Collins and Lyne's
Microbiological Methods
Eighth Edition

C. H. Collins, MBE, MA, DSc, FRCPath, FIBiol, FIBMS
The Ashes, Hadlow, Kent, UK

Patricia M. Lyne, CBiol, MIBiol
The Ashes, Hadlow, Kent, UK

J. M. Grange, MD, MSc
Visiting Professor, Centre for Infectious Diseases and International Health,
Royal Free and University College Medical School,
Windeyer Institute for Medical Sciences, London, UK

J. O. Falkinham III, PhD
Professor, Fralin Biotechnology Center, Virginia
Polytechnic Institute and State University, Blacksburg, Virginia, USA

ARNOLD

A member of the Hodder Headline Group
LONDON

First published in 1964 by Butterworth-Heinemann
Reprinted 1966
Second edition 1967
Reprinted 1970
Third edition 1970
Fourth edition 1976
Reprinted with additions 1979
Fifth edition 1984
Revised reprint 1985
Reprinted 1987
Sixth edition 1989
Revised reprint 1991
Seventh edition 1995
Reprinted 1997, 1998

This edition published in 2004 by
Arnold, a member of the Hodder Headline Group,
338 Euston Road, London NW1 3BH

http://www.arnoldpublishers.com

Distributed in the United States of America by
Oxford University Press Inc.,
198 Madison Avenue, New York, NY10016
Oxford is a registered trademark of Oxford University Press

Whilst the advice and information in this book are believed to be true and accurate at the date of going to press, neither the authors nor the publisher can accept any legal responsibility or liability for any errors or omissions that may be made. In particular (but without limiting the generality of the preceding disclaimer) every effort has been made to check drug dosages; however it is still possible that errors have been missed. Furthermore, dosage schedules are constantly being revised and new side-effects recognized. For these reasons the reader is strongly urged to consult the drug companies' printed instructions before administering any of the drugs recommended in this book.

British Library Cataloguing in Publication Data
A catalogue record for this book is available from the British Library

Library of Congress Cataloging-in-Publication Data
A catalog record for this book is available from the Library of Congress

ISBN 0 340 80896 9

1 2 3 4 5 6 7 8 9 10

Commissioning Editor: Serena Bureau
Development Editor: Layla Vandenbergh
Project Editor: Anke Ueberberg
Production Controller: Deborah Smith
Cover Design: Stewart Larking

Typeset in 10 on 13 pt Sabon by Phoenix Photosetting, Chatham, Kent
Printed and bound in the UK by Butler & Tanner Ltd

What do you think about this book? Or any other Arnold title?
Please send your comments to **feedback.arnold@hodder.co.uk**

and Lyne's

ethods

Edition

Contents

List of contributors

L. Berg, BSc, CLS, West Hills Hospital and Medical Center, West Hills, California, USA

J. S. Brazier, MSc, PhD, CBiol, MIBiol, FIBMS, Anaerobe Reference Laboratory, National Public Health Service of Wales, University Hospital of Wales, Cardiff, UK

E. Y. Bridson, MPhil, PhD, CBiol, FIBiol, FIBMS, Camberley, Surrey, UK

C. K. Campbell, MSc, PhD, Mycology Reference Laboratory and National Collection of Pathogenic Fungi, Health Protection Agency, Specialist and Reference Microbiology Division, South West HPA, Bristol, UK

C. H. Collins, MBE, MA, DSc, FRCPath, FIBiol, FIBMS, Hadlow, Kent, UK

Janet E. L. Corry, BSc, MSc, PhD, CBiol, MIBiol, FIST, Division of Farm Animal Food Science, Department of Clinical Veterinary Medicine, University of Bristol, Bristol, UK

Kate Davy, MSc, FIBMS, Mycology Reference Laboratory and National Collection of Pathogenic Fungi, Health Protection Agency, Specialist and Reference Microbiology Division, South West HPA, Bristol, UK

T. J. Donovan, PhD, CBiol, FIBiol, FIBMS, Ashford, Kent, UK

J. O. Falkinham III, PhD, Fralin Biotechnology Center, Virginia Polytechnic and State University, Blacksburg, Virginia, USA

J. M. Grange, MD, MSc, Centre for Infectious Diseases and International Health, Royal Free and University College Medical School, Windeyer Institute for Medical Sciences, London, UK

Elizabeth Johnson, BSc, PhD, Mycology Reference Laboratory and National Collection of Pathogenic Fungi, Health Protection Agency, Specialist and Reference Microbiology Division, South West HPA, Bristol, UK

D. A. Kennedy, MA, PhD, CBiol, MIBiol, FIOSH, Surbiton, Surrey, UK

S. F. Kinghorn-Perry, BSc, PhD, Health and Safety Executive, Biological Agents Unit, Bootle, Merseyside, UK

Christine L. Little, BSc, LLM, PhD, MIFST, Health Protection Agency, Environmental Surveillance Unit, Communicable Diseases Surveillance Centre, London, UK

Patricia M. Lyne, CBiol, MIBiol, The Ashes, Hadlow, Kent, UK

M. O. Moss, FLS, BSc, PhD, DIC, ARCS, School of Biomedical and Molecular Sciences, University of Surrey, Guildford, Surrey, UK

I. Ogden, HND, Department of Medical Microbiology, University of Aberdeen, Aberdeen, UK

D. N. Petts, MSc, PhD, CBiol, MIBiol, FIBMS, Microbiology Department, Basildon Hospital, Basildon, Essex, UK

T. L. Pitt, MPhil, PhD, Laboratory of Health Care and Associated Services, Health Protection Agency, London, UK

P. Silley, BSc, PhD, CBiol, FIBiol, Don Whitley Scientific Ltd, Shipley, West Yorkshire, UK

P. Taylor, BTech, MSc, MPhil, FIBMS, Department of Microbiology, Royal Brompton and Harefield NHS Trust, London, UK

Preface to the eighth edition

Forty years have passed since this book was first published and during its seven previous editions there have been many changes, not the least being its evolution from single authorship to multiple contributors and editors. Nevertheless, no single contributor is responsible for any one of the chapters. The system of peer review and revision has been adopted. Although the newer, automated and 'kit' methods have been accommodated, the basic techniques, essential to good practice in microbiology, have been retained and in some cases improved. These are still applicable in areas where automation is not yet possible and, we think, necessary for both teaching and research. One important change is the omission of the many formulae for the preparation of culture media. Very few laboratories, anywhere in the world, now make their own media because excellent, quality-controlled materials are now generally available.

Changes in nomenclature are inevitable but in some cases the older terminology, in addition to the latest, has been retained, on the 'custom and usage' principle.

We are indebted to our contributors for both their specialist knowledge and their general observations, as well as for allowing us to retain the style and format of the earlier editions, which appears to have been appreciated by readers and those who use the volume as a 'bench book'.

We also acknowledge the help that we have received from *Topley and Wilson's Microbiology and Microbial Infections* and *Cowan and Steel's Manual for the Identification of Medical Bacteria*.

C. H. Collins
Patricia M. Lyne Collins
J. M. Grange
J. O. Falkinham

I

Safety in microbiology

During the twentieth century some 4000 laboratory workers became infected with micro-organisms in the course of their work and some of them died. Unfortunately, laboratory-acquired infections still occur, although most of them are clearly preventable (see Collins and Kennedy, 1999).

To avoid becoming infected, therefore, laboratory workers should:

- be aware of the potential hazards of the materials and organisms they handle, especially those containing blood and those received from tropical and subtropical countries
- be aware of the routes by which pathogenic micro-organisms can enter the human body and cause infections
- receive proper instruction and thereafter practise good microbiological technique (GMT), which effectively 'contains' micro-organisms so that they do not have access to those routes.

CLASSIFICATION OF MICRO-ORGANISMS ON THE BASIS OF RISK

Micro-organisms vary in their ability to infect individuals and cause disease. Some are harmless, some may be responsible for disease with mild symptoms, others can cause serious illnesses, and a few have the potential for spreading in the community and causing serious epidemic disease. Experience and research into laboratory-acquired infections enabled investigators to classify micro-organisms

and viruses into four categories, also known as 'Classes', 'Risk Groups' or 'Hazard Groups' (in this book the last term is used). These are numbered 1–4 on the basis of the hazards that they present to laboratory workers and to the community should they 'escape' from the laboratory. These categories take into account the seriousness of the associated disease, the availability of prophylactic and therapeutic agents, the routes of infection and the history of laboratory infections caused by the organisms.

A number of classifications have been devised since the 1970s. They all agree in principle and are summarized in Table 1.1, although there are slight differences in wording and emphasis. These classifications originally applied to laboratory workers, but the increasing use of biological agents and materials in industry has led to their application to other workers (termed 'employees' in the most recent official publications).

Lists of organisms in Hazard Groups 2, 3 and 4 have been compiled by a number of states and organizations. There will inevitably be variations between countries according to their geography, reflecting differences in the incidence of the micro-organisms, their vectors, the associated diseases concerned and levels of hygiene. A comparison of these variations was made by the European Federation of Biotechnology (Frommer *et al.*, 1989), but they are under continual revision. Readers should therefore consult their current national lists.

In this book, the UK lists (Advisory Committee on Dangerous Pathogens or ACDP, 2001) are used. The differences between these and those in the USA reflect the variations noted above. Most of the

Table 1.1 Summary of classifications of micro-organisms on the basis of hazard

Hazard Group 1	Hazard Group 2	Hazard Group 3	Hazard Group 4
LOW RISK	$\rightarrow$	$\rightarrow$	*HIGH RISK*
Low to individual and to community	Moderate to individual, low to community	High to individual, low to community	Very high to individual *and* to community

organisms are in Hazard Groups 1 and 2. Cautions, in bold italic type, are given for those in Hazard Group 3. No Hazard Group 4 agents (which are all viruses) are discussed in the text. In the USA the lists are published by the Centers for Disease Control and Prevention (CDC, 1999) and the American Society for Microbiology (Fleming *et al.*, 1995). These publications, and also that of Collins and Kennedy (1999), also give 'agent summary statements' which provide information about the organisms and their histories in relation to laboratory-acquired infections.

ROUTES OF INFECTION

There are at least four routes by which micro-organisms may enter the human body.

Through the lungs

Aerosols, i.e. very small droplets (< 0.5–5.0 μm) containing micro-organisms, are frequently released during normal microbiological manipulations such as work with loops and syringes, pipetting, centrifuging, blending, mixing and homogenizing, pouring, and opening culture tubes and dishes. Inhalation of these has probably caused the largest number of laboratory-acquired infections.

Through the mouth

Biological agents may be ingested during mouth pipetting and if they are transferred to the mouth by contaminated fingers or articles, such as food, cigarettes and pencils, that have been in contact with areas or objects contaminated by spills, splashes or settled droplets.

Through the skin

The skin of hands, etc. is rarely free from (often minute) cuts and abrasions, through which organisms can enter the bloodstream. Infections may also arise as a result of accidental inoculation with infected hypodermic needles, broken glass and other sharp objects.

Through the eyes

Splashes of infected material into the eyes and transfer of organisms on contaminated fingers to the eyes are not uncommon routes of infection.

LEVELS OF CONTAINMENT

The Hazard Group number allocated to an organism indicates the appropriate Containment Level (CL) in the UK or Biosafety Level (BSL) in the USA, i.e. the necessary accommodation, equipment, techniques and precautions. Thus, there are four of these levels (Table 1.2). In the World Health Organization's (WHO's) classification the Basic Laboratory equates with Levels 1 and 2, Containment Laboratory with Level 3 and the Maximum Containment Laboratory with Level 4.

The bold italic type '*Caution*' at the head of sec-

Table 1.2 Summary of biosafety/containment levels

Level	Facilities	Laboratory practice	Safety equipment
1	Basic	GMT	None. Open bench
2	Basic	GMT plus protective clothing. Biohazard signs	Open bench + safety cabinet for aerosol potential
3	Containment	Level 2 + special clothing and controlled access	Safety cabinet for all activities
4	Maximum containment	Level 3 plus air lock entry, shower exit, special waste disposal	Class III safety cabinet, pressure gradient, double-ended autoclave

GMT, good microbiological technique

tions dealing with Group 3 organisms indicates the appropriate containment level and any additional precautions.

Full details are given by the WHO (1993), the CDC (1999), Fleming *et al.* (1995), the ACDP (2001) and HSAC (2003). A summary is presented here and in Table 1.2. More detailed measures are given in Table 1.3.

Levels 1 and 2: the Basic laboratories

These are intended for work with organisms in Hazard Groups 1 and 2. Ample space should be provided. Walls, ceilings and floors should be smooth, non-absorbent, easy to clean and disinfect, and resistant to the chemicals that are likely to be used. Floors should also be slip resistant. Lighting and heating should be adequate. Hand basins, other than laboratory sinks, are essential. Bench tops should be wide, the correct height for work at a comfortable sitting position, smooth, easy to clean and disinfect, and resistant to chemicals. Adequate storage facilities should be provided.

If agents in Hazard Group 2 are handled, e.g. in clinical laboratories, a microbiological safety cabinet (see Chapter 3) should be provided. Access should be restricted to authorized individuals.

Level 3: the Containment laboratory

This is intended for work with organisms in Hazard Group 3. All the features of the Basic laboratory should be incorporated, plus the following:

- The room should be physically separated from other rooms, with no communication (e.g. by pipe ducts, false ceilings) with other areas, apart from the door, which should be lockable, and transfer grilles for ventilation (see below).
- Ventilation should be one way, achieved by having a lower pressure in the Containment laboratory than in other, adjacent rooms and areas. Air should be removed to atmosphere (total dump, not recirculated to other parts of the building) by an exhaust system coupled to a microbiological safety cabinet (see p. 30) so that, during working hours, air is continually extracted by the cabinet or directly from the room and airborne particles cannot be moved around the building. Replacement air enters through transfer grilles.
- Internal windows should be provided, e.g. in the door, with internal mirrors if necessary in case of accidents to the occupants.
- Access should be strictly regulated. The international biohazard warning sign, with appropriate wording, should be fixed to the door (Figure 1.1).

Table 1.3 Summary of safety measures for work at containment levels 2, 3 and 4*

	Level 2	Level 3	Level 4
Sites, buildings, engineering and facilities			
Laboratories separated from other activities in same building	If justified by risk assessment	Yes	Yes, preferably separate building
External windows	Closed during work	Lockable	Sealed, watertight and unbreakable
Internal windows so that occupants may be observed	Not necessary	Recommended	Yes
Handbasins	Yes	In each room	In each room
Showers	Not necessary	Recommended	Yes
Surfaces impermeable to water and cleaning agents	All work surfaces	All work surfaces and floors	All exposed surfaces
Surfaces impermeable to acids, bases, solvents and disinfectants	Recommended	Yes	Yes
Laboratories maintained at negative pressure relative to atmosphere	Not necessary	Yes	With gradient
HEPA filtration of intake and extract air	Not necessary	Extract air	Intake and extract air
Airlock access to laboratories	Not necessary	Recommended	Yes
Room sealable for decontamination	Not necessary	Yes	Yes
Separate effluent treatment	Not necessary	If indicated by risk assessment	Yes
Two-way intercom system	Not necessary	Yes	Yes
Laboratories to have own dedicated equipment	Not necessary	Yes	Yes
Microbiological safety cabinets	If indicated by risk assessment	Classes I and/or II	Class III
Autoclaves	In building	In room	Pass-through, double ended
Centrifuges	Available	In room	In room
Operational measures			
Initial and annual tests for containment capability	Not necessary	Yes	Yes
Biohazard sign displayed	On all doors	On all doors	On all doors
Authorized access	Yes	Limited	Strictly limited
Validated disinfection and disposal methods	Yes	Yes	Yes
Emergency plans for spillages of infectious material	Yes	Yes	Yes
Safe storage of biological agents	Yes	In Level 3 room only	In Level 4 room only
Effective rodent and other vector control	Yes	Yes	Yes
Protective clothing to be worn	Standard overalls	Standard overalls	Full protection
Gloves to be worn	If indicated by risk assessment	Yes	Yes
Medical supervision	By occupational health service	By unit physician	By unit physician
Immunization	As for general public	As for general public and for agents used	As for general public and for agents used
Base line sera	Not necessary	Recommended	Yes
Accident reporting	Yes	Yes	Yes
Two-person rule	Not necessary	Recommended	Yes
Supervision of maintenance and service personnel	Yes	Accompanied by staff member	Accompanied by staff member, after decontamination of room
Technical procedures			
Work with infectious materials	On open bench except if aerosol release is possible	In Class I or II microbiological safety cabinet	In Class III microbiological safety cabinet
Centrifugation	In closed containers	In sealable safety buckets	In sealable safety buckets
Use of 'sharps'	As authorized	Avoid	Banned except when especially authorised

*From Collins and Kennedy (1999). Sources: Frommer *et al.* (1989), WHO (1993), ACDP (1995a). Health Canada (1996), CDC/NIH (1999). See also HSAC (2003).

Figure 1.1 International biohazard sign

Level 4: the Maximum Containment laboratory

This is required for work with Hazard Group 4 materials and is outside the scope of this book. Construction and use generally require government licence or supervision.

PREVENTION OF LABORATORY-ACQUIRED INFECTIONS: A CODE OF PRACTICE

The principles of containment involve good micro-biological technique, with which this book is particularly concerned, and the provision of:

1. primary barriers around the organisms to prevent their dispersal into the laboratory
2. secondary barriers around the worker to act as a safety net should the primary barriers be breached
3. tertiary barriers to prevent the escape into the community of any organisms that are not contained by the primary and secondary barriers.

This code of practice draws very largely on the requirements and recommendations published in other works (CDC, 1999; WHO, 1993, 1997; ACDP, 1995a, 1998, 2001; Fleming *et al.*, 1995; Collins and Kennedy, 1999; HSE, 2002; HSAC,

2003). (See also Approved Codes of Practice in the Appendix to this book.)

Primary barriers

These are techniques and equipment designed to contain micro-organisms and prevent their direct access to the worker and their dispersal as aerosols:

- All body fluids and pathological material should be regarded as potentially infected. 'High Risk' specimens, i.e. those that may contain agents in Hazard Group 3 or originate from patients in the AIDS and hepatitis risk categories, should be so labelled (see below).
- Mouth pipetting should be *banned in all circumstances*. Pipetting devices (see p. 38) should be provided.
- No article should be placed in or at the mouth, including mouthpieces of rubber tubing attached to pipettes, pens, pencils, labels, smoking materials, fingers, food and drink.
- The use of hypodermic needles should be restricted. Cannulas are safer.
- Sharp glass Pasteur pipettes should be replaced by soft plastic varieties (see p. 38).
- Cracked and chipped glassware should be replaced.
- Centrifuge tubes should be filled only to within 2 cm of the lip. On centrifugation in fixed angle centrifuges, the liquid level is perpendicular to the direction of the forces.
- All Hazard Group 3 materials should be centrifuged in sealed centrifuge buckets (see p. 28).
- Bacteriological loops should be completely closed, with a diameter not more than 3 mm and a shank not more than 5 cm, and be held in metal, not glass, handles.
- Homogenizers should be inspected regularly for defects that might disperse aerosols (see p. 39). Only the safest models should be used. Glass Griffith's tubes and tissue homogenizers should be held in a pad of wadding in a gloved hand when operated.

- All Hazard Group 3 materials should be processed in a Containment laboratory (see 'Tertiary barriers' below) and in a microbiological safety cabinet (see p. 30) unless exempted by agreed national regulations.
- A supply of suitable disinfectant (see p. 50) at the use dilution should be available at every work station.
- Benches and work surfaces should be disinfected regularly and after spillage of potentially infected material.
- Discard jars for small objects and re-usable pipettes should be provided at each work station. They should be emptied, decontaminated and refilled with freshly prepared disinfectant daily (see p. 56).
- Discard bins or bags supported in bins should be placed near to each work station. They should be removed and autoclaved daily (see p. 56).
- Broken culture vessels, e.g. after accidents, should be covered with a cloth, Suitable disinfectant should be poured over the cloth and the whole left for 30 min. The debris should then be cleared up into a suitable container (tray or dustpan) and autoclaved. The hands should be gloved and stiff cardboard used to clear up the debris.
- Infectious and pathological material to be sent through the post or by airmail should be packed in accordance with government, post office and airline regulations, obtainable from those authorities. Full instructions are also given on p. 9.
- Specimen containers should be robust and leak proof.
- Discarded infectious material should not leave the laboratory until it has been sterilized or otherwise made safe (see p. 55).

- Proper overalls and/or laboratory coats should be worn at all times and fastened. They should be kept apart from outdoor and other clothing. When agents in Hazard Group 3 containing, or material containing or suspected of containing, hepatitis B or the AIDS virus are handled, plastic aprons should be worn over the normal protective clothing. Surgeons' gloves should also be worn.
- Laboratory protective clothing should be removed when the worker leaves the laboratory and not worn in any other areas such as canteens and rest rooms.
- Hands should be washed after handling infectious material and always before leaving the laboratory.
- Any obvious cuts, scratches and abrasions on exposed parts of the worker's body should be covered with waterproof dressings.
- There should be medical supervision in laboratories where pathogens are handled.
- All illnesses should be reported to the medical or occupational health supervisors. They may be laboratory associated. Pregnancy should also be reported, because it is inadvisable to work with certain micro-organisms during pregnancy.
- Any member of the staff who is receiving steroids or immunosuppressive drugs should report this to the medical supervisor.
- Where possible, and subject to medical advice, workers should be immunized against likely infections (Wright, 1988; WHO, 1993; Collins and Kennedy, 1999; HSAC, 2003).
- In laboratories where tuberculous materials are handled the staff should have received BCG or have evidence of a positive skin reaction before starting work. They should have a pre-employment chest radiograph (Wright, 1988).

Secondary barriers

These are intended to protect the worker should the primary barriers fail. They should, however, be observed as strictly as the latter:

Tertiary barriers

These are intended to offer additional protection to the worker and to prevent escape into the community of micro-organisms that are under investigation in the laboratory.

They concern the provision of accommodation and facilities at the Biosafety/Containment levels appropriate for the organisms handled, and of microbiological safety cabinets.

For more information on laboratory design, see British Occupational Hygiene Society (BOHS, 1992) and Collins and Kennedy (1999).

PRECAUTIONS AGAINST INFECTION WITH HEPATITIS B (HBV), THE HUMAN IMMUNODEFICIENCY VIRUS (HIV) AND OTHER BLOOD-BORNE PATHOGENS

Laboratory workers who may be exposed to these viruses in clinical material are advised to follow the special precautions described by one or more of the following: ACDP (1990, 1995b, 2001); Occupational Health and Safety Agency (OHSA, 1991); WHO (1991); Fleming *et al.* (1995); Collins and Kennedy (1997), HSAC (2003).

PRECAUTIONS IN LABORATORIES WORKING WITH RECOMBINANT MICRO-ORGANISMS

It seems generally to be agreed that the potential risks associated with recombinant micro-organisms are of the same order as those offered by naturally occurring micro-organisms – with reference, of course, to the hazard groups of the hosts and inserts (Organization for Economic Co-operation and Development or OECD, 1986; Frommer *et al.*, 1989; WHO, 1993).

Nevertheless, work of this nature is closely regulated. As it is outside the scope of this book, readers are referred to the following: Collins and Beale (1992); National Institutes of Health (NIH, 1994); Advisory Committee on Genetic Modification (ACGM, 1997).

PLANT PATHOGENS

Very few, if any, plant pathogens can cause disease in humans. Laboratory precautions and containment are therefore aimed at preventing the escape of these agents to the environment. The subject has received little attention, however, except from the European Federation of Biotechnology (Frommer *et al.*, 1992). In the USA, laboratories must obtain permission from the US Department of Agriculture before they work on plant pathogens.

THE SAFE COLLECTION, RECEIPT AND TRANSPORT OF DIAGNOSTIC SPECIMENS AND 'INFECTIOUS SUBSTANCES'

Fears are sometimes expressed that micro-organisms might, as a result of poor packaging or accidental damage, 'escape' from specimens and cultures while they are being transported to or between laboratories. There are no recorded cases, however, of illness arising from such an 'escape' (WHO, 2003).

A distinction is drawn between diagnostic specimens and 'infectious substances' largely on the basis of risk assessment.

Diagnostic specimens

A diagnostic specimen is defined (WHO, 2003) as:

> . . . any human or animal material, including but not limited to excreta, blood and its components, tissue or tissue fluids, collected for the purpose of diagnosis but excluding live animals.

As a general rule, such specimens do not offer much risk, but if they are collected during an outbreak of an infectious disease of unknown aetiology they should be regarded as 'infectious substances' – see below.

Specimen containers

These containers should be robust, preferably screw capped, and unlikely to leak. Plastic bags should be provided to receive the container after the specimen has been collected. Request forms should not be placed in the same plastic bag as the specimen. Unless the source and laboratory are in the same premises, robust outer containers with adequate absorbent material should be provided. Various packages are available commercially.

Transport within hospitals and institutions

Leak-proof trays or boxes should be provided for the transport of diagnostic specimens from wards to the laboratory. These should be able to withstand autoclaving or overnight exposure to disinfectant. Suitable trays and boxes are readily available. Some are known as kitchen 'tidies' or 'tidy boxes', and are sold cheaply by chain stores, hardware and tool shops for cutlery, kitchen utensils, shoe-cleaning equipment and handyman's tools. Ordinary hospital ward or surgical trays are also widely used. Some purpose-made containers are made of stainless steel, others of propylene ethylene co-polymer. They should be deep enough for specimens to be carried upright to avoid leakage and spillage. These trays and boxes should not be used for any other purposes.

Organs in glass jars or plastic buckets, and 24-h urines in 2-litre screw-capped bottles or the standard plastic carboys should be placed in disposable plastic (or paper) bags for transport for both hygienic and aesthetic reasons.

Surface transport between and to hospitals

Diagnostic specimens are frequently conveyed from one hospital to another because of the centralization of laboratory services. The usual means of transport is by hospital van or hospital car, i.e. the vehicle is under the control of the hospital authorities. Similarly, specimens are transported from local authority offices or clinics to laboratories by the authorities' own vehicles. Rarely, such material may be carried by taxi cabs or by messengers on public service vehicles.

The UK Health Services Advisory Committee (HSAC, 2003) recommends that secure transport boxes with lids be used. These must be capable of withstanding autoclaving and prolonged exposure to disinfectants. Such boxes are in common use among food inspectors for conveying food, milk and water samples to public analysts and public health laboratories. Suitable boxes, used by some laboratories, included those that can be fitted to the carriers of motor cycles and scooters, and the plastic 'Coolkeepers' or 'Chilly Bins' used by motorists for picnics. The boxes should be marked quite clearly with the name and address of the laboratory that owns them. In case they are lost or the vehicle is involved in an accident, they should also carry warning labels 'Danger of Infection, do not open' as well as the international biohazard sign, the meaning of which is rarely clear to laypeople. The labels should request the finder to call the nearest hospital or police station. The boxes should be inspected daily for evidence of spillage, and should be decontaminated and then washed out at least weekly. Again, the drivers or messengers should be warned that the material in the boxes may be infectious. If the boxes are suitable for the job and are packed with care, the hazards are minimal and there should be no need for disinfectants to be carried.

Some authorities send specimens to laboratories by hand on public service vehicles. The same type of boxes should be used, but it may be advisable to label them less explicitly or transport employees and other passengers may be alarmed.

Inland mail

Within state boundaries the usual requirements are that the specimen is in a leak-proof primary container, securely cocooned in absorbent material and packed in a robust outer container, which is labelled 'Pathological Specimen' or 'Diagnostic Specimen'.

Specimens for mailing across a frontier/border

should be packaged as described below for infectious substances

Infectious substances

An infectious substance (WHO, 2003) is:

. . . a substance containing a viable micro-organism such as a bacterium, virus, rickettsia, parasite or fungus that is known or reasonably believed to cause disease in humans or animals.

Airmailing infectious substances

Until recently there have been problems here, because several different organizations are involved (Collins and Kennedy, 1999). These have now been resolved by international agreement and the WHO has issued an excellent guidance document (WHO, 2003).

The internationally agreed 'triple packaging system' must be used. This must meet the requirements of the UN class 2 specifications and packaging instruction PI(602) (UN, 1996) which impose strict tests, including surviving, undamaged, a 9-metre drop on to a hard surface, as well as puncture tests.

1. The material is placed in a watertight, leak-proof receptacle – the 'primary container'. This is wrapped in enough absorbent material to absorb all fluid in case of breakage.
2. The primary container is then packed in a watertight 'secondary container' with enough absorbent material to cushion the primary container.
3. The secondary container is placed in an outer shipping package that protects its contents from damage during transit.

An example of a triple package that satisfies the requirements of the UN packaging instruction is shown in Figure 1.2

A certain amount of documentation must accompany the package. This is explained in detail by the WHO (2003).

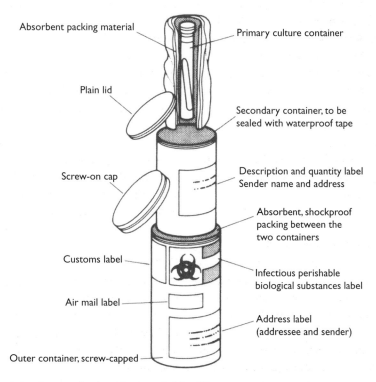

Absorbent packing material
Primary culture container
Plain lid
Secondary container, to be sealed with waterproof tape
Screw-on cap
Description and quantity label
Sender name and address
Absorbent, shockproof packing between the two containers
Customs label
Infectious perishable biological substances label
Air mail label
Address label (addressee and sender)
Outer container, screw-capped

Figure 1.2 Packing infectious substances for the overseas post (airmail)

RECEIPT OF SPECIMENS IN THE LABORATORY

Specimens from local sources, i.e. within state boundaries, may be unpacked only by trained reception staff who wear laboratory protective clothing. The specimens should be left in their plastic bags pending examination. Specimens from across frontiers/borders, which may contain exotic biological agents, should always be unpacked only by trained laboratory staff under appropriate containment conditions.

Instruction in safety

This instruction should be part of the general training given to all microbiological laboratory workers and should be included in what is known as good microbiological technique. In general, methods that protect cultures from contamination also protect workers from infection, but this should not be taken for granted. Personal protection can be achieved only by good training and careful work. Programmes for training in safety in microbiology have been published (International Association of Medical Technology or IAMLS, 1991; WHO, 1993; Collins and Kennedy, 1999).

Bioterrorism

We consider that it would be inappropriate to suggest that 'routine' laboratories (the target of the book) should test samples for agents of bioterrorism. Such samples should be submitted to a national laboratory that has the necessary expertise. States that do not have one should negotiate with, for example, the UK, the USA and Russia, who have the necessary facilities. See also the Appendix to this book.

For a review of the history, incidence, causes and prevention of laboratory-acquired infection see Fleming *et al.* (1995) and Collins and Kennedy (1999).

REFERENCES

Advisory Committee on Dangerous Pathogens (ACDP) (1990) *HIV, the Causative Agent of AIDS and Related Conditions.* London: HMSO.

ACDP (1995a) *Categorisation of Pathogens According to Hazard and Categories of Containment,* 2nd edn. London: HMSO.

ACDP (1995b) *Protection against Blood-borne Infections in the Workplace.* London: HMSO.

ACDP (2001) *The management, design and operation of microbiological containment laboratories.* Advisory Committee on Dangerous Pathogens. Sudbury: HSE Books.

ACDP (2002) Second Supplement to the ACDP guidance on the: *Categorisation of Pathogens According to Hazard and Categories of Containment.* Sudbury: HSE Books.

Advisory Committee on Genetic Modification (ACGM) (1997) *Compendium of Guidance on Good Practice for people Working with Genetically-modified Micro-organisms.* ACGM/HAS/DE Note No. 7. London: The Stationery Office.

British Occupational Hygiene Society (1992) Laboratory Design Issues. *British Occupational Hygiene Society Technical Guide No. 10.* Leeds: H & Scientific Consultants.

Centers for Disease Control and National Institutes of Health (1999) *Biosafety in Microbiological and Biomedical Laboratories,* 4th edn, Washington DC: CDC.

Collins, C. H. and Beale, A. J. (1992) *Safety in Industrial Microbiology and Biotechnology.* Oxford: Butterworth–Heinemann.

Collins, C. H. and Kennedy, D. A. (eds) (1997) *Occupational Blood-borne Infections.* Wallingford: CAB International.

Collins, C. H. and Kennedy, D. A. (1999) *Laboratory-acquired Infections,* 4th edn. Oxford: Butterworth–Heinemann.

Fleming, D. O., Richardson, J. H., Tulis, J. I. and Vesley, D. (eds) (1995) *Laboratory Safety: Principles and practice.* Washington DC: ASM Press.

Frommer, W.C. and a Working Party on Safety in Biotechnology of the European Federation of

Biotechnology (1989) Safe Biotechnology III. Safety precautions for handily microorganisms of different risk classes. *Applied Microbiology and Biotechnology* **30**: 541–552.

Frommer, W.C. and a Working Party on Safety in Biotechnology of the European Federation of Biotechnology (1992) Safe Biotechnology (4). Recommendations for safety levels for biotechnological operations with microorganisms that cause diseases in plants. *Applied Microbiology and Biotechnology* **38**: 139–140.

Health and Safety Executive (HSE) (2002) *Control of Substances Hazardous to Health: Approved Codes of Practice*, 4th edn. Sudbury: HSE Books.

HSAC, Health Services Advisory Committee (2003) *Safe Working and the Prevention of Infection in Clinical Laboratories*. Sudbury: HSE Books.

International Association of Medical Technology, Stockholm (1991) A Curriculum for Instruction in Laboratory Safety. *Medical Technical International* 1. Stockholm: IAMLS.

National Institutes of Health (1994) Guidelines for research involving recombinant DNA molecules. *Federal Register* 5 July, Part IV.

Occupational Health and Safety Agency (OHSA) (1991) Blood-borne Pathogens Standard. *Federal Register* **56**: 64175–64182.

Organization for Economic Co-operation and Development (1986) *Recombinant DNA Safety Considerations*. Paris: OECD.

United Nations (1996) Committee of Experts on the Transport of Dangerous Goods. *Division 2. Infectious Substances*, 10th edn. Revised New York: UN.

World Health Organization (1991) *Biosafety Guidelines for Diagnostic and Research Laboratories Working with HIV. AIDS Series* 9. Geneva: WHO.

WHO (1993) *Laboratory Biosafety Manual*, 2nd edn. Geneva: WHO.

WHO (1997) Safety *in Health-Care Laboratories*. Geneva: WHO.

WHO (2003) *Guidelines for the Safe Transport of Infectious Substances and Diagnostic Specimens.* WHO/EMC/97.3. Geneva: WHO.

Wright, A. E. (1988) Health care in the laboratory. In: Collins C. H. (ed.), *Safety in Clinical and Biomedical Laboratories*. London: Chapman & Hall.

2

Quality assurance

The information given in this chapter is intended to provide users with sufficient information to implement a quality assurance programme. For a comprehensive treatise of the subject see Snell *et al.* (1999). Three terms require definition: quality assurance, quality control and quality assessment. Quality assurance is the total process that guarantees the quality of laboratory results and encompasses both quality control and quality assessment. The quality of a microbiological service is assured by quality control, i.e. a continual monitoring of working practices, equipment and reagents. Quality assessment is a system by which specimens of known but undisclosed content are introduced into the laboratory and examined by routine procedures. Quality control and quality assessment are therefore complementary activities and one should not be used as a substitute for the other (European Committee for Clinical Laboratory Standards or ECCLS, 1985). Essentially, with a quality system, we are seeking to ensure that we are doing the right things, for the right people, at the right time and doing them right every time (Crook, 2002).

THE NEED FOR QUALITY

There is increasing reliance on the quality of laboratory results and confidence in them is essential to the organization of the laboratory. Quality control of laboratory tests cannot be considered in isolation: the choice, collection, labelling and transport of specimens are all subject to error and require regular monitoring for quality and suitability. Additional recognition of the fact that the quality of microbiology results is only a component within

the chain of patient care (Figure 2.1) is demonstrated by the introduction of policies in the UK that are aimed at embedding continuous quality improvement (CQI) at all levels and across all parts of the National Health Service (NHS) (Heard *et al.,* 2001). The key goal of these policies is to achieve changes in practice, thereby improving patient outcomes. The term 'clinical governance' is used to describe the links between clinical and organizational responsibility for the quality of care. Clinical governance is a system through which NHS organizations are accountable for continuously improving the quality of their services and safeguarding high standards of care by creating an environment in which excellence in clinical care will flourish (Campbell *et al.,* 2002).

Cost-effectiveness is increasingly relevant in all areas of microbiology. In the health services, speed of response may affect the outcome of health care and the length and cost of a patient's stay in hospital, and it is therefore as much an element of quality as the accuracy of the test result.

An important contributory factor to high standards of quality is the competence and motivation of staff. These aspects can be achieved and maintained by continued education and, above all, by communication within and between laboratories and with those using the services.

PRACTICAL QUALITY CONTROL WITHIN THE LABORATORY

Practical, realistic and economic quality control programmes should be applied to all procedures,

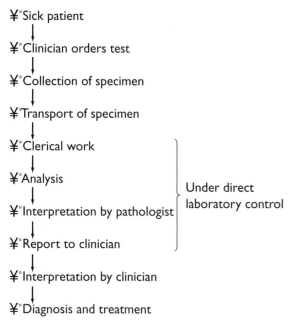

¥°Sick patient
↓
¥°Clinician orders test
↓
¥°Collection of specimen
↓
¥°Transport of specimen
↓
¥°Clerical work
↓
¥°Analysis
↓
¥°Interpretation by pathologist
↓
¥°Report to clinician
↓
¥°Interpretation by clinician
↓
¥°Diagnosis and treatment

Under direct laboratory control

Figure 2.1   Chain of patient care

irrespective of the size of the laboratory. Schedules should be based on the known stability of each item and its importance to the accuracy of the investigation as a whole. It is unnecessary to monitor all reagents and methods on every working day: the frequency of monitoring should be in direct proportion to the frequency with which errors are discovered. Nevertheless, to ensure success all the components of a quality programme must be in place and operating before good quality laboratory results can be produced.

Each laboratory must develop its own programme because the time spent on quality control will vary according to the size of the laboratory and the range of investigations. For its success, a senior member of staff should supervise the programme throughout the laboratory. This supervisory role has indeed been formalized by the fact that it is a requirement of the UK Clinical Pathology Accreditation (CPA) and the American Society of Clinical Pathologists (ASCP) that a quality manager be appointed. This person will act as a focal point for all quality issues within the laboratory and be responsible for ensuring that the quality management system (QMS) works.

Internal quality control can be divided into four main areas: collection of specimens; preventive monitoring and maintenance of equipment (see Chapter 3); manual procedures; and monitoring of media and reagents.

Collection of specimens

In clinical microbiology, results of studies on specimens taken from sites containing indigenous or colonizing bacteria, e.g. from clinical sites, or those likely to be contaminated during collection, are often difficult to interpret. Paradoxically, specimens containing the largest numbers of microbial species, with consequent high laboratory cost, are less likely to provide useful results than those yielding fewer types of organisms. Guidelines for the evaluation of the quality of specimens and the appropriateness of requests should be established by the medical and laboratory staff (Bartlett et al., 1978). The usefulness of rapid reporting of preliminary results should also be assessed (Bartlett, 1982) (see also Chapter 11).

A revolution is occurring in the detection, identification and characterization of pathogens (Borriello, 1999), such that the scenario of taking a clinical sample and, within an hour, knowing whether a pathogen is present and its susceptibility to antimicrobial agents may soon be a reality. Such developments, particularly with regard to near patient testing, will have important implications for the delivery of health care. Although there are several advantages of near patient testing, there are also disadvantages, not least of which is the reduced ability for internal and external quality assurance, which in turn may lead to misdiagnosis. To establish widespread use of such methodologies, there must be confidence in the operators as well as in the kits. Those responsible for health care must be confident that the correct specimen was obtained, the kit was used appropriately by properly qualified individuals, the result was interpreted correctly, and any machine used in the process has been appropriately maintained and used.

Preventive monitoring and maintenance of equipment

This is vitally important because many investigations are conducted with complex automated equipment. Regular monitoring and maintenance avoid costly repairs and long periods when the equipment is out of action ('down time'). Prevention of down time is particularly important when an uninterrupted 24-h service is being provided. Unique preventive maintenance systems, which are adaptable to the needs of an individual laboratory, have been described (Wilcox *et al.*, 1978). The following components belong to all preventive maintenance programmes.

Routine service tasks

The tasks needed to keep each item of equipment calibrated, running and clean should be defined in the standard operating procedures manual (see below).

Interval of service

The frequency with which defined tasks are performed is assessed according to the type of equipment and its usage. Although regularly used equipment will require more frequent cleaning and servicing than that used infrequently, minimum standards of preventive maintenance must be defined for the latter. In some areas service representatives will call annually to perform routine service and calibration work on instruments (e.g. balances and centrifuges) if there are enough of them on the same site.

Accountability

The concept of accountability must be introduced, and members of staff should be assigned specific responsibilities for maintenance tasks.

Staff

A list of trained people who are readily available for trouble-shooting, verification tasks and repairs should be compiled. Although laboratory staff may perform minor uncomplicated maintenance work,

advice with respect to complex equipment and procedures must be sought from the manufacturer's service department. The cost of a service contract must be weighed against the possibility of even more expensive repair costs and the consequences of prolonged 'down time'.

Implementation

Equipment maintenance is not a 'one-off' affair. Once tasks have been defined, and staff trained, the programme should continue regularly and uninterrupted (see also Chapter 3).

Documentation

A system of recording is essential and may be adapted to the specific needs of a laboratory, e.g. index cards or a computerized system. Computer records are ideal because they may easily be updated and made available.

Standard operating procedures manual

The standard operating procedures (SOP) manual is the most important document for the improvement and maintenance of the quality of the service. Such a manual can be in an electronic format and therefore more readily available and less likely to be misplaced. An out-of-date manual or one from another laboratory may be used as a guide to writing but, as each laboratory is unique, site-specific manuals must be developed. If there are no written methods and new staff members are taught by 'word of mouth', techniques will soon 'drift'. The headings standard to any SOP manual are listed and discussed by Fox *et al.* (1992).

Authoritative

A senior member of staff who has practical experience of the methods must write the manual.

Realistic

The manual should avoid theoretical ideals and provide clear, concise and explicit descriptions of procedures. Unnecessarily complicated procedures will quickly lead to loss of respect and to the possibility of staff introducing unofficial short cuts. A well-structured manual, organized and indexed, will allow a microbiologist from one laboratory to perform any of the tests described in another.

Up to date

The updating of manuals is vital and should be a continuous and formal process. The manual should be revised annually and amended when necessary. An electronic manual is therefore ideal. Only designated individuals should be permitted to modify a procedure, and no new or modified procedure should be used until approval has been obtained from the quality manager.

Availability

The manual must be freely available to all relevant personnel. It is the responsibility of senior staff to ensure that it is used and strictly adhered to. If it is not in electronic format it need not be a bulky bound item; loose-leaf systems have the advantage that sections may be separated and kept on relevant benches. Each section should have a unique and clear code number for identification and for cross-referencing.

Culture media and reagents

Most laboratories now use dehydrated media, which, if purchased from a reputable manufacturer, will have undergone extensive quality control testing before sale. Repetition of these tests should be unnecessary but, as dehydrated media will have to be reconstituted, heated and possibly supplemented with additives during preparation, it is essential to control the effect of these processes. Overheating of these additives may reduce their effectiveness and/or release toxic substances. These are good arguments for using media preparators, in which temperatures are strictly controlled instead of autoclaves (see Chapters 3, 4 and 5). Media must fulfil four main parameters: sterility; support

of bacterial growth; selective inhibition of growth, in some cases; and appropriate response of biochemical indicators.

All batches of media must be tested for sterility. Methods are usually a compromise because the sterility of every plate or tube of medium cannot be ensured without incubation of the whole batch, with the concomitant risk of deterioration of its ability to support microbial growth. A representative portion of each new batch should be incubated overnight to check for sterility before use. For batches of 100 ml or less, a 3–5% minimum sample should be tested; for larger batches, 10 plates or tubes taken randomly will suffice (Nicholls, 1999). After overnight incubation, media should be left at room temperature or in a refrigerator for a further 24 h as a check for contamination by psychrophiles. Low-level surface contamination of plates may not be detected on the controls but is usually recognizable during use by, for example, the observation of growth part way through a streaked population.

Selective media, being inhibitory to many organisms, pose particular problems: even gross contamination may not be detected by the above methods but inoculation of a few millilitres of the final product into 10 times its volume of sterile broth (to dilute inhibitory substances) will permit detection of heavy contamination.

Tests for performance of media

The ability of each batch of new medium to support growth and exhibit the required differential and selective qualities should be determined before use. This may be done by the efficiency of plating technique described in Chapter 5.

The organisms used should be stock strains from a recognized culture collection such as the National Collection of Type Cultures (NCTC) or American Type Cultures (ATCC). As characteristics of bacteria may alter when maintained by serial subculture, it is preferable to use stock strains stored at $-70°C$, freeze-dried or in gelatin discs (see Chapter 6). Selected media are tested with organisms expected to grow and those expected to be inhibited. Biochemical indicator media are tested with two or more organisms to demonstrate positive and negative reactions. A practical approach based on a minimum number of strains has been described (Westwell, 1999) (see also Chapters 5 and 6 of this book).

Reagents

Most pure chemicals and stains have greater stability than biological reagents. All reagents must be labelled as to contents, concentration, date of preparation and expiry date or shelf-life. They must be stored according to manufacturers' instructions and discarded when out of date. Freshly prepared reagents must be subjected to quality control before use; working solutions of stains are tested weekly and infrequently used reagents and stains are tested for their required purpose before use. Chemical reagents, e.g. Kovac's reagent, should be checked for performance against new batches of media. Some chemicals deteriorate unless stored correctly, e.g. certain stains are light sensitive. Storage should therefore be in accordance with the manufacturers' data sheets.

QUALITY ASSESSMENT

The concepts and statistical methods for the implementation of quality assessment were largely pioneered in clinical biochemistry laboratories. Schemes in microbiology laboratories are less easily organized because of the complexity, expense and labour intensity of the work. The diversity of microbiological investigations makes it difficult to produce control materials that cover all areas, and several technical difficulties have to be overcome in the preparation of quality assessment specimens (DeMello and Snell, 1985).

Quality assessment may be organized locally within one laboratory (internal quality assessment) (Gray *et al.*, 1995) or on a wider geographical basis (external quality assessment) (World Health Organization or WHO, 1994). A senior member of staff within the laboratory should organize an internal scheme to follow which arrangements are made so that examinations of specimens or samples are repeated, thereby enabling direct comparisons with previously obtained results. An obvious advantage with this approach is that the specimens or samples are real, rather than simulated, ones. False identifications may be given to repeated specimens so that staff – unaware that they are quality control specimens – do not give them special attention. Care should be taken with the use of stored materials because deterioration may prevent the reproduction of previous results.

A central laboratory, often on a national basis, usually organizes an external quality assessment scheme, as, for example, that of the ASCP. Simulated samples of known but undisclosed content are dispatched to participants who examine them and report their findings to the organizing laboratory. The organizer compares the results from all participating laboratories and sends an analysis to each participant (Figure 2.2), thereby enabling them to gain an insight into their routine capabilities. If internal quality control procedures are functioning correctly, external assessment should reveal few unpleasant surprises.

Organization

Participation in the UK quality assessment scheme is voluntary and strictly confidential. Confidentiality is achieved by allocating each laboratory a unique identification number known only to the organizer and participant. This number is used

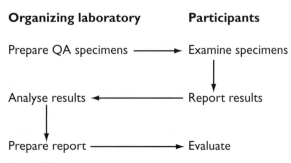

Figure 2.2 External quality assessment scheme

in all routine transactions and individual participant's results are not revealed to other parties except under previously agreed and strictly defined circumstances.

Difficulties arise in the organization of external schemes in countries where the size and function of participant laboratories vary widely. There may also be differences in the type of laboratory report issued, particularly in the field of clinical microbiology. Thus some laboratories report only the results of tests whereas others edit and interpret the results. External quality assessment schemes must take account of these differences by offering basic and comprehensive options (Gavan, 1974). Other schemes reflect the philosophy that, although there may be differences in the types of laboratories, the same standards should be applied to their assessment. Although international schemes do exist, their aims differ from those of national schemes because they are principally used to assess selected national laboratories in developing countries as part of a wider support programme. International schemes may be used to compare the standards of major laboratories in different countries. These laboratories can then use the expertise and knowledge gained to initiate their own national schemes (Snell and Hawkins, 1991).

Specimens

External quality assessment specimens should resemble, as closely as possible, materials usually examined by the participant laboratories. In some instances this is easily achieved, as with specimens for assay of antibodies in serum. In other cases, sufficient quantities of natural material may not be available and it is necessary to produce artificial specimens. As external quality assessment should test a laboratory's routine capabilities, specimens must to a great extent reflect the routine workload. Thus, as most specimens examined in a clinical laboratory do not yield pathogenic organisms, some of the simulated specimens should contain only non-pathogens. Commonly isolated organisms must be represented in quality assessment specimens but, as

most laboratories experience little difficulty in isolating and identifying, e.g. *Staphylococcus aureus*, there is little to be learned from the inclusion of too many such specimens. Less frequently encountered organisms, such as *Corynebacterium diphtheriae*, must also be distributed because they play an important role in the education of junior members of staff.

To gain and maintain credibility, participants must have full confidence in the specimens being distributed. The homogeneity and stability of specimens must be established and controlled to ensure their comparability. Specimens designed to test isolation procedures must have the same number of organisms in each vial; this is most easily achieved by freeze-drying (DeMello and Snell, 1985). Specimens prepared for the detection of antigen or antibody must react in a normal and predictable manner in a full range of commonly used assays. If participants are to evaluate their performance, they need to know if their results are incorrect. Thus, in general, specimens should yield clear-cut results. In reality, this can be quite difficult, e.g. it is not easy to select strains of bacteria that are unequivocally susceptible or resistant to a range of relevant antibiotics. The use of strains giving 'borderline' results may, however, highlight the failure of laboratories to use optimal methods.

Microbial susceptibility tests

Quality control is vital here (see Chapter 12).

Benefits of quality assessment

The success of a quality assessment scheme is judged ultimately by its usefulness, the extent of which is primarily determined by the attitude of the participant. Quality assessment is a management tool and information gained from the scheme will be rendered ineffective if not used correctly. The whole philosophy of quality assessment is based on the assumption that the results obtained should reflect what is happening in day-to-day practice. For this to be true, quality assessment specimens must be examined by the same staff and with the same methods and reagents as specimens in the normal workload.

Expenditure of extra effort on quality assessment specimens, as a conscious attempt to 'cheat' or as a subconscious desire to excel in a challenge of ability (La Motte et al., 1977), negates the usefulness of the assessment and fails to reveal any unsuspected problems. Pressures on laboratory staff to perform well in quality assessment schemes, especially if success is seen as contributing to the economic viability of the laboratory by providing proof of excellence to customers, are understandable but should be resisted.

Accreditation

Accreditation is a valuable component of quality assurance in clinical microbiology (Batstone, 1992). The purpose is simple: the audit by an external agency of an applicant's organizational and quality assurance programme to see that certain defined standards are met. An accreditation scheme should be comprehensive, covering all aspects of the laboratory, including organization and administration, staff development and education, facilities and equipment, and policies and procedures. Participation in all relevant quality assessment schemes is a requirement for accreditation. There are several accreditation schemes available and before a laboratory embarks on the process consideration should be given to the appropriateness of the various schemes.

REFERENCES

Bartlett, R. C. (1982) Making optimum use of the microbiology laboratory. *Journal of the American Medical Association* **247**: 857–859.

Bartlett, R. C., Heard, S. R., Schiller, G. *et al.* (1978) Quality assurance in clinical microbiology. In

Horn, S. L. (ed.) *Quality Assurance Practices for Health Laboratories* Washington: American Public Health Association, 1978: 871–1005.

Batstone, F. (1992) Medical audit in clinical pathology. *Journal of Clinical Pathology* **45**: 284–287.

Borriello, S. P. (1999) Science, medicine, and the future: Near patient microbiological tests. *British Medical Journal* **319**: 298–301.

Campbell, S. M., Sheaf, R., Sibbald, B. *et al.* (2002) Implementing clinical governance in English primary care groups/trusts: reconciling quality improvement and quality assurance. *Quality and Safety in Health Care* **11**: 9–14.

Crook, M. (2002) Clinical governance and pathology. *Journal of Clinical Pathology* **55**: 177–179.

DeMello, J. V. and Snell, J. J. S. (1985) Preparation of simulated clinical material for bacteriological examination. *Journal of Applied Bacteriology* **59**: 421–436.

European Committee for Clinical Laboratory Standards (1985) *Standard for Quality Assurance. Part 1: Terminology and general principles*, Vol 2. Lund: ECCLS: 1–13.

Fox, M. C., Ward, K. and Kirby, J. (1992) First steps to accreditation. *IMLS Gazette* August: 412.

Gavan, T. L. (1974) A summary of the bacteriology portion of the 1972 (CAP) quality evaluation program. *American Journal of Clinical Pathology* **61**: 971–979.

Gray, J. J., Wreggitt, T. G., McKee, T. A. *et al.* (1995) Internal quality assurance in a clinical virology laboratory. I Internal quality assessment. *Journal of Clinical Pathology* **48**: 168–173.

Heard, S. R., Schiller, G., Shef, R. *et al.* (2001) Continuous quality improvement: educating towards a culture of clinical governance. *Quality in Health Care* **10**: 70–78.

La Motte, L. C., Guerrant, G.O., Lewis, D.S. *et al.* (1977) Comparison of laboratory performance with blind and mail- distributed proficiency testing samples. *Public Health Report* **99**: 554–560.

Nicholls, E. (1999) Quality control of culture media. In: Snell, J. J. S., Brown, D. F. J. and Roberts, C. (eds), *Quality Assurance Principles and Practice in the Microbiology Laboratory*. London: Public Health Laboratory Service, 119–140.

Snell, J. J. S. and Hawkins, J. M. (1991) Quality assurance – achievements and intended directions. *Reviews in Medical Microbiology* 3: 28–34.

Snell, J. J. S., Brown, D. F. J. and Roberts. C. (eds) (1999) *Quality Assurance Principles and Practice in the Microbiology Laboratory*. London: Public Health Laboratory Service.

Westwell, A. (1999) Quality control of bacteriological characterization tests. In: Snell, J. J. S., Brown, D. F. J. and Roberts, C. (eds) *Quality Assurance Principles and Practice in the Microbiology Laboratory*. London: Public Health Laboratory Service, pp. 141–150.

Wilcox, K. R., Baynes, T.E., Crabbe, J.V. *et al.* (1978) Laboratory management. In: Horn, S. L. (ed.), *Quality Assurance Practices for Health Laboratories*. Washington DC: American Public Health Association, pp. 3–126.

WHO (1994) Practice of quality assurance in laboratory medicine in developing countries. In: Sharma K. B., Agarwell, D. S., Bullock, D. G. *et al.* (eds), *Health Laboratory Services in Support of Primary Health Care in Developing Countries*. New Delhi: Regional Publication, South-East Asia Regional Office, pp. 77–137.

3

Laboratory equipment

BASIC PRINCIPLES OF EQUIPMENT MANAGEMENT

Without equipment, there can be no microbiological methods and no practical microbiology. Experience has shown that, to be effective, equipment has to be properly managed. Such management should be an important part of the laboratory's overall quality assurance (QA) system. Equipment management is therefore an important topic for the microbiologist (see Chapter 2).

LABORATORY EQUIPMENT IN A MODERN LABORATORY

Taking a large modern clinical microbiology laboratory as an example, the range of capital equipment to be found in use would include the following:

- Microscopes of different types
- Centrifuges of different types
- Incubators of different types
- Water baths and heating blocks
- Refrigerators and freezers
- Autoclaves
- Microbiological safety cabinets
- Laminar flow clean air cabinets
- Colony counters.
- Freeze-drying apparatus
- Shakers and mixers
- Multi-point inoculators
- Micro-organism detection and identification systems
- Automatic pipettors

- Antimicrobial susceptibility test instruments
- Disc dispensers
- Anaerobic work stations and anaerobic jars
- pH meters
- Glassware washers
- Drying cabinets
- Plate pouring and other culture preparation systems.

In addition to this there are the basic tools and consumables:

- Inoculating loops, wires and spreaders
- Glassware items that are intended to be recycled, such as slides, bottles, tubes, Petri dishes and pipettes
- Pipetting devices and pipette tips
- Petri dish and pipette drums
- Racks and baskets for tubes and bottles
- Disposable plastic items, including Petri dishes, pipette tips
- Specimen and sample containers of glass or plastic.

IN VITRO DIAGNOSTIC MEDICAL DEVICES

By far the largest proportion of the workload of a clinical and public health laboratory is the examination of human samples and the provision of results that are intended to help in making clinical diagnoses or in clinical management. Accordingly, these laboratories will be using some equipment, together with dedicated consumable items, that falls within the coverage of the European Union's

Directive on *in vitro* diagnostic medical devices (IVDs) (EU, 1998).

According to the Directive '*in vitro* diagnostic medical device' means any medical device that is a reagent, reagent product, calibrator, control material, kit, instrument, apparatus, equipment or system, whether used alone or in combination, intended by the manufacturer to be used *in vitro* for the examination of specimens, including blood, solely or principally for the purpose of providing information:

- concerning a physiological or pathological state
- concerning a congenital abnormality
- to determine the safety and compatibility with potential recipients
- to monitor therapeutic measures.

The first of these includes the identification of a cause of infection. Monitoring therapeutic measures, e.g. an antimicrobial susceptibility test or an antibiotic assay, are everyday activities for clinical microbiology laboratories.

It is important to note that the Directive indicates that specimen containers, if specifically intended by their manufacturers for the primary containment and preservation of specimens derived from the human body for the purposes of *in vitro* diagnostic examination, are considered to be IVDs. However, the Directive emphasizes that items of general laboratory equipment are not IVDs unless their manufacturer specifically intends them to be used for *in vitro* diagnostic examination.

After 7 December 2003, manufacturers of IVDs must comply with national legislation that implements the requirements of the Directive, i.e. before they can be marketed IVDs must meet the relevant *essential requirements* that are laid down in the Directive. These focus on design and manufacturing requirements aimed at ensuring reliable results and safety in use.

When satisfied that essential requirements are met, after 7 December 2003, the manufacturer has to affix to the product the CE mark of conformity. However, distributors of IVDs may continue to sell until 7 December 2005 products that are not CE marked, but which are still in the supply chain.

CE marking involves different levels of regulatory control of different IVDs. The level of control reflects the risk category of the type of IVD. The most highly regulated IVDs include tests for markers of HIV-1 and -2, HTLV-I and -II, and hepatitis B virus (HBV), HCV and HDV.

Member states of the EU are required by the Directive to take all necessary steps to ensure that IVDs may be placed on the market and/or put into service only if they comply with the requirements of the Directive when duly supplied and properly installed, maintained and used in accordance with their intended purpose. This raises an additional important point, i.e. users clearly have a role to play in the proper use of an IVD if it can be expected to produce reliable results. Indeed, under the UK's *In Vitro* Diagnostic Medical Device Regulations (Statutory Instrument, 2000), if a user reassigns the intended use of a CE-marked IVD or modifies the instructions for use, without the agreement of the manufacturer, he or she becomes liable for any resulting performance failure that arises from such use of the product.

Against this background, with reference to the manufacturer's intended application, e.g. as stated in advertisements, sales literature and on packaging and labelling, examples of IVDs that might be of interest to a clinical microbiology laboratory would include dedicated equipment for antimicrobial sensitivity testing, blood culture, HIV and HBV testing, and specimen containers intended for blood, urine and faeces. On the other hand, a laboratory centrifuge, microscope, mixing machine, pipette or test tube that is not marketed for a specific application would not fall within the scope of the IVD Directive.

Finally, a hospital laboratory may be responsible for the management of IVDs that are used outside the laboratory by nursing or medical staff to provide on-the-spot results, i.e. in a situation that is known, among other things, as point of care testing (POCT). Other terms used to describe POCT are near patient testing, bedside testing, extra-laboratory testing and disseminated laboratory testing.

Examples that may be relevant to the clinical microbiology laboratory are rapid test systems for infectious disease markers and urinalysis. Some issues in the general management of IVDs and point of care testing IVDs are dealt with below.

PRINCIPLES OF EQUIPMENT MANAGEMENT

The equipment management principles outlined in this chapter are based on guidance that has been issued by the UK Department of Health's Medical Devices Agency (MDA) and its predecessor the Medical Devices Directorate (MDD, 1991). The purpose is to ensure, in a cost-effective manner, that equipment:

- is suitable for its intended purpose
- is understood by the users
- is in a safe and serviceable condition
- meets current safety and quality requirements
- where relevant, has a data output that is compatible with the laboratory's existing or envisaged data management system.

The essential features of equipment management are:

- justification of need
- selection
- acceptance upon delivery
- training in use
- maintenance, repair and modification
- replacement (if necessary) and disposal.

Justification of need

It is recommended that the laboratory have a procedure in place that requires the purchaser to make out a case in support of the need for the equipment. Where the equipment has never been used before in the laboratory, the case should indicate the job of work to be done by the new item and, where a replacement for an existing item of equipment is being sought, the need for replacement should be justified. The justification of need should take into account the current work pattern and likely future trends, including possible changes in workload and availability of trained staff.

Selection

Experience has shown that, unless good quality equipment is purchased in the first place, no subsequent activity will effect improvement. It is therefore of the utmost importance that there are arrangements for providing expert advice when needed and for carrying out all the necessary pre-purchasing investigations. In the UK, expert advice is available from the MDA, which organizes an evaluation programme and publishes reports on IVDs and other items used in the laboratory. This provided independent third party assessment of performance, sometimes carried out as comparative evaluation between different products with the same intended purpose. The MDA operated a voluntary adverse incident reporting programme and published safety notices that are of interest to those involved in the purchase of laboratory equipment. On 1 April 2003, however, the MDA merged with the Medicines Control Agency to form the Medicines and Health Care Products Regulatory Agency, and it is not clear what kind of advice this new body will provide.

In any case, before new equipment is purchased, it is always advisable to obtain the personal advice of microbiologists who have had experience in its use. It is bad policy to rely entirely on advertisements, catalogues, extravagant claims of representatives, and the opinions of purchasing officers who are mainly concerned with balancing budgets. The best is not always the most expensive, but it is rarely the cheapest.

A detailed specification should be drawn up for the equipment. This should itemize the performance requirements with reference to its intended immediate use. Although care should be taken not to specify a performance better than needed, the longer-term use of the equipment in the light of possible future test requirements should be taken into account. Especially when the purchase of expensive capital that is available from more than one supplier is contemplated, there is a need for an objective demonstration of how the specification is met by each of them. The emphasis should be on safety and quality, the latter being defined as fitness for purpose. Furthermore, the equipment should have been designed specifically for use in microbiological laboratories. That designed for other disciplines or other purposes is rarely satisfactory.

Although nowadays it is unlikely that laboratory equipment offered for sale will not meet the current safety requirements, especially if it is manufactured for the European or American markets, nevertheless compliance with safety standards should be investigated. Fortunately, if an item of equipment from a reputable manufacturer carries the European CE marking or claims of compliance with a relevant standard are made, a reasonable presumption can be made that the relevant safety standards are likely to be met. Nevertheless, it is advisable to include safety checks as part of the acceptance procedure (see below). A list of British Standards (BS and BS EN) for laboratory equipment is given in the Addendum to this chapter (see p. 43).

Although the EU Directive applies only to states within the EU and the MDA operates only in the UK, the principle of both could be adopted, with advantage, in other countries.

Other factors to consider when selecting equipment are the availability, throughout the planned life of the equipment, of adequate maintenance and repair facilities, availability of spare parts and consumables, and training for staff. The overall cost of equipment is a highly significant factor when making a purchasing decision. In assessing the cost of equipment, the whole life figure should be considered rather than simply the initial capital expenditure. Over the lifetime of the equipment, recurring revenue costs can often amount to more than the initial purchase price. Indeed, manufacturers often rely on the recurring sale of consumables for their profits. Other revenue costs include those for maintenance and repair, spare parts, installation and training.

It is recommended that, wherever possible, the selection process should be a team effort, the aim being that the final purchasing decision is a consensus following consultation with the intended users and with full consideration being given to their views.

Given that they will have to work with the equipment that is chosen, perhaps for a significant part of a working day, if there is a choice of more than one supplier at the end of the line, it is probably best that the deciding vote should go to the users. Some fields of investigation that might be pursued are set out below.

Acceptance procedure

A formal acceptance procedure is needed to ensure that the entry of all equipment into service is properly controlled. This procedure will include the following:

- an initial inspection with safety checks for electrical equipment
- incorporation into the equipment management record system
- formal commissioning.

Initial inspection

On receipt of a new acquisition, an initial inspection should be carried out to ensure that the equipment is complete with all the accessories and users' operating manuals, and that it is undamaged and in proper working order.

Records

It is recommended that a record be maintained and kept at all times with the equipment. This record should contain brief descriptions of all inspections, maintenance, repair and any modification. The first entry should be of the acceptance of the equipment into service. The record should be used to note minor faults that are found during normal use so that they can be brought to the attention of the service technician at the next scheduled planned preventive maintenance visit (see below). Periods of time that the equipment is out of service because of breakdown should also be noted in the record.

Commissioning

Some types of equipment, e.g. autoclaves and biological safety cabinets, will require, after installation, a formal commissioning procedure against objective test criteria to demonstrate that they are functioning safely and effectively before routine use. This will need to be recorded.

Staff training

No new laboratory equipment should be allowed to enter service until potential users have had adequate training in its use, user servicing and emergency action in the event of malfunction. It is recommended that those unfamiliar with the equipment should be forbidden to operate it unless supervised or until they are considered competent in its use.

The content of the training and who delivers the training will depend on the complexity of the equipment, but as a general rule it is recommended that, if a supplier offers a training course for a new item of equipment, this should be taken up. It is important that the members of staff who attend the course will be the ones who will actually use it. Furthermore, it should be recognized that trained staff will move on to other posts and their places may be taken by those who may not have seen or operated the equipment. Reliance should not be placed on them being able to pick up how to operate equipment from established operators; this approach often perpetuates bad habits and dangerous short-cuts. It is important that replacement staff receive full and systematic training according to a syllabus. Those for centrifuges and microbiological safety cabinets are given on pp. 29 and 35, respectively.

This is particularly important where user servicing is concerned because experience has shown that operational malfunctions and dangerous occurrences, such as fires or explosions, have sometimes occurred immediately after faulty user servicing (see below).

Maintenance, repair and modification

If equipment is to remain safe, serviceable and reliable throughout its working life, planned controlled maintenance is essential. In addition, arrangements should be in place to ensure that emergency repairs and any modifications that have to be carried out to improve the performance or safety of equipment are carried out promptly and effectively.

Decontamination

Equipment that has to be inspected, maintained, repaired, modified or scrapped (see below) should be in a condition that makes it safe to be handled by all those who may come into contact with it during transit or subsequent handling (MDA, 1998). Before any such work is done on microbiology laboratory equipment, to reduce the risk of transmission of infection, or fear of infection, it must be cleaned and decontaminated in accordance with the manufacturer's specific instructions, if provided. In the absence of specific instructions for decontamination, the manufacturer should be consulted about compatibility with intended agents to ensure, as far as possible, that they are compatible with the equipment and do not damage it (see under Microbiological safety cabinets and Chapter 4).

If the equipment is to be sent out of the laboratory for the work to be done, e.g. it is to be sent to the manufacturer's factory, or the work is to be carried out on site by staff other than those employed by the laboratory, such as by a visiting engineer, a decontamination certificate should be issued under the authority of the head of the laboratory.

Maintenance

There are two levels of maintenance to consider:

1. Basic scheduled maintenance, i.e. user servicing, which the manufacturer intends should be carried out by the actual user of the equipment; sometimes this amounts to nothing more than the laboratory equivalent of good housekeeping, e.g. cleaning and simple adjustments,

2. Planned preventive maintenance (PPM), which the manufacturer intends should be carried out to ensure long-term reliable performance in accordance with his specification. This often involves comprehensive inspection, changing components and subassemblies that require regular replacement, calibration, performance tests and a final safety check. This amounts to the laboratory equivalent of car servicing.

Responsibility for day-to-day supervision of user servicing should rest with the head of the laboratory who may wish to delegate this function to a senior member of the staff. It is essential that the work is done only by trained staff strictly in accordance with a written schedule, and that complete and up-to-date records are kept. A daily checklist will usually suffice.

Although contract PPM will usually be available from the manufacturer or an appointed agent, it may also be available from a third-party commercial source or though an in-house maintenance scheme, e.g. as is sometimes operated by hospital engineering departments in the UK. Setting aside economic considerations, the head of the laboratory will need to ensure that whoever carries out the PPM can demonstrate that they have readily available:

- sufficient technical information and spare parts
- sufficient trained staff, including supervisors
- necessary tools and test equipment.

Repair

If user servicing and PPM are being carried out properly and in accordance with the manufacturer's schedule, breakdowns should be minimized. However, random failures can occur even with well-maintained equipment, and the head of the laboratory should ensure that arrangements are in place so that repairs can be effected speedily. In addition to the requirements detailed for the provision of PPM, whoever is given the responsibility of attending to a request for a repair should be able to provide a guarantee that they can respond within an agreed timescale.

Modification

Uncontrolled modification of laboratory equipment can adversely affect performance and can be dangerous. Modification should be carried out only if formal authorization has been given by the head of the laboratory or a senior member of staff to whom authority has been delegated. It is preferable that modification be carried out by the manufacturer in order to ensure that essential safety features are not impaired. Manufacturers under contract for the provision of PPM will sometimes offer modifications that have been developed to improve the safety or performance of their equipment. One of the disadvantages of third-party contract or in-house provision of PPM is that the availability of such modifications may not become apparent.

Replacement and disposal

In due course, equipment may have to be replaced for one or more of the following reasons. The following can happen to the equipment:

- It may be worn out beyond economic repair.
- It may be damaged beyond economic repair.
- It may be unreliable.
- Results produced are clinically obsolete or the methodology for producing them is technologically obsolete.
- Spare parts are no longer available.
- Consumables are no longer available.
- More cost-effective equipment is available.

It is good practice to keep all existing equipment under review so that an item of equipment that will need replacement can be identified in good time, thereby facilitating the necessary financial and systematic procurement arrangements for replacement. In particular, the head of the laboratory should resist any drive that would lead to the hurried or unstructured purchase of replacement equipment at the end of a financial year because of fear that 'the money will be lost otherwise'. Such activity has often been associated with poor value for money or the purchase of equipment that is not really needed.

However unwanted equipment is disposed of, even if it is destined for the scrapyard, it should never pass out of the control of the laboratory without it being cleaned and decontaminated and accompanied by a signed decontamination certificate.

MANAGEMENT OF IVDS

Purchasers of IVDs will need to be aware that the essential requirements of the IVD Directive are that products must achieve the claims that are made for them by manufacturers in terms of analytical sensitivity, diagnostic sensitivity, analytical specificity, diagnostic specificity, accuracy, repeatability and reproducibility, including control of known relevant interference and limits of detection. Furthermore, the manufacturer must be able to demonstrate with data that the performance claims have been met. The Directive also states that the product must not compromise the health or safety of users.

In short, this means that a customer who purchases a CE-marked product from a reputable manufacturer should be able to have a high degree of confidence that it is capable of working properly if used as specified, and will be safe to use. However, it is important to emphasize that the Directive does not require a manufacturer to demonstrate clinical utility as a condition of placing an IVD on the market. This means that the onus is on the purchaser to assess whether the product is suitable for the use to which the results will be put in the clinical management of patients.

Another aspect to consider is that the IVD Directive exists to remove technical barriers to trade within the EU. Thus, it can be expected that manufacturers from across the expanding EU will be offering IVDs that, although they may have been developed originally for a specific home market, will be widely exported. This may necessitate translations of the original instructions. Good clear instructions have a crucial role in the safe and effective use of IVDs. In the light of these points, the laboratory is advised to ensure (MDA, 2002a) that:

- The users of IVDs and the clinicians who will use the results produced by them should be consulted when making purchasing decisions.
- Performance claims made by manufacturers and arrangements for provision of training and maintenance and the compatibility of a candidate system's output with the laboratory's data management system should be considered at the selection stage.
- Before an IVD is introduced into routine use, it should be subjected to a thorough evaluation in the hands of the laboratory staff, as a means of validation.
- The evaluation should include an investigation to establish whether existing specimen collection devices, e.g. blood sample tubes and bacteriological swabs, can adversely affect the results produced by the IVD.
- The instructions for use must be understandable by all the intended users – this is especially important when the instructions that have been issued are translations.
- Adequate levels of staff training are maintained.
- Arrangements are in place for maintenance and repair.

Point of care testing IVDs

Although today's POCT IVDs are generally more reliable and less prone to error than those of previous generations, the successful implementation of POCT service is still dependent on the effective organization and management of staff.

Users of IVDs for POCT should have a sound understanding of the relevant analytical principles, the importance of QA and the interpretation of test results. It is therefore important that the users of POCT IVDs have access to clear guidance on these issues. In the light of these points, the laboratory that is responsible for providing the clinical microbiology service for the place where a POCT IVD is to be used is advised (MDA, 2002b) to ensure that it:

- is involved at the earliest opportunity in all aspects of the provision of a POCT service involving a microbiological test
- oversees all training, updating and monitoring of all staff involved in the POCT service, and should be prepared to act as an initial 'trouble-shooter' for all problems involving POCT IVDs
- oversees the preparation of all standard operating procedures involving the POCT service
- oversees the integration of the POCT service into its overall quality system.

CAPITAL EQUIPMENT

Equipment for which there are British Standards are marked with an asterisk(*) (see the Addendum to this chapter on p. 43).

Microscopes

Laboratory workers should, of course, be familiar with the general mechanical and optical principles of microscopes but a detailed knowledge of either is unnecessary and, apart from superficial cleaning, maintenance should be left to the manufacturers, who will arrange for periodic visits by their technicians. Most manufacturers publish handbooks containing useful explanations and information.

Wide-field compensating eyepieces should be fitted; they are less tiring than other eyepieces and are more convenient for spectacle wearers. For low-power work, $\times 5$ and $\times 10$ objectives are most useful and, for high-power (oil immersion) microscopy, $\times 50$ and $\times 100$ objectives are desirable.

Careful attention to critical illumination, centring and the position of diaphragms is essential for adequate microscopy.

Low-power microscopes
Low-power, binocular, dissecting microscopes are better than hand lenses for examining colony mor-

phology. Experience in their use may save much time and expense in pursuing unnecessary tests (see p. 85). They also facilitate the subculture of small colonies from crowded Petri dishes, thus avoiding mixed cultures.

Incubators

Incubators are available in various sizes. In general, it is best to obtain the largest possible models that can be accommodated, although this may create space problems in laboratories where several different incubation temperatures are employed. The smaller incubators suffer wider fluctuations in temperature when their doors are opened than the larger models, and most laboratory workers find that incubator space, like refrigerator space, is subject to 'Parkinson's law'.

Although incubators rarely develop faults, it is advisable, before choosing one, to ascertain that service facilities are available. The circuits are not complex but their repair requires expert technical knowledge. Transporting incubators back to the manufacturers is most inconvenient.

In medical and veterinary laboratories, incubators are usually operated only at 35–37°C. The food or industrial laboratory usually requires incubators operated at 15–20°C, 28–32°C and 55°C. For temperatures above ambient there are no problems, but lower temperatures may need a cooling as well as a heating device.

CO$_2$ incubators
Incubation in an atmosphere of 5–8% CO_2 is preferable for the cultivation of many bacteria of medical importance. Incubators that provide automatic control of CO_2 and humidity are now available.

Water used to maintain humidity may become contaminated with fungi, especially aspergilli, causing problems with cultures that require prolonged incubation.

Cooled incubators

For incubation at temperatures below the ambient, incubators must be fitted with modified refrigeration systems with heating and cooling controls. These need to be correctly balanced.

Automatic temperature changes

These are necessary when cultures are to be incubated at different temperatures for varying lengths of time, as in the examination of water by membrane filtration, and when it is not convenient to move them from one incubator to another. Two thermostats are required, wired in parallel, and a time switch wired to control the thermostat set at the higher temperature. These are now built in to some models.

Portable incubators

These are useful in field work, e.g. environmental microbiology and water examination (see p. 258).

Incubator rooms

These should be installed by specialist suppliers. Metal shelving is preferable to wood: if the humidity in these rooms is high the growth of mould fungi may be encouraged by wood. Solid shelving should be avoided and space left at the rear of the shelves to allow for the circulation of air.

If, on the other hand, low humidity is a problem cultures may be prevented from drying out by placing them along with pieces of wet filter paper in plastic boxes such as those sold for the storage of food.

Centrifuges*

Over the years there have been a number of accidents involving centrifuges. Some of these involved the ejection into the laboratory of centrifuge buckets and other parts. Although some of these accidents were caused by faulty components, others were the result of misuse (Kennedy, 1988). The high kinetic energy of parts ejected from a centrifuge poses a serious risk of injury. In addition, if

an accident occurs while bodily fluids are being centrifuged, an infection risk arises from microbially contaminated environmental surfaces and aerosols (Collins and Kennedy, 1999).

Mechanical, electrical and microbiological safety are important considerations in the purchase and use of centrifuges. They should conform to national and international standards, e.g. those listed in the Addendum (see p. 43), and also the International Federation of Clinical Chemistry (IFCC, 1990) and the International Electrical Commission (IEC, 1990).

Although modern centrifuges are designed to meet high safety standards (including those of BS EN, 1995, see Addendum) and are less likely to lend themselves to misuse, it is strongly recommended that all potential users of a centrifuge are trained in its safe use before they are permitted to start using it. It is important that users receive training in every type of centrifuge that they will be using, including operations involving sealed buckets if these are to be used. A model set of instructions and precautions for the safe use of centrifuges are offered below.

The ordinary laboratory centrifuge is capable of exerting a force of up to 3000 g and this is the force necessary to deposit bacteria within a reasonable time.

For general microbiology, a centrifuge capable of holding 15-ml and 50-ml buckets and working at a maximum speed of 4000 rev./min is adequate. The swing-out head is safer than the angle head because it is less likely to distribute aerosols if uncapped tubes are used (Collins and Kennedy, 1999).

For maximum microbiological safety, sealed buckets should be fitted. These confine aerosols if breakage occurs and are safer than sealed rotors. Several companies supply sealed buckets ('safety cups'). Windshields offer no protection. There should be an electrically operated safety catch so that the lid cannot be opened when the rotors are spinning.

Centrifuge buckets are usually made of stainless steel and are fitted with rubber buffers. They are paired and their weight is engraved on them. They are always used in pairs, opposite to one another, and it is convenient to paint each pair with different coloured patches so as to facilitate recognition. If the buckets fit in the centrifuge head on trunnions, these are also paired.

Centrifuge tubes to fit the buckets are made of glass, plastic, nylon or spun aluminium. Conical and round-bottomed tubes are made. Although the conical tubes concentrate the deposit into a small button, they break much more readily than the round-bottomed tubes. When the supernatant fluid, not the deposit, is required, round-bottomed tubes should be used.

Instructions for using the centrifuge

1. Select two centrifuge tubes of identical lengths and thickness. Place liquid to be centrifuged in one tube and water in the other to within about 2 cm of the top.
2. Place the tubes in paired centrifuge buckets and place the buckets on the pans of the centrifuge balance. This can be made by boring holes large enough to take the buckets in small blocks of hard wood, and fitting them on the pans of a simple balance.
3. Balance the tubes and buckets by adding 70% alcohol to the lightest bucket. Use a Pasteur pipette and allow the alcohol to run down between the tube and the bucket.
4. Place the paired buckets and tubes in diametrically opposite positions in the centrifuge head.
5. Close the centrifuge lid and make sure that the speed control is at zero before switching on the current. (Many machines are fitted with a 'no-volt' release to prevent the machine starting unless this is done.)
6. Move the speed control slowly until the speed indicator shows the required number of revolutions per minute.

Precautions

1. Make sure that the rubber buffers are in the buckets, otherwise the tubes will break.
2. Check the balancing carefully. Improperly balanced tubes will cause 'head wobble' and spin-off accidents, and wear out the bearings.

3. Check that the balanced tubes are really opposite one another in multi-bucket machines.
4. Check that the buckets are properly located on the centrifuge head.
5. Never start or stop the machine with a jerk.
6. Observe the manufacturer's instructions about the speed limits for the various loads.
7. Load and open sealed centrifuge buckets in a microbiological safety cabinet.

Maintenance

The manufacturers will, by contract, arrange periodic visits for inspection and maintenance in the interests of safety and efficiency.

Water baths

The contents of a test tube placed in a water bath are raised to the required temperature much more rapidly than in an incubator. These instruments are therefore used for short-term incubation. If the level of water in the bath comes to one-half to two-thirds of the way up that of the column of liquid in the tube, convection currents are caused, which keep the contents of the tube well mixed and hasten reactions such as agglutination.

All modern water baths are equipped with electrical stirrers and in some the heater, thermostat and stirrer are in one unit, easily detached from the bath for use elsewhere or for servicing. Water-baths must also be lagged to prevent heat loss through the walls. A bath that has not been lagged by the manufacturers may be insulated with slabs of expanded polystyrene.

Water-baths should be fitted with lids in order to prevent heat loss and evaporation. These lids must slope so that condensation water does not drip on the contents. To avoid chalky deposits on tubes and internal surfaces, only distilled water should be used, except, of course, in those baths that operate at or around ambient temperature, which are connected to the cold water supply and fitted with a constant level and overflow device.

This kind of bath offers the same problems as the low-temperature incubators unless the water is supplied at a much lower temperature than that required in the water bath. In premises that have an indirect cold water supply from a tank in the roof, the water temperature in summer may be near or even above the required water bath temperature, giving the thermostat no leeway to operate. In this case, unless it is possible to make direct connection with the rising main, the water supply to the bath must be passed through a refrigerating coil. Alternatively, the water bath may be placed in a refrigerator and its electricity supply adapted from the refrigerator light or brought in through a purpose-made hole in the cabinet. The problem is best overcome by fitting laboratory refrigeration units designed for this purpose.

Metal block heaters

These give better temperature control than water baths and do not dry out, but they can be used only for tubes and thin glass bottles that must fit neatly into the holes in the steel or aluminium blocks.

It is advisable to use these blocks for work with tubes containing Group 3 agents that might leak into and contaminate the water in water baths.

Refrigerators and freezers

Although domestic appliances are usually satisfactory for laboratory work they may pose a hazard if they have unsealed electrical contacts and if chemicals with flammable or explosive vapours are stored in them. If there is a need to store such material in a domestic-type refrigerator, it should be a small amount in a sound container with a leak-proof closure and preferably packaged inside a larger container.

Ideally all sources of ignition, e.g. thermostat contacts and light switch mechanisms, should be outside the refrigeration chamber. Intrinsically safe refrigerators are available commercially but they are expensive.

Refrigerators and freezers that are not intended for the storage of flammable liquids should be labelled as such (Kennedy, 1988).

Sterilization equipment

For details about this equipment, see Chapter 4.

Microbiological safety cabinets*

Microbiological safety cabinets (MSCs), which are also known as biological safety cabinets (BSCs), are intended to capture and retain infected airborne particles released in the course of certain manipulations and to protect the laboratory worker from infection that may arise from inhaling them.

There are three kinds: Classes I, II and III. In Europe Class I and Class II cabinets are used in diagnostic, Level 2 and Containment, Level 3 laboratories for work with Hazard Group 3 organisms. Class III cabinets are used for Hazard Group 4 viruses. Some other countries use only Class II and Class III cabinets.

A Class I cabinet is shown in Figure 3.1. The operator sits at the cabinet, works with the hands inside, and sees what he or she is doing through the glass screen. Any aerosols released from the cultures are retained because a current of air passes in at the front of the cabinet. This sweeps the aerosols up through the filters, which remove all or most of the organisms. The clean air then passes through the fan, which maintains the airflow, and is

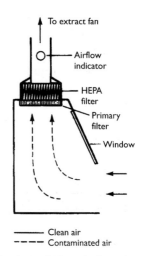

Figure 3.1 Class I microbiological safety cabinet

exhausted to atmosphere, where any particles or organisms that have not been retained on the filter are so diluted that they are no longer likely to cause infection if inhaled. An airflow of between 0.7 and 1.0 m/s must be maintained through the front of the cabinet, and modern cabinets have airflow indicators and warning devices. The filters must be changed when the airflow falls below this level.

The Class II cabinet (Figure 3.2) is more complicated and is sometimes called a laminar flow cabinet, but as this term is also used for clean air cabinets that do not protect the worker it should be avoided (see p. 37). In the Class II cabinet about 70% of the air is recirculated through filters, so that the working area is bathed in clean (almost sterile) air. This entrains any aerosols produced in the course of the work, which are removed by the filters. Some of the air (about 30%) is exhausted to atmosphere and is replaced by a 'curtain' of room air, which enters at the working face. This prevents the escape of any particles or aerosols released in the cabinet.

There are other types of Class II cabinets, with different airflows and exhaust systems, which may also be used for toxic chemicals and volatile chemicals. Manufacturers should be consulted.

Class III cabinets are totally enclosed and are tested under pressure to ensure that no particles can leak from them into the room. The operator works with gloves that are an integral part of the cabinet. Air enters through a filter and is exhausted to atmosphere through one or two more filters (Figure 3.3).

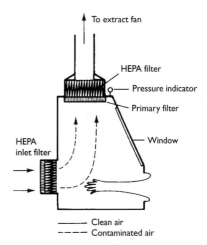

Figure 3.3 Class III microbiological safety cabinet

Purchasing standards for safety cabinets

Safety cabinets should comply with national standards, e.g. British Standards Institution (see Addendum, p. 43) and the (US) National Sanitation Foundation (NSF, 1992).

Use of safety cabinets

These cabinets are intended to protect the worker from airborne infection. They will not protect the worker from spillage and the consequences of poor techniques. The cabinet should not be loaded with unnecessary equipment or it will not do its job properly. Work should be done in the middle to the rear of the cabinet, not near the front, and the worker should avoid bringing hands and arms out of the cabinet while working. After each set of manipulations and before withdrawing the hands, the worker should wait for 2–3 min to allow any aerosols to be swept into the filters. The hands and arms may be contaminated and should be washed immediately after ceasing work. Bunsen burner

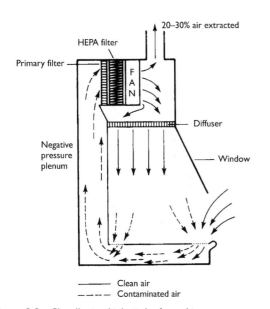

Figure 3.2 Class II microbiological safety cabinet

burners and even micro-incinerators should not be used because they disturb the airflow. Disposable plastic loops (see p. 37) are recommended.

Siting

The efficiency of a safety cabinet depends on correct siting and proper maintenance. Possible sites for cabinets in a room are shown in Figure 3.4. Site A is a poor site, because it is near to the door and airflow into the cabinet will be disturbed every time the door is opened and someone walks into the room and past the cabinet. Site B is not much better, because it is in almost a direct line between the door and window, although no one is likely to walk past it. Its left side is also close to the wall and airflow into it on that side may be affected by the 'skin effect', i.e. the slowing down of air when it passes parallel and close to a surface. Air passing across the window may be cooled, and will meet warm air from the rest of the room at the cabinet face, when turbulence may result.

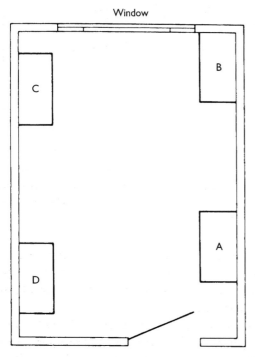

Figure 3.4 Possible sites for safety cabinets in relation to cross draughts from door and window and movement of staff. A is bad; B is poor; C is better; D is best

Site C is better and site D is best of all. If two cabinets are required in the same room, sites C and D would be satisfactory, but they should not be too close together, or one may disturb the airflow of the other. We remember a laboratory where a Class I cabinet was placed next to a Class II cabinet. When the former was in use it extracted air from the latter and rendered that cabinet quite ineffective.

Care must also be taken in siting any other equipment that might generate air currents, e.g. fans and heaters. Mechanical room ventilation may be a problem if it is efficient, but this can be overcome by linking it to the electric circuits of the cabinets so that either, but not both, is extracting air from the room at any one time. Alternatively, baffles may be fitted to air inlets and outlets to avoid conflicting air currents near to the cabinet. Tests with smoke generators (see below) will establish the directions of air currents in rooms.

None of these considerations needs to apply to Class III cabinets, which are in any case usually operated in more controlled environments.

Testing airflows

The presence and direction of air currents and draughts are determined with 'smoke'. This may be generated from burning material in a device such as that used by bee keepers, but is usually a chemical that produces a dense, visible vapour. Titanium tetrachloride is commonly used. If a cottonwool-tipped stick, e.g. a throat swab, is dipped into this liquid and then waved in the air, a white cloud is formed which responds to quite small air movements.

Commercial airflow testers are more convenient. They are small glass tubes, sealed at each end. Both ends are broken off with the gadget provided and a rubber bulb fitted to one end. Pressing the bulb to pass air through the tube causes it to emit white smoke. These methods are suitable for ascertaining air movements indoors.

To measure airflow an anemometer is required. Small vane anemometers, timed by a stopwatch, are useful for occasional work but for serious activities electronic models, with a direct reading scale, are essential. The electronic vane type has a diameter of about 10 cm. It has a satisfactory time constant and responds rapidly enough to show the changes in velocity that are constantly occurring when air is passing into a Class I safety cabinet. Hot wire or thermistor anemometers may be used but they may show very rapid fluctuations and need damping. Both can be connected to recorders.

It is necessary to measure the airflow into a Class I cabinet in at least five places in the plane of the working face (Figure 3.5). At all points this should be between 0.7 and 1.0 m³/s. No individual measurement should differ from the mean by more than 20%. If there is such a difference there will be turbulence within the cabinet. Usually some piece of equipment, inside or outside the cabinet, or the operator, is influencing the airflow.

velocity of air through this is measured and the average velocity over the whole face calculated from this figure and the area of the working face. The inflow at the working face should be not less than 0.4 m/s. A check should also be made with smoke to ensure that air is in fact entering the cabinet all the way round its perimeter and not just at the lower edge.

The downward velocity of air in a Class II cabinet should be measured with an anemometer at 18 points in the horizontal plane 10 cm above the top edge of the working face (Figure 3.6). The mean downflow should be between 0.25 and 0.5 m/s and no individual reading should differ from the mean by more than 20%.

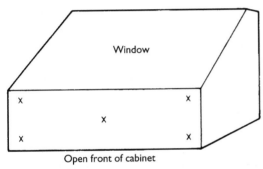

Figure 3.5 Testing airflow into a Class I safety cabinet. Anemometer readings should be taken at five places, marked with an X, in the plane of the working face with no-one working at the cabinet

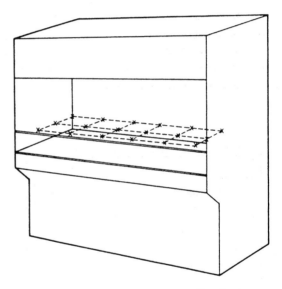

Figure 3.6 Testing the vertical airflow in a Class II safety cabinet. Anemometer readings should be taken at points marked X on an imaginary grid 6 inches within the cabinet walls and just above the level of the bottom of the glass window

Although airflows into Class I cabinets are usually much the same at all points on the working face, this is not true for Class II cabinets, where the flow is greater at the bottom than at the top. The average inward flow can be calculated by measuring the velocity of air leaving the exhaust and the area of the exhaust vent. From this the volume per minute is found and this is also the amount entering the cabinet. The average velocity is obtained by dividing this volume by the area of the working face. A rough and ready way, suitable for day-to-day use, requires a sheet of plywood or metal that can be fitted over the working face. In the centre of this an aperture is made which is 2 × 2.5 cm. The inward

It is usual, when testing airflows with an anemometer, to observe or record the readings at each position for several minutes, because of possible fluctuations.

To measure the airflow through a Class III cabinet the gloves should be removed and readings taken at each glove port. Measurements should also be taken at the inlet filter face when the gloves are attached.

Decontamination

As safety cabinets are used to contain aerosols, which may be released, during work with Hazard Group 3 micro-organisms, the inside surfaces and the filters will become contaminated. The working surface and the walls may be decontaminated on a day-to-day basis by swabbing them with disinfectant. Glutaraldehyde is probably the best disinfectant for this purpose because phenolics may leave sticky residues and hypochlorites may, in time, corrode the metal (*Caution* – see below).

Glutaraldehyde vapour is a cause of occupational asthma, irritation of the eyes, nose and throat. It is also a skin sensitizer and may cause allergic dermatitis. In the UK the Health and Safety Executive sets a maximum exposure limit of 0.05 ppm.

For thorough decontamination, however, and before filters are changed, formaldehyde (*Caution* – see below) should be used and precautions should be taken before use. The installation should be checked to ensure that none of the gas can escape into the room or other rooms. The front closure (night door) of the cabinet should seal properly on to the carcase, or masking tape should be available to seal it. Any service holes in the carcase should be sealed and the high efficiency particulate air (HEPA) filter seating examined to ensure that there are no leaks. If the filters are to be changed and the primary or roughing filter is accessible from inside the cabinet, it should be removed and left in the working area. The supply filter on a Class III cabinet should be sealed with plastic film.

Formaldehyde is severely irritating to the eyes and the respiratory tract. Inhalation may cause lung oedema. It is possibly carcinogenic to humans. The International Labour Organization advises an occupational exposure limit of 0.3 ppm. Mixtures of 7.75% (v/v) formaldehyde and dry air are explosive: the ignition point is 430°C.

Formaldehyde is generated by:

- boiling formalin, which is a 40% solution of the gas in water

- heating paraformaldehyde, which is its solid polymer.

Boiling formalin

The volume used is important. Too little will be ineffective; too much leads to deposits of the polymer, which is persistent and may contribute to the natural blocking of the filters. The British Standard (BS, 1992 – see Addendum, p. 43) specifies a concentration of 50 mg/m³. To achieve this 60 ml formalin, mixed with 60 ml water, is required for each cubic metre of cabinet volume (this volume is specified by the manufacturer).

The formalin may be boiled in several ways. Some cabinets have built-in devices in which the liquid is dripped on to a hot-plate. Other manufacturers supply small boilers. A laboratory hot-plate, connected to a timing clock, is adequate. The correct amount of formalin and water is placed in a flask.

Heating paraformaldehyde

Tablets (1 g) are available from chemical manufacturers. Three or four tablets are adequate for an average size cabinet. They may be heated on an electric frying pan connected to a time clock, or placed in a dish and heated by an electric hair drier.

Exposure time

It is convenient to start decontamination in the late afternoon and let the gas act overnight. In the morning the fan should be switched on and the front closure 'cracked' open very slightly to allow air to enter and purge the cabinet. Some new cabinets have a hole in the front closure, fitted with a stopper that can be removed for this purpose. With Class III cabinets the plastic film is taken off the supply filter to allow air to enter the cabinet and purge it of formaldehyde. After several minutes the gloves may be removed. The fan is then allowed to run for about 30 min, which should remove all formaldehyde.

Summary of decontamination procedure

The prefilter should be removed from inside the cabinet (where possible) and placed on the working surface. It is usually pushed or clipped into place and is easily removed, but as it is likely to be contaminated gloves should be worn. Formalin 60 ml plus water 60 ml, per cubic metre of cabinet volume, should be placed in its container on the heater in the cabinet or the reservoir. The front closure is then put in place and sealed if necessary. The heater is switched on and the formalin boiled away. After switching off the heater the cabinet is left closed overnight. The next morning the cabinet fan is switched on, and the front closure is opened very slightly to allow air to pass in and purge the cabinet of formaldehyde. After several minutes the front closure is removed and the cabinet fan allowed to run for about 30 min. Any obvious moisture remaining on the cabinet walls and floor may then be wiped away and the filters changed.

Changing the filters and maintenance

The cabinet must be decontaminated before filters are changed and any work is done on the motor and fans. If these are to be done by an outside contractor, e.g. the manufacturer's service engineer, the front closure should be sealed on again after the initial purging of the gas and a notice – *Cabinet decontaminated but not to be used* – placed on it awaiting the engineer's arrival. He or she will also require a certificate stating that the cabinet has been decontaminated.

The primary, or prefilters, should be changed when the cabinet airflow approaches its agreed local minimum. Used filters should be placed in plastic or tough paper bags, which are then sealed and burned. If the airflow is not restored to at least the middle of the range, then arrangements should be made to replace the HEPA filter. This is usually done by the service engineer, but may be done by laboratory staff if they have received instruction. Unskilled operators often place the new filter in upside down or fail to set it securely and evenly in its place. Used filters should be placed in plastic bags, which are then sealed for disposal. They are not combustible. Some manufacturers accept used filters and recover the cases, but this is not usually a commercial practice – no refund is given. When manufacturers replace HEPA filters they may offer a testing service.

Training operators

We recommend that all potential users of microbiological safety cabinets be trained in the proper use of every type of cabinet that they are likely to use, especially in the following:

- Classification
- Appropriate and inappropriate use
- Mode of operation and function of all controls and indicators
- Limitations of performance
- How to work safely with them
- How to decontaminate them after use
- Principles of airflow and operator protection tests (see below).

Testing: further information

This is beyond the scope of this book and the reader is referred to national standards and other publications (Kuhne *et al.* 1995; Collins and Kennedy, 1999; CDC/NIH, 2000).

Although a cabinet may be designed and constructed satisfactorily and be *capable* in the factory of meeting the performance claims made for it, unless it is properly installed in the laboratory it may not provide adequate operator protection. It was for this reason that BS 5726: Part 3, 1992 specified that an operator protection factor test, among other performance tests, must be carried out after installation. This is a very sensitive test that actually measures the ability of a cabinet to retain an aerosol. Unfortunately, the European Standard (designated BS EN 12469:2000 in the UK) that has replaced BS 5726:1992 does not mandate an operator protection factor test (which it calls a retention test) after installation, but makes it an option. It is recommended that an operator protection factor test is included in the commissioning procedure.

Autoclaves

For information about autoclaves, see Chapter 4.

Anaerobic work stations and anaerobic jars

For information about these, see Chapter 6.

Freeze-drying apparatus

For information about this, see Chapter 6.

Blenders and shakers

These range from instruments suitable for grinding and emulsifying large samples, e.g. food, to glass and plastic devices for homogenizing small pieces of tissue (see 'Tissue grinders and homogenizers', below). Most are electrically driven (see BS EN, 1996 in the Addendum). They suffer from the disadvantage of requiring a fresh, sterile cup for each sample. These cups are expensive and it takes time to clean and re-sterilize them.

Some of the devices may release aerosols during operation and when they are opened. A heavy Perspex or metal cover should be placed over them during use. This should be decontaminated after use. All blenders, etc. should be opened in a microbiological safety cabinet because the contents will be warm and under pressure, and aerosols will be released.

The Stomacher Lab-Blender overcomes these problems. Samples are emulsified in heavy-duty sterile plastic bags by the action of paddles. This is an efficient machine, capable of processing large numbers of samples in a short time without pauses for washing and re-sterilizing. There is little risk of aerosol dispersal. The bags are automatically sealed while the machine is working.

Vortex mixers are useful for mixing the contents of single bottles, e.g. emulsifying sputum, but should always be operated in a microbiological safety cabinet.

Shaking machines

Conventional shaking machines are sold by almost all laboratory suppliers. They are useful for mixing and shaking cultures and for serological tests, but should be fitted with racks, preferably of polypropylene, which hold the bottles or tubes firmly. These machines should be covered with a stout Perspex box when in use to prevent the dispersal of aerosols from containers that might leak. Alternatively, or in addition, we believe that all bottles that contain infected material and are shaken in any machine should be placed inside individual self-sealing or heat-sealed plastic bags, in order to minimize the dispersal of infected airborne particles.

The better models allow bottles, flasks, etc. to be rotated at varying speeds as well as for shaking in two dimensions.

Spiral platers

For information about these, see Chapter 6.

Micro-organism detection and identification systems

For information about these, see Chapter 6.

Antimicrobial susceptibility test instruments

For information about these, see Chapters 8 and 12.

Disc dispensers

These allow antimicrobial sensitivity and other paper discs to be easily and rapidly dispensed, singly or in sets on media in Petri dishes.

Glassware washing machines

These are useful in large laboratories that use enough of any one size of tube or bottle to give an economic load.

Glassware drying cabinets

A busy wash-up room requires a drying cabinet with wire-rack shelves and a 3-kW electric heater in the base, operated through a three-heat control. An extractor fan on the top is a refinement; otherwise the sides near the top should be louvered.

Laminar flow, clean air work stations

These cabinets are designed to protect the work from the environment and are most useful for aseptic distribution of certain media and plate pouring. A stream of sterile (filtered) air is directed over the working area, either horizontally into the room or vertically downwards when it is usually re-circulated. They are particularly useful for preventing contamination when distributing sterile fluids and hand pouring into Petri dishes.

Laminar flow clean air work stations are NOT microbiological safety cabinets and must NOT used for manipulations with micro-organisms or tissue culture cells. The effluent air, which may be contaminated, is blown into the face of the operator.

BENCH EQUIPMENT, GLASSWARE AND PLASTICS

Equipment for which there are British Standards are marked with an asterisk(*) (see Addendum, p. 43).

Bunsen burner burners and loop incinerators

The usual Bunsen burner with a bypass is satisfactory for most work, but for material that may spatter or that is highly infectious a hooded Bunsen burner should be used. There are several versions of these, but we recommend the Kampff and Bactiburner types, which enclose the flame in a borosilicate tube.

'Electric Bunsen burners' or electric incinerator burners are available (e.g. the Bacti-Cinerator, made by Oxford Labware, St Louis, MO, USA). A loop or inoculating wire (see below) is inserted into the heated ceramic tube and after about 6 seconds organic material is incinerated and sterilized at a temperature of about 815°C. Any spatter is contained within the heated tube. These devices are suitable for use in Class I and Class II MSCs, provided that the cabinets are free of any flammable material.

Inoculating loops and wires

These are usually made of 25 SWG Nicrome wire, although this is more springy than platinum iridium. They should be short (not more than 5 cm long) in order to minimize vibration and therefore involuntary discharge of contents. Loops should be small (not more than 3 mm in diameter). Large loops are also inclined to empty spontaneously and scatter infected airborne particles. They should be completely closed, otherwise they will not hold fluid cultures. This can be achieved by twisting the end of the wire round the shank, or by taking a piece of wire 12 cm long, bending the centre round a nail or rod of appropriate diameter, and twisting the ends together in a drill chuck. Ready-made loops of this kind are available.

Loops and wires should not be fused into glass rods because these may shatter in a Bunsen burner flame. Aluminium holders are sold by most laboratory suppliers.

Plastic disposable loops are excellent. There are

two sizes – 1 and 10 µl – and both are useful. They are sold sealed in packs of about 25, sterilized ready for use. Pointed plastic rods, for use in place of 'straight wire', for subculturing are now available but some are too coarse for picking up very small colonies.

Spreaders

Many workers prefer these to loops for inoculating Petri dishes with more than one loopful of material. To make them, cut a glass rod of 3–4 mm diameter into 180-mm lengths and round off the cut ends in a flame. Hold each length horizontally across a bat's wing flame so that it is heated and bends under its own weight, approximately 36 mm from one end, and an L shape is obtained. Plastic spreaders are on the market, sterilized ready for use.

Microscope slides and cover glasses

Unless permanent preparations are required, the cheaper microscope slides are satisfactory. Slides with ground and polished edges are much more expensive. Most microscope slides are sold in boxes of 100 slides. They should not be washed and reused but discarded.

Cover glasses are sold in several thicknesses and sizes. Those of thickness grade No. 1, 16 mm square, are the most convenient. They are sold in boxes containing about 100 glasses. Plastic cover glasses are available.

Pasteur pipettes*

Glass Pasteur pipettes are probably the most dangerous pieces of laboratory equipment in unskilled hands. They are used, with rubber teats, to transfer liquid cultures, serum dilutions, etc.

Very few laboratories make their own Pasteur pipettes nowadays; they are obtainable, sterile and ready for use, from most laboratory suppliers and are best purchased in bulk. Long and short forms are available and most are made of 6–7 mm diameter glass tubing. Some are supplied.

Safer Pasteur pipettes with integral teats, made of low-density polypropylene, are now available and are supplied ready sterilized.

Pasteur pipettes are used once only. Attempts to recycle them may result in cuts and pricks to fingers. There is a British Standard for glass disposable Pasteur pipettes (BS 5732:1985 – see Addendum).

Graduated pipettes

Straight-sided blow-out pipettes, 1–10 ml capacity are used. They must be plugged with cottonwool at the suction end to prevent bacteria entering from the pipettor or teat and contaminating the material in the pipette. These plugs must be tight enough to stay where they are during pipetting, but not so tight that they cannot be removed during cleaning. About 25 mm of non-absorbent cottonwool is pushed into the end with a piece of wire. The ends are then passed through a Bunsen burner flame to tidy them. Wisps of cottonwool that get between the glass and the pipettor or teat may permit air to enter and the contents to leak.

Pipettes are sterilized in the hot air oven in square-section aluminium containers, similar to those used for Pasteur pipettes (square-section containers that do not roll have supplanted the time-honoured round-section boxes). A wad of glass wool at the bottom of the container prevents damage to the tips.

Disposable 1-ml and 10-ml pipettes are available. Some firms supply them already plugged, wrapped and sterilized.

Pipetting aids

'Mouth pipetting' is, properly, banned in all microbiological laboratories that handle pathogens or potential pathogens.

Rubber teats and pipetting devices provide alternatives to mouth pipetting.

Rubber teats

Choose teats with a capacity greater than that of the pipettes for which they are intended, i.e. a 1-ml teat for Pasteur pipettes, a 2-ml teat for a 1-ml pipette; otherwise the teat must be used fully compressed, which is tiring. Most beginners compress the teat completely, then suck up the liquid and try to hold it at the mark while transferring it. This is unsatisfactory and leads to spilling and inaccuracy. Compress the teat just enough to suck the liquid a little way past the mark on the pipette. Withdraw the pipette from the liquid, press the teat slightly to bring the fluid to the mark and then release it. The correct volume is now held in the pipette without tiring the thumb and without risking loss. To discharge the pipette, press the teat slowly and gently, and then release it in the same way. Violent operation usually fails to eject all the liquid; bubbles are sucked back and aerosols are formed.

'Pipettors'

A large number of devices that are more sophisticated than simple rubber teats are now available. Broadly speaking there are four kinds:

1. Rubber bulbs with valves that control suction and dispensing
2. Syringe-like machines that hold pipettes more rigidly than rubber bulbs and have a plunger operated by a rack and pinion or a lever
3. Electrically operated pumps fitted with flexible tubes in which pipettes can be inserted
4. Mechanical plunger devices that take small plastic pipette tips and are capable of repeatedly delivering very small volumes with great accuracy.

It is extremely difficult to give advice on the relative merits of the various devices. That which suits one operator, or is best for one purpose, may not be suitable for others. The choice should be made by the operators, not by managers or administrators who will not use them. None of those in categories 1, 2 and 4 above is expensive, and it should be possible for several different models to be available. What is necessary is some system of instruction in their use and in their maintenance. None should be expected to last forever.

In the USA, a number of manufacturers offer cleaning and calibration services at a reasonable cost.

Tissue grinders and homogenizers

Hand-held devices are used for work with small pieces of tissue or other material. They are available in several sizes. A heavy glass tube is constricted near to its closed end and a pestle, usually made of glass covered with polyfluorotetraethylene (PTFE) or of stainless steel, is ground into the constriction. With the pestle in place, the tissue and some fluid are put into the tube. The pestle is rotated by hand. The tissue is ground through the constriction and the emulsion collects in the bottom of the tube.

These grinders present some hazards. The tubes, even though they are made of borosilicate glass, may break and disperse infected material. They should be used inside a microbiological safety cabinet and be held in the gloved hand in a wad of absorbent material.

Recently, single-use, plastic, disposable tissue grinders have become available. Although they are less likely to break than the glass models, they do have a minor disadvantage that small shards of sharp-edged plastic end up in the ground tissue, although these are easily identifiable.

Small pieces of soft tissue, e.g. curettings, may be emulsified by shaking in a screw-capped bottle with a few glass beads and 1 or 2 ml broth on a vortex mixer.

Glassware and plastics*

For ordinary bacteriological work, soda-glass tubes and bottles are satisfactory. In assay work, the more expensive borosilicate glass might be justified. An important consideration is whether glassware should be washed or discarded. Purchasing cheaper glassware in bulk and using plastic disposable Petri dishes and culture tubes may, in some circumstances, be more economical than employing labour to clean them. Plastics fall into two categories:

1. Disposable items, such as Petri dishes, specimen containers and plastic loops, which are destroyed when autoclaved and cannot be sterilized by ordinary laboratory methods (but see Chapter 4)
2. Recoverable material, which must be sterilized in the autoclave.

Non-autoclavable plastics

These include polyethylene, styrene, acrylonitrile, polystyrene and rigid polyvinyl chloride.

Autoclavable plastics

These withstand a temperature of 121°C and include polypropylene, polycarbonate, nylon, PTFE (polytetrafluororethylene or Teflon), polyallomer, TPX (methylpentene polymer), Viken and vinyl tubing.

It is best to purchase plastic apparatus from specialist firms who will give advice on the suitability of their products for specific purposes. Most recoverable plastics used in microbiology can be washed in the same way as glassware. Some plastics soften during autoclaving and may become distorted unless packed carefully.

Petri dishes*

Disposable plastic Petri dishes (see BS 1990 in the Addendum) are now used in most laboratories in developed countries. They are supplied already sterilized and packed in batches in polythene bags. They can be stored indefinitely, are cheap when purchased in bulk, and are well made and easy to handle. Two kinds are available: vented and unvented. Vented dishes have one or two nibs that raise the top slightly from the bottom and are to be

preferred for anaerobic and carbon dioxide cultures.

Glass Petri dishes are, however, still popular in some areas. In general, two qualities are obtainable. The thin blown dishes, usually made of borosilicate glass, are pleasant to handle, their tops and bottoms are flat and they stack safely. They do not become etched through continued use but are fragile, must be washed with care and are expensive. The thick pressed glass dishes are often convex, cannot always be stacked safely, and are easily etched and scratched with use and washing. On the other hand, they are cheap and not fragile. Aluminium, stainless steel and plastic 'lids' are available, which doubles the number of glass dishes.

Glass Petri dishes are usually placed in aluminium drums and sterilized by hot air.

Test tubes and bottles, plugs, caps and stoppers

Culture media may be distributed in either test tubes or small bottles. Apart from personal choice, test tubes are easier to handle in busy laboratories and take up less space in storage receptacles and incubators, although bottles are more convenient in the smaller workroom where media are kept for longer periods before use. Media in test tubes may dry up during storage. Screw-capped test tubes are available.

The most convenient sizes of test tube are: 127 × 12.5 mm, holding 4 ml of medium; 152 × 16 mm, holding 5–10 ml; 152 × 19 mm, holding 10–15 ml; and 178 × 25 mm, holding 20 ml. Rimless test tubes of heavy quality are made for bacteriological work. The lipped, thin, glass chemical test tubes are useless and hazardous.

Cottonwool plugs have been used for many years to stopper test-tube cultures, but have largely been replaced by metal or plastic caps, or in some laboratories by soft, synthetic, sponge bungs.

Aluminium test-tube caps were introduced some years ago but they have a limited tolerance and, in spite of alleged standard specifications of test tubes, a laboratory very soon accumulates many tubes that will not fit the caps. Caps that are too loose are useless. These caps are cheap, last a long time, available in many colours, and save a great deal of time and labour. Those held in place by a small spring have a wider tolerance and fit most tubes. Polypropylene caps in several sizes and colours and which stand up to repeated autoclaving can also be obtained. Rubber stoppers of the orthodox shape are useless because they are blown out of the tubes in the autoclave, although synthetic sponge rubber stops remain *in situ*. Temporary closures can be made from kitchen aluminium foil.

Aluminium capped test tubes are sterilized in the hot air oven in baskets. Polypropylene capped tubes must be autoclaved.

Several sizes of small culture bottles are made for microbiological work. Some of these bottles are of strong construction and are intended for reuse. Others, although tough enough for safe handling, are disposable. The most useful sizes are: those holding 7, 14 and 28 ml; and the 'Universal Container', 28 ml, which has a larger neck than the others and is also used as a specimen container. These bottles usually have aluminium screw caps with rubber liners. The liners should be made of black rubber; some red rubbers are thought to give off bactericidal substances. Polypropylene caps are also used; they need no liners but in our experience they may loosen spontaneously during long incubation or storage. The medium dries up. They are satisfactory in the short term

All of these bottles can be sterilized by autoclaving.

Media storage bottles*

'Medical flats' or 'rounds' with screw caps are made in sizes from 60 ml upwards. The most convenient sizes are 110 ml, holding 50–100 ml of medium, and 560 ml, for storing 250–500 ml. The flat bottles are easiest to handle and store but the round ones are more robust.

Specimen containers

The screw-capped glass or plastic 'Universal Containers', which hold about 28 ml, are the most popular. There are many other containers, mostly made of plastic. There are too many different containers, many of plastic: some are satisfactory, others not; some leak easily and others do not stand up to handling by patients. Only screw-capped containers with more than one-and-a-half threads on their necks should be used. Those with 'push-in' or 'pop-up' stoppers are dangerous. They generate aerosols when opened. Waxed paper pots should not be used because they invariably leak. For larger specimens, there is a variety of strong screw-capped jars.

Before any containers are purchased in bulk, it is advisable to test samples by filling them with coloured water and standing them upside down on blotting paper for several days after screwing the caps on moderately well. Patients and nurses may not screw caps on as tightly as possible. Similar bottles should be sent through the post, wrapped in absorbent material in accordance with postal regulations (see p. 8). Leakage will be evident by the staining of the blotting paper or wrapping. More severe tests include filling the containers with coloured water and centrifuging them upside down on a wad of blotting paper. Specimen container problems and tests are reviewed by Collins and Kennedy (1999).

Sample jars and containers

Containers for food samples, water, milk, etc. should conform to local or national requirements. In general, large screw-capped jars are suitable but are rather heavy if many samples are taken, and should not be used in food factories. Plastic containers may be used because leakage and spillage are not such a problem as with pathological materials. Strong plastic bags are useful, particularly if they are of the self-sealing type. Otherwise 'quick-ties' may be used.

Racks and baskets

Test-tube and culture bottle racks should be made of polypropylene or of metal covered with polypropylene or nylon so that they can be autoclaved. These racks also minimize breakage, which is not uncommon when metal racks are used. Wooden racks are unhygienic. Some metal racks may rust and should not be used in water baths.

Aluminium trays for holding from 10 to 100 bottles, according to size, are widely employed in the UK. They can be autoclaved and are easily taken apart for cleaning.

The traditional wire baskets are unsafe for holding test tubes. They contribute to breakage hazards and do not retain spilled fluids. For non-infective work, those baskets in which the wire is covered with polypropylene or nylon are satisfactory, but autoclavable plastic boxes of various sizes are safer for use with cultures.

Other bench equipment

A hand magnifier, forceps and a knife or scalpel should be provided, and also a supply of tissues for mopping up spilled material and general cleaning. Swab sticks, usually made of wood and about 6 inches (15 cm) long, are useful for handling some specimens, and wooden throat spatulas are useful for food samples. These can be sterilized in large test tubes or in the aluminium boxes used for Pasteur pipettes.

Some form of bench 'tidy' or rack is desirable to keep loops and other small articles together.

Discard jars and disinfectant pots are considered in Chapter 4.

REFERENCES

British Standards (BS and BS EN) are listed separately in the Addendum (see below).

Centers for Disease Control and National Institutes of Health (CDC/NIH) (2000) *Primary Containment for Biohazards Selection, Installation and Use of Biological Safety Cabinets*. Washington: Government Printing Office.

Collins, C.H. and Kennedy, D.A. (1999) *Laboratory Acquired Infections*, 4th edn. Oxford: Butterworth–Heinemann.

European Union (1998) Directive 98/79/EC of the European Parliament and of the Council of 27 October 1998 on *in vitro* diagnostic medical devices. *Official Journal of the European Communities* L331, 7.12.98, pp. 1–37.

International Electrical Commission (1990) 1010 Part 1. *Safety Requirements – Electrical Equipment for Measuring and Control and Laboratory Use*. Geneva: IEC.

International Federation of Clinical Chemistry (1990) *Guidelines for the Selection of Safe Laboratory Centrifuges and their Safe Use*. Copenhagen: IFCC.

Kennedy, D. A. (1988) Equipment-related hazards. In: Collins, C. H. (ed.), *Safety in Clinical and Biomedical Laboratories*. London: Chapman & Hall Medical, pp. 11–46.

Kuhne, R. W. Chatigny, M. A., Stainbrook, B. W. *et al.* (1995) Primary barriers and personal protective equipment in biomedical laboratories. In: Fleming, D. O., Richardson, J. H., Tulis, J. I. *et al.* (eds), *Laboratory Safety: Principles and Practices*, 2nd edn. Washington: ASM Press.

Medical Devices Agency (MDA) (1998) Medical device and equipment management, for hospital and community-based organisations. *Device Bulletin*, MDA DB 9801, January 1998.

MDA (2002a) Management of *in vitro* diagnostic medical devices. *Device Bulletin*, MDA DB 2002 (02), March 2002.

MDA (2002b) Management and use of IVD point of care test devices. *Device Bulletin*, MDA DB 2002 (03), March 2002.

Medical Devices Directorate (MDD) (1991) Management of medical equipment and devices, *Health Equipment Information* 98: November 1990.

National Sanitation Foundation (1992) *Standard 49. Class II (Laminar Flow) Biohazard Cabinetry*. Ann Arbor: NSF.

SI (2000) *In Vitro* Diagnostic Medical Device Regulations 2000, Statutory Instrument 2000 No. 1315.

ADDENDUM: BRITISH STANDARDS FOR LABORATORY EQUIPMENT

Autoclaves, sterilizers and disinfectors

BS EN 61010-1:1993. Safety requirements for electrical equipment for measurement, control and laboratory use. General requirements.

BS EN 61010-2-041:1997. Safety requirements for electrical equipment for measurement, control and laboratory use. Particular requirements for autoclaves using steam of the treatment of medical materials and for laboratory processes.

BS EN 61010-2-042:1997. Safety requirements for electrical equipment for measurement, control and laboratory use. Particular requirements for autoclaves and sterilizers using toxic gas for the treatment of medical materials and for laboratory processes.

BS EN 61010-2-043:1998. Safety requirements for electrical equipment for measurement, control and laboratory use. Particular requirements for dry heat sterilizers using either hot air or hot inert gas for the treatment of medical materials and for laboratory processes.

BS EN 61010-2-045:2001. Safety requirements for electrical equipment for measurement, control and laboratory use. Particular requirements for washer disinfectors used in medical, pharmaceutical, veterinary and laboratory fields.

BS 2646-2:1990 [1996]. Autoclaves for sterilization in laboratories. Guide to planning and installation.

BS 2646-4:1991 [1998]. Autoclaves for sterilization in laboratories. Guide to maintenance.

BS 2646-1:1993 [2000]. Autoclaves for sterilization in laboratories. Specification for design, construction, safety and performance.

BS 2646-3:1993 [2000]. Autoclaves for sterilization in laboratories. Guide to safe use and maintenance.

BS 2646-5:1993 [2000]. Autoclaves for sterilization in laboratories. Methods of test for function and performance.

Cell disruptors

BS EN 12884:1999. Biotechnology. Performance criteria for cell disruptors.

Centrifuges

BS EN 61010-2-020:1995. Safety requirements for electrical equipment for measurement, control and laboratory use. Particular requirements for laboratory centrifuges.

BS EN 12884:1994. Biotechnology. Performance criteria for centrifuges.

Drying ovens

BS 2648:1955 [2000]. Performance requirements for electrically-heated laboratory drying ovens.

Glassware

BS EN ISO 4796-1:2001. Laboratory glassware. Screw-neck bottles.

BS EN ISO 4796-2:2001. Laboratory glassware. Conical-neck bottles.

BS EN ISO 4796-3:2001. Laboratory glassware. Aspirator bottles.

BS ISO 4798:1997. Laboratory glassware. Filter funnels.

BS ISO 4800:1998. Laboratory glassware. Separating funnels and dropping bottles.

BS 5732:1985 [1997]. Specification for glass disposable Pasteur pipettes.

Microbiologicial safety cabinets

BS EN 12469:2000. Biotechnology. Performance criteria for microbiological safety cabinets.

BS 5726:Part 1:1992. Microbiological safety cabinets. Part 1. Specification for design, construction and performance prior to installation.

BS 5726:Part 2:1992. Microbiological safety cabinets. Part 2. Recommendations for information to be exchanged between purchaser, vendor and installer and recommendations for installation.

BS 5726:Part 3:1992. Microbiological safety cabinets. Part 3. Specification for performance after installation.

BS 5726:Part 4:1992. Microbiological safety cabinets. Part 4. Recommendations for selection use and maintenance.

Mixers and stirrers

BS EN 61010-2-051:1996. Safety requirements for electrical equipment for measurement, control and laboratory use. Particular requirements for laboratory equipment for mechanical mixing and stirring.

pH meters

BS 3145:1978 [1999]. Specification for laboratory pH meters.

Plastics ware

BS 5404-1: 1976 [1994]. Plastics laboratory ware. Beakers.

BS 5404-2:1977 [1994]. Plastics laboratory ware. Graduated measuring cylinders.

BS 5404-3:1977 [1994]. Plastics laboratory ware. Filter funnels.

BS 5404-4:1977 [1994]. Plastics laboratory ware. Wash bottles.

BS 611-2:1990. Specification for plastics Petri dishes for single use.

4

Sterilization, disinfection and the decontamination of infected material

DEFINITION OF TERMS

- *Sterilization* implies the complete destruction of all micro-organisms, including spores. It is accomplished by physical methods.
- *Disinfection* implies the destruction of vegetative organisms that might cause disease or, in the context of the food industries, that might cause spoilage. It usually employs chemicals and does not necessarily kill spores.

These two terms are not synonymous.

- *Decontamination* is the preferred term in microbiological laboratories for rendering materials safe for use or disposal.

STERILIZATION

The methods commonly used in microbiological laboratories are:

- red heat (flaming)
- dry heat (hot air)
- steam under pressure (autoclaving)
- steam not under pressure (tyndallization)
- filtration
- inspissation.

Incineration is also a method of sterilization but as it is applied, usually outside the laboratory, for the ultimate disposal of laboratory waste, it is considered separately (see p. 59).

Red heat

Instruments such as inoculating wires and loops are sterilized by holding them in a Bunsen flame until they are red hot. Hooded Bunsens (see p. 37) are recommended for sterilizing inoculating wires contaminated with highly infectious material (e.g. tuberculous sputum) to avoid the risk of spluttering contaminated particles over the surrounding areas.

Micro-incinerators, electrically operated and intended to replace Bunsens, are becoming increasingly popular.

Dry heat

This is applied in electrically heated ovens, also known as dry-heat sterilizers, which are thermostatically controlled and fitted with a large circulating fans to ensure even temperatures in all parts of the load. Modern equipment has electronic controls which can be set to raise the temperature to the required level, hold it there for a prearranged time and then switch off the current. A solenoid lock is incorporated in some models to prevent the oven being opened before the cycle is complete. This safeguards sterility and protects the staff from accidental burns. There is a British Standard (BS 2648) for these appliances (see p. 43).

Materials that can be sterilized by this method include glass Petri dishes, flasks, pipettes and metal objects. Various metal canisters and drums that conveniently hold glassware during sterilization and keep it sterile during storage are available from laboratory suppliers. Some laboratory workers still prefer the time-honoured method of wrapping pipettes, etc. individually in brown (Kraft) paper.

Loading

Air is not a good conductor of heat so oven loads must be loosely arranged, with plenty of space to allow the hot air to circulate.

Holding times and temperatures

When calculating processing times for hot air sterilizing equipment, there are three time periods that must be considered:

1. *The heating-up period*, which is the time taken for the entire load to reach the sterilization temperature; this may take about 1 h.
2. *The holding periods* at different sterilization temperatures recommended in the UK are 160°C for 2 h or 180°C for 30 min (*British Pharmacopoeia*, 1993; Russell, 1999).
3. *The cooling-down period*, which is carried out gradually to prevent glassware from cracking as a result of a too rapid fall in temperature; this period may take up to 2 h.

Control of hot air sterilizers

Tests for electrically operated fan ovens are described in the British Standard (BSI, 1998).

Hot air sterilization equipment should be calibrated with thermocouples when the apparatus is first installed and checked with thermocouples afterwards when necessary.

Ordinary routine control can be effected simply with commercial chemical indicator tubes.

Steam under pressure: the autoclave

The temperature of saturated steam at atmospheric pressure is approximately 100°C. Temperature increases with pressure, e.g. at 1 bar (about 15 lb/in² – some older autoclaves use Imperial units) it is 121°C. Bacteria are killed by autoclaving at this temperature for 15–20 min. Air has an important influence on the efficiency of autoclaving. The above relationship holds well only if no air is present. If about 50% of the air remains in the autoclave the temperature of the steam/air mixture will be only 112°C. If air is present in the autoclave load, it will also adversely affect penetration by the steam.

All the air that surrounds and permeates the load must first be removed before sterilization can commence. In some autoclaves this is done by vacuum pumps.

Loads in autoclaves

As successful autoclaving depends on the removal of all the air from the chamber and the load, the materials to be sterilized should be packed loosely. 'Clean' articles may be placed in wire baskets, but contaminated material (e.g. discarded cultures) should be in solid-bottomed containers not more than 20 cm deep (see 'Disposal of infected waste' below). Large air spaces should be left around each container and none should be covered. If large volumes of fluid are to be sterilized, then the time should be lengthened to ensure that this reaches the appropriate temperature.

Types of autoclave

Only autoclaves designed for laboratory work and capable of dealing with a 'mixed load' should be used. 'Porous load' and 'bottled fluid sterilizers' are rarely satisfactory for laboratory work. There are two varieties of laboratory autoclave:

1. pressure cooker types
2. gravity displacement models with automatic air and condensate discharge.

Laboratory bench autoclaves

These, as domestic pressure cookers, are still in use in many parts of the world. The more modern type has a metal chamber with a strong metal lid that can be fastened and sealed with a rubber gasket. It has an air and steam discharge tap, pressure gauge and safety valve. There is an electric immersion heater in the bottom of the chamber (Figure 4.1).

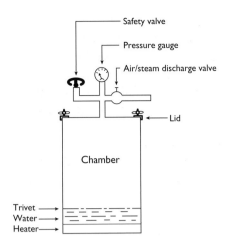

Figure 4.1 'Pressure cooker' laboratory autoclave

Operating instructions

There must be sufficient water inside the chamber. The autoclave is loaded and the lid is fastened down with the discharge tap open. The safety valve is then adjusted to the required temperature and the heat is turned on.

When the water boils, the steam will issue from the discharge tap and carry the air from the chamber with it. The steam and air should be allowed to escape freely until all the air has been removed. This may be tested by attaching one end of a length of rubber tubing to the discharge tap and inserting the other end into a bucket or similar large container of water. Steam condenses in the water and the air rises as bubbles to the surface; when all of the air has been removed from the chamber, bubbling in the bucket will cease. When this stage has been reached, the air–steam discharge tap is closed and the rubber tubing removed. The steam pressure then rises in the chamber until the desired pressure and temperature are reached and steam issues from the safety valve.

When the load has reached the required temperature (see 'Testing autoclaves' below), the pressure is held for 15 min. At the end of the sterilizing period, the heater is turned off and the autoclave is allowed to cool.

The air and steam discharge tap is opened very slowly after the pressure gauge has reached zero (atmospheric pressure). If the tap is opened too soon, while the autoclave is still under pressure, any fluid inside (liquid media, etc.) will boil explosively and bottles containing liquids may even burst. The contents are allowed to cool. Depending on the nature of the materials being sterilized, the cooling (or 'run-down') period needed may be several hours for large bottles of agar to cool to 80°C, when they are safe to handle.

Gravity displacement autoclaves

These may be relatively simple in their construction and operation, as shown diagrammatically in Figure 4.2, or extremely sophisticated pieces of engineering in which air and, finally, steam are removed by vacuum pumps and in which the whole sterilization cycle may be programmed.

The jacket surrounding the autoclave consists of an outer wall enclosing a narrow space around the chamber, which is filled with steam under pressure to keep the chamber wall warm. The steam enters the jacket from the mains supply, which is at high pressure, through a valve that reduces this pressure to the working level. The working pressure is measured on a separate pressure gauge fitted to the jacket. This jacket also has a separate drain for air and condensate to pass through.

The steam enters the chamber from the same source that supplies steam to the jacket. It is introduced in such a way that it is deflected upwards and fills the chamber from the top downwards,

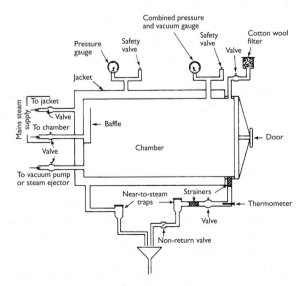

Figure 4.2 Gravity displacement autoclave

thus forcing the air and condensate to flow out of the drain at the base of the chamber by gravity displacement. The drain is fitted with strainers to prevent blockage by debris. The drain discharges into a closed container (not shown in the diagram) so that there is a complete air break that prevents backflow. There is also a filter to ensure that aerosols are not released into the room.

The automatic steam trap or 'near-to-steam' trap is designed to ensure that only saturated steam is retained inside the chamber, and that air and condensate, which are at a lower temperature than saturated steam, are automatically discharged. It is called a 'near-to-steam' trap because it opens if the temperature falls to about 2°C below that of saturated steam and closes within 2°C or nearer to the saturated steam temperature. The trap operates by the expansion and contraction of a metal bellows that operates a valve.

There may be a thermometer probe in the drain but, as this registers the temperature of steam at that point and not that in the load within the chamber, this may be misleading, e.g. the temperature in the drain may reach 121°C whereas that in the load is only 50°C (Collins and Kennedy, 1999).

In modern autoclaves flexible thermocouple probes are fitted in the chamber so that temperatures in various parts of the load may be recorded. In older models the thin thermocouple leads may be safely taken in through the door seals.

There are usually interlocking devices that prevent the opening of the door before the temperature in the chamber has fallen to 80°C. This does not imply that the temperature in the load has also fallen to a safe level. In large, sealed bottles it may still be over 100°C, when the contents will be at a high pressure. Sudden cooling may cause the bottles to explode. The autoclave should not be opened until the temperature in the load has fallen to 80°C or below. This may take a very long time, and in some autoclaves there are locks that permit the door to be opened only fractionally to cool the load further before it is finally released. This is a good reason for not using these autoclaves for the preparation and melting of culture media (see Chapter 5).

Operation of a gravity displacement autoclave

If the autoclave is jacketed, the jacket must first be brought to the operating temperature. The chamber is loaded, the door is closed and the steam valve is opened, allowing steam to enter the top of the chamber. Air and condensate flow out through the drain at the bottom (Figure 4.2). When the drain thermometer reaches the required temperature a further period must be allowed for the load to reach the temperature. This should be determined initially and periodically for each autoclave as described below. Unless this is done the load is unlikely to be sterilized. The autoclave cycle is then continued for the holding time. When it is completed the steam valves are closed and the autoclave allowed to cool until the temperature dial reads less than 80°C. Not until then is the autoclave safe to open. It should first be 'cracked' or opened very slightly and left in that position for several minutes to allow steam to escape and the load to cool further (see 'Protection of the operator' below).

Time/temperature cycles

For most purposes the time/temperature cycles in Table 4.1 will ensure sterilization of a properly packed load (see p. 46) (MDA, 1993/1996).

These are the holding times at temperature (HTAT), determined as described below. The usual HTAT in microbiology laboratories is 15 min at 121°C.

Material containing, or suspected of containing, the agents of spongiform encephalopathies should, however, be autoclaved for 18 min or for six consecutive cycles of 3 min at 134°C (Advisory Committee on Dangerous Pathogens or ACDP, 1994).

Table 4.1 Time/temperature cycles

Temperature (°C)	Holding period (min)
121–124	15
134–138	3

Testing autoclaves

The time/temperature cycles should be tested under 'worst load' conditions, e.g. a container filled with

5 ml screw-capped bottles. This should be placed in the centre of the chamber and, if space is available, other loaded containers may be placed around it. Thermocouple leads are placed in the middle of the load and at other places. The cycle is then started and timed.

There are three periods:

1. *Warming up*, until the temperature in the middle of the load is 121°C.
2. *Sterilization*, i.e. HTAT, in which the temperature in the load is maintained at 121°C for 15 min.
3. *Cooling down*, after the steam valve has been closed and the temperature in the load has fallen to 80°C.

If the autoclave is to be operated manually these times should be noted and displayed. Automated autoclaves may be programmed.

Monitoring

During daily use or from time to time the recording thermometers may be supplemented with indicators. There are three types. Two give immediate results, the other is retrospective:

1. *Bowie-Dick autoclave tape* (Bowie et al., 1963): this is tape impregnated with a chemical and placed in the load. The colour changes if there is adequate steam penetration. These tapes should be reserved for vacuum-type autoclaves.
2. *Chemical indicators*: these are usually in sealed tubes or sachets, and change colour if the correct time and temperature combination has been achieved.
3. *Biological indicators*. Spores of two organisms are used: *Bacillus stearothermophilus* (NCTC 1007, ATCC 7935) and *Clostridium sporogenes* (NTCC 8596, ATCC 7955). They are used as suspensions or absorbed on carriers such as filter paper strips. Laboratory preparations may be unreliable because the heat resistance of the spores depends on the culture media used. It is best to purchase commercial products. Strips are placed at various locations in the load and, after autoclaving, are added to tubes of broth, which are then incubated. Turbidity indicates unsuccessful processing.

There are British Standards for the design, installation, maintenance and testing of laboratory autoclaves (BSI, 1997a, 1997b, 1996, 1998, 2000a, 2000b, 2000c – see Addendum to Chapter 3).

For further information on autoclaving see Kennedy (1988), Gardner-Abbate (1998) and Russell *et al.* (1999).

Protection of the operator

Serious accidents, including burns and scalds to the face and hands have occurred when autoclaves have been opened, even when the temperature gauge is read below 80°C and the door lock has allowed the door to be opened. Liquids in bottles may still be over 100°C and under considerable pressure. The bottles may explode on contact with air at room temperature.

When autoclaves are being unloaded operators should wear full-face visors of the kind that cover the skin under the chin and throat. They should also wear thermal protective gloves (see Kennedy, 1988).

Steam at 100°C ('tyndallization')

This process, named after the Irish physicist cum bacteriologist John Tyndall, employs a Koch or Arnold steamer, which is a metal box in the bottom of which water is boiled by a gas burner, electric heater or steam coil. The articles to be processed are placed so that they rest on a perforated rack just above the water level. The lid is conical so that the condensation runs down the sides instead of dripping on the contents. A small hole in the top of the lid allows air and steam to escape.

This method is used to sterilize culture media that might be spoiled by exposure to higher temperatures, e.g. media containing easily hydrolysed carbohydrates or gelatin. These are steamed for 30–45 min on each of 3 successive days. On the first occasion, vegetative bacteria are killed; any spores that survive will germinate in the nutrient medium overnight, producing vegetative forms that are killed by the second or third steaming.

Filtration

Micro-organisms may be removed from liquids by passing them through filters with very small pores that trap bacteria but, in general, not mycoplasmas or viruses. The method is used for sterilizing serum for laboratory use, antibiotic solutions and special culture media that would be damaged by heat. It is also used for separating the soluble products of bacterial growth (e.g. toxins) in fluid culture media.

Filters of historical interest, now rarely used, include the Berkefeld (made of Kieselguhr), the Chamberland (unglazed porcelain), the Seitz (asbestos) and sintered glass. Details of these may be found in older bacteriology textbooks. They have been superseded by membrane filters.

Membrane filters

These filters are made from cellulose esters (cellulose acetate, cellulose nitrate, collodion, etc.). A range of pore sizes is available. Bacterial filters have a pore size of less than 0.75 µm. The membranes and the devices that hold them for use may be sterilized by autoclaving.

For use, a sterile membrane is mounted, using aseptic precautions, on a perforated platform, usually made of stainless steel, which is sealed together between the upper and lower funnels. Filtration is achieved by applying either a positive pressure to the entrance side of the filter or a negative pressure to the exit side.

Small filter units for filtering small volumes of fluid (e.g. 1–5 ml) are available. The fluid passes through the filter by gravitational force in a centrifuge or is forced through a small filter from a syringe.

Membrane filters and filter holders to suit different purposes are obtainable and several manufacturers publish useful booklets or leaflets about their products.

CHEMICAL DISINFECTION

Some disinfectants present health hazards (Table 4.2). It is advisable to wear eye and hand protection when making dilutions.

Many different chemicals may be used and they are collectively described as disinfectants or (micro)biocides. The former term is used in this book. Some are ordinary reagents and others are special formulations, marketed under trade names. Disinfectants should not be used to 'sterilize' materials or when physical methods are available. In some circumstances, e.g. in food establishments, cleaning with detergents is better. The effects of time, temperature, pH, and the chemical and physical nature of the article to be disinfected and of the organic matter present are often not fully appreciated.

Types and laboratory uses of disinfectants

There is an approximate spectrum of susceptibility of micro-organisms to disinfectants. The most susceptible are vegetative bacteria, fungi and lipid-containing viruses. Mycobacteria and non-lipid-containing viruses are less susceptible and spores are generally resistant.

Consideration should be given to the toxicity of disinfectants and any harmful effects that they may have on the skin, eyes and respiratory tract.

Only those disinfectants that have a laboratory application are considered here. For further information and other applications, see Gardner-Abbate (1998), Russell *et al.* (1999) and Ayliffe *et al.* (1999).

The most commonly used disinfectants in laboratory work are clear phenolics and hypochlorites. Aldehydes have a more limited application, and alcohol and alcohol mixtures are less popular but deserve greater attention. Iodophors and quaternary ammonium compounds (QACs) are more popular in the USA than in the UK, whereas mercurial compounds are the least used. The properties of these disinfectants are summarized below and in Table 4.2. Other substances, such as ethylene oxide and propiolactone, are used commercially in the preparation of

Table 4.2 Properties of some disinfectants

| | Active against | | | | | | | Inactivated by | | | | | | Toxicity | | |
	Fungi	Bacteria G+	Bacteria G−	Myco-bacteria	Spores	Lipid viruses	Non-lipid viruses	Protein	Natural materials	Synthetic materials	Hard water	Deter-gent	Skin	Eyes	Lungs
Phenolics	+++	+++	+++	++	−	+	v	+	++	++	+	C	+	+	−
Hypochlorites	+	+++	+++	+	++	+	+	+++	+	+	+	C	+	+	+
Alcohols	−	+++	+++	+++	−	+	v	+	+	+	+	−	−	+	−
Formaldehyde	+++	+++	+++	+++	+++ᵃ	+	+	+	+	+	+	−	+	+	+
Glutaraldehyde	+++	+++	+++	+++	+++ᵇ	+	+	NA	+	+	+	−	+	+	+
Iodophors	+++	+++	+++	+++	+	+	+	+++	+	+	+	A	+	+	−
QACs	+	+++	++	−	−	−	−	+++	++	+++	+++	A(C)	+	+	−

From Collins and Kennedy (1999)

+++, good; ++, fair; +, slight; −, nil; v, depends on virus; ᵃ > 40°C; ᵇ > 20°C; C, cationic; A, anionic; NA, not applicable; QACs, quarternary ammonium compounds.

sterile equipment for hospital and laboratory use. They are not normally used in laboratories but are used in some hospitals for medical equipment. Commercial equipment is available for this purpose.

Clear phenolics

These compounds are effective against vegetative bacteria (including mycobacteria) and fungi. They are inactive against spores and non-lipid-containing viruses. Most phenolics are active in the presence of considerable amounts of protein but are inactivated to some extent by rubber, wood and plastics. They are not compatible with cationic detergents. Laboratory uses include discard jars and disinfection of surfaces. Clear phenolics should be used at the highest concentration recommended by the manufacturers for 'dirty situations', i.e. where they will encounter relatively large amounts of organic matter. This is usually 2–5%, as opposed to 1% for 'clean' situations where they will not encounter much protein. Dilutions should be prepared daily and diluted phenolics should not be stored for laboratory use for more than 24 h, although many diluted clear phenolics may be effective for more than 7 days. *Skin and eyes should be protected.*

Hypochlorites

The activity is caused by chlorine, which is very effective against vegetative bacteria (excluding mycobacteria), spores and fungi. Hypochlorites are considerably inactivated by protein and to some extent by natural non-protein material and plastics, and they are not compatible with cationic detergents. Their uses include discard jars and surface disinfection, but as they corrode some metal care is necessary. They should not be used on the metal parts of centrifuges and other machines that are subjected to stress when in use.

The hypochlorites sold for industrial and laboratory use in the UK contain 100 000 ppm available chlorine. They should be diluted as follows:

- Reasonably clean surfaces 1 : 100 giving 1000 ppm
- Pipette and discard jars 1 : 40 giving 2500 ppm
- Blood spillage 1 : 10 giving 10 000 ppm

Some household hypochlorites (e.g. those used for babies' feeding bottles) contain 10 000 ppm and should be diluted accordingly. Household 'bleaches' in the UK and the USA contain 50 000 ppm available chlorine and dilutions of 1 : 20 and 1 : 5 are appropriate.

Hypochlorites decay rapidly in use, although the products as supplied are stable. Diluted solutions should be replaced after 24 h. The colouring matter added to some commercial hypochlorites is intended to identify them: it is not an indicator of activity. *Hypochlorites may cause irritation of skin, eyes and lungs.*

Sodium dichloroisocyanate (NaDCC) is a solid chlorine-releasing agent. The tablets are useful for preparing bench discard jars and the powder for dealing with spillages, especially of blood.

Aldehydes

Formaldehyde (gas) and glutaraldehyde (liquid) are good disinfectants. They are active against vegetative bacteria (including mycobacteria), spores and fungi. They are active in the presence of protein and not very much inactivated by natural or synthetic materials, or detergents.

Formaldehyde is not very active at temperatures below 20°C and requires a relative humidity of at least 70%. It is not supplied as a gas, but as a solid polymer, paraformaldehyde, and a liquid, formalin, which contains 37–40% formaldehyde. Both forms are heated to liberate the gas, which is used for disinfecting enclosed spaces such as safety cabinets and rooms. Formalin diluted 1 : 10, to give a solution containing 4% formaldehyde, is used for disinfecting surfaces and, in some circumstances, cultures. Solid, formaldehyde-releasing compounds are now on the market and these may have laboratory applications although they have not yet been evaluated for this purpose. Formaldehyde is used mainly for decontaminating safety cabinets (see p. 34) and rooms.

Some glutaraldehyde formulations need an activator, which is supplied with the bulk liquid. Most activators contain a dye so that the user can be sure that the disinfectant has been activated. Effectiveness and stability after activation vary with the product and the manufacturer's literature should be consulted.

Aldehydes are toxic. Formaldehyde is particularly unpleasant because it affects the eyes and causes respiratory distress. Special precautions are required (see below).

Glutaraldehyde is moderately toxic and is also an irritant, especially to eyes, skin and the upper respiratory tract.

Alcohol and alcohol mixtures

Ethanol and propanol, at concentrations of about 70–80% in water, are effective, albeit slowly, against vegetative bacteria. They are not effective against spores or fungi. Protein and other material or detergents do not especially inactivate them.

Effectiveness is enhanced by the addition of formaldehyde, e.g. a mixture of 10% formalin in 70% alcohol, or hypochlorite to give 2000 ppm of available chlorine.

Alcohols and alcohol mixtures are useful for disinfecting surfaces and, with the exception of alcohol–hypochlorite mixtures, for balancing centrifuge buckets.

They are relatively harmless to skin but may cause eye irritation.

Quaternary ammonium compounds

These are cationic detergents known as QACs or quats, and are effective against vegetative bacteria and some fungi, but not against mycobacteria or spores. They are inactivated by protein, by a variety of natural and plastic materials, and by anionic detergents and soap. Their laboratory uses are therefore limited, although they have the distinct advantages of being stable and not corroding metals. They are usually employed at 1–2% dilution for cleaning surfaces, and are very popular in food hygiene laboratories because of their detergent nature.

QACs are not toxic and are harmless to the skin and eyes.

Iodophors

As with chlorine compounds these iodines are effective against vegetative bacteria (including mycobacteria), spores, fungi, and both lipid-containing and non-lipid-containing viruses. They are rapidly inactivated by protein and to a certain extent by natural and plastic substances, and they are not compatible with anionic detergents. For use in discard jars and for disinfecting surfaces they should be diluted to give 75–150 ppm iodine but, for hand-washing or as a sporicide, diluted in 50% alcohol to give 1600 ppm iodine. As sold, iodophors usually contain a detergent and they have a built-in indicator: they are active as long as they remain brown or yellow. They stain the skin and surfaces but stains may be removed with sodium thiosulphate solution.

Iodophors are relatively harmless to skin although some eye irritation may be experienced.

Mercurial compounds

Activity against vegetative bacteria is poor and mercurials are not effective against spores. They do have an action on viruses at concentrations of 1 : 500 to 1 : 1000 and a limited use, as saturated solutions, for safely making microscopic preparations of mycobacteria.

Their limited usefulness and highly poisonous nature make mercurials unsuitable for general laboratory use.

Precautions in the use of disinfectants

As indicated above, some disinfectants have undesirable effects on the skin, eyes and respiratory tract. Disposable gloves and safety spectacles, goggles or a visor should be worn by anyone who is handling strong disinfectants, e.g. when pouring from stock and preparing dilutions for use.

Testing disinfectants

Several countries and organizations have 'official' tests for disinfectants. They are used by manufacturers and have no place in clinical or public health laboratories that might use them only occasionally, because reproducibility is then a problem. Some, e.g. the Rideal–Walker and the Chick–Martin tests, have fallen into disrepute because they have been used to compare unlikes. Those that use *Salmonella typhi*, even strains that are claimed to be avirulent, should be abandoned altogether. Details of current 'official' tests will not, therefore, be given here – only the principle.

Suspensions containing known numbers of colony-forming units (cfu)/millilitre of *Staphylococcus aureus* and/or *Pseudomonas aeruginosa* are prepared and added to various dilutions of the test product. After predetermined contact times, a known volume of the mixture is removed to a neutralizing fluid that deactivates the disinfectant. The number of surviving colony-forming units is then counted. A satisfactory disinfectant might be expected to reduce the number of colony-forming units by 5 log units within an hour.

Laboratory tests

Although the standard or official tests are best left to manufacturers, there is one, the 'in use' test, that is simple and useful.

The 'in-use' test

Samples of liquid disinfectants are taken from such sources as laboratory discard jars, floor-mop buckets, mop wringings, disinfectant liquids in which cleaning materials or lavatory brushes are stored, disinfectants in central sterile supply departments, used instrument containers and stock solutions of diluted disinfectants. The object is to determine whether the fluids contain living bacteria, and in what numbers. The test (Maurer, 1972) is described here in detail for use in any laboratory, because meaningful results can be obtained only in the light of local circumstances.

1. Take a 1-ml sample of the disinfectant solution from each pot or bucket with separate sterile pipettes.
2. Add to each sample 9 ml diluent in a sterile universal container or a 25-ml screw-capped bottle and mix well. Select the diluent according to the group to which the disinfectant belongs (Table 4.3).
3. Within 4 h use separate Pasteur pipettes to withdraw small volumes of the disinfectant/diluent and to place 10 drops, separately, on the surface of each of two, well-dried, nutrient agar plates.
4. Incubate one plate at 32 or 37°C for 3 days and the other at room temperature for 7 days. The optimum temperature for most pathogenic bacteria is 37°C but those that have been damaged by disinfectants often recover more readily at 32°C.
5. After incubation, examine the plates.

The growth of bacterial colonies on one or both of a pair of plates is evidence of the survival of bacteria in the particular pot from which the sample is taken. One or two colonies on a plate may be ignored – a disinfectant is not a sterilant and the presence of a few live bacteria in a pot is to be expected.

The growth of five or more colonies on one plate should arouse suspicion that all is not well. The relationship between the number of colonies on the plate and the number of live bacteria in the pot may easily be calculated, because the disinfectant sample is diluted 1 in 10 and the 50-drop pipette delivers 50 drops/ml. If five colonies are grown from ten drops of disinfectant/diluent, then five live bacteria were present in one drop of disinfectant and 250 live bacteria were present in 1 ml disinfectant.

Table 4.3 'In-use' test neutralizing diluents

Diluent	Disinfectant group
Nutrient broth	Alcohols
	Aldehydes
	Hypochlorites
	Phenolics
Nutrient broth + Tween 80, 3% w/v	Hypochlorites + detergent
	Iodophors
	Phenolics + detergent
	QACs
QACs, quarternary ammonium compounds	

DECONTAMINATION AND DISPOSAL OF INFECTED MATERIALS

It is a cardinal rule that no infected material shall leave the laboratory.

This principle (Collins *et al.*, 1974) still holds well. Laboratory waste that contains micro-organisms is technically clinical waste (Health Services Advisory Committee or HSAC, 1999). It is clearly the responsibility of the laboratory management to ensure that no waste containing viable micro-organisms leaves the laboratory premises, when it may offer hazards to other people. The waste should therefore be *made safe* on site before final disposal. The best way of doing this is to autoclave it. This procedure offers no problems in properly equipped and well-managed laboratories.

Disinfectants alone should not be used. We have always believed that disinfection is a first-line defence, and for, discarded bench equipment, for example, it is a temporary measure, to be followed as soon as possible by autoclaving.

Table 4.4, adapted from Collins and Kennedy (1993), lists items that should be regarded as infectious and should therefore be autoclaved. (If incineration is contemplated, see p. 59.)

Containers for discarded infected material

In the laboratory there should be five important types of receptacles for discarded infected materials:

1. Colour-coded discard bins or plastic bags for specimens and cultures
2. Discard jars to receive slides, Pasteur pipettes and small disposable items
3. Pipette jars for graduated (recoverable) pipettes
4. Colour-coded plastic bags for combustibles such as specimen boxes and wrappers that might be contaminated
5. Colour-coded 'sharps' containers for hypodermic needles and syringes.

Table 4.4 Proposed classification of clinical and biomedical laboratory waste

Disposables other than sharps
Specimens or their remains (in their containers) submitted for tests: containing blood, faeces, sputum, urine, secretions, exudates, transudates, other normal or morbid fluids but not tissues
All cultures made from these specimens, directly or indirectly
All other stocks of micro-organisms that are no longer required
Used diagnostic kits (which may contain glass, plastics, chemicals and biologicals)
Used disposable transfer loops, rods, plastic Pasteur pipettes
Disposable cuvettes and containers used in chemical analyses
Biologicals, standards and quality control materials
Food samples submitted for examination in outbreaks of food poisoning
Paper towels and tissues used to wipe benches and equipment and to dry hands
Disposable gloves and gowns
Sharps
Hypodermic needles (with syringes attached if custom so requires)
Disposable knives, scalpels, blades, scissors, forceps, probes
Glass Pasteur pipettes; slides and cover glasses
Broken glass, ampoules and vials
Tissues and animal carcases
Bedding from animal cages

Adapted from Collins and Kennedy (1993)

Recommended colours in the UK (HSAC, 1999) are:

- *Yellow*: for incineration
- *Light blue or transparent with blue inscription*: for autoclaving (but may subsequently be incinerated)
- *Black*: normal household waste – local authority refuse collection
- *White or clear plastic*: soiled linen (e.g. laboratory overalls).

Discard bins and bags

Discard bins should have solid bottoms that should not leak, otherwise contaminated materials may escape. To overcome steam penetration problems, these containers should be shallow, not more than 20 cm deep and as wide as will fit loosely into the autoclave. They should never be completely filled. Suitable plastic (polypropylene) containers are available commercially, although not specifically designed for this purpose. Some authorities prefer stainless steel bins.

Plastic bags are popular but only those made for this purpose should be used. They should be supported in buckets or discard bins. Even the most reliable may burst if roughly handled. It is possible that, under new EU legislation, rigid containers will supersede plastic bags.

For safe transmission to the preparation room, the bins may require lids and the bags may be fastened with wire ties.

Discard bins and bags should be colour coded, i.e. marked in a distinctive way so that all workers recognize them as containing infected material.

Discard jars

The jars or pots of disinfectant that sit on the laboratory bench and into which used slides, Pasteur pipettes and other rubbish are dumped have a long history of neglect and abuse. In all too many laboratories these jars are filled infrequently with unknown dilutions of disinfectants, are overloaded with protein and articles that float, and are infrequently emptied. The contents are rarely properly disinfected.

Choice of container

Old jam jars and instant coffee jars are not suitable. Glass jars are easily broken and broken glass is an unnecessary laboratory hazard, especially if it is likely to be contaminated. Discard jars should be robust and autoclavable and the most serviceable articles are 1-litre polypropylene beakers or screw-capped polypropylene jars. These are deep enough to hold submerged most of the things that are likely to be discarded, are quite unbreakable and survive many autoclave cycles. They go dark brown in time, but this does not affect their use. Screw-capped polypropylene jars are better because they can be capped after use and inverted to ensure that the contents are all wetted by the disinfectants, and air bubbles that might protect objects from the fluid are removed.

Correct dilution

A 1-litre discard jar should hold 750 ml of diluted disinfectant and leave space for displacement without overflow or the risk of spillage when it is moved. A mark should be made at 750 ml on each jar, preferably with paint (grease pencil and felt-pen marks are less permanent). The correct volume of neat disinfectant to be added to water to make up this volume for 'dirty situations' can be calculated from the manufacturers' instructions. This volume is then marked on a small measuring jug, e.g. of enamelled iron, or a plastic dispenser is locked to deliver it from a bulk container. The disinfectant is added to the beaker and water added to the 750-ml mark.

Sensible use

Laboratory supervisors should ensure that inappropriate articles are not placed in discard jars. There is a reasonable limit to the amount of paper or tissues that such a jar will hold, and articles that float are unsuitable for disinfectant jars, unless these can be capped and inverted from time to time to wet all the contents.

Large volumes of liquids should never be added to dilute disinfectants. Discard jars, containing the usual volume of neat disinfectant, can be provided for fluids such as centrifuge supernatants, which should be poured in through a funnel that fits into the top of the beaker. This prevents splashing and aerosol dispersal. At the end of the day water can be added to the 750-ml mark and the mixture left overnight. Material containing large amounts of protein should not be added to disinfectants but should be autoclaved or incinerated.

Regular emptying

No material should be left in disinfectant in discard jars for more than 24 h, or surviving bacteria may grow. All discard jars should therefore be emptied once daily, but whether this is at the end of the day or the following morning is a matter for local choice. Even jars that have received little or nothing during that time should be emptied.

'Dry discard jars'

Instead of jars containing disinfectants there is a place in some laboratories for the plastic containers used for discarded disposable syringes and their needles. These will accommodate Pasteur pipettes, slides, etc. and can be autoclaved or incinerated.

Pipette jars

Jars for recoverable pipettes should be made of polypropylene or rubber. These are safer than glass. The jars should be tall enough to allow pipettes to be completely submerged without the disinfectant overflowing. A compatible detergent should be added to the disinfectant to facilitate cleaning the pipettes at a later stage. Tall jars are inconvenient for short people, who tend to place them on the floor. This is hazardous. The square-

based rubber jars may be inclined in a box or on a rack that is convenient and safer.

Plastic bags for combustibles

These should be colour coded (yellow in the UK, see p. 56).

Treatment and disposal procedures

There are three practical methods for the treatment of contaminated, discarded laboratory materials and waste:

1. Autoclaving
2. Chemical disinfection
3. Incineration.

The choice is determined by the nature of the material: if it is disposable or recoverable; if the latter, if it is affected by heat. With certain exceptions none of the methods excludes the others. Disinfection alone is advisable only for graduated, recoverable pipettes. It will be seen from Figure 4.2 that incineration alone is advised only if the incinerator is under the control of the laboratory staff.

Organization of treatment

The design features of a preparation room for dealing with discarded laboratory materials should include autoclaves, a sluice, a waste disposal unit plumbed to the public sewer, deep sinks, glassware washing machines, drying ovens, sterilizing ovens and large benches.

These should be arranged to preclude any possible mixing of contaminated and decontaminated materials. The designers should therefore work to a flow, or critical pathway chart, provided by a professional microbiologist.

Such a chart is shown in Figure 4.3. The contaminated materials arrive in colour-coded containers onto a bench or into an area designated and used for that purpose only. They are then sorted according to their colour codes and dispatched to the

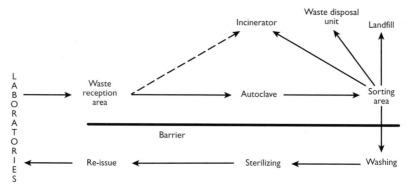

Figure 4.3 Design of preparation (utility) rooms. Flow chart for the disposal of infected laboratory waste and re-usable materials

incinerator or loaded into the autoclave. Nothing bypasses this area. After autoclaving, the containers are taken to the sorting bench where the contents are separated:

- Waste for incineration, which is put into colour-coded containers
- Waste for the rubbish tip, which is also put into different colour-coded containers
- Waste suitable for the sluice or waste disposal unit
- Recoverable material which is passed to the next section or room for washing and re-sterilizing. This room should have a separate autoclave. Contaminated waste and materials for re-use or re-issue should not be processed in the same autoclave.

Procedures for various items

Before being autoclaved the lids of discard bins should be removed and then included in the autoclave load in such a way that they do not interfere with steam penetration. Plastic bags should have the ties removed and the bags opened fully in the bins or buckets that support them.

Contaminated glassware

After autoclaving, culture media may be poured away or scraped out, and the tubes and bottles, etc. washed by hand or mechanically with a suitable detergent. The washing liquid or powder used will depend on the hardness of the water supply and the method of washing. The advice of several laboratory detergent manufacturers should be sought.

Busy laboratories require glassware washing machines. Before purchasing one of these machines, it is best to consider several and to ask other laboratories which models they have found satisfactory. A prerequisite is a good supply of distilled or deionized water.

If re-usable articles are washed by hand double sinks, for washing and then rinsing, are necessary, plus plastic or stainless steel bowls for final rinsing in distilled or deionized water. Distilled water from stripper stills, off the steam line, is rarely satisfactory for bacteriological work.

Rubber liners should be removed from screw caps and the liners and caps washed separately and reassembled. Colanders or sieves made of polypropylene are useful for this procedure.

New glassware, except that made of borosilicate or similar material, may require neutralization. When fluids are autoclaved in new soda-glass tubes or bottles, alkali may be released and alter the pH. Soaking for several hours in 2–3% hydrochloric acid is usually sufficient, but it is advisable to test a sample by filling with neutral water plus a few drops of suitable indicator and autoclaving.

Discard jars

After standing overnight to allow the disinfectant to act, the contents of the jars should be poured carefully through a polypropylene colander and flushed down the sluice sink. The colander and its contents are then placed in a discard bin and autoclaved. Rubber gloves should be worn for these operations. The empty discard jars should be autoclaved before they are returned to the laboratory for further use. There may be residual contamination.

Re-usable pipettes

After total immersion overnight in disinfectant (e.g. hypochlorite, 2500 ppm available chlorine – an anionic detergent may be added), the pipettes should be removed with gloved hands.

Before they are washed, the cottonwool plugs must be removed. This can be carried out by inserting the tip into a piece of rubber tubing attached to a water tap. Difficult plugs can be removed with a small crochet hook. Several excellent pipette washing machines are manufactured that rely on water pressure and/or a siphoning action, but the final rinse should be in distilled or deionized water.

24-hour urines

Although rare in microbiology laboratories other pathology departments may send them for disposal on the grounds that they may contain pathogens. Ideally they should be processed in the department concerned as follows.

Sufficient disinfectant, e.g. hypochlorite, should be added to the urine to give the use dilution. After standing overnight the urine should be poured carefully down the sink or sluice to join similar material in the public sewer. The containers, which are usually plastic, may then be placed in colour-coded bags for incineration.

Incineration

The problems with this method of disposal of infected waste that has not been autoclaved are in ensuring that the waste actually reaches the incinerator, and that if it does it is effectively sterilized and none escapes, either as unburned material or up the flue. Incinerators are rarely under the control of laboratory staff. Sometimes they are not even under the control of the staff of the hospital or institution, but are some distance away, and contaminated and infectious material has to be sent to them on the public highway. Some older incinerators are inefficient.

Unburned material may be found among the ash and, from its appearance, it may be deduced that it may not have been heated enough to kill micro-organisms. We have recovered unconsumed animal debris, including fur and feathers and entrails, from a laboratory incinerator. The updraught of air may carry micro-organisms up the flue and into the atmosphere if the load is too large or badly distributed.

In the UK there are strict regulations about the incineration of clinical waste, which includes infected laboratory waste (see Collins and Kennedy, 1999) and inefficient incinerators are being phased out. The ACDP (1995) states that all waste must be made safe before disposal or removal to the incinerator but the HSAC (1991) permits it to be transported to an incinerator provided that 'it is securely packaged' and the incinerator is 'under the supervision of an operator who is fully conversant with appropriate safe working procedures for handling contaminated waste'. As infected laboratory waste, especially discarded cultures, is the most hazardous of all clinical waste, and as all reputable microbiological laboratories possess autoclaves, this practice should be resisted and the material made safe by autoclaving it before it leaves the laboratory (Collins et al., 1974; Collins and Kennedy, 1993, 1999; Collins, 1994; ACDP, 1995).

There is a British Standards Institution Code of Practice (BS 3316) for the incineration of hospital waste.

REFERENCES

Advisory Committee on Dangerous Pathogens (ACDP) (1994) *Precautions for Work with Human and Animal Transmissible Spongiform Encephalopathies*. London: HMSO.

Advisory Committee on Dangerous Pathogens (1995) *Categorization of Pathogens according to Hazard and Categories of Containment*. London: HMSO.

Ayliffe, G. A. J., Babb, J. R. and Taylor, L (1999) *Hospital Acquired Infections*, 3rd edn. Oxford, Butterworth-Heinemann.

Bowie, J. W., Kelsey, J. C. and Thompson, G. R. (1963) The Bowie and Dick autoclave tape test. *Lancet* i: 586–587.

British Pharmacopoeia (1993). London: HMSO.

British Standards Institution (BSI) (1987) BS 3316 Part 4. Code of Practice for the design, specification and commissioning of incineration plant for the destruction of hospital waste. London: British Standards Institution.

British Standards Institution (BSI) (1998) BS EN 61010-2-043:1998. Safety requirements for electrical equipment for measurement, control and laboratory use. Particular requirements for dry heat sterilizers using either hot air or hot inert gas for the treatment of medical materials and for laboratory processes. London: British Standards Institution.

Collins, C. H. (1994) Infected laboratory waste. *Letters in Applied Microbiology* 19: 61–62.

Collins, C. H. and Kennedy, D. A. (1993) *Treatment and Disposal of Clinical Waste*. Leeds: Science Reviews.

Collins, C. H. and Kennedy, D. A. (1999) *Laboratory Acquired Infections*, 4th edn. Oxford: Butterworth-Heinemann.

Collins, C. H., Hartley, E. G. and Pilsworth, R. (1974) *The Prevention of Laboratory-Acquired Infections*. PHLS Monograph No. 6. London: HMSO.

Gardner-Abbate, S. (1998) *Introduction to Sterilization and Disinfection*, 3rd edn. London: Churchill Livingstone.

Health Services Advisory Committee (HSAC) (1991) *Safe Working and the Prevention of Infection in Clinical Laboratories*. London: HMSO.

HSAC (1999) *Safe Disposal of Clinical Waste*. London: HMSO.

Kennedy, D. A. (1988) Equipment-related hazards. In: Collins, C. H. (ed.), *Safety in Clinical and Biomedical Laboratories*. London: Chapman & Hall, pp. 11–46.

Maurer, I. M. (1972) The management of laboratory discard jars. In: Shapton, D. A. and Board, R. G. (eds), *Safety in Microbiology* Society for Applied Bacteriology Technical Series No. 6, London: Academic Press, pp. 53–59.

Medical Devices Agency (MDA) (1993/1996) *Sterilization, Disinfection and Cleaning of Medical Equipment*. London: HMSO.

Russell, A. D. (1999) Microbial susceptibility and resistance to chemical and physical agents. In: Collier, L., Balows, S. and Sussman, M. (eds), *Topley and Wilson's Microbiology and Microbial Infections*, 9th edn, Vol. 2, *Systematic Bacteriology*. London: Arnold, Chapter 7.

Russell, A. D., Hugo, W. B. and Ayliffe, G. A. J. (1999) *Principles and Practice of Disinfection, Preservation and Sterilization*, 3rd edn. Oxford: Blackwell Scientific.

5

Culture media

In developed countries it is unusual for medical microbiology laboratories to make their own culture media from raw materials. Furthermore, even the reconstitution of dehydrated media takes place less and less in the individual laboratory as bottles, tubes and plates of ready-to-use media become available from central suppliers and/or commercial sources.

Although this loss of skill in the preparation of media may be regretted, the growing weight of *in vitro* diagnostics legislation makes independent laboratory manufacture a legal minefield. For every person making a product, two more staff will be required to check activity, labels, instructions for use, shelf-life limits and safety precautions in use. One effect of centralization of supplies is an inevitable reduction in the variety of media formulations available. Commercial manufacturers annually review their list of media and monitor the sales of each medium. They may decide to cease manufacture when sales fall below specified figures. Central non-commercial suppliers will operate from a limited list of consensus decisions about the variety of media that can be supplied to the receiving laboratories. This means that the days have long gone since Sir Graham Wilson once stated that a UK microbiologist would rather borrow a colleague's toothbrush than use his or her particular culture medium.

Nevertheless, there will always be culture media for special purposes that are not available commercially. Formulae and methods for those that are mentioned in this book are therefore included (see p. 72). Laboratory workers who do decide to make their own media can find the formulae, etc. in the commercial catalogues and handbooks and in older editions of this book.

All culture media must now display labels describing the formulation, with the amounts of the individual constituents present, normally expressed as grams per litre. The labels of the same medium from different suppliers will appear to be identical but each label will also state that some variation in formulation may be made to meet performance standards. This caveat allows formula manipulation to accommodate variation in the raw materials used to achieve standard performance results. Some culture media have commercially protected names and, although such media may have the same formulation between different manufacturers; others cannot use the protected name.

Further details about the formulation and performance of culture media may be found in the following: MacFaddin (1985), Baird *et al.* (1987), Barrow and Feltham (1993) and Bridson (1994). Also the current editions of BBL, Difco, LabM and Oxoid manuals and catalogues are useful sources of information.

IN VITRO CULTIVATION OF BACTERIA

The work of Pasteur (Gieson, 1955) and his colleagues in France, together with that of Robert Koch (Brock, 1988) and his team in Germany, provided the foundation of *in vitro* cultivation of micro-organisms in the second half of the nineteenth century. Cultivation was essential because, apart from the morphology of individual cells, these extremely small units of life could not be separately studied to determine their metabolism and biochemistry, and to classify their separate identities, without using millions of cells – a situation that, despite advances in molecular biology, still largely prevails today. It was this essential groundwork that provided the proof of the causation of infectious disease.

All the organisms of importance in medical microbiology are described as chemo-organotrophs because they obtain energy from organic molecules. The constituents of early culture media for their cultivation were undefined, complex, organic mixtures. This is another characteristic of medical microbiology that remains to this day. Pasteur started with simple organic solutions and then moved on to urine and blood, finally using live animals. Koch started with a meat stew (meat extract) for the isolation of his first pathogenic organism – *Bacillus anthracis*. He did not consider this to be a universal culture medium and he also experimented with blood, serum and egg. Koch and his colleagues in Berlin then went on to isolate and identify the bacterial causes of all the major infectious diseases in Europe. This work was completed over a period of 20 years – 1880–1900.

This great achievement came about because of two important developments. First, Koch created a solid culture medium format from his meat extract solution, the 'Poured Plate', using at first gelatin and later agar as the solidifying agent. He established the principle of one cell forming one colony of identical cells, which was a great improvement over Pasteur's dilution to extinction method using fluid cultures. Second, his colleague Loeffler supplemented Koch's meat extract with peptone (hydrolysed protein). The full significance of this supplement only became apparent in the closing years of the twentieth century.

The major work in microbiology in the first half of the twentieth century was the study of the metabolism in prokaryote cells: the catabolic chemical pathways producing energy and the anabolic pathways of biosynthesis consuming it. At the same time, study of the dynamics of bacterial growth revealed the lag, exponential growth, and the stationary and decline phases of growth that are typical in any closed system of bacterial cultivation. In the second half of the twentieth century, antibiotics were developed and the DNA/RNA genetic code was cracked to reveal the complexities of the molecular control of life.

THE COMMON COMPONENTS OF CULTURE MEDIA FOR MEDICAL MICROBIOLOGY

An examination of the formulae of culture media commonly shows a few key nutrient ingredients, and several other components added for the selection and/or identification of separate groups or species of bacteria.

All life is dependent on the presence of water and all the nutrients from which micro-organisms synthesize cell material and obtain energy must be dissolved in water. Simple chemical analysis of bacterial cells reveals 11 macroelements – C, H, O, N, S, P, K, Na, Ca, Mg, Fe – and many trace elements that are often found as impurities in the macroelements, e.g. Mn, Mb, Zn, Cu, Co, Ni, V, B, Se, Si, W. This simple analysis is not very helpful in the construction of specific culture media and it is better to divide the constituents into one or more functional parts.

Amino-nitrogen nutrients

Protein hydrolysates (peptones) infusions or extracts: these nutrients often contain sufficient energy-rich molecules and trace elements to become the sole ingredient necessary for the growth of common medical bacteria.

Energy source

This is usually glucose but other easily used carbohydrates may be substituted. Peptones can also provide carbon energy, especially those derived from plants.

Growth factors

Some more fastidious organisms appear to require heat-labile supplements to stimulate growth, e.g. blood, serum and vitamin complexes. Further work has shown that, with *Bordetella*, *Neisseria*, *Campylobacter* and *Legionella* species, the essential role of blood is to act as a protective agent against toxic oxygen radicals rather than act as a supplementary nutrient.

Buffer salts

Soluble sodium or magnesium phosphate, acetate or citrate salts are commonly added to culture media containing carbohydrates, to maintain pH stability when fermentative organisms are growing. These buffer salts, however, can chelate essential metals, especially Fe^{2+} ions (Munro, 1968). Some growth factors are toxic (e.g. fatty acids) and must be added in a form or composition (e.g. with albumin or as a detergent) to permit growth and survival.

Mineral salts and metals

Supplements of salts and metals are normally restricted to synthetic (defined) culture media. Undefined media should provide an adequate supply unless extra is required to overcome a chelating substance in the formulation or an excess of NaCl is added to select halophilic organisms.

Selective agents

Toxic chemicals, antibiotics and inhibitory dyes have been used separately or in combination in culture media. It is essential that the selective agents are at the correct strength for the particular medium formulation and the selected organism. There is an element of compromise for all selective media, i.e. not all unwanted organisms will be inhibited and not all desired organisms will grow (Miller and Banwart, 1965; Bridson, 1978). The active concentration of an antibacterial agent (e.g. Hg^{2+}, antibiotic) may be influenced by the components of the medium (e.g. by albumin).

Indicator dyes

Dyes such as phenol red, neutral red and bromocresol purple are added to culture media to indicate changes in pH value, during and after growth. Some dyes are colourless (e.g. triphenyltetrazolium chloride) and act as electron acceptors, and their reduced forms (e.g. triphenylformazan) are pigmented (red) and insoluble, resulting in dye deposition around colonies in agar medium or red colour in broths as an indication of growth. Fermentative carbohydrates are normally added to the formulation. The dyes used can be toxic to sensitive or stressed cells and this fact needs to be kept in mind, if growth is absent.

Gelling agents

Agar is the most common gelling agent used in culture media because it has natural advantages over gelatin and alginate. It is not an inert constituent of culture media; it can contribute metals, minerals, sulphate and pyruvate. Agar can also bind water and inhibit growth of organisms (see 'Poured plates', p. 69).

DEFINED (SYNTHETIC) VERSUS UNDEFINED (COMPLEX) CULTURE MEDIA

The great success of Koch and his colleagues in the last 20 years of the nineteenth century was based on totally undefined and (as we now know) highly complex soups and stews made in the laboratory kitchen. With the exception of *Mycobacterium tuberculosis*, a wide variety of pathogenic organisms were isolated with the same nutrient broth/agar medium first described by Koch and Loeffler. With small modifications, this formula has continued to serve bacteriologists to the present day. There was a period, however, when microbiologists became ashamed of their apparent culinary art. In the 1950s, the biochemical neighbours of bacteriologists began to surge ahead with new knowledge of energy-yielding and synthetic pathways of metabolism, which they combined with rapid, automated, microanalytical techniques. Bacteriologists became aware that, if Koch or Pasteur were to enter the average clinical bacteriology laboratory, it would soon feel like home to them. Thus began a 25-year search for defined (synthetic) culture media to replace the soup and stew. There were two sound reasons for the development of defined media at that time:

1. Existing (largely manufactured by individual laboratories) undefined media could give widely variable results between and within laboratories.
2. Increasing exploration of microbial metabolism, using defined media, suggested that reproducible results could come only from synthetic culture media.

Over the next 25 years, active research was reported in a large number of published papers, which mostly demonstrated hope rather than reality. Nutrient agar/broth can grow practically all the genera of clinically important organisms and with one or two supplements will grow most of the others. This includes all the species and subvariants of the genera. None of the synthetic media formulations published could meet this standard. Most workers started with the Synthetic Mineral-Salts Medium of Davis and Mingioli (1950). This contained:

Carbon source	2 g
K_2HPO_4	7 g
KH_2PO_4	3 g
$MgSO_4 \cdot 7H_2O$	0.1 g
$(NH4)_2SO_4$	1 g
Na citrate·$3H_2O$	0.5 g
Water	1 litre
(Agar)	(15 g)
pH	7.0

This formula allowed the growth of *Escherichia coli* from a generous inoculum. To grow more demanding organisms supplementation with amino acids, nucleotides, vitamins, trace metals, etc. was necessary The list of supplements increased considerably for organisms such as pathogenic neisserias. One such defined medium for *Neisseria gonorrhoeae* required 10 multi-component stock solutions plus 5 separated components, a total of 50 components. Catlin's (1973) paper describing this medium showed that reproducible defined media cannot be prepared by weighing all the dried ingredients into water, boiling to dissolve them and

finally autoclaving the mixture. A very large, unpredictable and highly complex interaction of the components would result, which would not be repeatable with every batch of the same medium. Furthermore, even the stock solutions had to be protected from photo- and chemo-oxidation. Experience showed that simpler defined media, designed around a less demanding genus, would seldom show growth of other genera and often would not grow all the species of the single genus, without further supplementation.

The search for defined (synthetic) media that could be used in general clinical bacteriology slowly ebbed away in the mid-1970s. There were two major reasons for abandoning the search:

1. By the mid-1970s most bacteriological laboratories were purchasing commercially prepared, dehydrated and ready-to-use media. The gross disparity of quality in culture media between and within laboratories had gone.
2. In 1966, Ellner *et al.* published the results of their work on Columbia agar. These workers had postulated that, if a one-peptone medium showed good results with many organisms, then more than one peptone would be better. They finally incorporated four different peptones and formulated a general-purpose medium for medical microbiology, which has still not been eclipsed in performance. The explanation for its astonishing success lies in the complex mixture of polypeptides produced from different proteolytic enzymes digesting different proteins. Subsequently, it was realized that the Columbia formula could recover growth from small inocula of stressed organisms.

STRESSED ORGANISMS AND RESUSCITATION

In the early days of clinical bacteriology, when surgeons were delighted to see 'laudable' pus from their patients, it was considered that all pathogenic organisms were aggressive, vigorous cells. The idea that such organisms might be stressed or need special nurture would have seemed ludicrous to bacteriologists in those by-gone days.

Preservation of food by drying, salting, heating to scalding temperature, chilling with ice or making acid with vinegar and/or the use of spices evolved empirically, long before the existence of micro-organisms was known. Pasteur rationalized the heating process now called 'pasteurization' and he was able to explain why the new 'autoclave' canning of food worked to overcome putrefaction on storage at room temperature. It is unlikely that the food preservation steps described above could destroy all the viable micro-organisms in food. However, those that just survived were probably damaged to a degree that fermentation or putrefaction was long delayed. Such damaged or stressed organisms may also not survive a period of acid digestion in the stomachs of those eating the food. Those micro-organisms that did escape alive into the small intestine would meet other defence mechanisms. However, in an imperfect world not all consumers have the same immune systems and food poisoning is still a hazard today. Food microbiologists, aware that their products could contain damaged (stressed) micro-organisms, were the first to recognize that special steps needed to be taken to recover them. Quantitative counts of organisms in food samples, before and after processing, were essential facts required to make judgements about the duration and temperatures required to process foods safely.

The recovery of stressed cells by the process of resuscitation

The traditional test to detect sublethal injury in bacteria caused by inhibitory agents (chemical, physical or biological) is recognition of failure to grow on a selective medium that normally supports its growth, although growth occurs in the absence of the inhibitory agent (Mossel and Ratto, 1970; Hackney et al., 1979). Although this selective/non-selective medium test can be helpful to detect injured cells, it is not entirely reliable. Damaged

cells may not be able to grow at all until the recovery (repair) process has taken place (Walker, 1966; Dukan et al., 1997). In the 1960s it was recognized that the addition of pyruvate or catalase to culture media yielded higher colony counts. Slowly, over the next 30 years, a full appreciation of the effects of oxidation/reduction in culture media and their effects on microbial growth was made. Oxidation has been considered as the cause of 'viable but non-culturable' (VBNC) organisms (Bloomfield et al., 1998) but this hypothesis has been contested (Barer and Harwood, 1999).

An unexpected but essential factor in the recovery of stressed bacteria was the discovery of a wide variation in the length of lag phases of growth associated with the numbers of organisms present in the inoculum. Stephens et al. (1997) used an automated growth analysis system to measure times and growth of heat-stressed salmonellas in a resuscitation broth, in which the inocula varied from one cell to 10^7 cells. Single stressed cell inocula produced lag phases in different recovery broths, which varied from 6 to 36 h. The lag phases of non-heat-injured single cells varied from 6 to 8 h in the same recovery broths. At 100-cell inocula, both heat-injured and non-heat-injured cells produced similar lag phases of 6–8 h. There is some protective action when numbers of stressed cells attempt recovery of growth that is not available to the single stressed cell. Reducing agents are beneficial because, in later work, Stephens (1999) demonstrated that the most inhibitory broths had higher levels of peroxide present and that the addition of reducing agents to the inhibitory media overcame the long lag phase with single stressed cells.

Is there a single optimum recovery medium for pre-enrichment or resuscitation of bacteria that have been sublethally injured by various methods? Experience has shown that complex undefined media are preferable to chemically defined media or buffer salt solutions. Key factors seem to be the presence of polypeptides and a reduced Eh medium. A review of the literature made in 1977 listed the recovery agents that can be added to media to assist the recovery of stressed organisms (Mossel and Corry, 1977):

1. Neutralizing substances to counter the effects of damaging residual chemicals, e.g. β-lactamases, sulphydral compounds, wetting agents
2. Lysozyme to help germination of 'dead' spores
3. Catalase or whole blood to remove peroxides
4. Pyruvate and other citric acid cycle intermediates to improve recovery from membrane damage
5. Magnesium to stabilize RNA and cell membranes
6. General recovery agents: oxygen radical scavengers, whole blood, activated charcoal.

SELECTIVE AND ELECTIVE CULTURE MEDIA

The concept of selective enrichment of microorganisms was developed early in the history of culture media. Bile and bile salts were found to select *Salmonella typhi* in particular and coliform bacteria in general. Conjugated bile salts (combined with taurine or glycine) are less toxic than free bile acids (deoxycholic and cholic acids). The toxic activity of selective agents has to be accurately determined. The amount of agent added to the formulation is calculated and checked by testing each new batch with a standard reference batch of the same formulation.

Chemical selective agents

Typical chemical selective agents are:

1. Inorganic salts: azide, bismuth, lithium, selenite, tellurite, thallium
2. Dyes: brilliant green, malachite green, crystal violet
3. Surface active agents: bile, cetrimide, lauryl sulphate, Tergitol (Bridson, 1990) and Tween 80
4. Others: dichloran, irgasan, phenylethyl alcohol.

Chemical selective agents are relatively crude in action and in most circumstances a line must be drawn in a trial to define an optimum concentration to be used. An ideal concentration that will destroy 100% of unwanted organisms and allow 100% of the desired organisms to survive is unobtainable. From a limited range of concentrations (predetermined by experiment) that can allow a predominance of the desired organism(s) and maximum suppression of the unwanted organisms, the best concentration is selected for use with the chosen medium. Further details on chemical selective agents can be found in the Oxoid publication (Bridson, 1994).

Antibiotic selective agents

Fleming carried out the first demonstration of the selective action of antibiotics in culture media in 1929, using his crude penicillin preparation to isolate *Haemophilus* species. This work could not be repeated until commercial supplies of penicillin became available in the 1950s. Lacey (1954) returned to Fleming's technique using penicillin to isolate *H. pertussis*. By that date, however, the incidence of penicillin resistance was sufficiently common to require other selective agents to be used with it.

Although not the first of the antibiotic-selective media to be used, the medium of Thayer and Martin (1966) for the isolation of *Neisseria* species was the most widely known early example. These media opened the door for opportunities to isolate 'new' and 'difficult' organisms. A host of exotic organisms could now be found and cultivated in any general microbiology laboratory, e.g. *Campylobacter* (*Helicobacter*), *Legionella*, *Gardnerella*, *Listeria*, *Mycoplasma* and *Yersinia* species, and *Clostridium difficile*.

Early work with these media, using single antibiotics, was promising, e.g. novobiocin for salmonellas and neomycin for anaerobic organisms, but mixtures of two or more antibiotics gave better long-term prospects against antibiotic-resistant contaminating organisms. Although there are very many antibiotics, their basic mechanisms of action are few:

- alteration of cell membranes
- inhibition of protein synthesis
- inhibition of nucleic acid synthesis
- inhibition of cell wall synthesis
- anti-metabolic activity or competitive antagonism.

By incorporating at least two of these five inhibitory actions in the antibiotic mixture, the likelihood of resistance developing is much reduced.

Commercial media suppliers were requested to supply antibiotic selective media. Only two antibiotics, chloramphenicol and kanamycin, were sufficiently stable to be incorporated into dehydrated culture media. However, the hazards of inhaling antibiotic-containing dust by customers and manufacturing staff were considered to be too great. Incorporating labile antibiotics into prepared media meant that the individual activities of the antibiotics would fall from the day of preparation until the media were used. The best solution was to have separate vials of sterile, lyophilized antibiotic mixtures, ready for rehydration and adding to prescribed volumes of pre-sterilized, cooled media. It was Skirrow's medium for the isolation of *Campylobacter* species that started the commercial production of lyophilized antibiotic supplements in 1977 (Skirrow, 1977). In a relatively short period of time, campylobacters arose from obscurity to become identified as the most common cause of food poisoning in the UK. Since that time, another 60–70 antibiotic mixtures have been made available to microbiological laboratories.

Elective media

A selective medium contains a surfeit of nutrients and inhibitory agents are added to restrict growth of undesired organisms and make it selective for a particular genus/species. By contrast, an elective medium is designed to have minimal nutritional ingredients, enabling only a limited group of micro-organisms to grow. Elective media are not used in medical microbiology because they are ineffective for resuscitation of stressed organisms. Reuter (1985) has described elective media for lactic acid bacteria and gives examples of selection and election.

THE SIGNIFICANCE OF COMPLEX POLYPEPTIDES IN UNDEFINED CULTURE MEDIA

It was Loeffler's contribution of peptone to Koch's system of cultivating bacteria that ensured 120 years later that the original nutrient broth/agar medium formulation has scarcely changed. Throughout most of this time, the role of peptone was seen to be as a nutritional supplement but its complex structure of peptides was little appreciated until the work of Ziska (see Bridson 1994) was published in 1967–68. This, combined with the work of Ellner *et al.* (1966), raised a possibility that polypeptides might have a special role in the recovery of stressed organisms. Columbia Agar, with its wide spectrum of peptides ranging in size from 50 to 2 amino acids in length plus numerous free amino acids, was superior in performance to culture media containing one peptone only. The success of undefined polypeptides over defined mixtures of amino acids may be attributed to one or more of the following hypotheses.

Polypeptides may act as redox agents

It is known that larger numbers of stressed organisms are easier to recover than single or few cells. This protective effect may be a reduced Eh in the immediate cell environment. Polypeptides may have a similar protective effect in lowering the Eh of the medium. Wright (1933) demonstrated that 'toxic' (oxidized) peptones could be reduced by heating with meat particles. He found that such reduced peptones would then recover small numbers of fastidious strains of *Streptococcus pneumoniae*.

Use of peptides may be more efficient than amino acids

The formation of proteins in the living cell involves assembly of specific amino acids into strings and then folding of the amino acid sequence into secondary, tertiary or quarternary structures. It is the three-dimensional structure of the sequence, assisted by protein chaperones, that determines whether the protein will have a structural or a catalytic function. Protein formation is a heavy burden for the stressed microbial cell; the creation and polymerization of amino acids consume 75% of the available cell energy (Stouthamer, 1973). It also requires a fully functional cascade of enzymes that may not be present in the stressed cell. For the stressed cell, a supply of appropriate polypeptides may enable the cell to overcome these constraints.

Transport of polypeptides

Transport of amino acids and peptides into the cell costs energy. It requires one proton to transport a single amino acid or a peptide. Peptide transport has greater energy advantages over amino acid transport.

How large polypeptides (greater than a tetrapeptide) cross cell membranes is still a matter of conjecture. Chaperone molecules could capture, transport and, if required, refold large peptides (Clarke, 1996). Stressed cells produce increased amounts of chaperone molecules as heat shock proteins (Hsp), as well as other SOS proteins, and these may have a critical transport role (Martin and Hartl, 1997).

Polypeptide mixtures and ribozymes

Existing technology cannot reveal the complex chemical pathways involved in the resuscitation of a single cell. All currently described microbial biochemical pathways are based on mean value analyses of millions of cells. The fact remains, however, that a severely disabled cell has to create a large variety of enzymes and structural proteins to enable it to commence normal metabolism and division.

There are theories that RNA developed before DNA in the evolutionary pathway. These theories are based on the fact that nucleotide subunits of DNA are modified RNA molecules, and that many enzymes require coenzymes that are RNA molecules or close relatives. In 1982, RNA molecules with catalytic powers (ribozymes) were discovered in protozoa and bacteria. Ribozymes are hybrid molecules, part peptide and part 'gene'. Molecules can be selected by the 'fitness' of the peptide arm and then amplified with the attached 'gene'. They are used to cut strands of DNA and they can be replicated in a synthesizer. When replicated using a synthesizer that allows 'errors' to occur in copying, the mutant enzymes can show increases in DNA cutting efficiency of 100%. This discovery fitted the three main pillars of Darwinian evolution:

- a mechanism to introduce mutations in molecules
- a selection or 'fitness' pressure that favours some mutations over others
- an amplification mechanism that encourages favoured mutants to multiply.

Exposing stressed cells to millions of polypeptide variations would allow cells to select polypeptides by 'fitness' of combination with the cell ribozymes. These RNA–polypeptide molecules may be used in a non-ribosomal pathway as 'starters' for the synthesis of key enzymic and structural proteins.

These speculative but intriguing hypotheses are more fully discussed in Bridson (1999).

STORAGE AND PROTECTION OF PREPARED CULTURE MEDIA FROM ENVIRONMENTAL FACTORS

It is a not uncommon paradox that much science and care can be expended in producing high-quality media, only to have the products stored under conditions in which they will deteriorate. Media supplied from commercial or central sources now have precise instructions about shelf-lives, storage temperature limits and light protection factors. Microbiological laboratories are expected to show evidence that they conform to these storage and use instructions. Like good wine, culture media solutions are best kept cool (2–8°C), not frozen, and in the dark (photo-oxidation spoils both wine and broth). All bottles should carry labels identifying the contents and show the preparation date, shelf-life, batch number and expiry date. Media should be used in strict date order and outdated material discarded. There are also a few elementary observations that can be carried out. Bottles with reduced clarity and darkening of colour (especially the upper layer of the liquid) should be rejected and the pH value checked for acid drift.

Great damage can be caused to culture media in the autoclave. If powdered culture media is added to the bottles, water simply poured on top to the prescribed volume and the bottles placed in the autoclave for sterilization, the problem can be seen when the autoclave is opened. There is a dark-brown concentrate of nutrients at the bottom of the bottle. When shaken, the colour becomes a uniform straw colour but the damage has been done. Superheating biological concentrates causes interactive complexity of the amino acids, phosphates and free metal ions. It reduces the nutritive value of the medium and reduces the shelf-life of the product. All prepared culture media are dynamic complexes of ingredients that can only deteriorate on storage. The best storage conditions are those that reduce the rate of deterioration to the minimum. Stirred media preparators are preferable to static autoclaves.

PROTECTION OF PREPARED PLATES

The most labile prepared culture medium that is widely used in microbiological laboratories is the nutrient gel slab residing in the Petri dish. What causes the rapid fall off in microbiological performance? Chemo- and photo-oxidation have a role but the most serious effect is the loss of surface moisture. Modern agars at 1.2% w/v hold water in a firm gel. The agarose polymer, however, has a natural contraction rate which squeezes water out of the gel; this is called syneresis. If the covered plate is left on the open bench, this water evaporates and the surface 'skin' of agar becomes less porous. The greater the water loss, the thicker the 'skin' of inhibitory agar on the surface. The two forces of syneresis and evaporation work together and can result in a thin layer of very hard agar at the surface of the dish. The microbial inoculum on the surface of the agar quickly uses the available nutrients in its immediate locality. It must now wait while nutrients diffuse up through the agar thickness, or from the periphery of the inoculum to replace nutrients taken up by the microcolony. Anything that slows down this flow of nutrients results in attenuated growth but the agar 'skin' is a poor diffuser of solute. All bacteria suffer in this situation and usually produce smaller colonies, but more serious is the fact that stressed cells are particularly vulnerable and may fail to multiply at all.

How much water must be lost before there is a significant effect on bacterial growth? Tests carried out in Oxoid laboratories many years ago, before stressed cells were fully appreciated, showed that if prepared plates lost more than 5% of their gross weight (Petri dish plus agar layer) on storage, when compared with their fresh weight, then a reduction in colony size of unstressed bacteria could be detected. This is quite a small amount of water but it is removed from the highly critical surface between organism and agar.

Armed with this knowledge, is it possible to compare the performance of current commercially prepared plates with those made in the individual laboratory 30 years or so ago? It would be tempting to say that current plates are much superior to those in the past. Although 7-day-old plates made in the past were poorer than present-day, commercially prepared plates of the same age, in those days, 7-day-old plates were not used in most laboratories. Plates were often poured in the morning and used the same day. These freshly prepared plates would probably be comparable in performance with most commercially/centrally poured plates distributed today.

Packaging is a very important factor but it is not the sole answer. It is possible to see moisture beading the lids and packaging of packs of prepared plates. This problem is caused by changes in temperature and great care is now paid to keeping the temperature of the plates as steady as possible with polystyrene packs. Unfortunately, once agar has lost water it cannot reabsorb it unless it is raised to the melting temperature of agar (80°C.) It is possible to improve the growth of micro-organisms by placing the inoculated dishes in closed containers, together with moisture-saturated air but little or none of the water can rehydrate the agar.

The most successful way to keep moisture in agar plates is to place a sterile plastic film over the surface of the agar. This will prolong the serviceable life of the plate by minimizing the water loss from the surface. A simple test to demonstrate this is to incubate two sterile nutrient agar plates at 35°C for 48 hours. Take them out of the incubator and, using a sterile scalpel blade, cut around the meniscus of agar of one plate. Replace it in the lid and slap the plate smartly on the bench. The agar slab will fall into the lid. Inoculate both plates lightly with a suspension of *Staphylococcus aureus* (Oxford strain) NTCC 6571. After 16 hours' incubation, the conventionally exposed agar surface will show poorer growth than the upside-down surface, which has been protected by the plastic dish. The poor growth represents a reduction in water activity and even moderately high water activity (a_w) reduction can totally inhibit the growth of stressed cells.

QUALITY CONTROL AND THE PERFORMANCE TESTING OF CULTURE MEDIA

Total quality management (TQM) of culture media is discussed on pages 134–146 of Bridson (1994), but clinical laboratories will require only the end-user control tests. These will be to check storage facilities and shelf-lives of culture media in the laboratory or to compare one supplier's media with another. Only quantitative tests should be used for statistical purposes but a great deal of essential microbiological detail is obtained from colonial qualitative characteristics, e.g. colour, translucence, surface reflectance of light, pigment diffusion, size variation, capsule formation, etc. For the identification of certain organisms, such characteristics may be quite critical and the lack or variation of these could fail a particular batch or another supplier's product, and constitute a failure to meet the user's quality standards.

When clinical laboratories made their own culture media, they tested them with laboratory stock cultures and with clinical specimens taken from patients. When commercial suppliers commenced manufacturing media for use in clinical diagnosis, they could either have requested that an outside hospital laboratory tested samples with clinical specimens or have begged for clinical material from local hospital laboratories. Both practices were carried out and both were equally difficult to control. Increasingly stringent safety rules in industry made the importation of clinical material into manufacturing premises very difficult. At this time the expansion of food microbiology led to demands from that industry that all culture media made for them should be tested with specified standard lyophilized strains from national collection centres, e.g. ATCC (American Type Culture Collection) or NCTC (National Collection of Type Cultures) in particular.

As food industry requirements included both spoilage and pathogenic organisms, media manufacturers turned their backs on clinical isolates and used these stable and reproducible organisms for all their testing. Unfortunately, this change opened a gulf between the manufacturer and the medical laboratory. The manufacturer could supply much detail about the quantitative recovery of standard organisms, but the clinical customer did not isolate standard organisms from patients. The manufacturer's task was to nurture very unhappy organisms, taken from a hostile environment of immune and antibiotic attacks, possibly in quite small numbers, i.e. a few highly stressed cells. The difference between these testing situations was exacerbated by the use of selective media. Bartl (1985) discussed this problem where he compared 'wild' versus 'culture collection' strains on selective culture media. He accepted, however, that uncertainties arose from using random non-reproducible 'wild' strains. A better solution was required. The answer seems to be to stress standard strains under strictly controlled conditions, e.g. exposing them to higher or lower temperatures for fixed periods of time, such as 20 min at 52°C and 30 min at –20°C or any variations between these extremes of temperature. There is still much work required to standardize the stress conditions for a wide variety of commercially available lyophilized cultures and to apply the appropriately stressed organisms to test a variety of culture media.

Meanwhile, various schedules of testing methods for culture media are in use. Appendix A of the 'QA and QC of microbiological culture media' (National Committee for Clinical Laboratory Standards or NCCLS, 1996) can be recommended.

- The NCCLS QA Standard for Commercially Prepared Microbiological Culture Media (NCCLS, 1996) describes the appropriate control organisms and test procedures that are to be used by manufacturers of culture media in the USA. If these standards are met customer laboratories in the USA are exempt from further testing. Two culture media formulations have failed to meet NCCLS standards (MacFaddin, 1985): Campylobacter Agar and selective media for pathogenic *Neisseria* species. When using these media, the customer laboratory must carry out appropriate quality control tests.
- The British Standard BS 12322:1999 (BSI, 1999) is a concise description and definition of terms used, performance evaluation criteria, control strains used, quality criteria and information to be supplied by the manufacturer.
- DD ENV ISO 110133-1:2000: (DD, 2000) is a draft publication of extensive terminology with dozens of headings but the small print underneath each heading is not as helpful as CEN/TC 275 (CEN, 2001).
- The CEN/TC 275 N489 (2001) on practical guidelines on performance testing of culture media is much more detailed than the two standards above, although both should be studied. The tables of commonly used culture media with test organisms, culture conditions and expected reactions are particularly useful.
- BS 6068-4.12:1998 (ISO 11731) (BSI, 1998) is concerned with water quality and the detection and enumeration of *Legionella* species in particular. It recommends using Buffered Charcoal Yeast-Extract Agar Medium (BCYE) with GVPC (glycine–vancomycin–polymyxin–cyclohexamide). A specific warning is given against using stock strains of *L. pneumophila* to check the growth properties of BCYE–GVPC medium. Known positive water samples containing 'wild' strains of *Legionella* are preferable.

Corry *et al.* (1995) have updated their 1987 publication on culture media for food microbiology. Each medium is fully described with a brief history plus physical properties, shelf-life, inoculation method, reading results and interpretation, quality assessment (productivity and selectivity) with appropriate references.

A closing comment should be made about culture media incubation temperature. It is a sad and persistent error that all European standards specify 37°C as the mean incubation temperature, rather than the US standard of 35°C. Walk-in incubators normally operate at ± 2°C of the stated mean temperature. Stressed organisms will recover and grow better between 33 and 37°C than they will between 35 and 39°C.

IS THERE A FUTURE FOR CONVENTIONAL CULTURE MEDIA?

This is a question that has been regularly asked over the last 30 years, as novel microbial detection and identification methods have come and gone. One weakness with many 'alternative' tests has been that, with the exception of methods based on nucleic acid amplification, high concentrations of organisms are required before they can be applied. Culture media and all their problems are still required before the novel method can be used. The novel part is then often very expensive compared with conventional culture media microbiology.

Rapid diagnostic tests for the presence of pathogenic organisms in patients can be justified, in spite of high costs, in certain circumstances:

- Where growth of the suspected organism is very slow or uncertain in culture media
- Where very early diagnosis or differentiation of agents causing identical symptoms is required to allow life-saving procedures to be applied, e.g. meningitis
- Where rapid screening of potential patients is required for barrier nursing or isolation from wards
- Where identification of very specific antigen-variant types of the pathogen is urgently sought.

Other situations could be listed if the costs were of less consequence. Unfortunately, health services consume money with the same avidity that 'black holes' consume matter. In the 'rob Peter to pay Paul' situation that is common in most health services, a best value solution is always sought.

There are good reasons why conventional methods of culturing micro-organisms will continue for a substantial period of time:

- Unsuspected or 'novel' micro-organisms that emerge (e.g. legionellas).
- The methods used are very economical and it can be envisaged that smaller culture, media-packaged formats will be produced on high-speed machines, thus improving the economy of the methods.
- There are large parts of the world where, even today, microbial diagnostic services are sadly lacking. Even simple pathology services would be a great advantage to the deprived populations.

- Most importantly, only broad-spectrum culture media are capable of nurturing and cultivating quite unexpected organisms. Until it is possible accurately to interpret tissue SOS signals, the pyrexial, hypotensive and very unwell patient requires an open mind and non-discriminate investigations, including conventional microbiological testing.

As science and medicine progress at an accelerating rate, it would be foolish to attempt to predict when a totally new and reliable system of identifying significant microbial infection in a patient will appear. The agar plate is about 120 years old and, despite many forecasts of its imminent demise, it is currently manufactured in larger numbers than ever. It still appears to be in its log phase of growth. Experience proves that technology does not become obsolete until it is superseded by something that is clearly superior. Microbiologists await the arrival of superior microbial diagnostic methods with great interest.

LABORATORY-PREPARED MEDIA

Most of the media mentioned in this book are available commercially but some are not and the formulae for these are given below. A few others are in appropriate chapters.

Arginine broth

This is useful in identifying some streptococci and Gram-negative rods:

Tryptone	5 g
Yeast extract	5 g
Dipotassium hydrogen phosphate	2 g
L-Arginine monohydrochloride	3 g
Glucose	0.5 g
Water	1000 ml

Dissolve by heating, adjust to pH 7.0; dispense in 5- to 10-ml amounts and autoclave at 115°C for 10 min.

For arginine breakdown by lactobacilli, use MRS (p. 347) broth in which the ammonium citrate is replaced with 0.3% arginine hydrochloride.

Carbohydrate fermentation and oxidation tests

Formerly known as 'peptone water sugars', some of these media are now more nutritious and do not require the addition of serum except for very fastidious organisms. Liquid, semisolid and solid basal media are available commercially and contain a suitable indicator, e.g. Andrade (pH 5–8), phenol red (pH 6.8–8.4) or bromothymol blue (pH 5.2–6.8).

Some sterile carbohydrate solutions (glucose, maltose, sucrose, lactose, dulcitol, mannitol, salicin) are available commercially. Prepare others as a 10% solution in water and sterilize by filtration. Add 10 ml to each 100 ml of the reconstituted sterile basal medium and tube aseptically. Do not heat carbohydrate media. Durham's (fermentation) tubes need to be added only to the glucose tubes; gas from other substrates is not diagnostically significant.

These media are not suitable for some organisms; variations are given below.

Ammonium salt 'sugars'

Pseudomonads and spore bearers produce alkali from peptone water and the indicator may not change colour. Use this basal synthetic medium instead.

Ammonium dihydrogen phosphate	1.0 g
Potassium chloride	0.2 g
Magnesium sulphate	0.2 g
Agar	10.0 g
Water	1000 ml

Dissolve by heating, add 4 ml 0.2% bromothymol blue and a final 1% sterile (filtered) carbohydrate solution (0.1% aesculin, 0.2% starch – these are sterilized by steaming).

'Anaerobic sugars'

Peptone water or serum peptone water sugar media need to have a low oxygen tension for testing the reactions of anaerobes, even under anaerobic conditions. Add a clean wire nail to each tube before sterilization. Do not add indicator until after growth is seen.

Baird-Parker's carbohydrate medium

For the sugar reactions of staphylococci and micrococci use Baird-Parker (1966) formula:

Yeast extract	1 g
Ammonium dihydrogen phosphate	1 g
Potassium chloride	0.2 g
Magnesium sulphate	0.2 g
Agar	12 g
Water	1000 ml

Steam to dissolve and adjust to pH 7.0. Add 20 ml of 2% bromocresol purple and bottle in 95-ml amounts. Sterilize at 115°C for 10 min. For use, melt and add 5 ml 10% sterile carbohydrate.

Lactobacilli fermentation medium

For sugar reactions of lactobacilli, make MRS base (see p. 347) without Lab-Lemco and glucose and adjust to pH 6.2–6.5. Bottle in 100-ml amounts and autoclave at 115°C for 15 min. Melt 100 ml, add 10 ml 10% sterile (filtered) carbohydrate solution and 2 ml 0.2% chlorophenol red. Dispense aseptically.

Robinson's serum water sugars

Some organisms will not grow in peptone water sugars. Robinson's serum water medium is superior to that of Hiss:

Peptone	5 g
Disodium hydrogen phosphate	1 g
Water	1000 ml

Steam for 15 min, adjust to pH 7.4 and add 250 ml horse serum. Steam for 20 min, add 10 ml

Andrade indicator and 1% of appropriate sugar (0.4% starch).

Unheated serum contains diastase and may contain a small amount of fermentable carbohydrate, so a buffer is desirable, particularly in starch fermentation tests.

Gluconate broth

For differentiation among enterobacteria:

Yeast extract	1.0 g
Peptone	1.5 g
Dipotassium hydrogen phosphate	1.0 g
Potassium gluconate	40.0 g
(Or sodium gluconate	37.25 g)
Water	1000 ml

Dissolve by heating, adjust to pH 7.0, filter if necessary, dispense in 5- to 10-ml amounts and autoclave at 115°C for 10 min.

Hugh and Leifson's medium

Peptone	2.0 g
Sodium chloride	5.0 g
Dipotassium hydrogen phosphate	0.3 g
Agar	3.0 g
Water	1000 ml

Heat to dissolve, adjust to pH 7.1 and add 15 ml 0.2% bromothymol blue and sterile (filtered) glucose to give a final 1% concentration. Dispense aseptically in narrow (< 1 cm) tubes in 8- to 10-ml amounts.

Baird-Parker (1966) modification for staphylococci and micrococci

Tryptone	10 g
Yeast extract	1 g
Glucose	10 g
Agar	2 g
Water	1000 ml

Steam to dissolve, adjust to pH 7.2. Add 20 ml 0.2% bromocresol purple, tube in 10-ml amounts in narrow (12 mm) tubes and sterilize at 115°C for 10 min.

Donovan's (1966) medium

Originally devised to differentiate klebsiellas, this is useful for screening lactose fermenting Gram-negative rods:

Tryptone	10 g
Sodium chloride	5 g
Triphenyltetrazolium chloride	0.5 g
Ferrous ammonium sulphate	0.2 g
1% aqueous bromothymol blue	3 ml
Inositol	10 g
Water	1000 ml

Dissolve by heating and adjust to pH 7.2. Distribute in 4-ml amounts and autoclave at 115°C for 15 min. Allow to set as butts.

Dorset's egg medium

Wash six fresh eggs in soap and water, and break into a sterilized basin. Beat with a sterile fork and strain through sterile cotton gauze into a sterile measuring cylinder. To three parts of egg add one part nutrient broth. Mix, tube or bottle, and inspissate in a sloped position for 45 min at 80–85°C. Glycerol 5% may be added if desired. (Griffith's egg medium uses saline instead of broth and is useful for storing stock cultures.)

Egg yolk salt broth and agar

For lecithinase tests:

Nutrient broth or blood agar base	10 ml
Egg yolk emulsion	1 ml
NaCl (5%)	0.2 ml

EMJ medium for leptospires

The formula is given by Waitkins (1985) but a commercial preparation is available.

Selective EMJH/5FU medium (Waitkins, 1985)

Dissolve 1 g 5-fluorouracil (5FU) in about 50 ml distilled water containing 1 mol/l NaOH by gentle heat (do not exceed 56°C). Adjust to pH 7.4–7.6 with 1 mol/l HCl, make up to 100 ml with distilled and sterilize through a membrane filter (0.45 μm). Dispense in 1-ml lots and store at –4°C.

Thaw 1 ml at 56°C and add it to 100 ml of the EMJ Medium (final concentration of 5FU is 100 μg/ml).

Modified alkaline peptone water

This is used for isolating *Vibrio cholerae*:

Sodium chloride	10.0 g
Magnesium chloride hexahydrate	4.0 g
Potassium chloride	4.0 g
Tryptone peptone	10.0 g

Adjust to pH 8.6, distribute and sterilize by autoclaving (Roberts *et al.*, 1995).

N medium (Collins, 1962)

This is used in the differentiation of mycobacteria:

NaCl	1.0 g
$MgSO_4 \cdot 7H_2O$	0.2 g
KH_2PO_4	0.5 g
$Na_2HPO_4 \cdot 12H_2O$	3.0 g
$(NH_4)_2SO_4$	10.0 g
Glucose	10.0 g

Phenylalanine agar

DL-Phenylalanine	2 g
Yeast extract	3 g

Na_2HPO	1 g
NaCl	5 g
Agar	20 g
Distilled water	1000 ml

Dissolve by heat; sterilize at 115°C for 10 min and tube as slopes.

Phenolphthalein phosphate agar

For the phosphatase test: add 1 ml of a 1% solution of phenolphthalein phosphate to 100 ml of nutrient agar or commercial blood agar base and pour plates.

Phenolphthalein sulphate agar

For the arylsulphatase test: dissolve 0.64 g potassium phenolphthalein sulphate in 100 ml water (0.01 mol/l). Dispense one of the Middlebrook Broth media in 2.7-ml amounts and add 0.3 ml of the phenolphthalein sulphate solution to each tube.

SPS (Sulphite Polymyxin Sulphadiazine agar)

Peptone (from casein)	15.0 g
Yeast extract	10.0 g
Ferric citrate	0.5 g
Sodium sulphite	0.5 g
Sodium sulphadiazine	0.12 g
Agar	13.9 g
pH at 25°C	7.0

From Angelotti *et al.* (1962).

Tyrosine agar: xanthine agar

To 100 ml melted nutrient agar, add 5 g tyrosine or 4 g xanthine and steam for 30 min. Mix well to suspend amino acid and pour plates.

'Sloppy agar' (Craigie tube)

Broth containing 0.1–0.2% agar will permit the migration of motile organisms but will not allow convection currents. This permits motile organisms to be separated from non-motile bacteria in the Craigie tube.

Dispense sloppy agar into 12-ml amounts in screw-capped bottles. To each bottle add a piece of glass tubing of dimensions 50×5 or 6 mm. There must be sufficient clearance between the top of the tube and the surface of the medium so that a meniscus bridge does not form, and the only connection between the fluid and the inner tube is through the bottom.

Starch agar

Prepare a 10% solution of soluble starch in water and steam for 1 h. Add 20 ml of this solution to 100 ml melted nutrient agar and pour plates.

Diluents

Saline and Ringer's solutions

Physiological saline, which has the same osmotic pressure as micro-organisms, is a 0.85% solution of sodium chloride in water. Ringer's solution, which is ionically balanced so that the toxic effects of the anions neutralize one another, is better than saline for making bacterial suspensions. It is used at one-quarter the original strength.

Sodium chloride (AnalaR)	2.15 g
Potassium chloride (AnalaR)	0.075 g
Calcium chloride, anhydrous (AnalaR)	0.12 g
Sodium thiosulphate pentahydrate	0.5 g
Distilled water	1000 ml

The pH should be 6.6.
Both may be purchased as tablets.

Calgon Ringer's

Alginate wool, used in surface swab counts, dissolves in this solution:

Sodium chloride	2.15 g
Potassium chloride	0.075 g
Calcium chloride	0.12 g
Sodium hydrogen carbonate	0.05 g
Sodium hexametaphosphate	10.00 g
pH	7.0

It is conveniently purchased as tablets.

Thiosulphate Ringer's

This may be preferred for dilutions of rinses where there may be residual chlorine. It is conveniently manufactured in tablet form.

Maximum recovery diluent (MRD)

The safest and least lethal of diluents is 0.1% peptone water.

Buffered peptone water

This is a useful pre-enrichment medium for the enrichment of salmonellas from foods.

Peptone	10.0 g
Sodium chloride	5.0 g
Disodium hydrogen phosphate	3.5 g
Potassium dihydrogen phosphate	1.5 g
Distilled water	1000 ml

The pH should be 7.2.

Phosphate-buffered saline

This is a useful general diluent at pH 7.3:

Sodium chloride	8.0 g
Potassium dihydrogen phosphate	0.34 g
Dipotassium hydrogen phosphate	1.21 g
Water	1000 ml

Adjustment of pH

The pH of reconstituted and laboratory-prepared media should be checked with a pH meter but, for practical purposes, devices such as the BDH Lovibond Comparator are good enough.

Pipette 10 ml of the medium into each of two 152 × 16 mm test tubes. To one add 0.5 ml 0.04% phenol red solution. Place the tubes in a Lovibond Comparator with the blank tube (without indicator) behind the phenol red colour disc, which must be used with the appropriate screen. Rotate the disc until the colours seen through the apertures match. The pH of the medium can be read in the scale aperture. Rotate the disc until the required pH figure is seen, add 0.05 mol/l sodium hydroxide solution or hydrochloric acid to the medium plus indicator tube and mix until the required pH is obtained.

From the amount added, calculate the volume of 1 or 5 mol/l alkali or acid that must be added to the bulk of the medium. After adding, check the pH again.

For example, 10 ml of medium at pH 6.4 requires 0.6 ml 0.05 mol/l sodium hydroxide solution to give the required pH of 7.2. Then the bulk medium will require:

$$[0.6 \times 100]/20 = 3.0 \text{ ml } 1 \text{ mol/l NaOH/litre}$$

Discs of other pH ranges are available.

All final readings of pH must be made with the medium at room temperature because hot medium will give a false reaction with some indicators. Note the volume of acid and alkali used and compare with records of previous batches. Major differences in pH or buffering capacity may indicate an error in calculation or changes in the content of raw materials.

GENERAL PRECAUTIONS

Ensure that all equipment, including autoclaves, preparators and distributors, is regularly serviced and maintained in good condition.

Do not overheat media or quality will be impaired. Do not autoclave concentrates before they are dissolved by prior heating or they will set at the bottom of the vessel. Do not hold partly processed media overnight.

Screw caps, aluminium or propylene closures are better than cottonwool plugs.

CONTAMINATION

This is a major cause of media loss and may be caused by inadequate heat processing or failure to sterilize the distributing apparatus. The appearance of a few colonies on plated media after storage may indicate environmental fall-out. This may be reduced or even eliminated by using automated plate pouring machines and/or by pouring media in laminar-flow clean air cabinets.

ISOLATION AND SELECTIVE MEDIA: EFFICIENCY OF PLATING TECHNIQUE

New batches of culture media may vary considerably and should be tested in the following way.

Prepare serial tenfold dilutions, e.g. 10^{-2}–10^{-7}, of cultures of various organisms that will grow on or be inhibited by the medium. For example, when testing DCA use *S. sonnei*, *S. typhi*, *S. typhimurium*, several other salmonellas and *E. coli*. Use at least four well-dried plates of the test medium for each organism and at least two plates each of a known satisfactory control medium and a non-selective medium, e.g. nutrient agar. Do Miles and Misra drop counts with the serial dilutions of the organisms on these plates so that each plate is used for several dilutions (Figure 5.1) (see p. 148).

Count the colonies, tabulate the results and compare the performances of the various media. These may suggest that in a new batch of a special medium it is necessary to alter the proportion of certain ingredients.

Stock cultures will, however, frequently grow on

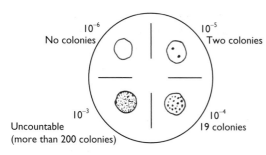

Figure 5.1 Miles and Misra count

media that will not support the growth of 'damaged' organisms from natural materials. It is best, therefore, to use dilutions or suspensions of such material.

With some culture media, these drop counts do not give satisfactory results because of reduced surface tension. An alternative procedure is recommended.

Make cardboard masks to fit over the tops of Petri dishes and cut a square 25 × 25 mm in the centre of each. Place a mask over a test plate and drop one drop of the suspension through the square on the medium. Spread the drop over the area limited by the mask. This method has the additional advantage of making colony counts easier to perform but more plates must, of course, be tested.

In addition to these EOP tests, ordinary plating methods should be used to compare colony size and appearance.

IDENTIFICATION MEDIA

Maintain stock cultures (see p. 86) of organisms known to give positive or negative reactions with the identification tests in routine use. The same organism may often be used for a number of tests. When a new batch of any identification medium is anticipated, subculture the appropriate stock strain so that a young, active culture is available. Test the new medium with this culture.

Keep stock cultures of the organisms likely to be encountered, preferably from a type culture collection and maintained freeze-dried until required.

Subculture these cultures to test new batches of culture media.

Caution: bacteria in Hazard Group 3 should not be used unless the tests can be conducted in Level 3 laboratories.

REFERENCES

Angelotti, R., Hall, H. E., Foter, M. J. *et al.* (1962) Quantitation of *Clostridium perfringens* in foods. *Applied Microbiology* 10: 193–199.

Baird, R. M., Corry, J. E. L. and Curtis, G. D. W. (eds) (1987) Pharmacopoeia of culture media for food microbiology. *International Journal of Food Microbiology* 5: 187–229.

Baird-Parker, A. C. (1966) An improved diagnostic and selective medium for isolating coagulase-positive staphylococci. *Journal of Applied Bacteriology* 25: 12–19.

Barer, M. R. and Harwood, C. R. (1999) Bacterial viability and culturability. *Advances in Microbial Physiology* 41: 93–137.

Barrow, G. I. and Feltham, R. K. A. (eds) (1993) *Cowan & Steel's Manual for the Identification of Medical Bacteria*, 3rd edn. Cambridge: Cambridge University Press.

Bartl V. (1985) Choice of test strains. *International Journal of Food Microbiology* 2: 99–102.

Bloomfield, S. F., Stewart, G. S., Dodd, C. C. *et al.* (1998) The viable but non-culturable phenomenon explained? *Microbiology* 144: 1–3.

Bridson, E. Y. (1978) Problemen by de standaardisatie van selective stoffen voor isolatie media voor Enterobacteriaceae. In: *Van Monster tot Resultaat.* Wageningen: Nederlands Society for Microbiology, pp. 58–67.

Bridson, E. Y. (1990) Media in Microbiology. *Review of Medical Microbiology* 1: 1–9.

Bridson, E. Y. (1994) *The Development, Manufacture and Control of Microbiological Culture Media.* Basingstoke: Oxoid Ltd.

Bridson, E. Y. (1999) Quantal microbiology: the interactive effect of chaos and complexity on stressed cell recovery. PhD Thesis, University of the West of England.

British Standards Institution (BSI) (1998) British Standard 6068-4.12:1998 (ISO 11731:1998) Water quality – Part 4: Microbiological methods – Section 4.12: Detection and enumeration of *Legionella*. London: BSI.

BSI (1999) British Standard BS EN 12322:1999. *In vitro* diagnostic medical devices – Culture media for microbiology – Performance criteria for culture media. London: BSI.

Brock, T. (1988) *Robert Koch: A life in medicine and bacteriology*. Madison, WI: Science Technical Publications.

Catlin, B. W. (1973) Nutritional profiles of *Neisseria gonorrhoeae*, *N. meningitidis* and *N. lactamica* in chemically defined media and the use of growth requirements for gonococcal typing. *Journal of Infectious Disease* **128**: 178–181.

CEN (2001) CEN/TC 275 N489:2001. *Microbiology of food and animal feeding stuffs – Guidelines on preparation and production of culture media – Part 2: Practical guidelines on performance testing of culture media*. Brussels: EEC.

Clarke, A. R. (1996) Molecular chaperones in protein folding and translocation. *Current Opinions in Structural Biology* **6**: 43–60.

Collins, C. H. (1962) The classification of 'anonymous' acid fast bacilli from human source. *Tubercle* **43**: 293–298.

Corry, J. E. L. , Curtis, G. W. D. and Baird, R. M. (eds) (1995) Culture Media for Food Microbiology. In: *Progress in Industrial Microbiology*, Vol. 34. Oxford: Elsevier Science.

Davis, B. D. and Mingioli, E. S. (1950) Mutants of *Escherichia coli* requiring methionine or vitamin B$_{12}$. *Journal of Applied Bacteriology* **60**: 17–21.

DD (2000) DD ENV ISO 11133-1:2000. *Microbiology of food and animal feeding stuffs – Guidelines on quality assurance and performance testing of culture media. Part 1: General guidelines*. Brussels: EEC.

Donovan, T. J. (1966) A *Klebsiella* screening medium. *Journal of Medical Laboratory Technology* **23**: 11194–11196.

Dukan, S., Levi, Y. and Touati, D. (1997) Recovery of culturability of an HOCl-stressed population of *Escherichia coli* after incubation in phosphate buffer: resuscitation or regrowth? *Applied and Environmental Microbiology* **63**: 4204–3209.

Ellner, P. D., Stossel, C. I., Drakeford, E. *et al.* (1966) A new culture medium for medical microbiology. *American Journal of Clinical Pathology* **45**: 502–507.

Gieson, G. L. (1955) *The Private Science of Louis Pasteur*. Princeton: Princeton University Press.

Hackney, C. R., Ray, B., Speck, M. L. (1979) Repair detection procedure for enumeration of faecal coliforms and enterococci from seafoods and marine environments. *Applied and Environmental Microbiology* **37**: 947–953.

Lacey, B. W. (1954) A new selective medium for *Haemophilus pert*ussis containing diamidine, sodium fluoride and penicillin. *Journal of Hygiene* **52**: 273–303.

MacFaddin, J. F. (1985) *Media for the Isolation–Cultivation–Identification–Maintenance of Medical Bacteria*, Vol. 1. Baltimore, MA: Williams & Williams.

Martin, J. and Hartl, F. U. (1997) Chaperone-assisted protein folding. *Current Opinions in Structural Biology* **7**: 41–52.

Millar, V. R. and Banwart, G. J. (1965) Effect of various concentrations of brilliant green and bile salts on salmonella and other organisms. *Applied Microbiology* **13**: 77–80.

Mossel, D. A. A. and Corry, J. E. L. (1977) Detection and enumeration of sublethally injured pathogenic and index bacteria in foods and water processed for safety. *Alimenta* **16**: 19–34.

Mossel, D. A. A. and Ratto, M. A. (1970) Rapid detection of sublethally impaired cells of Enterobacteriaceae in dried foods. *Applied Microbiology* **20**: 273–275.

Munro, A. L. S. (1968) Measurement and control of pH values. In: Norris JR, Ribbons DW. (eds), *Methods in Microbiology*, Vol. 2. London: Academic Press, pp. 39–89.

National Committee for Clinical Laboratory Standards (NCCLS) (1996) *Quality Assurance for Commercially Prepared Microbiological Culture Media*, 2nd edn. Approved Standard. 1996. CCLS M22-A2, Vol. 16, No. 16. Villanova, PA: NCCLS.

Reuter, G .(1985) Elective and selective media for lactic acid bacteria. *International Journal of Food Microbiology* 2: 55–68.

Roberts, D., Hooper, W. and Greenwood, M. (eds) (1995) *Practical Food Microbiology*. London: Public Health Laboratory Service, p. 155.

Skirrow, M. B. (1977) Campylobacter enteritis – a 'new' disease. *British Medical Journal* ii: 9–11.

Stephens, P. J. (1999) Recovery of stressed salmonellas. PhD Thesis, University of Exeter.

Stephens, P. J., Joynson, J. A., Davies, K. W. *et al.* (1977). The use of an automated growth analyser to measure recovery times of single heat-injured salmonella cells. *Journal of Applied Microbiology* 83: 445–455.

Stouthamer, A. H. (1973) The search for a correlation between theoretical and experimental growth yields. In: Quayle JR (ed.), *International Review of Biochemistry and Microbiological Biochemistry*, Vol. 21. Baltimore, MA: University Park Press, pp. 1–47.

Thayer, J. D. and Martin, J. E. (1966) Improved medium selective for cultivation of *Neisseria gonorrhoea* and *N. meningitidis*. *Public Health Reports* 81: 559–562.

Waitkins, S. (1985) Leptospiras and leptospirosis. In: Collins, C. H. and Grange, A. M. (eds), *Isolation and identification of Micro-organisms of Medical and Veterinary Importance*. Society for Applied Bacteriology Technical Series No. 21. London: Academic Press, pp. 251–296.

Walker, G. C. (1966) The SOS response of *Escherichia coli*. In: Neidhardt FC (ed.), *Escherichia coli and Salmonella*, Vol. 1. Washington DC: ASM Press, pp. 1400–1416.

Willis, A. T. and Hobbs, B. (1959) A medium for the identification of clostridia producing opalescence in egg yolk emulsions. *Journal of Pathology and Bacteriology* 77: 299–300.

Wright, H. D. (1933) The importance of adequate reduction of peptone in the preparation of media for the pneumococcus and other organisms. *Journal of Pathology and Bacteriology* 37: 257–282.

6

Cultural methods

GENERAL BACTERIAL FLORA

Specimens and samples for microbiological examination may contain mixtures of many different micro-organisms. To observe the general bacterial flora of a sample of any material, use non-selective media. No one medium or any one temperature will support the growth of all possible organisms. It is best to use several different media, each favouring a different group of organisms, and to incubate at various temperatures aerobically and anaerobically. The pH of the medium used should approximate that of the material under examination.

For bacteria, examples are nutrient agar, glucose tryptone agar and blood agar and, for moulds and yeasts, malt agar and Sabouraud's agar.

Methods for estimating bacterial numbers are given in Chapter 10, and the dilution methods employed for that purpose can be used to make inocula for the following techniques. Three methods are given here for plating material. They depend on separating clumps or aggregates of bacteria so that each will grow into a separate colony (but see Chapter 10). Colonies resulting from these manipulations may then be subcultured for further examination.

Pour plate method

Melt several tubes each containing 15 ml medium. Place in water bath at 45–50°C to cool. Emulsify the material to be examined in 0.1% peptone water and prepare 10^{-1}, 10^{-2} and 10^{-3} dilutions in the same diluent (see p. 146). Add 1 ml of each dilution to 15 ml of the melted agar, mix by rotating the tube

between the palms of the hands, and pour into a Petri dish. Make replicate pour plates of each dilution for incubation at different temperatures. Allow the medium to set, invert the plates and incubate. This method gives a better distribution of colonies than that used for the plate count (see p. 146), but cannot be used for that purpose because some of the inoculum, diluted in the agar, remains in the test tube.

Spreader method

Dry plates of a suitable medium. Make dilutions of the emulsified material as described above. Place about 0.05 ml (1 drop) of dilution in the centre of a plate and spread it over the medium by pushing the glass spreader backwards and forwards while rotating the plate. Replace the Petri dish lid and leave for 1–2 h to dry before inverting and incubating as above. Plates that are wet may yield confluent growth in spots because of growth of cells in the liquid on the surface.

Looping-out method

Make dilutions of material as described above. Usually the neat emulsion and a 10^{-1} dilution will suffice. Place one loopful of material on the medium near the rim of the plate and spread it over the segment (Figure 6.1a). Flame the loop and spread from area A over area B with parallel streaks, taking care not to let the streaks overlap. Flame the loop and repeat with area C, and so on.

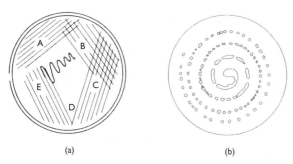

Figure 6.1 (a) 'Spreading' or 'looping-out' on a plate; (b) rotary plating

Each looping out dilutes the inoculum. Invert the plates and incubate.

When looping-out methods are used, two loops are useful: one is cooling while the other is being used.

These surface plate methods are said to give better isolation of anaerobes than pour plates.

Spiral or rotary plating

Spiral platers are used extensively in the food, pharmaceuticals, cosmetics and water industries to count bacteria. The benefits of spiral plating compared with conventional methods include minimal set-up time, the elimination of serial dilutions, up to 69% labour saving, a reduction in consumable costs and laboratory waste with up to a two-thirds reduction in incubator space requirements. Incubated plates can then be counted manually or automatically using image analysis systems with the requisite software.

Although the early spiral platers were purely mechanical the current generation of instruments use a microprocessor to control all aspects of the deposition of liquid sample on to the surface of a rotating agar plate. The stylus arm dispenses liquid samples in an Archimedes spiral, either uniformly across the plate, or as a continuous decreasing volume producing the equivalent of a thousandfold dilution across a single plate. Pre-programmed options allow liquid sample to be dispensed in a variety of ways, using up from 10 to 400 µl of

sample. One impressive feature of the WASP system (Don Whitley Scientific Ltd, UK) is the ability to load sample, inoculate a plate and then clean the stylus with a single keystroke. This instrument also has an automatic cleaning cycle, which uses sanitizing solution from an integral reservoir and sterile water contained in autoclavable vessels. To ensure that less experienced users can operate the instrument with the minimum risk of inaccurate results, the user-selectable parameters can be chosen and then locked. The microprocessor continually monitors the travel and relative positions of the syringe, carriage and turntable, and these can be interrogated by service engineers. The results are then compared with factory settings, generating valuable instrument calibration data, essential for today's quality standards.

Hand-operated platers (with electrically driven turntable) are also available. Place an open plate on the turntable. Touch the rotating plate with the charged loop at the centre of the medium and draw it slowly towards the circumference. A spiral of bacterial growth will be obtained after incubation which will show discrete colonies (see Figure 6.1b).

Multipoint inoculators

These are convenient tools when many replicate cultures are needed. They are fully or semi-automatic and spot inoculate large numbers of Petri dish cultures. A number of pins take up the inoculum from a single reservoir or different inocula from the wells in, for example, a microtitre plate and spot inoculate the medium on one or a succession of Petri dishes. Multipoint inoculators have heads for up to 36 pins for 90-mm dishes and 96 for larger dishes. A variety of these instruments is available commercially. They are particularly useful for combined antibiotic susceptibility, minimum inhibitory concentration tests, the quality control of culture media, the assessment of bacterial loads of large numbers of samples (e.g. urines, foods) and in some identification procedures, e.g. of staphylococci and enterobacteria.

The UK Public Health Laboratory Service has published an excellent guide to multipoint inoculation methods (Faiers *et al.*, 1991).

Shake tube cultures

These are useful for observing colony formation in deep agar cultures, especially of anaerobic or microaerophilic organisms.

Dispense media, e.g. glucose agar or thioglycollate agar, in 15- to 20-ml amounts in bottles or tubes 20–25 mm in diameter. Melt and cool to approximately 45°C. Add about 0.1 ml inoculum to one tube, mix by rotating between the palms, remove one loopful to inoculate a second tube, and so on. Allow to set and incubate. Submerged colonies will develop and will be distributed as follows: obligate aerobes grow only at the top of the medium and obligate anaerobes only near the bottom; microaerophiles grow near but not at the top; and facultative organisms grow uniformly throughout the medium. Banding of growth in response of oxygen concentration may be observed by reducing the agar concentration in the medium to 0.5–1.0% ('sloppy agar').

Stab cultures

These can be used to observe motility, gas production and gelatin liquefaction. Dispense the medium in 5-ml amounts in tubes or bottles of 12.5 mm diameter and inoculate by stabbing a loaded wire down the centre of the agar.

Subcultures

Place the plate culture under a low-power binocular microscope and select the colony to be examined. With a flamed straight wire (*not* a loop) touch the colony. There is no need to dig or scrape; too vigorous handling may result in contamination with adjacent colonies or with those beneath the selected colony.

With the charged wire, touch the medium on another plate, flame the wire and then use a loop to spread the culture to obtain individual colonies. This should give a pure culture but the procedure may need repeating.

To inoculate tubes or slopes, follow the same procedure. Hold the charged wire in one hand and the tube in the other in an almost horizontal position. Remove the cap or plug of the tube by grasping it with the little finger of the hand holding the wire. Pass the mouth of the tube through a Bunsen flame to kill any organism that might fall into the culture, and inoculate the medium by drawing the wire along the surface of a slope or touching the surface of a liquid. Replace the cap or plug and flame the wire. Several tubes may be inoculated in succession without recharging the wire.

ANAEROBIC CULTURE

Anaerobic jars

Modern anaerobic jars are made of metal or transparent polycarbonate, are vented by Schrader valves and use 'cold' catalysts. They are therefore safer than earlier models; internal temperatures and pressures are lower. More catalyst is used and the escape of small particles of catalyst, which can ignite hydrogen–air mixtures, is prevented. Nevertheless, the catalyst should be kept dry, dried after use and replaced frequently.

For ordinary clinical laboratory work the commercial sachets are a convenient source of hydrogen and hydrogen–carbon dioxide mixtures. A recent development is a sachet that achieves reliable anaerobic conditions without generating hydrogen. The final atmosphere contains less than 1% oxygen, supplemented with carbon dioxide.

Place the plates 'upside down', i.e. with the medium at the bottom of the vessel. If they are incubated the right way up, the medium sometimes falls into the lid when the pressure is reduced. Dry the catalyst capsule by flaming and replace the lid.

Some workers prefer to use pure hydrogen or a mixture of 90% hydrogen and 10% carbon dioxide. A non-explosive mixture of 10% hydrogen, 10% carbon dioxide and 80% nitrogen is safer (see Kennedy, 1988), but may be difficult to obtain in some countries. There is a common misconception about storing these cylinders horizontally: they may safely be stored vertically.

For the evacuation–replacement jar technique with a mixture containing 90% or more hydrogen attach a vacuum pump to the jar across a manometer and remove air to about −300 mmHg. Replace the vacuum with the gas mixture, disconnect both tubes and allow the secondary vacuum to develop as the catalyst does its work. Leave for 10 min and reconnect the manometer to check that there is a secondary vacuum of about −100 mmHg. This indicates that the seals of the jar are gas tight and that the catalyst is working properly. Replace the partial vacuum with gas to atmospheric pressure.

Note that if the gas mixture contains only 10% hydrogen the jar must be evacuated to at least −610 mmHg so that enough oxygen is removed for the catalyst and hydrogen to remove the remainder.

Indicators of anaerobiosis

'Redox' indicators based on methylene blue and resazurin are available commercially. There is a simple test to demonstrate that anaerobiosis is achieved during incubation: inoculate a blood agar plate with *Clostridium perfringens* and place on it a 5-µg metronidazole disc. This bacterium is not very exacting in its requirement for anaerobiosis and will grow in suboptimal conditions. A zone of inhibition around the growth indicates that conditions were sufficiently reduced for the drug to work and that anaerobiosis is adequate. There is little to commend the use of *Pseudomonas aeruginosa* as a 'negative' control. Even if a 'positive' control such as *C. tetani* fails to grow, it does not necessarily suggest inadequate anaerobic conditions.

Anaerobic cabinets

Although the development of anaerobic jars proved to be a major advance, it was not until the advent of anaerobic workstations that the microbiologist could easily process, culture and examine samples without exposing them to atmospheric oxygen. It is well established that anaerobe isolation rates are significantly increased through the use of anaerobic workstations. Wren (1977), working at the London Hospital, showed an increase in isolation rate of anaerobes from 9.7% when using anaerobic jars opened at both 24 and 48 hours for examination to 28.1% with jars left for 48 hours before being opened, and to 35.7% when using an anaerobic workstation. Wren concluded that 'uninterrupted anaerobic incubation for 48 h substantially increases the yield of anaerobic isolates'. Ten years later a study at Newcastle General Hospital by Sissons *et al.* (1987) showed an isolation rate of 43.6% using anaerobic jars, which was increased to 56.4% with an anaerobic workstation. The principal reason for these differences is straightforward: anaerobic workstations provide an immediate and sustained oxygen-free atmosphere throughout the sample manipulation and incubation period. Furthermore, anaerobic jars are frequently not anaerobic as a result of either a poisoned catalyst or the poor technique used when setting up the jar. It must of course be mentioned that fundamental to good anaerobic microbiology is the requirement to ensure that samples are not exposed to oxygen after sampling. Oxygen exposure will of course reduce culturability of the most stringent of anaerobes in the sample.

Culture under carbon dioxide

Some microaerophiles grow best when the oxygen tension is reduced. Capnophiles require carbon dioxide. For some purposes, candle jars are adequate, providing approximately 2.5% CO_2. Place the cultures in a can or jar with a lighted candle or night-light. Replace the lid. The light will go out when most of the oxygen has been removed and be replaced with CO_2.

Alternatively, place 25 ml 0.1 mol/l HC1 in a 100-ml beaker in the container and add a few marble chips before replacing the lid.

Commercial CO_2 sachets, hydrogen–CO_2 sachets or CO_2 from a cylinder may be used with an anaerobic jar. Remove air from the jar down to –76 mmHg and replace with CO_2 from a football bladder. This is the best method for capnophiles, which require 10–20% CO_2 to initiate growth, but, if extensive work of this nature is carried out, a CO_2 incubator is invaluable. A nitrogen–CO_2 mixture can be used instead of pure CO_2.

Many clinical laboratories now purchase 'CO_2 incubators' that automatically maintain a predetermined level of CO_2 and humidity (see p. 27).

Incubation of cultures

Incubators are discussed in Chapter 2. In most clinical laboratories the usual temperature to be maintained is 35–37°C. A dish of water should be placed on the floor of an incubator (especially if air is circulated to prevent the drying of media in Petri dishes). For culturing at 20–22°C incubators fitted with cooling coils may be necessary. Alternatively, cultures may be left on the bench, but a cupboard is better because draughts can be avoided. A maximum

and minimum thermometer is necessary. Ambient temperatures often fluctuate widely when windows are open or heating is switched off at night.

Psychrophiles are usually incubated at 4–7°C. Again, incubators need a cooling coil, but a domestic refrigerator may be used, provided that it is not opened too often. Thermophiles require 55–60°C.

Incubate Petri dish cultures upside down, i.e. medium uppermost (except in anaerobic vessels), otherwise condensation collects on the surface of the medium and prevents the formation of isolated colonies.

If it is necessary to incubate Petri dish cultures for several days, seal them with adhesive tape or place them in plastic bags or plastic food containers.

COLONY AND CULTURAL APPEARANCES

Although colony recognition is often now considered to be 'old-fashioned' much time and labour may be saved in the identification of bacteria if workers familiarize themselves with colonial appearances on specific media. These appearances and the terms used to describe them are illustrated in Figure 6.2.

Examine colonies and growths on slopes with a hand lens or, better, plate microscope. Observe pigment formation both on top of and under the surface of the colony and note whether pigment diffuses into the medium. In broth cultures, surface growth may occur in the form of a pellicle.

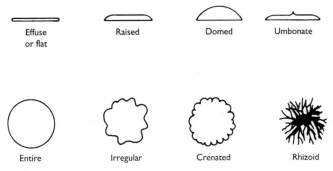

Effuse or flat   Raised   Domed   Umbonate

Entire   Irregular   Crenated   Rhizoid

Figure 6.2 Description of colonies

Odours

Some cultures develop characteristic odours, but caution is required before smelling them. Laboratory workers have acquired infections, e.g. with brucellas, as a result of sniffing cultures (Collins and Kennedy, 1999).

Nevertheless, odours may be helpful in noting the presence of some bacteria. Examples are the 'grapy' odour of *P. aeruginosa* and the 'fruity' odour of *Alcaligenes faecalis*, likened to green apples. Some other organisms with quite distinct odours are: *Proteus* and *Providencia* spp. (similar to *Proteus* spp., but somewhat sweeter), *Serratia odorifera* (usually described as musty, or compared with mouldy potatoes), *Streptococcus milleri* (like caramel), *Klebsiella* spp. (a fishy, amine odour), *Enterobacter* spp. (resembling mouldy oranges), *Nocardia* (earthy) and *Citrobacter* spp. (faecal and quite unpleasant). Some of these odours are described by Murray *et al.* (1999).

AUXANOGRAMS

These are used to investigate the nutritional requirements of bacteria and fungi, e.g. which amino acid or vitamin substance is required by 'exacting strains' or which carbohydrates the organisms can use as carbon sources.

Make a minimal inoculum suspension of the organism to be tested in phosphate-buffered saline. Centrifuge and re-suspend in fresh phosphate-buffered saline. Wash again in the same way and add 1 ml to several 15-ml amounts of mineral salts medium (see p. 64) with indicator. Pour into plates. On each plate place a filter-paper disc of 5-mm diameter and on the filter paper place one loopful of a 5% solution of the appropriate carbohydrate. Incubate and observe growth.

When testing for amino acid or vitamin requirements in this way, use the basal medium plus glucose. Several discs, each containing a different amino acid, can be used on a single plate.

Commercially prepared discs and auxanogram kits are available.

STOCK CULTURES

Most laboratories need stock cultures of 'standard strains' of micro-organisms for testing culture media and controlling certain tests, and so that workers may become familiar with cultural appearances and other properties. These must be subcultured periodically to keep them alive, with the consequent risk of contamination; more importantly, in laboratories engaged on assay work, this is for the involuntary selection of mutants. It is because of mutations in bacterial cultures that the so-called 'standard strains' or 'type cultures' so often vary in behaviour from one laboratory to another.

To overcome these difficulties, type culture collections of bacteria, fungi, yeasts and other micro-organisms of medical and industrial importance are maintained by government and other laboratories. In these establishments, the cultures are preserved in a freeze-dried state, and in the course of maintenance of the stocks they are tested to ensure that they conform to the official descriptions of the standard or type strains.

In the UK these collections, under the aegis of the UK National Culture Collection (UKNCC), are distributed among a number of institutions. The addresses of the various collections may be found on the website (www.ukncc.co.uk). In the USA the American Type Culture Collection (ATCC) is at PO Box 1549, Manassas, VA20108, USA.

Strains that are to be used occasionally, or rarely, should be obtained when required and not maintained in the laboratory. Strains that are in constant use may be kept for short periods by serial subculture, for longer periods by one of the drying methods and indefinitely by freeze-drying. When stocks of dried or freeze-dried cultures need renewing, however, it is advisable to start again with a culture from one of the National Collection rather than to perpetuate one's own stock.

Serial subculture

The principle is to subculture as infrequently as possible in order to keep the culture alive, and to arrest growth as early as possible in the logarithmic phase so as to avoid the appearance and possible selection of mutants. Useful media for keeping stock cultures of bacteria are Dorset egg, fastidious anaerobe medium, Robertson's cooked meat and litmus milk. Fungi may be maintained on Sabouraud's medium. These media must be in screw-capped bottles. It is difficult to keep cultures of some of the very fastidious organisms, e.g. gonococci, and it is advisable to store them above liquid nitrogen (see below).

Inoculate two tubes, incubate until growth is just obvious and active, e.g. 12–18 h for enterobacteria, other Gram-negative rods and micrococci, and 2 or 3 days for lactobacilli. Cool the cultures and store them in a refrigerator for 1–2 months.

Keep one of these cultures, A, as the stock, and use the other, B, for laboratory purposes.

$$\text{Stock} \quad A_1 \rightarrow A_2 \rightarrow A_3 \rightarrow A_4$$
$$\downarrow \quad \downarrow \quad \downarrow \quad \downarrow$$
$$\text{Use} \quad B \quad B \quad B \quad B$$

PRESERVATION OF CULTURES

Drying cultures

Two methods are satisfactory. To dry by Stamp's method, make a small amount of 10% gelatin in nutrient broth and add 0.25% ascorbic acid. Add a thick suspension of the organisms before the medium gels (to obtain a final density of about 10^{10} organisms/ml). Immerse paper sheets in wax, drain and allow to set. Drop the suspension on the waxed surface. Dry in a vacuum desiccator over phosphorous pentoxide. Store the discs that form in screw-capped bottles in a refrigerator. Petri dishes smeared with a 20% solution of silicone in light petroleum and dried in a hot-air oven can be used instead of waxed paper.

Rayson's method is slightly different. Use 10% serum broth or gelatin broth and drop on discs of about 10-mm diameter cut from Cellophane and sterilized in Petri dishes. Dry in a vacuum desiccator over phosphorous pentoxide. Store in screw-capped bottles in a refrigerator.

To reconstitute, remove one disc with sterile forceps into a tube of broth and incubate.

Freezing

Make thick suspensions of bacteria in 0.5 ml sterile tap water or skim milk in screw-capped bottles and place in a deep freeze at −40°C or, better, −70°C. Many organisms remain viable for long periods in this way.

Make heavy suspensions in 30% glycerol in 1% peptone water (preserving medium) and distribute in small vials. Freeze at −70°C (−20°C may suffice). There are commercial systems (e.g. Prolab, UK) in which suspensions are added to vials containing prepared beads. The excess fluid is removed and the vial frozen. One bead is removed to culture medium as required.

Suspensions may also be stored in ampoules in the vapour phase above liquid nitrogen in a Dewar vessel. Note that this should not be in the liquid nitrogen itself. If the liquid finds its way into an imperfectly sealed ampoule it will expand rapidly to its gaseous phase when the ampoule is removed from the Dewar vessel, causing it to explode and disperse its contents. Also, if an ampoule breaks the liquid nitrogen can become contaminated with the micro-organisms.

Freeze-drying

Suspensions of bacteria in a mixture of one part 30% glucose in nutrient broth and three parts commercial horse serum are frozen and then dried by evaporation under vacuum. This preserves the bacteria indefinitely and is known as lyophilization. Yeasts, cryptococci, nocardias and streptomycetes can be lyophilized by making a thick suspension in sterile skimmed milk fortified with 5% sucrose. This simple protective medium can be sterilized by autoclaving in convenient portions, e.g. 3 ml in 5-ml screw-capped bottles.

Several machines are available. The simplest consists of a metal chamber in which solid carbon dioxide and ethanol are mixed. The bacterial suspension is placed in 0.2-ml amounts in ampoules or small tubes together with a label, and these are held in the freezing mixture, which is at about −78°C, until they freeze. They are then attached to the manifold of a vacuum pump capable of reducing the pressure to less than 0.01 mmHg. Between the pump and the manifold, there is a metal container, usually part of the freezing container, in which a mixture of solid carbon dioxide and phosphorous pentoxide condenses the moisture withdrawn from the tubes. The tubes or ampoules are then sealed with a small blowpipe.

Simultaneous freezing and drying is carried out in centrifugal freeze-dryers in which bubbling of the suspension is prevented by centrifugal action.

REFERENCES

Collins, C. H. and Kennedy, D. A. (1999) *Laboratory Acquired Infections*, 4th edn. Oxford: Butterworth-Heinemann.

Faiers, M., George, R., Jolly, J. and Wheat, P (1991) *Multipoint Methods in the Clinical Laboratory*. London: Public Health Laboratory Service.

Kennedy, D. A. (1988) Equipment-related hazards. In: Collins C. H. (ed.), *Safety in Clinical and Biomedical Laboratories*. London: Chapman & Hall, pp. 11–46.

Murray, P. R., Baron, J. O., Pfaller, M. A. *et al.* (1999) *Manual of Clinical Microbiology*, 7th edn. Washington: ASM Press.

Sisson P. R., Ingham H. R. and Byrne P. O. (1987) Wise anaerobic station: an evaluation. *Journal of Clinical Pathology* 40: 286–291.

Wren, M. W. D. (1977) The culture of clinical specimens for anaerobic bacteria. *Journal of Medical Microbiology* 10: 195–201.

7

Identification methods

In general, the identification of micro-organisms requires both microscopic examination and cultural tests. It may also require serological tests. Pure cultures are essential and methods for obtaining them are described in Chapter 6. It may be necessary, however, to plate out subcultures repeatedly to ensure that only one kind of organism is present. General-purpose media should be used, provided that they support the growth of the organism in question. Colonies taken from selective media may contain organisms, the growth of which has been suppressed.

Cultures for identification should first be examined microscopically. Stained films may reveal mixed cultures.

MICROSCOPY

Preparation of films for microscopy

It is not easy to observe the morphology of most bacteria in wet, unstained preparations, even with light, dark-field or phase contrast microscopy. It is better to stain thin films of the organisms, prepared as follows.

Place a very small drop of saline or water in the centre of a 76×25 mm glass microslide. Remove a small amount of bacterial growth with an inoculating wire or loop, emulsify the organisms in the liquid, spreading it to occupy about 1 or 2 cm^2. For all stains it is important to be able to examine single cells, not cells in clumps. Cells in clumps may not display the correct stain reaction. This is why a minimal amount of material is transferred to the

slide surface. Allow to dry in air or by waving high over a Bunsen burner flame, taking care that the slide becomes no warmer than can be borne on the back of the hand. Pass the slide, film side down, once only through the Bunsen flame to 'fix' it. This coagulates bacterial protein and makes the film less likely to float off during staining. It may not kill all the organisms, however, and the film should still be regarded as a source of infection. Methods for staining are given below. After staining, drain, blot with fresh clean filter paper and dry by gentle heat over a Bunsen flame.

Most laboratories now buy stains in solution ready for use or as concentrated solutions, rather than as dry substances. The formulae of some are, however, given below. Make sure that any stain solution used is free of particulate matter. Filter if necessary.

In the USA only stains certified by the Biological Stains Commission should be used. In the UK there is no comparable system and stains prepared by one of the specialist organizations should be purchased.

A large number of stains, mostly basic aniline dyes, have been used and described. For practical purposes, however, one or two simple stains, the Gram method and an acid-fast stain are all that are required, plus one stain for fungi. A few other stains are described here, including methods for staining capsules and spores.

Methylene blue stain

Stain for 1 min with a 0.5% aqueous solution of methylene blue or Loeffler's methylene blue: mix 30 ml saturated aqueous methylene blue solution with 100 ml 0.1% potassium hydroxide. Prolonged storage, with occasional shaking, yields 'polychrome' methylene blue for M'Fadyean's reaction for *Bacillus anthracis* in blood films.

Fuchsin stain

Stain for 30 s with the following: dissolve 1 g basic fuchsin in 100 ml 95% ethanol. Stand for 24 h. Filter and add 900 ml water. *Or* use carbol-fuchsin (see 'Acid-fast stain') diluted 1 : 10 with phenol saline.

Gram stain

There are many modifications of this. This is Jensen's version:

1. Dissolve 0.5 g methyl violet in 100 ml distilled water.
2. Dissolve 2 g potassium iodide in 20 ml distilled water. Add 1 g finely ground iodine and stand overnight. Make up to 300 ml when dissolved.
3. Dissolve 1 g safranin or 1 g neutral red in 100 ml distilled water.

Stain with methyl violet solution for 20 s. Wash off and replace with iodine solution. Leave for 1 min. Wash off iodine solution with 95% alcohol or acetone, leaving on for a few seconds only. Wash with water. Counterstain with fuchsin or safranin for 30 s. Wash and dry.

Some practice is required with this stain to achieve the correct degree of decolourization. Acetone decolourizes much more rapidly than alcohol.

The Hucker method is commonly used in American laboratories.

1. Dissolve 2 g crystal violet in 20 ml 95% ethanol. Dissolve 0.8 g ammonium oxalate in 80 ml distilled water. Mix these two solutions, stand for 24 h and then filter.
2. Dissolve 2 g potassium iodide and 1 g iodine in 300 ml distilled water using the method described for Jensen's Gram stain.
3. Grind 0.25 g safranin in a mortar with 10 ml 95% ethanol. Wash into a flask and make up to 100 ml with distilled water.

Stain with the crystal violet solution for 1 min. Wash with tap water. Stain with the iodine solution for 1 min. Decolourize with 95% ethanol until no more stain comes away. Wash with tap water. Stain with the safranin solution for 2 min. Wash and dry.

Acid-fast stain

The Ziehl–Neelsen method employs carbol fuchsin, acid–alcohol and a blue or green counterstain. Colour-blind workers should use picric acid counterstain.

1. Basic fuchsin	5 g
Crystalline phenol (*caution*)	25 g
95% alcohol	50 ml
Distilled water	500 ml

Dissolve the fuchsin and phenol in the alcohol over a warm water bath, then add the water. Filter before use.

| 2. 95% ethyl alcohol | 970 ml |
| Conc. hydrochloric acid | 30 ml |

3. 0.5% methylene blue or malachite green, or 0.75% picric acid in distilled water.

Pour carbol–fuchsin on the slide and heat carefully until steam rises. Stain for 3–5 min but do not allow to dry. Wash well with water, decolourize with acid–alcohol for 10–20 s, changing twice, and counterstain for a few minutes with methylene blue, malachite green or picric acid.

Some workers prefer the original decolourizing procedure with 20% sulphuric acid followed by 95% alcohol.

When staining nocardias or leprosy bacilli, use 1% sulphuric acid and no alcohol.

Cold staining methods are popular with some workers. For the Muller–Chermack method add 1 drop Tergitol 7 (sodium heptadecyl sulphate) to 25 ml carbol–fuchsin and proceed as for the Ziehl–Neelsen stain.

For Kinyoun's method make the carbol–fuchsin as follows:

Basic fuchsin	4 g
Melted phenol (*caution*)	8 g
95% ethanol	20 ml
Distilled water	100 ml

Dissolve the fuchsin in the alcohol. Shake gently while adding the water. Then add the phenol. Proceed as for Ziehl–Neelsen stain but do not heat. Decolourize with 1% sulphuric acid in water.

Fluorescent stain for acid-fast bacilli

There are several of these. We have found this auramine–phenol stain adequate:

1. Phenol crystals (*caution*)   3 g
   Auramine   0.3 g
   Distilled water   100 ml
2. Conc. HCl   0.5 ml
   NaCl   0.5 g
   Ethanol   75 ml
3. Potassium permanganate 0.1% in distilled water.

Prepare and fix films in the usual way. Stain with auramine–phenol for 4 min. Wash in water. Decolourize with acid–alcohol for 4 min. Wash with potassium permanganate solution. Permanganate bleaches the colour, so be careful not to decolourize too much.

Staining corynebacteria

To show the barred or beaded appearance and the metachromatic granules of these organisms, Laybourn's modification of Albert's stain may be used:

1. Dissolve 0.2 g malachite green and 0.15 g toluidine blue in a mixture of 100 ml water, 1 ml glacial acetic acid and 2 ml 95% alcohol.
2. Dissolve 3 g of potassium iodide in 50–100 ml distilled water. Add 2 g finely ground iodine and leave overnight. Make up to 300 ml.

Stain with solution (1) for 4 min. Wash, blot and dry, and stain with solution (2) for 1 min. The granules stain black and the barred cytoplasm light and dark green.

Spore staining

Make a thick film, and stain for 3 min with hot carbol–fuchsin (Ziehl–Neelsen). Wash and flood with 30% aqueous ferric chloride for 2 min.

Decolourize with 5% sodium sulphite solution. Wash and counterstain with 1% malachite green.

Flemming's technique substitutes nigrosin for the counterstain. Spread nigrosin over the film with the edge of another slide.

Spores are stained red. Cells are green, or with Flemming's method are transparent on a grey background.

Capsule staining

Place a small drop of India ink on a slide. Mix into it a small loopful of bacterial culture or suspension. Place a cover glass over the drop avoiding air bubbles and press firmly between blotting paper. Examine with a high-power lens. (Dispose of blotting paper in disinfectant.)

For dry preparations mix one loopful of India ink with one loopful of a suspension of organisms in 5% glucose solution at one end of a slide. Spread the mixture with the end of another slide, allow drying and pour a few drops of methyl alcohol over the film to fix. Stain for a few seconds with methyl violet (Jensen's Gram solution No. 1). The organisms appear stained blue with capsules showing as haloes.

Giemsa stain may be used. Fix the films in absolute methanol for 3 min. Pour on Giemsa stain and leave until it has almost dried up. Wash rapidly in water and then in phosphate buffer (0.001 mol/l, pH 7.0). Blot and dry. Capsules are stained pink and bacterial cells blue.

Flagella stains

This is a modification of Ryu's method (Kodaka *et al.*, 1982).

1. Tannic acid powder   10 g
   Phenol 5% in water (*caution*)   50 ml
   Aluminium potassium sulphate·
      12H$_2$O, saturated solution   50 ml
2. Crystal violet   12 g
   Ethanol   100 ml

Mix 10 parts of (1) with 1 part of (2) and stand for 3 days before use.

Use a young culture and dilute it with water until it is barely turbid. Place a small drop on a very clean, flamed slide. Allow it to spread and dry in air. Stain for 15 min, wash and dry.

Diene's stain for mycoplasmas

Methylene blue	2.5 g
Azure II	1.25 g
Sodium bicarbonate	0.15 g
Maltose	10 g
Benzoic acid	0.2 g

Dissolve in 100 ml distilled water.

CULTURE PROCEDURES

There are three approaches to the identification of bacteria: 'traditional', 'kit' and 'automated'. As many of the traditional methods have now been replaced in most clinical laboratories by paper discs, strips and batteries of tests in kits, these will be considered first. Automated methods for identification are described in Chapter 8.

Controls

As indicated in Chapter 2, quality control is important. Stock cultures of known organisms should therefore be included in test procedures. Suitable organisms, with their NCTC and ATCC catalogue numbers, are listed in Table 7.1. They may be stored and resuscitated for use as described on p. 87.

Kits for bacteriology

These methods offer certain advantages, e.g. the saving of time and labour but, before an arbitrary choice is made between conventional tests and the newer methods, and between the several products available, the following factors should be considered.

1. There are no 'universal kits'. Some organisms cannot be identified by these methods.

Table 7.1 Bacterial strains for use in control tests in identification procedures

Strain	NCTC	ATCC
Acinetobacter		
calcoaceticus	7844	15308
lwoffi	5866	15309
Aeromonas hydrophila	8049	7966
Alkaligenes faecalis	415	19018
Bacillus		
cereus	10876	7464
subtilis	6633	10400
Clostridium		
histolyticum	19401	503
perfringens	13124	8237
Edwardsiella tarda	10396	19547
Enterobacter		
aerogenes	13048	10006
cloacae	10005	–
Enterococcus faecalis	29212	–
Escherichia coli	25922	10418
Mycobacterium		
fortuitum	10349	6841
kansasii	10268	14471
phlei	8151	19249
terrae	10856	15755
Nocardia		
braziliensis	11274	19296
otitidiscaviarum	1934	14629
Proteus		
mirabilis	10975	–
rettgeri	7475	–
Pseudomonas aeruginosa	27853	10662
Serratia marcescens	13880	10218
Staphylococcus		
epidermidis	12228	–
aureus	25923	6571
Streptococcus		
agalactiae	13813	8181
milleri	10708	–
pneumoniae	6303	–
salivarius	8618	7073

2. Caution is required in interpretation. A kit result may suggest unlikely organisms, e.g. *Yersinia pestis, Brucella melitensis,*

Pseudomonas mallei, and ill-considered reporting of these could cause havoc.

3. Kits may not employ media suitable for growth of novel micro-organisms. If one choses to use different media, be sure that the results and interpretation of the results are consistent with another method for measuring the same characteristic or activity.

4. Conventional, paper strip and disc methods enable the user to make his or her own choice of tests. Kit methods do not; they give 10–20 test results whether the user needs them or not.

5. Identification of organisms by conventional, paper strip or disc methods usually requires the judgement of the user. With kit methods, the results are often interpreted by numerical charts or computers.

6. If final identification depends on serology, a few screening tests are usually adequate. If identification depends on biochemical tests, the more tests used the more reliable will be the results. Kits may then be the methods of choice, especially with unusual organisms and in epidemiological investigations when the kit manufacturers' services are particularly useful.

7. Some techniques may be more hazardous than others. Those that involve pipetting increase the risk of dispersing aerosols and of environmental contamination. Those employing syringe and needle work increase the risk of self-inoculation.

8. Some methods are much more expensive per identification than others.

9. Kits devised for identifying bacteria of medical importance may not give the correct answers if they are used to identify organisms of industrial significance

It follows that choices should be made according to the nature of the investigations, the professional knowledge and skill of the workers, and the size of the laboratory budget. The relative merits of the individual products mentioned below are not assessed here but assistance may be obtained from the survey of an advisory group (Bennett and Joynson, 1986) in which 10 commercial kits were tested with over 1000 bacterial strains. Kit manufacturers will supply other references on request.

CONVENTIONAL TESTS

Only those tests mentioned in this book are included here.

Aesculin hydrolysis

Inoculate aesculin medium or Edwards' medium and incubate overnight. Organisms that hydrolyse aesculin blacken the medium.

Controls: positive: *Serratia marcescens*; negative: *Edwardsiella tarda*.

Ammonia test

Incubate culture in nutrient or peptone broth for 5 days. Wet a small piece of filter paper with Nessler reagent and place it in the upper part of the culture tube. Warm the tube in a water bath at 50–60°C. The filter paper turns brown or black if ammonia is present.

Arginine hydrolysis

Incubate the culture in arginine broth for 24–48 h and add a few drops of Nessler reagent. A brown colour indicates hydrolysis. (For lactobacilli, see p. 73.)

Controls: positive: *Enterobacter cloacae*; negative: *Proteus rettgeri*.

CAMP test

This test is named after the initials of its originators (Christie, Atkins and Munch-Petersen, 1944).

Inoculate a plate of sheep blood agar (preferably layered over nutrient agar) with a single streak of *Staphylococcus aureus* that produces β-toxin (NCTC 7428) across its diameter. Streak the test organisms at right angles to this but do not touch the staphylococcus streak. Incubate aerobically overnight at 37°C and examine immediately after removal from the incubator. A positive CAMP reaction is shown by a clear zone where the inocula meet.

Carbohydrate fermentation and oxidation

Liquid, semisolid and solid media are described on p. 73. They contain a fermentable carbohydrate, alcohol or glucoside, and an indicator to show the production of acid. To demonstrate gas production, a small inverted tube (Durham's tube, gas tube) is placed in the fluid media. In solid media (stab tube, see p. 83), gas production is obvious from the bubbles and disruption of the medium. Normally, gas formation is recorded only in the glucose tube and any that occurs in tubes of other carbohydrates results from the fermentation of glucose formed during the first part of the reaction.

- For most bacteria, use peptone or broth-based media.
- For *Lactobacillus* spp., use MRS base (see p. 347) without glucose or meat extract and adjust to pH 6.2–6.5. Add chlorphenol red indicator.
- For *Bacillus* spp. and *Pseudomonas* spp., use ammonium salt 'sugar' media (see p. 73). These organisms produce ammonia from peptones and this may mask acid production.
- For *Neisseria* spp. and other fastidious organisms use solid or semisolid media. Neisserias do not like liquid media. Enrich with Fildes' extract (5%) or with rabbit serum (10%) Horse serum may give false results because it contains fermentable carbohydrates. Robinson's buffered serum

sugar medium (see p. 73) overcomes this problem and is the medium of choice for *Corynebacterium* spp.

Treat all anaerobes as fastidious organisms. The indicator may be decolourized during incubation, so add more after incubation. A sterile iron nail added to liquid media may improve the anaerobic conditions.

Tests may be carried out on solid medium containing indicator. Inoculate the medium heavily, spreading it all over the surface and place carbohydrate discs on the surface. Acid production is indicated by a change of colour in the medium around the disc. We do not recommend placing more than four discs on one plate.

Casein hydrolysis

Use skimmed milk agar and observe clearing around colonies of casein-hydrolysing organisms. The milk should be dialysed so that acid production by lactose fermenters does not interfere. To detect false clearing caused by this, pour a 10% solution of mercuric chloride (*caution*) in 20% hydrochloric acid over this medium. If the cleared area disappears, casein was not hydrolysed.

Controls: positive: *Neisseria braziliensis*; negative: *N. otitidiscaviarum*.

Catalase test

There are several methods:

1. Emulsify some of the culture in 0.5 ml of a 1% solution of Tween 80 in a screw-capped bottle. Add 0.5 ml 20-vol (6%) hydrogen peroxide (*caution*) and replace the cap. Effervescence indicates the presence of catalase. Do not do this test on an uncovered slide because the effervescence creates

aerosols. Cultures on low-carbohydrate medium give the most reliable results.

2. Add a mixture of equal volumes 1% Tween 80 and 20-vol (6%) hydrogen peroxide to the growth on an agar slope. Observe effervescence after 5 min.

3. Test mycobacteria by Wayne's method (see p. 389). Both for mycobacteria and for other bacteria, for bubbles to form the hydrogen peroxide must enter cells because catalase is an intracellular enzyme.

4. Test minute colonies that grow on nutrient agar by the blue slide test. Place one drop of a mixture of equal parts methylene blue stain and 20-vol (6%) hydrogen peroxide on a slide. Place a cover slip over the colonies to be tested and press down firmly to make an impression smear. Remove the cover slip and place it on the methylene blue–peroxide mixture. Clear bubbles, appearing within 30 s, indicate catalase activity.

Controls: positive: *Staphylococcus epidermidis*; negative: *Enterobacter faecalis*.

Citrate utilization

Inoculate solid medium (Simmons) or fluid medium (Koser) with a straight wire. Heavy inocula may give false-positive results. Incubate at 30–35°C and observe growth.

Controls: positive: *P. rettgeri*; negative: *S. epidermidis*.

Coagulase test

Staphylococcus aureus and a few other organisms coagulate plasma (see p. 330).

Controls: positive: *S. aureus*; negative: *S. epidermidis*.

Decarboxylase tests

Falkow's method may be used for most Gram-negative rods but Moeller's method gives the best results with *Klebsiella* spp. and *Enterobacter* spp. Commercial Falkow medium allows only lysine decarboxylase tests, whereas commercial Moeller media permit lysine, ornithine and arginine tests to be used.

Inoculate Falkow medium and incubate for 24 h at 37°C. The indicator (bromocresol purple) changes from blue to yellow as a result of fermentation of glucose. If it remains yellow, the test is negative; if it then changes to purple, the test is positive.

When using Moeller medium, include a control tube that contains no amino acid. After inoculation, seal with liquid paraffin to ensure anaerobic conditions and incubate at 37°C for 3–5 days. There are two indicators: bromothymol blue and cresol red. The colour changes to yellow if glucose is fermented. Decarboxylation is indicated by a purple colour. The control tube should remain yellow.

Controls. Arginine: positive: *E. cloacae*; negative: *E. aerogenes*. Lysine: positive, *Serratia marcescens*; negative: *P. rettgeri*. Ornithine: positive: *S. marcescens*; negative: *P. rettgeri*.

DNase test

Streak or spot the organisms heavily on DNase agar. Incubate overnight and flood the plate with 1 mol/l hydrochloric acid, which precipitates unchanged nucleic acid. A clear halo around the inoculum indicates a positive reaction.

Controls: positive: *S. marcescens*; negative: *E. aerogenes*.

Gelatin liquefaction

1. Inoculate nutrient gelatin stabs, incubate at room temperature for 7 days and observe

digestion. For organisms that grow only at temperatures when gelatin is fluid, include an uninoculated control and after incubation place both tubes in a refrigerator overnight. The control tube should solidify. This method is not entirely reliable.

2. Inoculate nutrient broth. Place a denatured gelatin charcoal disc in the medium and incubate. Gelatin liquefaction is indicated by the release of charcoal granules, which fall to the bottom of the tube.

3. For a rapid test, use 1 ml 0.01 mol/l calcium chloride in saline. Inoculate heavily (whole growth from slope or Petri dish). Add a gelatin charcoal disc and place in a water bath at 37°C. Examine at 15-min intervals for 3 h.

4. Inoculate gelatin agar medium. Incubate overnight at 37°C and then flood the plate with saturated ammonium sulphate solution. Haloes appear around colonies of organisms producing gelatinase.

Controls: positive: *Aeromonas hydrophila*; negative: *E. coli*.

Gluconate oxidation

Inoculate gluconate broth (see p. 74) and incubate for 48 h. Add an equal volume of Benedict's reagent and place in a boiling water bath for 10 min. An orange or brown precipitate indicates gluconate oxidation.

Controls: positive: *E. cloacae*; negative: *P. rettgeri*.

Hippurate hydrolysis

Inoculate hippurate broth, incubate overnight and add excess 5% ferric chloride. A brown precipitate indicates hydrolysis.

Controls: positive: *Streptococcus agalactiae*; negative: *S. salivarius*.

Hugh and Leifson test (oxidation fermentation test)

This is also known as the 'oxferm' test. Some organisms metabolize glucose oxidatively, i.e. oxygen is the ultimate hydrogen acceptor and culture must therefore be aerobic. Others ferment glucose, when the hydrogen acceptor is another substance; this is independent of oxygen and cultures may be aerobic or anaerobic.

Heat two tubes of medium (see p. 74) in boiling water for 10 min to drive off oxygen, cool and inoculate; incubate one aerobically and the other either anaerobically or seal the surface of the medium with 2 cm melted Vaseline or agar to give anaerobic conditions.

- Oxidative metabolism: acid in aerobic tube only.
- Fermentative metabolism: acid in both tubes.

For testing staphylococci and micrococci, use the Baird-Parker modification (see p. 73).

Controls: oxidation: *Acinetobacter calcoaceticus*; fermentation: *E. coli*; no action: *Alkaligenes faecalis*.

Hydrogen sulphide production

1. Inoculate a tube of nutrient broth. Place a strip of filter paper impregnated with a lead acetate indicator in the top of the tube, holding it in place with a cottonwool plug. Incubate and examine for blackening of the paper.

2. Inoculate one of the iron or lead acetate media. Incubate and observe blackening.

TSI (Triple Sugar Iron) agar and similar media do not give satisfactory results with sucrose-fermenting organisms. The indicator paper (lead acetate) is the most sensitive method and the ferrous chloride media the least sensitive method, but the latter is probably the method of choice for identifying salmonellas and for differentiation in other groups where the amount of hydrogen sulphide produced varies with the species.

Controls: positive: *E. tarda*; negative: *P. rettgeri*.

Indole formation

Grow the organisms for 2–5 days in peptone or tryptone broth.

1. Ehrlich's method

Dissolve 4 g *p*-dimethylaminobenzaldehyde in a mixture of 80 ml concentrated hydrochloric acid and 380 ml ethanol (do not use industrial spirit – this gives a brown instead of a yellow solution). To the broth culture add a few drops of xylene and shake gently. Add a few drops of the reagent. A rose-pink colour indicates indole.

2. Kovac's method

Dissolve 5 g *p*-dimethylaminobenzaldehyde in a mixture of 75 ml amyl alcohol and 25 ml concentrated sulphuric acid. Add a few drops of this reagent to the broth culture. A rose-pink colour indicates indole.

3. Spot test (Miller and Wright, 1982)

Dissolve 1 g *p*-dimethylaminocinnamaldehyde (DMAC) (*caution*) in 100 ml 10% hydrochloric acid. Moisten a filter paper with it and smear on a colony from blood or nutrient agar. A blue-green colour is positive and pink negative. This test may be applied directly to colonies but is unreliable in the presence of carbohydrates, e.g. on MacConkey or CLED media.

Controls: positive: *P. rettgeri*; negative: *S. marcescens*.

Lecithinase activity

1. Inoculate egg yolk, salt broth and agar (see p. 74) and incubate for 3 days. Lecithinase-producing colonies are surrounded by zones of opacity.

2. Inoculate egg yolk salt broth (see p. 74) and incubate for 3 days. Lecithinase producers make the broth opalescent. Some organisms (e.g. *Bacillus cereus*) give a thick turbidity in 13 h.

Controls: positive: *B. cereus*; negative: *B. subtilis*.

Levan production

Inoculate nutrient agar containing 5% sucrose. Levan producers give large mucoid colonies after incubation for 24–48 h.

Controls: positive: *Streptococcus salivarius*; negative, *S. milleri*.

Lipolytic activity

Inoculate tributyrin agar. Incubate for 48 h at 25–30°C. A clear zone develops around colonies of fat-splitting organisms. These media can be used for counting lipolytic bacteria in dairy products.

Controls: positive: *S. epidermidis*; negative: *Proteus mirabilis*.

Malonate test

Inoculate one of the malonate broth media and incubate overnight. Growth and a deep-blue colour indicate malonate utilization.

Controls: positive: *E. cloacae*; negative: *P. rettgeri*.

Motility

Hanging drop method

Place a very small drop of liquid bacterial culture in the centre of a 16-mm square No. 1 cover glass, with the aid of a small (2-mm) inoculating loop. Place a small drop of water at each corner of the cover glass. Invert, over the cover glass, a microslide with a central depression – a 'well slide'. The cover glass will adhere to the slide and when the slide is inverted the hanging drop is suspended in the well.

Instead of using 'well slides' a ring of Vaseline may be made on a slide. This is supplied in collapsible tubes or squeezed from a hypodermic syringe.

Bring the edge of the hanging drop, or the air bubbles in the water seal, into focus with the 16-mm lens before turning to the high-power dry objective to observe motility.

Bacterial motility must be distinguished from brownian movement. There is usually little difficulty with actively motile organisms, but feebly motile bacteria may require prolonged observation of individual cells.

Careful examination of hanging drops may indicate whether a motile organism has polar flagella – a darting zigzag movement – or peritrichate flagella – a less vigorous and more vibratory movement.

To examine anaerobic organisms for motility grow them in a suitable liquid medium. Touch the culture with a capillary tube 60- to 70-mm long and bore of about 0.5–1 mm. Some culture will enter the tube. Seal both ends of the tube in the Bunsen flame and mount on Plasticine on the microscope stage. Examine as for hanging drops.

Craigie tube method

Craigie tubes (see p. 76) can be used instead of hanging drop cultures. Inoculate the medium in the inner tube. Incubate and subculture daily from the outer tube. Only motile organisms can grow through the sloppy agar.

Nagler test

This tests for lecithinase activity (see p. 97) but the word has come to mean the half antitoxin plate test

for *Clostridium perfringens*. Egg yolk agar containing Fildes' enrichment or Willis and Hobbs medium is satisfactory.

Control: *C. perfringens* – no activity in presence of its antitoxin.

Nitrate reductase test

1. Inoculate nitrate broth medium and incubate overnight.
2. Grow the organisms in a suitable broth. Add a few drops of 1% sodium nitrate solution and incubate for 4 h.

Acidify with a few drops of 1 mol/l hydrochloric acid and add 0.5 ml each of a 0.2% solution of sulphanilamide and 0.1% N-naphthylethylenediamine hydrochloride (*caution*) (these two reagents should be kept in a refrigerator and freshly prepared monthly). A pink colour denotes nitrate reductase activity. As some organisms further reduce nitrite, however, if no colour is produced add a very small amount of zinc dust. Any nitrate present will be reduced to nitrite and produce a pink colour, i.e. a pink colour in this part of the test indicates no nitrate reductase activity and no colour indicates that nitrates have been completely reduced. This latter point is important because the absence of nitrite does not necessarily mean that there is no nitrate reductase. Rather, the absence of nitrite could be the result of rapid reduction of nitrite by nitrite reductase. Assimilative nitrate reductase may not be produced in the presence of ammonia or amino nitrogen, and dissimilatory nitrate reductase may not be produced under aerobic conditions.

Controls: positive: *S. marcescens*; negative: *Acinetobacter lwoffii*.

ONPG (*o*-nitrophenylgalactoside) test

Lactose is fermented only when β-galactosidase and permease are present. Deficiency of the latter gives late fermentation. True non-lactose fermenters do not possess β-galactosidase.

Place an ONPG disc in 0.1 ml 0.85% sodium chloride (physiological saline) in which a colony of the test organism has been emulsified. Incubate for 6 h. If β-galactosidase is present, a yellow colour is formed as the result of o-nitrophenol liberation. If there is no colour, incubate overnight to detect lactose fermenters.

Controls: positive: *S. marcescens*; negative: *P. rettgeri*.

Optochin test

Pneumococci are sensitive but streptococci are resistant to optochin (ethylhydrocupreine hydrochloride).

Streak the organisms on blood agar and place an optochin disc on the surface. Incubate overnight and examine the zone around the disc.

Controls: positive: *S. pneumoniae*; negative: *S. milleri*.

Oxidase test (cytochrome oxidase test)

1. Soak small pieces of filter paper in 1% aqueous tetramethyl-*p*-phenylenediamine dihydrochloride or oxalate (which keeps better). Some filter papers give a blue colour and these must not be used. Dry or use wet. Scrape some of a fresh young culture with a clean platinum wire or a glass rod (dirty or Nichrome wire gives false positives) and rub on the filter paper. A blue colour within 10 s is a positive oxidase test. Old cultures are unreliable. Tellurite inhibits oxidase as do fermentable carbohydrates. Organisms that have produced acid from a carbohydrate should be subcultured to a sugar-free medium.

2. Incubate cultures on nutrient agar slopes for 24–48 h at the optimum temperature for the strain concerned. Add a few drops each of freshly prepared 1% aqueous *p*-amino-dimethylaniline oxalate and 1% α-naphthol in ethanol. Allow the mixture to run over the growth. A deep blue colour is a positive reaction.

Controls: positive: *Pseudomonas aeruginosa*; negative: *A. lwoffii*.

Oxidative or fermentative metabolism of glucose

See Hugh and Leifson test above.

Phenylalanine test (PPA or PPD test)

Inoculate phenylalanine agar and incubate overnight. Pour a few drops of 10% ferric chloride solution over the growth. A green colour indicates deamination of phenylalanine to phenylpyruvic acid. Among the enterobacteria only strains of *Proteus* and *Providencia* have this property.

Controls: positive: *P. rettgeri*; negative: *E. cloacae*.

Phosphatase test

Some bacteria, e.g. *S. aureus*, can split ester phosphates. Inoculate a plate of phenolphthalein phosphate agar (see p. 75) and incubate overnight. Expose the culture to ammonia vapour (*caution*). Colonies of phosphatase producers turn pink. Phosphate may prevent the synthesis of phosphatase.

Controls: positive: *S. aureus*; negative: *S. epidermidis*.

Proteolysis

Inoculate cooked meat medium and incubate for 7–10 days. Proteolysis is indicated by blackening of the meat or digestion (volume diminishes). Tyrosine crystals may appear.

Controls: positive: *C. histolyticum*; negative: *Cl. perfringens*.

Starch hydrolysis

Inoculate starch agar (see p. 76). Incubate for 3–5 days, then flood with dilute iodine solution. Hydrolysis is indicated by clear zones around the growth. Unchanged starch gives a blue colour.

Controls: positive: *B. subtilis*; negative: *E. coli*.

Sulphatase test

Some organisms (e.g. certain species of mycobacteria) can split ester sulphates. Grow the organisms in media (Middlebrook 7H9 for mycobacteria) containing 0.001 mol/l potassium phenolphthalein disulphate for 14 days. Add a few drops of ammonia solution (*caution*). A pink colour indicates the presence of free phenolphthalein. Sulphate may prevent the synthesis of sulphatase.

Controls: positive: *Mycobacterium fortuitum*; negative: *M. phlei* (3-day test only).

Tellurite reduction

Some mycobacteria reduce tellurite to tellurium metal (see p. 389).

Controls: positive: *M. fortuitum*; negative: *M. terrae*.

Tween hydrolysis

Some mycobacteria can hydrolyse Tween (polysorbate) 80, releasing fatty acids that change the colour of an indicator. This test is used mostly with mycobacteria (see p. 389).

Controls: positive: *M. kansasii*; negative: *M. fortuitum*.

Tyrosine decomposition

Inoculate parallel streaks on a plate of tyrosine agar (see p. 75) and incubate for 3–4 weeks. Examine periodically under a low-power microscope for the disappearance of crystals around the bacterial growth.

Controls: positive: *N. braziliensis*; negative: *N. otitidiscaviarum*.

Urease test

Inoculate one of the urea media heavily and incubate for 3–12 h. The fluid media give more rapid results if incubated in a water bath. If urease is present, the urea is split to form ammonia, which changes the colour of the indicator from yellow to pink.

Controls: positive: *P. rettgeri*; negative: *S. marcescens*.

Voges Proskauer reaction (VP)

This tests for the formation of acetyl methyl carbinol (acetoin) from glucose. This is oxidized by the reagent to diacetyl, which produces a red colour with guanidine residues in the media.

Inoculate glucose phosphate broth and incubate for 5 d at 30°C. A very heavy inoculum and 6 h incubation may suffice. Test by one of the following methods.

1. Add 3 ml 5% alcoholic α-naphthol solution and 3 ml 40% potassium hydroxide solution (Barritt's method).
2. Add a 'knife point' of creatinine and 5 ml 40% potassium hydroxide solution (O'Meara's method).
3. Add 5 ml of a mixture of 1 g copper sulphate (blue) dissolved in 40 ml saturated sodium hydroxide solution plus 960 ml 10% potassium hydroxide solution (APHA method).

A bright pink or eosin red colour appearing in 5 min is a positive reaction. For testing *Bacillus* spp., add 1% NaCl to the medium.

Controls: positive: *S. marcescens*; negative: *P. rettgeri*.

Xanthine decomposition

Inoculate parallel streaks on a plate of xanthine agar (see p. 75) and incubate for 3–4 weeks. Examine periodically under a low-power microscope for the disappearance of xanthine crystals around the bacterial growth.

Controls: positive: *N. otitidiscaviarum*; negative: *N. braziliensis*.

AGGLUTINATION TESTS

These tests are performed in small test tubes (75×9 mm). Dilutions are made in physiological saline with graduated pipettes controlled by a rubber teat or (for a single-row test), with Pasteur pipettes marked with a grease pencil at approximately 0.5 ml. Automatic pipettes or pipettors are useful for doing large numbers of tests.

Standard antigen suspensions and agglutinating sera can be obtained commercially. Antigen suspensions are used mainly in the serological diagnosis of enteric fever, which may be caused by several related organisms, and of brucellosis.

Standard agglutinating sera are used to identify unknown organisms.

Testing unknown sera against standard *H* and *O* antigen suspensions

To test a single suspension

Prepare a 10^{-1} dilution of serum by adding 0.2 ml to 1.8 ml saline. Set up a row of seven small tubes. Add 0.5 ml saline to tubes 2–7 and 0.5 ml of the 10^{-1} serum to tubes 1 and 2. Rinse the pipette by sucking in and blowing out saline several times. Mix the contents of tube 2 and transfer 0.5 ml to tube 3. Rinse the pipette. Continue with doubling dilutions but discard 0.5 ml from tube 6 instead of adding it to tube 7. Rinse the pipette between each dilution. The dilutions are now 1 : 10 to 1 : 320.

Add 0.5 ml of standard suspension to each tube. The last tube, containing no serum, tests the stability of the suspension. The dilutions are now 1 : 20 to 1 : 640.

To test with several suspensions

In some investigations, e.g. in enteric fevers, a number of suspensions will be used. Set up six large tubes (152×16 mm). To tube 1 add 9 ml saline, to tubes 2–6 add 5 ml saline and to tube 1 add 1 ml serum. Mix and transfer 5 ml from tube 1 to tube 2, and continue in this way, rinsing the pipette between each dilution.

Set up a row of seven small tubes (75×9 mm) for each suspension to be tested. Transfer 0.5 ml from each large tube to the corresponding small tube. Work from right to left, i.e. weakest to strongest dilution to avoid unnecessary rinsing of the pipette. The final dilutions are now 1 : 20 to 1 : 640.

Incubation temperature and times

Incubate tests with *O* suspensions in a water bath at 37°C for 4 h, then allow to stand overnight in refrigerator.

Incubate *H* tests for 2 h in a water bath at 50–52°C (*not* 55–56°C, as the antibody may be partly destroyed). The level of water in the bath should be adjusted so that only about half of the liquid in the tubes is below the surface of the water. This encourages convection in the tubes, mixing the contents.

Brucella agglutinations

To avoid false-negative results resulting from the prozone phenomenon, double the dilutions for at least three more tubes, i.e. use 1 : 20 to 1 : 5120.

Reading agglutinations

Examine each tube separately from right to left, i.e. beginning with the negative control. Wipe dry and use a hand lens. If the tubes are scratched, dip them in xylene. The titre of the serum is that dilution in which agglutination is easily visible with a low-power magnifier.

Testing unknown organisms against known sera

Preparation of O suspensions

The organism must be in the smooth phase. Grow on agar slopes for 24 h. Wash off in phenol–saline and allow lumps to settle. Remove the suspension and dilute it so that there are approximately 1×10^9 bacteria/ml by the opacity tube method (see p. 145). Heat at 60°C for 1 h. If this antigen is to be stored, add 0.25% chloroform.

If a K antigen is suspected, heat the suspension at 100°C in a water bath for 1 h (although the B antigen is thermostable).

Preparation of H suspensions

Check that the organism is motile and grow it in nutrient broth for 18 h, or in glucose broth for 4–6 h. Do not use glucose broth for the overnight culture because the bacteria grow too rapidly. Add formalin to give a final concentration of 1.0% and leave for 30 min to kill the organisms. Heat at 50–55°C for 30 min (this step may be omitted if the suspension is to be used at once). Dilute to approximately 1×10^9 organisms/ml by the opacity tube method.

Agglutination tests

Use the same technique as that described under 'Testing unknown sera . . .' but as most sera are issued with a titre of at least 1 : 250 (and labelled accordingly) it is not necessary to dilute beyond 1 : 640. In practice, as standard sera are highly specific and an organism must be tested against several sera, it is usually convenient to screen by adding one drop of serum to about 1 ml suspension in a 75 × 9 mm tube. Only those sera that give agglutination are then taken to titre.

Slide agglutination

This is the normal procedure for screening with O sera.

Place a loopful of saline on a slide and, next to it, a loopful of serum. With a straight wire, pick a colony and emulsify in the saline. If it is sticky or granular, or autoagglutinates, the test cannot be done. If the suspension is smooth, mix the serum in with the wire. If agglutination occurs it will be rapid and obvious. Dubious slide agglutination should be discounted.

Only O sera should be used. Growth on solid medium is used for slide agglutination and this is not optimum for the formation of flagella. False-negative results may be obtained with H sera unless there is fluid on the slope.

Suspensions of organisms in the R phase are agglutinated by 1 : 500 aqueous acriflavine solution.

FLUORESCENT ANTIBODY TECHNIQUES

The principle of fluorescent antibody techniques is that proteins, including serum antibodies, may be labelled with fluorescent dyes by chemical combination without alteration or interference with the biological or immunological properties of the proteins. These proteins may then be seen in microscopic preparations by fluorescence microscopy.

The preparation is illuminated by ultraviolet or ultraviolet blue light. Any fluorescence emitted by the specimen passes through a barrier filter above the object. This filter transmits only the visible fluorescence emission. Microscopes suitable for this purpose are now readily available, and it is a simple matter to convert standard microscopes for fluorescence work. Fluorescent dyes are used instead of ordinary dyes because they are detectable in much smaller concentrations. They are available in a form that simplifies the conjugation procedure considerably. The fluorochromes in frequent current use are fluorescein isothiocyanate (FITC) and Lissamine rhodamine B (RB200). Of these, FITC is the most commonly used. It gives an apple-green fluorescence.

Fluorescence labelling is often used in microbiology and immunology along with or instead of traditional serological tests. Immune serum globulin conjugated with fluorochrome is usually employed to locate the corresponding antigen in microbiological investigations.

Reagents, accompanied by technical instructions, are available commercially.

TYPING MICRO-ORGANISMS

Reasons for typing micro-organisms

The principal reason for typing or fingerprinting micro-organisms is to provide information to permit the tracing of epidemics of infection or to identify the source of an outbreak (Maslow *et al.*, 1993), e.g. fingerprints of isolates from a cluster of patients can be used to determine whether they share the same or different clones of a particular species. If isolates from different individuals share the same fingerprint or type, it suggests that they were infected from a single source. In outbreaks of disease that are thought to be food related, typing can be used to identify the source of the infectious micro-organism. Also, fingerprinting methods (e.g. multilocus enzyme electrophoresis or MLEE) can be used to describe the population genetics of micro-organisms (Selander *et al.*, 1986, 1987).

Typing targets

Typing or fingerprinting methods fall into two categories based on the target molecule. One category uses phenotypic features of the micro-organism such as cell surface antigens, e.g. proteins or polysaccharides (i.e. serotyping), profiles of enzymatic activities (biotyping), profiles of enzyme polymorphisms (MLEE), or patterns of susceptibility to bacteriophage (i.e. phage typing) or antibiotics (i.e. antibiograms). Unfortunately, expression of these characteristics can be influenced by growth conditions and they are thus subject to variation. It is also common that some strains are untypable for reasons other than the absence of the particular typing target, e.g. if a typing method employs agglutination with antisera, cells that spontaneously clump cannot be typed.

In the second category of typing methods DNA is the target. Repeated sequences, e.g. a transposable genetic element such as an insertion sequence (IS) or a reiterated sequence, are commonly employed as targets. DNA-based fingerprinting can also be performed by comparison of large genomic fragments generated by cleavage with restriction endonucleases and separated by pulsed-field gel electrophoresis (PFGE). As all cells contain DNA, growth conditions do not necessarily influence the results, unless they result in cells resistant to lysis and release of DNA. It is also important to remember that DNA-fingerprinting methods involving the use of restriction endonucleases (i.e. restriction fragment length polymorphism or RFLP) can be influenced by the presence of DNA-modifying enzymes which can prevent cleavage of specific DNA sequences.

Requirements of targets for typing or fingerprinting

Excellent descriptions of the principles of typing or fingerprinting (Maslow *et al.*, 1993) and interpreting typing data (Tenover *et al.*, 1995) have been published. First, it should be pointed out that typing or fingerprinting should always be validated with a large number of isolates. Further, in cases of outbreak investigations, a large number of isolates should be tested. The results from small samples can be influenced by chance and thus may not reflect the entire population. If the objective of the typing is to identify clusters (implying a common source of infection), small samples may omit members of clusters and the extent of clustering would be underestimated (Murray, 2002).

A primary requirement of a target is that all isolates must express or have that target. There cannot be isolates that lack the particular target (i.e. untypable). There must be a diversity of patterns or profiles exhibited by the population. However, one should be careful to choose a target with sufficient diversity to provide meaningful results, e.g. if there were little variation in the target among isolates, then all isolates would be thought to belong to a limited number of different types. Use of such a target might lead to a false clustering of isolates or linkage to a possible source. Such targets might be useful as species markers rather than as typing targets. At the other end of the spectrum, it is possible to use a target that displays too much variation within the studied population. Typing or fingerprinting with such a target would result in a false absence of clustering or linkage to a possible source. This would be the case if a transposable genetic element were subject to a very high rate of transposition. In this case, almost all isolates will be thought to belong to individual groups, to the extreme of one isolate per group.

Another way of stating the requirement that there be a diversity of typing or fingerprinting patterns is that a useful target provides a high level of discrimination. A discrimination index can be calculated using Simpson's index of diversity, originally developed for the description of species diversity within the habitat or an organism (Simpson, 1949). Using that index, it was possible to compare the discriminatory power of different methods used for typing isolates of *Candida albicans* (Hunter and Gaston, 1988). Clearly, the number of types directly influences the discrimination index. A greater number of different types are associated with a higher dis-

criminatory index. Fortunately, this method takes into account the reduction in discriminatory index when many isolates belong to a single group, e.g. there is a non-random distribution of sites of insertion of IS6110 in *Mycobacterium tuberculosis* which is the result of the existence of sites ('hotspots') favoured for integration (McHugh and Gillespie, 1998).

Typing methods should be simple, rapid, inexpensive and within the technical skill of personnel. Typing methods, as with other methods employed in the laboratory, should include quality control and quality assessment using reference strains. For methods involving electrophoretic separation of either proteins (e.g. MLEE) or DNA fragments (e.g. RFLP, PFGE or arbitrary primed polymerase chain reaction), the criteria for determining whether an individual band's mobility is the same or different in two isolates should be established by repeated measurement, e.g. it is not uncommon to find that mobility of identical bands in different gels varies by about ± 5%. Thus, if the mobilities of bands in two different isolates are within 5%, they will be counted as identical in the analysis. Finally, there should be a straightforward method of analysis that provides meaningful information to users (e.g. clinicians or public health personnel). Basically, the object is to determine whether isolates belong to a single clone. If not members of the same clone, the relatedness of clones must be determined. One method of estimating the relatedness between isolates is to calculate the Dice (1945) coefficient of similarity. That value equals twice the number of shared bands divided by the total number of bands in both isolates (Dice, 1945).

Serotyping

Serotyping is one of the oldest methods for typing micro-organisms and has had wide use in tracing the sources of food-associated infections. Antisera are raised against different isolates of a particular species and used to agglutinate cells. Targets for serotyping can be cell surface polysaccharides or proteins. Provided that there is sufficient variation in the targets and a large number of isolates are employed, a panel of antisera can be identified that can distinguish between categories (i.e. serotypes) of isolates. To prevent cross-reactions, sera can be mixed with isolates of different types to leave only antibodies reactive against targets that are unique to the type used to generate the antiserum (adsorbed sera). The widespread applicability of serotyping is limited by the need to produce and store antisera. Growth conditions can lead to an absence or modification of the targets of the antisera (false type). If evidence of the presence of a particular target is based on agglutination, cells of isolates that can spontaneously agglutinate cannot be typed.

Phage typing

Phage typing followed the discovery that bacteria were susceptible to infection by a variety of different viruses (e.g. bacteriophage) and those viruses recognized different cell surface molecules (i.e. receptors) for attachment. For phage typing, lawns of individual isolates are spotted with virus-containing lysates and susceptibility (i.e. lysis of the lawn) scored for each virus. As is the case for all typing methods, there must be a large number of different viruses recognizing different receptors and sufficient variation in the cell surface receptors (present or absent) to allow sufficient discrimination between isolates. As is the case for serotyping, growth conditions can lead to an absence of a particular receptor and assignment of an isolate to a false type.

Multilocus enzyme electrophoresis

Multilocus enzyme electrophoresis relies on the variation in amino acid composition of cellular proteins (Selander *et al.*, 1986). It has been used to characterize the populations of *E. coli* (Whittam *et al.*, 1983), *Legionella pneumophila* (Selander *et al.*,

1985), and a variety of other bacteria (Selander *et al.*, 1987). Populations of micro-organisms contain mutant individuals expressing an active enzyme the amino acid composition of which is different. In some instances the difference in amino acid results in a change in electrophoretic mobility. The mobility of enzymes is measured by electrophoretic separation and detection of *in situ* enzyme activity (e.g. production of an insoluble coloured product). The discriminatory power of MLEE is dependent on the existence of mutant variants with different mobilities in a population. Further, measuring the mobility of a large number of different enzymes increases the discrimination index. MLEE requires that growth conditions be chosen to ensure expression of a wide variety of different enzymes, the activities of which can be detected in gels (Selander *et al.*, 1986). Although time-consuming, labour intensive and requiring sophisticated equipment and technical skill, MLEE is a powerful method for typing isolates.

Pulsed-field gel electrophoresis

Pulsed-field gel electrophoresis enables the separation of large fragments of DNA by gel electrophoresis (Schwartz *et al.*, 1983; Carle *et al.*, 1986). The fragments are generated by cleavage of genomic DNA with restriction endonucleases that recognize few sequences (e.g. 8- to 12-base recognition sequences) and generate fragments of typically between 100 and 1000 kb (Allardet-Servent *et al.*, 1989). Methods for isolation and purification of large DNA-restriction fragments are presented in Selden and Chory (1987). Variations in size and, hence, mobility of restriction fragments are based on mutations leading to acquisition or loss of a recognition site for the enzyme. As such large-sized fragments of DNA would be subject to fragmentation by shearing in solution, cells are immobilized in agarose blocks and followed sequentially by cell lysis and restriction endonuclease digestion of released DNA to reduce shear forces. The agarose blocks are inserted into wells of agarose and PFGE performed. PFGE can be used for typing virtually every microbial species. Unfortunately, in spite of its wide applicability, the number of required number of steps makes PFGE time-consuming. Further PFGE requires expensive equipment and sophisticated technical assistance.

It is possible to calculate the expected average size and number of fragments generated by a specific restriction endonuclease and the guanine + cytosine percentage (%G + C) of the micro-organism. Choosing restriction enzymes that produce a large number of fragments results in greater discrimination. Care should be taken to ensure that differences in patterns are not the result of the presence of a modifying methylase, which prevents cleavage of a recognition site. A guide for interpretation of PFGE patterns has been published (Tenover *et al.*, 1995). Criteria used to determine whether bands are identical or different must be established by repeated measurements of digests of a limited number of strains. In addition to simple visual inspection of the fragment patterns, Dice coefficients of similarity can be calculated and phylogenetic trees constructed (Fitch and Margoliash, 1967; Felsenstein, 1985). There are also a number of commercial software programs that can scan gels and calculate relatedness between isolates.

Restriction fragment length polymorphism

A variety of targets has been chosen for RFLP analysis including insertion sequences and repeated sequences. For RFLP analysis, DNA is recovered from a variety of isolates and subjected to restriction endonuclease digestion. The restriction fragments are separated by gel electrophoresis and a labelled nucleic acid probe is used to detect bands containing the target sequence by hybridization. Methods for DNA isolation – restriction endonuclease digestion, gel electrophoresis, blotting (i.e. southern transfer) and hybridization – are all described in detail in Sambrook *et al.* (1989). In cases where fingerprinting is widely used, e.g. in

tuberculosis epidemiology, a standard methodology has been published (Van Embden *et al.*, 1993).

The common feature of useful targets is that they are present in multiple copies in the genome (e.g. > 10). One way of detecting a genetic element of high copy number is to subject genomic DNA to cleavage with a variety of restriction endonucleases and separate the fragments by gel electrophoresis. Often, repeated sequences appear as more intense bands when stained with ethidium bromide and they can be cloned for sequencing. As is the case for PFGE, the variation in bands is caused by mutations in restriction endonuclease recognition sequences flanking or within a repeated genetic element. Variation is also a consequence of the transposition of an insertion sequence.

As with PFGE, the target for RFLP analysis is DNA. Few isolates are found to be untypable and this is because of the inability to recover sufficient amounts of DNA. RFLP and PFGE are both amenable to quality control and quality assurance programmes. The criteria for interpretation of RFLP patterns is no different from those used for PFGE, as is the method of analysis and presentation of data. As a result of the need for separate steps of electrophoresis, blotting and hybridization, RFLP analysis requires time, is rather expensive and demands technical sophistication.

Polymerase chain reaction fingerprinting methods

PCR amplification of sequences between insertion sequences

It is possible to amplify DNA between insertion sequences or repetitive genetic elements using the polymerase chain reaction (PCR). To be successful, there must be a high number of target sequences in the genome (i.e. > 10 and preferably 20). One example of a repeated sequence that is highly conserved among many different bacteria is the repetitive extragenic palindromic (REP) sequence (Versalovic *et al.*, 1991). Primers are chosen to amplify the target sequence (or related target sequences), and amplification of the sequence between targets occurs by a primer binding to one element and the second primer binding to a second element. If the two targets are not too far apart to prevent successful PCR amplification (e.g. < 5 kb), amplified products will be produced. One advantage of this approach is that the number of steps necessary for fingerprinting is reduced and the required level of DNA purity is reduced. One disadvantage of this approach is that it is less discriminatory than RFLP analysis. In a comparison of IS1245 fingerprinting in *Mycobacterium avium*, fewer amplified fragments were produced compared with the number of bands seen in RFLP (Picardeau and Vincent, 1996). PCR-based fingerprinting based on consensus primers for the REP sequence has been used for epidemiological studies of a wide range of bacteria, including *Citrobacter diversus* (Woods *et al.*, 1992), *Neisseria gonorrhoeae* (Poh *et al.*, 1996) and *Acinetobacter* spp. (Snelling *et al.*, 1996). A commercial kit for performing REP-PCR is available (Bacterial BarCodes, Inc., Houston, TX).

Arbitrary primed PCR and random amplified polymorphic DNA

Micro-organisms can be typed using the PCR-based technique that employs single primers to amplify sequences in DNA (Welsh and McClelland, 1990; Williams *et al.*, 1990). The method is called either arbitrary primed PCR (AP-PCR) or random amplified polymorphic DNA (RAPD). The advantage of this approach is that primers can be screened without knowledge of the genome's characteristics and thus the technique can be employed with a wide variety of micro-organisms. Screening chooses single primers that reproducibly yield multiple amplified products. Detailed methods that also provide examples of the application of AP-PCR to fingerprinting of a variety of micro-organisms have been published (Williams *et al.*, 1993). Using isolates of *E. coli*, it was shown that AP-PCR was more sensitive than MLEE in distinguishing isolates (Wang *et al.*, 1993). Data from AP-PCR (i.e. band patterns) can be treated as bands from PFGE or RFLP analysis and the data

can be used to estimate nucleotide sequence divergence (Clark and Lanigan, 1993). As is characteristic of all PCR-based techniques, extreme care must be taken to avoid amplification of contaminating DNA and amplification of non-target, but related sequences.

Conclusion

One of the distinct advantages of PFGE and RFLP analysis is that gels can be scanned, phylogenetic trees constructed and coefficients of relatedness calculated by computer. Such data can be available to clinicians in a hospital or other investigators at distant locations.

It is possible to select the level of discrimination desired in any epidemiological study. Care must be taken to avoid choosing a method that provides too little discrimination leading to false grouping of isolates. In contrast, a fingerprinting method with a high discrimination index may lead to separation of isolates that truly belong to the same group. Using MLEE data, it is assumed that isolates with the same mobility for a particular enzyme share the same allele (i.e. DNA sequence). However, this is not necessarily the case; the isolates may have different alleles. Therefore, it is not surprising that AP-PCR can be more discriminatory than MLEE for isolates of *E. coli* (Wang *et al.*, 1993). One should not conclude that the most discriminatory method is the 'best' for typing. One method, e.g. MLEE, may provide evidence of large related groups that can be found world wide (Whittam *et al.*, 1983). Such patterns would be missed by a more discriminatory method.

REFERENCES

Allardet-Servent, A. N., Bouziges, M. J., Carles-Nurit, G. *et al.* (1989) Use of low-frequency-cleavage restriction endonucleases for DNA analysis in epidemiological investigations of nosocomial bacterial infections. *Journal of Clinical Microbiology* 27: 2057–2061.

Bennett, C. H. N. and Joynson, D. H. M. (1986) Kit systems for identifying Gram- negative aerobic bacilli: report of the Welsh Standing Specialist Advisory Working Party in Microbiology. *Journal of Clinical Pathology* **39**: 666–671.

Carle, G. F., Frank, M. and Olson, M. V. (1986). Electrophoretic separations of large DNA molecules by periodic inversion of the electric field. *Science* **232**: 65–68.

Christie, R., Atkins, N. E. and Munch-Petersen, E. (1944) A note of the lytic phenomenon shown by Group B streptococci. *Australian Journal of Experimental Biology and Medical Science* **22**: 197.

Clark, A. G. and Lanigan, C. M. S. (1993) Prospects for estimating nucleotide divergence with RAPDs. *Molecular and Biological Evolution* **10**: 1096–1111.

Dice, L. R. (1945) Measures of the amount of ecological association between species. *Ecology* **26**: 297–302.

Felsenstein, J. (1985) Confidence limits on phylogenies: an approach using the bootstrap. *Evolution* **39**: 783–791.

Fitch, W. M. and Margoliash, E. (1967) Construction of phylogenetic trees. *Science* **155**: 279–284.

Hunter, P. R. and Gaston, M. A. (1988) Numerical index of the discriminatory ability of typing systems: an application of Simpson's index of diversity. *Journal of Clinical Microbiology* **26**: 2465-2466.

Kodaka, H., Armfield, A. Y., Lombard, G. and Dowell, V. R. (1982) Practical procedure for demonstrating bacterial flagella. *Journal of Clinical Microbiology* **16**: 948.

McHugh, T. D. and Gillespie, S. H. (1998) Non-random association of IS*6110* and *Mycobacterium tuberculosis*: implications for molecular epidemiological studies. *Journal of Clinical Microbiology* **36**: 1410–1413.

Maslow, J. N., Mulligan, M. E. and Arbeit, R. D. (1993) Molecular epidemiology: application of contemporary techniques to the typing of microorganisms. *Clinical Infectious Disease* **17**: 153–164.

Miller, J. M. and Wright, J. W. (1982) Spot indole test:

evaluation of four methods. *Journal of Clinical Pathology* **15**: 589–592.

Murray, M. (2002) Sampling bias in the molecular epidemiology of tuberculosis. *Emerging Infectious Diseases* **8**: 363–369.

Picardeau, M. and Vincent, V. (1996) Typing of *Mycobacterium avium* isolates by PCR. *Journal of Clinical Microbiology* **34**: 389–392.

Poh C. L., Ramachandran, V. and Tapsall, J. W. (1996) Genetic diversity of *Neisseria gonorrhoeae* IB-2 and IB-6 isolates revealed by whole-cell repetitive element sequence-based PCR. *Journal of Clinical Microbiology* **34**: 292–295.

Sambrook, J., Fritsch, E. F. and Maniatis, T. (1989) *Molecular Cloning: A laboratory manual*, 2nd edn. New York: Cold Spring Harbor Laboratory Press.

Schwartz, D. C., Saffran, W. J. Welsh, W. J. *et al.* (1983) New techniques for purifying large DNAs and studying their properties and packaging. *Cold Spring Harbor Symposium on Quantitative Biology* **47**: 189–195.

Selander, R. K., McKinney, R. M., Whittam, T. S. *et al.* (1985) Genetic structure of populations of *Legionella pneumophila*. *Journal of Bacteriology* **163**: 1021–1037.

Selander R. K., Caugant, D. A. Ochman, H. *et al.* (1986) Methods of multilocus enzyme electrophoresis for bacterial population genetics and systematics. *Applied and Environmental Microbiology* **51**: 873–884.

Selander, R. K., Musser, J. M., Caugant, D. A. *et al.* (1987) Population genetics of pathogenic bacteria. *Microbiology and Pathology* **3**: 1–7.

Selden, R. F. and Chory, J. (1987) Isolation and purification of large DNA restriction fragments from agarose gels In: Ausubel, F. M., Brent, R., Kingston, R.E. *et al.* (eds), *Current Protocols in Molecular Biology*. New York: Greene Publishing and Wiley-Interscience, pp. 2.6.1–2.6.8.

Simpson, E. H. (1949). Measurement of diversity. *Nature* **163**: 688.

Snelling, A. M., Gerner-Smidt, P., Hawkey, P. M. *et al.* (1996) Validation of use of whole-cell repetitive extragenic palindromic sequence-based PCR (REP-PCR) for typing strains belonging to the *Acinetobacter calcoaceticus–Acinetobacter baumanii* complex and application of the method to the investigation of a hospital outbreak *Journal of Clinical Microbiology* **34**: 1193–1202.

Tenover, F. C., Arbeit, R. D., Goering, R. V. *et al.* (1995) Interpreting chromosomal DNA restriction patterns produced by pulsed-field gel electrophoresis: criteria for bacterial strain typing. *Journal of Clinical Microbiology* **33**: 233–2239.

Van Embden, J. D. A., Cave, M. D., Crawford, J. T. *et al.* (1993) Strain identification of *Mycobacterium tuberculosis* by DNA fingerprinting: recommendations for standardized methodology. *Journal of Clinical Microbiology* **31**: 406–409.

Versalovic, J., Koeuth, T., McCabe, E. R. *et al.* (1991) Distribution of repeated DNA sequences in eubacteria applied to fingerprinting of bacterial genomes. *Nucleic Acid Research* **19**: 6823–6824.

Wang, G. T., Whittam, S., Berg, C. M. *et al.* (1993) RAPD (arbitrary primer) PCR is more sensitive than multilocus enzyme electrophoresis for distinguishing related bacterial strains. *Nucleic Acids Research* **21**: 930–5933.

Welsh, J. and McClelland, M. (1990) Fingerprinting genomes using PCR with arbitrary primers. *Nucleic Acids Research* **18**: 7213–7218.

Whittam, T. S., Ochman, H. and Selander, R. K. (1983) Multilocus genetic structure in natural populations of *Escherichia coli*. *Proceedings of the National Academy of Science of the USA* **80**: 1751–1755.

Williams, J. G. K., Kubelik, A. R., Livak, K. J. *et al.* (1990) DNA polymorphisms amplified by arbitrary primers are useful as genetic markers. *Nucleic Acids Research* **18**: 6531–6535.

Williams, J. G. K., Hanafey, M. K., Rafalski, J. A. *et al.* (1993) Genetic analysis using random amplified polymorphic DNA markers. *Methods in Enzymology* **218**: 704–740.

Woods, C. R., Versalovic, J., Koeuth, T. *et al.* (1992) Analysis of relationships among isolates of *Citrobacter diversus* by using fingerprints generated by repetitive sequence-based primers in the polymerase chain reaction. *Journal of Clinical Microbiology* **30**: 2921–2929.

8

Automated methods

In medical, agricultural and industrial microbiology, there is a continued need for 'rapid methods' that will shorten the time between receipt of a specimen or sample and the issue of a report. This has led to the development of instrumentation, usually known as automation, which has the additional advantage that it does not require constant attention, can proceed overnight and has therefore earned the title 'walk-away'. Such automation may be described as the operation of an instrument by mechanical, electrical or electronic procedures alone or in combination, thereby removing the need for direct human action.

The equipment described here can be truly considered as automation and is widely available. Clearly, not all such equipment can be included. Nor can details be given of use and operation; these may be found in the manufacturers' instruction manuals.

IDENTIFICATION AND ANTIMICROBIAL SUSCEPTIBILITY

A pure culture is a prerequisite. Thereafter, the user has a wide choice. Most systems incorporate antimicrobial agent testing options.

VITEK

Initially introduced as the automated system in 1976 by BioMerieux (France), this is based on the detection of microbial growth in microwells within plastic cards. The original cards were designed for the identification of urine isolates and used the most probable number (MPN) system to determine significant bacteriuria. Additional cards became available for the identification of pure cultures and for antimicrobial susceptibility testing. VITEK is a completely automated instrument that offers rapid results with an average 2- to 6-hour, same-day, turnaround time. The system allows for random or batch processing and has a built-in quality control module with a complete data management system. There are five VITEK models available, all of which provide the same analysis with rapid results and only differ in the number of cards they process: 32, 60, 120, 240 or 480.

Results from the identification cards are interpreted automatically. For antimicrobial susceptibility testing, over 40 antimicrobial agents are available and the results from each test include an interpolated minimum inhibitory concentration (MIC) as well as the National Committee of Clinical Laboratory Standards (NCCLS) breakpoints of susceptible, intermediate and resistant (NCCLS, 2000).

The VITEK Test Card is the size of a playing card and made up of 30 or 45 microwells, containing either identification substrates or antibiotics. VITEK offers a wide range of cards for the identification and susceptibility testing of most organisms encountered in the laboratory. The Test Card is a sealed container, so no aerosols are produced, thereby reducing the likelihood of personal contamination. Disposable waste is apparently reduced by more than 80% over microtitre methods. The current range of tests is summarized in Table 8.1.

The VITEK is an integrated modular system con-

Table 8.1 VITEK: range of tests

Product card	Organisms identified
GNI + Card V1311	Enterobacteria, vibrios, *P. aeruginosa* and other non-fermenters
GPI Card V1305	*Streptococcus, Staphylococcus, Enterococcus, Listeria* and *Corynebacterium* spp.
YBC Card V1303	Most clinically relevant yeasts
ANI Card V1309	> 80 anaerobes
NHI Card V1308	*Neisseria, Haemophilus* spp. and other fastidious organisms
UID-3 V1102	Nine urinary pathogens
UID-1 V1106	Nine urinary pathogens

sisting of a filler–sealer unit, reader–incubator, computer and printer, and it may be interfaced with other laboratory information management systems (LIMSs). The inoculum is prepared from a predetermined medium and, after adjustment to either MacFarlane 1 or 2 standard, the suspension is automatically transferred to the Test Card during the vacuum cycle of the filling module. The cards are then incubated in trays at 35°C. The optical density is monitored hourly; the first reading establishes a baseline, after which the light reduction caused by growth or biochemical reaction is recorded.

The VITEK has been introduced as a fully automated system that performs identification and susceptibility testing analyses from a standardized inoculum. The system combines both complementary and highly advanced skills to transform the antibiotic susceptibility test into a performance tool that detects and helps in the fight against bacterial resistance.

VITEK2 is targeted at the clinical market and claims to reduce time to identification and antibiotic susceptibility results. It offers a more extensive analysis menu than the original VITEK, also benefiting from a miniaturized card-format consumable item (10 cm × 6 cm × 0.5 cm), which consists of 64 wells for increased analytical capabilities.

Identification is based on fluorescent technology that provides broad profiles for the reliable identification of the most clinically relevant organisms.

Data analysis uses algorithms to look at a range of parameters and test conditions to ensure accurate test results; it is combined with an Advanced EXPERT Rules System for quality assurance purposes.

After primary isolation a suspension of the organism is prepared in a tube of saline and verified with a densitometer. The inoculum tube is then placed into a rack. The sample identification number is entered into the Smart Carrier via a barcode or keypad and electronically linked to the supplied barcode on each test code. Identification (ID) and antimicrobial susceptibility testing (AST) Test Cards can be mixed and matched in the cassette to meet specific laboratory needs. All information entered at the bench is transported to the instrument in a memory chip attached to the cassette. This provides positive text tracking from the bench to the report. From start to finish, all processing steps are completely autonomous, standardized, controlled and checked.

The optical system reads all 64 wells every 15 min. This kinetic monitoring provides an extended database for analysis and interpretation allowing VITEK2 to provide rapid results.

As AST becomes increasingly requested and complex as a result of emerging and low-level microbial resistances, the EXPERT system provides various levels of security, validations and in-depth analysis of results to predict the clinical outcome of therapy most accurately.

The EXPERT System allows for:

1. Biological validation of the quality of the results from a technical point of view.
2. Result interpretation to facilitate correction for improved clinical outcomes.
3. Systematic addition of recommendations that the laboratory wishes to communicate to physicians. These may be selected from footnotes pertinent to the national standards to which the laboratory is working, e.g. the NCCLS (2000), the British Society for

Antimicrobial Chemotherapy (BSAC: Phillips *et al.*, 1999) or specific recommendations defined by the user's laboratory.

VITEK have introduced DataTracas, an information management system that oversees the test data generated by the VITEK or VITEK2. The information is gathered and formatted into a variety of reports. In addition data can be exported for analysis into other PC-based software programs. Addition of the CAR program allows the VITEK and VITEK2 user to compose rules easily, listing specific conditions which, when met, suppress the reporting of toxic, inappropriate or more costly antimicrobial agents on the patient report.

The quality control module provides the laboratory with the ability to store the results of all VITEK cards, thereby allowing the quality of control performance to be tracked. Cumulative quality control reports, which list any results with the exceptions flagged or exceptions only, are easily generated.

The Bi-directional Computer Interface (BCI) package allows VITEK to receive patient information directly from the LIMS and transmit microbiology test data directly to it. This minimizes keystrokes and data transcription errors while automatically uploading test data into the LIMS.

The system has been evaluated by many workers. Jorgensen *et al.* (2000) showed that VITEK2 provided rapid, reliable susceptibility category determinations with a range of challenge and clinical isolates of *Streptococcus pneumoniae*. The mean time for generation of susceptibility results is 8.1 h. Hansen *et al.* (2002) used VITEK GN1+ and GNS-GA cards to identify and perform susceptibility testing on positive blood cultures from the BACTEC system. They concluded that the direct VITEK method could correctly report identifications and susceptibility patterns within 6 h, making same-day reporting possible for almost two-thirds of bacteraemic episodes with Gram-negative bacilli.

Joyanes *et al.* (2001) evaluated VITEK2 for identification and susceptibility testing of *Pseudomonas aeruginosa*, *Acinetobacter baumannii* and *Stenotrophomonas maltophilia*, and concluded that it allowed rapid identification of the *S. maltophilia* strains and most of the *P. aeruginosa* and *A. baumannii* isolates. It also performed reliable susceptibility testing of many of the antimicrobial agents used against these organisms.

In a multicentre evaluation of the VITEK2 Advanced Expert System for interpretive reading of antimicrobial resistance tests, Livermore *et al.* (2002) demonstrated the capacity of the system to detect and interpret resistance mechanisms with a high level of accuracy and standardization. Only 64 of 963 (6.6%) interpretations at 10 European centres were discrepant, with half of these concerning *S. pneumoniae* isolates that were intermediately resistant to penicillin by previous methods. In a similar evaluation of the Advanced Expert System, Cantón *et al.* (2001) showed it to be a reliable tool for the detection of extended spectrum β-lactamases or inhibitor-resistant TEM β-lactamases produced by Enterobacteriaceae.

Sensititre

The Sensitive System (Trek Diagnostic Systems, Inc.) is made up of a number of 'mix-and-match' components and provides walk-away automation for identification and susceptibility testing. The modular system is made up of an auto-inoculator, plate reader, the fully automated Automated Reading and Incubation System (ARIS) and the Sensititre Automated Management System (SAMS) software.

The Sensititre system uses fluorescence-based technology. Substrates are linked to a fluorophore such as 4-methylumbelliferone (4MU) or 7-aminomethylcoumarin (7AMC). The substrate fluorophore is normally non-fluorescent but, in the presence of specific enzymes, the substrates are cleaved from the fluorophore, when the unbound 4MU or 7AMC fluoresces under ultraviolet (UV) light.

The test media contain specifically designed probes for a number of reactions. Carbohydrate use is detected by monitoring the pH shift with a fluorescent indicator. When the pH in the test well is alkaline, the indicator is fluorescent but, as the carbohydrate is oxidized or fermented, producing acid, the fluorescence decreases, indicating a positive reaction. Carbon-source use, and decarboxylase and urease reactions also depend on monitoring pH changes with fluorescence indicators. Bacterial enzyme tests detect the production of various enzymes, including peptidase, pyranosidase, phosphatase and glucuronidase. Enzymes are detected by their ability to cleave a quenched fluorophore, resulting in fluorescence, e.g. aesculin, a fluorescent glycoside, breaks down to the non-fluorescent aescletin and glucose; non-fluorescence therefore indicates a positive result. Tryptophan deamination by the catabolic enzyme tryptophase results in a coupled reaction in which the formation of a dark colour suppresses the fluorescent signal and thus indicates a positive result.

A real benefit of a fluorescence endpoint is that emission may be detected with greater sensitivity and a reaction can therefore be observed much earlier than is possible with colour changes or turbidity readings. Fluorescence also overcomes some of the problems associated with conventional approaches such as loss of resolution caused by pigment formation in cultures, scanty or variable growth, and the use of opaque or pigmented supplements.

The system is based on the standard 96-well microtitre tray and consists of an auto-inoculator, plate reader and data handler. Inocula are prepared, with the aid of the built-in nephelometer, by transferring colonies from the isolation media to sterile distilled water to equal a 0.5 McFarlane standard. The inoculator dispenses 50 µl into each test well. The test plate is read after incubation. The light source is a broad-band xenon lamp (360 nm) which generates microsecond pulses of high-peak-power light. This passes through interference filters and a beam-splitting cube with wavelength-selective coatings to lenses that focus the light on a test well and the detector. The detector is a photomultiplier tube that transmits raw fluorescence data to a computer. Approximately 30 s are required to read a plate and the biocode generated

is matched to the Sensititre database. There may be additional test prompts and a reincubate prompt also occurs at 5 h if there is no, or a low, probability of identification. The result of each test is printed and the quality of each identification is determined on the basis of calculated probability values. ARIS has a 64-plate capacity, which allows the user to load any MIC, breakpoint or identification plates for a combination of 192 possible tests in a single instrument.

Stager and Davis (1992) reviewed the performance reports of the Sensititre and stated that the limitations were the requirement for off-line incubation and the necessity for reincubation of some panels. The development of systems for the rapid identification of Gram-positive bacteria would extend its usefulness.

One of the real developments in automated instrumentation has been the introduction of data management software. Sensititre is no exception and has introduced the SAMS software, which provides for a wide range of data-tracking and reporting capabilities. Extensive survey facilities allow a variety of customizable and automatically formatted reports, including epidemiology, statistics and workload data, to be produced. The QC Module automatically checks quality control ranges, showing Pass or Fail on reports. The Expert System flags when predefined NCCLS standards are violated. Automated Backup allows scheduling of automatic nightly data back-up and the Automated Interface allows scheduling of automatic initiation of results.

Biolog

A number of system options are available although it is the OmniLog ID system that is fully automated. It uses Biolog's proprietary carbon source use test methodology in a convenient 96-well format. When an organism is introduced to the wide variety of pre-selected carbon sources, it rapidly produces a characteristic pattern or 'fingerprint', which is compared with an extensive database. Unlike other systems, the Biolog results are based on 95 reactions from 6 to 8 different classes of carbon sources. The Biolog technology can detect mixed or contaminated cultures, and is also able to identify environmental and fastidious organisms. The sensitivity and precision of the technology have allowed Biolog to build extensive databases that include environmental organisms, veterinary organisms, plant pathogens and human pathogens.

A unique feature of the OmniLog System is user-defined incubation temperature. The OmniLog ID System was designed to identify a wide variety of organisms, which must be incubated at their optimal temperature to obtain an accurate identification. Other automated systems that incubate and read identification test panels do so at one set temperature, usually 37°C. As such they are not always able to identify bacteria such as those isolated when doing environmental monitoring. For those organisms that may require special incubation environments (e.g. elevated CO_2 or extreme temperatures), the system allows the user to incubate the MicroPlates 'off line' before placing them into the Reader for interpretation.

The OmniLog System simultaneously incubates, reads and interprets the Biolog MicroPlates (Table 8.2). It continuously processes samples but allows the user complete access at any time during a sample run. Samples can be loaded when ready and removed when complete without disturbing other samples still in process. Inside the Reader there are 25 trays. Each tray holds 2 MicroPlates, giving the Reader a total capacity to incubate and read 50 isolates.

Before the user inoculates the appropriate MicroPlates, they log the MicroPlate information into the OmniLog software. By simply following the software's instructions, the user then opens the door of the OmniLog Reader and places each MicroPlate in the appropriate tray slot indicated by the software. Once all the MicroPlates are loaded and the door of the Reader is closed, the OmniLog software takes over the responsibility for incubating, reading, saving and printing the results.

The OmniLog ID System begins reading the MicroPlates 4 h after they have been placed into the Reader. The pattern is compared with the identification database and an ID is called if enough positive reactions have developed. This result is then displayed on the Read menu screen, stored in the computer's memory, and an identification report is printed out. If no result is obtained after 6 h, the instrument automatically continues to incubate the MicroPlate and begins reading again after 16 h and up to 24 h.

The OmniLog ID System uses Windows-based software. The status of all the samples can be observed by simply looking at the Read menu screen of the OmniLog ID System software. All of the information for a specific MicroPlate is contained on a single line of this screen. Once the system has determined a final identification for a specific MicroPlate, it indicates this status with a check mark icon. A clock icon is used for those samples that are still incubating because an identification result has not yet been determined.

The 'Plus' version of the OmniLog adds the additional capability of identifying yeasts, anaerobes and filamentous fungi (Table 8.3).

Table 8.2 The OmniLog System Microbe Identification Process

Step 1	Isolate a pure culture on Biolog media
Step 2	Do a Gram stain and determine testing protocol
Step 3	Prepare inoculum at specified cell density
Step 4	Inoculate MicroPlates and place in Reader
Step 5	Obtain ID results from printer

Table 8.3 Omnilog and Omnilog Plus: range of tests

	OmniLog	OmniLog Plus
Gram-negative bacteria 500 species	✔	✔
Gram-positive bacteria 500 species	✔	✔
Yeasts 250 species		✔
Anaerobic bacteria 350 species		✔
Filamentous fungi 500 species		✔

In today's regulated environments control and validation of the microbial identification system and control are two crucial elements of regulatory compliance. In conjunction with validation experts, Biolog developed a comprehensive validation package for the OmniLog System. The Validation Protocol Manual provides the framework for performing the Installation Qualification and the Operation Qualification for each of the system elements. The Validation Package also includes test strains that can be run on the OmniLog to provide validation of the system performance. Biolog also provides optional software features to control operator access and to control access for building or modifying data files. This new capability is indicative of how Biolog is consistently responding to its customers' ongoing needs.

The operator can use one of the Biolog databases or compile a user-defined file. An unknown biochemical profile may then be compared with either the databases or a combination of the two. Other features of the software include on-line information about any species in the library, cluster analysis programs in the form of dendrograms and two- or three-dimensional plots to demonstrate the relatedness of strains or species, and the separation of the Gram-negative database into clinical and environmental sectors.

It is the last of these that gives some distinction to the Biolog because it identified a niche in the market that other systems had failed to fill. It includes non-clinical isolates as a feature of its initial database, thereby addressing the particular needs of environmental and research workers. This is best illustrated by considering the ES Microplate which is designed for characterizing and/or identifying different strains of *Escherichia coli* and *Salmonella* spp., for characterizing mutant strains and for quality control tests on *E. coli* and *S. typhimurium* strains carrying recombinant plasmids.

Phoenix

The Phoenix System is a new, automated, short incubation system for the rapid identification and susceptibility testing of clinically relevant bacteria. Phoenix can test up to 100 ID and AST combination panels at a time. The Phoenix system uses an optimized colorimetric redox indicator for AST and a variety of colorimetric and fluorimetric indicators for bacterial ID. A sealed and self-inoculating moulded polystyrene tray with 136 microwells serves as the Phoenix ID/AST Combination disposable.

The Phoenix System includes an inoculation station for panel set-up/ inoculation and an instrument with an Incubator/Reader carousel module. The carousel houses four horizontal tiers of 26 panel carriers to accommodate a tier-specific Normalizer and 25 Phoenix Panels.

The Phoenix Panel uses up to 51 microwells for identification and up to 85 microwells for susceptibility testing of 16 antimicrobial agents at five concentrations, or more than 25 agents at three concentrations per agent.

An optimized Mueller–Hinton broth base with a redox indicator and an optimized Inoculum Fluid is used for AST and ID inoculum preparations, respectively.

A bacterial inoculum concentration, approximately equivalent to a 0.5 McFarland Standard, is required for the identification of either Gram-negative or Gram-positive bacteria. Susceptibility testing is performed with an inoculum concentration of $3–7 \times 10^5$ colony-forming units (cfu)/ml with a colorimetric redox indicator. Kinetic measurements of bioreactivity within individual microwells via red, green, blue and fluorescence readings are collected and comparatively analysed with the Phoenix database.

The average time to results is 6 h for Gram-positive and 6–12 h for Gram-negative organisms. The instrument will provide interpreted test results with an integrated expert system (BDXpert). The system will offer an optional workstation (Epicenter) to complement Phoenix results with data management features, including epidemiology, user-defined reporting and Advanced BDXpert capability.

Midi microbial identification system

The Midi (Microbial ID Inc., USA) is a fully automated, computerized gas chromatography (GC) system that can separate, distinguish and analyse more than 300 C_9–C_{20} fatty acid methyl esters. The value of this type of analysis is that the fatty acid composition is a stable genetic trait, which is highly conserved within a taxonomic group.

The system software includes operational procedures, automatic peak naming, data storage and comparison of unknown profiles with the database, using pattern recognition algorithms. The database contains more than 100 000 strain profiles, including representatives of Enterobacteriaceae and *Pseudomonas*, *Staphylococcus* and *Bacillus* spp. It also includes mycobacteria, anaerobes and yeasts.

Representative isolates must be subcultured in defined media before preparation of the methyl ester, after which the whole process is automated. The system is calibrated by a mixture of straight-chain fatty acids. This is necessary after analysis of 10 samples and corrects for any changes in sample injection volume and variability in gas flow rates. Calibration samples can be set up automatically. The autosampler will take up to 2 days of analytical capacity and may therefore be considered as truly walk-away.

Gas chromatography analysis is a well-established tool in the anaerobe laboratory and extensive work has been done in this field, notably at the Anaerobe Laboratory, Virginia Polytechnic Institute and State University. McAllister *et al.* (1991) reported that the Midi system correctly identified 97% of the anaerobes tested.

The Midi database consists of more than 100 000 analyses of strains obtained from experts and from culture collections. The cultures were collected from around the world to avoid potential geographical bias. Where possible, 20 or more strains of a species or subspecies were analysed to make the entry. When fatty-acid profile subgroups were found within a taxon, more strains were obtained to delineate each group.

The system will analyse about 45 samples per day (with a dual tower, 90 samples per day can be analysed). A single technician can extract about 75 samples per day and so the operator time per sample averages at about 6 min.

Assessment of identification systems

Stager and Davis (1992) reviewed five studies in which the accuracy of either two or three of the systems was compared: percentage accuracy varied from as low as 35% to as high as 99.2%. This type of study is of somewhat limited value, however, because the companies are continually improving their systems, particularly with respect to the quality of the databases, which are, of course, fundamental to correct identification. This is best seen in the studies of Kelly *et al.* (1984), Stevens *et al.* (1984) and Truant *et al.* (1989), who all presented evidence that variation in biotypes of individual species from different geographical areas may in part be responsible for performance variability. This needs to be considered both by the manufacturers and by the users, especially by the latter when they are considering a purchase.

BLOOD CULTURE INSTRUMENTATION

BACTEC

The BACTEC series of instruments (Becton Dickinson, USA) is generally associated with blood cultures but has other uses. Samples of blood, spinal, synovial, pleural and other normally sterile body fluids are injected into the BACTEC vials. Growth of organisms is detected by the production of CO_2 from culture media. As the first of the blood culture systems, it is worth considering the development of this product line. Users should be aware of the fact that cross-contamination could occur between samples, should there be inadequate sterilization of the sample needles that are used to collect samples.

BACTEC 460

This early BACTEC system used the makers' culture medium containing ^{14}C-labelled substrate. The instrument takes 60 vials, which are inoculated with 5 ml of the test fluid. It then automatically measures the radioactivity in the headspace gas for ^{14}CO$_2$ each hour and prints the result. If the activity exceeds a preset threshold the blood culture is considered to be positive.

BACTEC NR-660, NR730 and NR-860

The NR series replaced the radioactive substrate by employing infrared (IR) spectrometry to detect the CO$_2$. Infrared light is absorbed by the CO$_2$ in the test cell and the amount passing through to the detector is registered. The amplifier converts this measurement of conductivity to voltage, which is converted to a read-out value. There is an inverse relationship between the amount of CO$_2$ generated and the amount of IR light detected. This is accounted for in the calculations that generate the growth value (GV).

During the test a pair of needles penetrates the vial septum and the headspace gas is drawn through one of them into the detector. The other needle is connected to an external gas cylinder so that the appropriate gas for aerobiosis or anaerobiosis can be replaced. The various models in the NR series use different IR systems. In the NR-860 a positive culture is flagged if the GV, or the difference between two consecutive measurements, exceeds a predetermined threshold. The vial headspace is also measured and when that exceeds a default setting the sample is flagged as potentially positive. Vial testing is automatic and an automated tray transport mechanism shuttles the trays between the incubator and the IR sensor unit. A barcode scanner reads the sample label. The NR-860 can store 480 samples in eight drawers. The two bottom drawers, containing the most recent aerobic cultures, are mounted on an orbital shaker.

The working capacity is 48 new cultures per day with a 5-day test protocol. When positive samples are identified they are removed for further investigation.

BACTEC 460/TB for mycobacteria

The BACTEC 460 instrument was modified to the 460/TB for the detection and susceptibility testing of mycobacteria. In this the cover was replaced by as hood with a forced, recirculating air supply and HEPA (high efficiency particulate air) filters to avoid dispersion of any aerosols into the laboratory atmosphere. This instrument, subsequently modified to detect CO$_2$ by IR spectrometry, instead of radioactivity, has proved to be very useful in the early detection of tubercle bacilli in sputum cultures and of antimicrobial resistance of mycobacteria.

BACTEC 9000 Series

This 9000 series instruments are fluorescence-based, walk-away detection systems reputedly offering increased sensitivity; as they are non-invasive, they offer a high level of built-in microbiological safety. As no venting is required the samples may be collected in Vacutainer blood-collecting sets. A dye in the sensor reacts with CO$_2$, thereby modulating the amount of light that is absorbed by a fluorescent material in the sensor. The photodetectors measure the fluorescence, which is related to the amount of CO$_2$ released in the culture. On initiation of a test the machine carries out a diagnostic routine, after which it starts continuous and automated on-line testing. A test cycle of all racks is continually being monitored every 10 min and positive cultures are flagged by a light on the front of the instrument and displayed on the monitor.

The BACTEC 9240 can monitor 240 culture vials, arranged in six racks. The vials are incubated at 35°C with agitation. The working capacity is 24 new culture sets per day with a 5-day test protocol.

Also available is the BACTEC 9120, which monitors 120 culture vials. The BACTEC 9050 has been introduced for laboratories that process fewer than 150 blood culture sets each month. The compact size of the BACTEC 9050 minimizes space requirements, yet it still features fully automated, non-invasive, continuous fluorescent monitoring of growth, but occupies only 4.5 square feet of bench space (less than half a square metre). Software and operating systems are totally built in, so there is no need for an external computer to take up more space and complicate operations. The instrument is designed for laboratories processing up to five blood cultures per day. The technician simply opens the door of the instrument, presses the 'Vial Entry' key, scans the vial barcode at the fixed barcode reader and places the vial into the assigned position. Once the door is shut, processing automatically begins.

In addition to the ability of networking up to 20 BACTEC instruments for large or group laboratories, Becton Dickinson have introduced BACTEC VISION Software, a data control centre for the BACTEC 9000 series of blood culture instruments.

The software allows for extensive data management, total reporting and sorting flexibility in a Windows environment.

Other BACTEC 9000 Data Management Systems

- BACTEC Level I: with bar code icon interface to increase work flow efficiency. Monitors instrument status, tracks and reports in protocol specimens.
- BACTEC Level II: includes Level I plus connectivity to all interfaces written to ASTM standards. Provides bidirectional capability to download patient demographics directly to core computer, saving technician time and speeding reporting of results. Monitors instrument status, tracks and reports in-protocol specimens, and records patient and specimen demographics.

Note: BACTEC VISION is a Level III System and includes all items in Levels I and II.

BacT/Alert

The BacT/Alert system (Biomerieux) was introduced in 1990–91 as a fully automated system and therefore as an alternative to semi-automated radiometric and IR systems. Like the BACTEC, it uses the production of CO_2 and employs a novel colorimetric sensor in the base of the aerobic and anaerobic blood culture bottles. The sensor changes colour as CO_2 is produced; the rate of change is detected by a reflectometer and the data are passed to a computer whose algorithm can distinguish between the constant CO_2 production by blood cells and the accelerating CO_2 production from a positive blood culture. The sensor is covered by an ion-exclusion membrane, which is permeable to CO_2 but not to free hydrogen ions, media components or whole blood. Indicator molecules in the water-impregnated sensor are dark green in their alkaline state, changing progressively to yellow as the pH decreases. CO_2 from a positive culture passes through the membrane, giving the following reaction:

$$CO_2 + H_2O \rightarrow HCO_3^- + H^+$$

The free hydrogen ions react with the indicator molecules causing the sensor to change colour. This is detected by a red light-emitting diode and measured by a solid-state detector.

Aerobic and anaerobic culture bottles are available and are maintained at negative pressure for ease of inoculation. The recommended adult sample is 5–10 ml per bottle. It is necessary to vent the aerobic bottles occasionally. Cultures taken outside normal laboratory hours may be incubated externally and entered into the system on the following day. The system holds up to 240 bottles, each with its own detector. Up to four data-detection units may be interfaced with the same computer, giving a maximum throughput for one system of 960 bottles at any one time. A single detection unit has 10 blocks of 24 wells, each block operating independently, thereby allowing easy removal and routine servicing. A Reflectance Standard Kit is supplied with each system, which allows the user to calibrate and test for quality control. The PC analyses all data, interpreting both positive and negative growth curves. It also allows complete patient files to be kept and will interface with existing in-house systems.

The system has been evaluated against the radiometric BACTEC 460 and shown to give comparable results (Thorpe *et al.*, 1990). Wilson *et al.* (1992) also concluded that the BacT/Alert and BACTEC 660/730 non-radiometric systems were comparable for recovering clinically significant micro-organisms from adult patients with bacteraemia or fungaemia. The BacT/Alert, however, detected microbial growth earlier than the BACTEC and gave significantly fewer false-positive results. In addition to the standard Bact/Alert system, MB/BacT has been introduced as an easier, safer and more cost-effective method for detection of mycobacteria than conventional or radiometric methods. It is a fully automated system for continuous non-invasive monitoring of non-blood specimens and, based on the unique colorimetric technology, it offers the same benefits in terms of speed, accuracy, ease of use and cost-effectiveness.

A further development is BacT/ALERT 3D which is a test system that combines blood, body fluid and mycobacteria specimen testing in one instrument. Its flexible modular design consists of one to six Incubator Modules, each capable of accommodating 240 bottles, and is directed by a touchscreen-activated Control Module. This allows for a text-free user interface for more direct random rapid loading and unloading of the test samples. The control Module supervises the reading of the Incubator Module's sensors and controls the decision-making algorithms, to determine whether or not a given bottle is positive.

Bact/VIEW is a complete data management program for use with BacT/ALERT and MB/BacT. It has an easy to learn and operate touchscreen interface. Collecting patient data, launching routine tasks, loading and unloading bottles, and customizing reports are very straightforward matters. There is no need to memorize commands or routines. A single computer system with BacT/VIEW can control up to nine microbiological units. In addition BacT/VIEW may be connected directly to an LIMS using BacT/LINK.

ESP

ESP (Trek Diagnostic Systems) is a fully automated culture system for routine blood culture, mycobacteria detection and *Mycobacterium tuberculosis* susceptibility testing on a single instrument platform. It is a closed system design, thereby allowing safe handling of all specimens throughout the laboratory.

The instrument design can be configured to allow for a system that efficiently accommodates a range of test volumes: 1920 test sites may be monitored by one PC, thereby reducing costs and improving efficiency.

ESP detects a wide range of micro-organisms by measuring both consumption and production of all gases. The technology does not rely on forced production of CO_2 for micro-organism detection. The non-invasive measurement of pressure changes in the headspace of the culture bottle provides an effective and rapid detection of microbial growth. Aerosol-free venting of the culture bottle occurs automatically before entry and after removal of a positive specimen. A hydrophobic membrane in the connector prevents aerosols from escaping into the laboratory environment.

Proprietary media support the growth of the widest variety of micro-organisms and are effective for all specimen types without any additives. Included essential growth factors allow a much smaller amount of blood to be drawn. The American Food and Drugs Administration (FDA) has cleared the use of volumes as low as 0.1 ml to accommodate difficult draws. Paediatric bottles are therefore not necessary with ESP. The optimized blood : broth dilution ratio neutralizes effects of antibiotics and other inhibitory substances, meaning that resins and fan media are not required.

A non-radiometric *M. tuberculosis* detection system is provided for. The ESP Culture System II combines a liquid culture medium, a growth supplement and, for potentially contaminated specimens, an antibiotic supplement, with a detection system that automatically incubates and continually monitors culture bottles inoculated with specimens suspected of containing mycobacteria. ESP detects mycobacterial growth by automatically monitoring (every 24 min) the rate of consumption of oxygen within the headspace of the culture bottle and reports that growth response with a visible positive sign.

An ESP Myco Susceptibility Kit consisting of three first-line anti-tuberculosis drugs – rifampin, isoniazid and ethambutol – is available. The total ESP susceptibility system is made up of specific lots of ESP media that are qualified to be used with each ESP Myco Susceptibility Kit. The primary drugs are injected into supplemented ESP Myco bottles. An inoculum of the M. tuberculosis isolate to be tested can be prepared from colonies grown on solid media or from a seed bottle. The inoculum is added to the drug-containing bottles and to a drug-free control bottle, used as a positive control. The presence of growth is determined automatically by the ESP instrument. If the organism is susceptible to a drug, microbial growth will not be detected or will be significantly delayed compared with the drug-free control bottle. If the organism is resistant to a drug, microbial growth will be detected within 3 days of the control bottle.

Safety has been addressed by the use of a disposable plastic connector with recessed probe that eliminates risk and liability of injury from venting probes and provides safe, non-invasive venting. There is a 0.2-μm hydrophobic membrane within the connector which prevents the escape of dangerous aerosols into the laboratory. The use of a non-radiometric mycobacteria broth eliminates potential hazards of radioactive components and eliminates the requirement for radioactive disposal or licensing fees.

As with all the current generation of instrumentation, a data management system has been developed to extend the functionality of the culture system. It is a fully featured Windows NT-based data management system using the Microsoft Access database. Custom reports can easily be integrated with standard reports, providing a wealth of information about epidemiology, quality of results, workplace efficiency, and cost containment. An interface allows download of patient demographics during specimen entry and automatic transfer of test results to the LIMS. ESP data can be retrieved from any computer on the network using a compatible version of Microsoft Access.

URINE EXAMINATION

There have been many attempts to automate urine analysis, none of which has had significant commercial success. Zaman et al. (2001) considered the UF-100 urine flow cytometer (Merck, Eurolab) and concluded that the technology does not accurately predict the outcome of urine cultures, and that this system is therefore unsuitable for the safe screening of urine samples for urinary tract infections.

Trek Diagnostic Systems, Inc. have introduced the Cellenium System which is claimed to provide the first rapid, fully automated approach to urine screening. Using proprietary advanced robotic operation, measured amounts of each urine sample are dispensed into a Cellenium cassette and fluorescent nucleic acid stains are added. The cassette is automatically coupled with a Cellenium membrane. Vacuum filtration results in a monolayer of stained micro-organisms on the membrane, which are examined using computerized fluorescent microscopy imaging. Results are equivalent to the colony-forming units per millilitre of a standard culture.

This approach can eliminate the culturing of negative samples and can detect and enumerate bacteria in urine, as well as providing Gram stain information. In positive specimens, bacteria are noted as Gram-positive cocci, Gram-negative rods or Gram-positive rods. It can also be used to detect yeast and white blood cells, and provides Gram stain morphological information.

The instrument is able to detect, enumerate, and classify pathogenic micro-organisms in urine within 30 min, thus eliminating the requirement to culture negative specimens. It consequently provides the ability to report negative results a day earlier than traditional culture methods.

Patient specimen data may be entered either by barcode or by manual entry through the instrument's input station. An intuitive Windows NT-based software guides the loading of reagents and allows laboratory personnel to monitor the test process at any time. The proprietary reagents are supplied in a ready-to-use form to eliminate any requirement for special preparation. The instrument is capable of continuous operation (24 h/day) analysing up to 70 specimens per h.

Reporting can be carried out in several ways, including a hard copy printout, and a graphic map that classifies specimens as positive, negative or inconclusive, or through an interface to the LIMS.

IMPEDANCE INSTRUMENTATION

Four systems are generally available: Bactometer (BioMerieux, France), Malthus (Radiometer, Denmark), RABIT (Don Whitley, UK) and Bactrac (Sy-Lab, Austria). All operate on the same principle.

Impedance may be defined as resistance to flow of an alternating current as it passes through a conductor (for detailed theory, see Eden and Eden, 1984; Kell and Davey, 1990). When two metal electrodes are immersed in a conducting fluid the system behaves either as a resistor and capacitor in series or as a conductor and capacitor in parallel (Kell and Davey, 1990). When it is in series combination, the application of an alternating sinusoidal potential will produce a current that is dependent on the impedance, Z, of the system which, in turn, is a function of its resistance, R, capacitance, C, and applied frequency, F. Thus:

$$Z = \sqrt{R^2 + \left(\tfrac{1}{2}\pi FC\right)^2}$$

Any increase in conductance, defined here as the reciprocal of resistance, or capacitance results in a decrease in impedance and an increase in current. The AC equivalent of conductance is admittance, defined as the reciprocal and the units of measurement are siemens (S).

Microbial metabolism usually results in an increase in both conductance and capacitance, causing a decrease in impedance and a consequent increase in admittance. The concepts of impedance, conductance, capacitance and resistance are only

different ways of monitoring the test system and are interrelated. An important factor, however, is that all are dependent on the frequency of the alternating current and all are complex quantities, in that they contain both real and imaginary parts (Kell and Davey, 1990). In practical terms the user must be aware that the electrical response is frequency dependent, has a conductive and capacitive component, and is temperature dependent. Temperature control in any impedance system is of critical importance. A temperature increase of 1°C will result in an average increase of 0.9% in capacitance and 1.8% in conductance (Eden and Eden, 1984), and from this it can be calculated that a less than 100 millidegree temperature drift can result in a false detection. It is interesting to note that three of the four systems mentioned above have adopted a different method of temperature control: Bactometer, a hot-air oven; Malthus, a water bath; and RABIT and BacTrac a solid block heating system.

These systems provide a measurement of net changes in conductivity of the culture medium. Tests are monitored continually and, when the rate of change in conductivity exceeds the user-defined pre-set criterion, the system reports growth. The time required to reach the point of detection is the 'time to detection' (TTD) and is a function of the size of the initial microbial population, its growth kinetics and the properties of the culture medium. For a given protocol, the TTD is inversely proportional to the initial microbial load of the sample. At the point of detection it is generally considered that there will be approximately 10^6 cfu/ml of the test organism present in the sample. This will vary according to the type of organism, growth media, etc., but will be constant for any one organism growing under defined test conditions.

The attributes of the principal systems are outlined in Table 8.4, which shows that there are differences in the measuring modes. The Bactometer has the option of measuring either conductance or capacitance, the Malthus measures conductance and the principal signal component in the RABIT is conductance, although there is a also a capacitance component. The benefits of capacitance are

Table 8.4 Comparison of the three principal impedance systems/ Bactometer, Malthus and RABIT

	Bactometer	Malthus 2000	RABIT
Test capacity	512	480	512
Modular system	Yes	No	Yes
Temperature options per system	8	1	16 max
Cell volume	2 ml max	2 and 5 ml	2–10 ml
Cell format	Block of 16	Single	Single
Reusable/disposable	Disposable	Reusable and disposable	Reusable
Media availability	Yes	Yes	Yes
Measuring mode	Conductance or capacitance	Conductance	Impedance
Indirect mode	No	Yes	Yes
Full access to cell throughout test period	No	No	Yes
Temperature control	Hot-air oven	Water-bath	Dry heating block

best exemplified in attempts to detect organisms in which normal metabolism results in no net increase in medium conductivity. This does not preclude the Malthus and RABIT from such applications, however, because both can utilize the indirect technique (Owens *et al.*, 1989) in which an electrical signal is generated by CO_2 production. The technique also has the advantage of being fully compatible with samples and test media that have a high salt content, which is not the case with the Bactometer. The indirect technique also allows the user a greater choice of culture media because these do not have to be optimized for electrical response but simply for growth of the target organism.

ATP MEASUREMENT

Stewart and Williams (1992) have reviewed the mechanism of bioluminescence assay, mediated by the luciferin/luciferase reaction, for adenosine triphosphate (ATP). The luciferin/luciferase complex reacts specifically with the ATP in living cells and emits a light signal proportional to the amount of ATP present. There is a linear relationship between the ATP and the plate count (Stannard and Wood, 1983). The ATP in the sample originates from both living cells and other sources (the

somatic ATP), and the two must be differentiated if the technique is to be used in microbiological testing (although it is not necessary for rapid hygiene monitoring, as in the food industry). The two commonly used methods are separation of the organisms from the sample before extracting the ATP, and prior destruction of the somatic ATP.

Bactofoss

Bactofoss (Foss Electric, Denmark) is a unique, fully automated instrument that combines the principles mentioned above to remove sample 'noise'. All functions are monitored by the central processing unit and messages are presented on the screen. Bactofoss is a simple push-button instrument which provides a result within 5 min.

The test sample is taken up automatically and deposited in a temperature-controlled funnel where several filtrations and pre-treatments remove somatic ATP. The micro-organisms are left on a filter paper which is then positioned in an extraction chamber. Extraction reagent is added to extract microbial ATP, the amount of which is determined by measurement of luminescence after the addition of luciferin and luciferase. All reagents, etc. are supplied by the manufacturer and the instrument can perform 20 analyses per hour. Limond and Griffiths (1991) have shown that it is applicable to meat and milk samples, with detection sensitivities of $3 \times 10^4 - 3 \times 10^8$ cfu/g and $1 \times 10^4 - 1 \times 10^8$ cfu/ml for milk.

Direct epifluorescence techniques

The direct epifluorescence technique (DEFT) was originally developed for counting bacteria in raw milk (Pettipher *et al.*, 1980, 1989). It takes less than 30 min and uses membrane filtration and epifluorescence microscopy. Pre-treatment of samples may be necessary to facilitate rapid filtration and distribution of bacteria on the membrane. A suitable fluorescent stain shows organisms that are easily distinguished from debris by an epifluorescence microscope, which is linked to an automatic counting system.

The direct epifluorescence technique has been used in manual or semi-automatic modes for several years, but the new Cobra system (France) is fully automated.

Bactoscan

Bactoscan (Foss Electric, Denmark) is fully automated and employs the DEFT principle, but is dedicated to the quality control of milk by making direct bacterial counts.

A 2.5-ml sample of milk is treated with lysing fluid to dissolve protein and somatic cells. The bacteria are then separated from the milk by gradient centrifugation. The bacterial suspension is mixed with a protease enzyme to dissolve particulate protein and is then stained with acridine orange. The stained organisms are differentiated from debris on the basis of dye uptake and are counted by a continuously functioning epifluorescence microscope with the results displayed on screen. Eighty samples per hour can be screened and the system is flushed with rinsing fluid between each sample. When testing is finished the machine is automatically cleaned

The protease solution must be prepared daily but the other reagents need be made up only every other day. The start-up procedure takes about 30 min and involves a calibration routine, lens checking and setting the discrimination level of the counting module. Milk samples are placed in special racks, heated to 40°C and loaded on the instrument. The sample identification can be keyed in manually or entered by a bar code reader. Results are displayed on a screen. Bactoscan counts correlate well with standard plate counts and may therefore be converted to equivalent colony-forming units per millilitre.

FLOW CYTOMETRY

Flow cytometry, widely used in research, has now been developed for routine industrial use. The technique allows cell-by-cell analysis of the test samples and, coupled with a range of fluorescent labels, provides a rapid and automated method for detecting microbes and examining their metabolic state. The sample is injected into a 'sheath' fluid that passes under the objective lens via a hydrodynamic focusing flow cell. The sheath fluid passes continuously through the flow cell, thereby focusing the sample stream into a narrow, linear flow. The sample then passes through a light beam that causes the labelled cells to emit fluorescent pulses. Each pulse is detected and subsequent analysis allows pulses to be recognized as separate counts and graded in terms of fluorescence intensities.

Systems currently available include the Chemunex (AES Laboratoire) range of products, which can now detect at the single cell level. The results may be printed as hard copy or transferred to a computer. The labelling and counting take less than 30 min per sample. Other systems from companies such as Coulter Electronics and Becton Dickinson are available.

The ability of flow cytometry to analyse and sort cells into defined populations on the basis of cell size, density and discriminatory labelling is a powerful new microbiological tool.

PYROLYSIS MASS SPECTROMETRY

Pyrolysis mass spectrometry (PMS) has considerable potential for the identification, classification and typing of bacteria (Magee *et al.*, 1989; Freeman *et al.*, 1990; Sisson *et al.*, 1992). In the RAPyD-400 benchtop automated PMS (Horizon Instruments, UK) samples are spread on V-shaped Ni–Fe pyrolysis foils held in pyrolysis tubes. Samples are then heated by Curie-point techniques in a vacuum, which causes pyrolysis in a controlled and reproducible manner. The gas produced is passed through a molecular beam and analysed by a rapid scanning quadruple mass spectrometer to produce a fingerprint of the original sample over a mass range of 12–400 Da. Data are analysed by onboard multivariate statistical routines, providing principal component analysis, discriminant function analysis, cluster analysis and factor spectra. The system is entirely automated, with sample loading, extraction, indexing to next sample and data collection controlled by a computer. Routine servicing and maintenance are minimal. Routine analysis takes approximately 90 s per sample with a batch of 300 samples.

It is important that the organisms are originally cultured on media that do not impose stress and consequent alteration of phenotype expression, which may obscure the relatedness of isolates. It may be desirable to facilitate expression of certain phenotypes before analysis, thereby allowing the technique to differentiate between unrelated strains, e.g. differentiation between toxigenic and non-toxigenic strains (Sisson *et al.*, 1992). Experience in UK public health laboratories suggests that speed, low running costs and versatility of PMS makes it suitable for the initial screening in outbreaks of infection.

MOLECULAR MICROBIOLOGY

Advances in molecular microbiology have added to diagnostic techniques, mainly with DNA probes or the polymerase chain reaction (PCR).

Probe-based technology

A DNA probe is a piece of single-stranded DNA that can recognize and consequently hybridize with a complementary DNA sequence. The probe also carries a label that 'lights up' the hybrid. Probes can be prepared from total cellular DNA, short-chain oligonucleotides or cloned DNA fragments and their specificity measured. There are many assay formats, although all follow the same principle whereby the strands of the DNA double helix of the target organism are separated by heating or alkali, and immobilized on a membrane. Immobilization prevents the complementary strands re-hybridizing and allows access for the probe.

After challenge with the probe the membrane is washed to remove any unbound probe and the resulting hybridization is visualized. The dot–blot assay immobilizes target DNA on a membrane and visualizes hybridization as a coloured spot. Colony–blot assays take up cells from a plate on to a sterile membrane, thereby forming a replica of that plate. The cells are then lysed to release the DNA, which is immobilized and challenged by the probe. Liquid hybridization formats use a probe that carries a chemifluorescent label, which is added directly to lysed cells. Adding a proprietary reagent breaks down non-hybridized probe. The amount of fluorescence measured is directly proportional to the number of target organisms. Paddle and bead formats carry the probe on their solid surfaces. After hybridization the unbound probe is washed off and detection is visualized. Several commercial kits are available for this type of examination (e.g. Gen-Probe Inc., San Diego, CA).

DNA fingerprinting

This is considered elsewhere in the book (see p. 103) because it is not yet fully automated.

Otherwise known as restriction fragment length polymorphism (RFLP) analysis, this combines probe and restriction enzyme technologies. The assay format is generally known as Southern blotting. Restriction enzymes cut DNA at sites characterized by short sequences of certain bases, resulting in a number of fragments of different lengths. The number and sizes of the fragments vary considerably from one organism to another, even within a species. The fragments can be separated according to size and transferred directly to a membrane by electrophoretic techniques. The immobilized fragments are then probed, resulting in a DNA fingerprint that is characteristic of the source DNA.

PCR technology: amplification-based technology

The best-established DNA amplification method is the polymerase chain reaction. This is based on the repetitive cycling of three simple reactions: denaturation of double-stranded DNA, annealing of single-stranded complementary oligonucleotides and extension of oligonucleotides to form a DNA copy. The conditions for these reactions vary only in incubation temperatures, all occurring in the same tube in a cascade manner. The repetitive cycling is therefore self-contained and can be automated in a programmable thermocycler. The first step is the heat denaturation of native DNA which melts as the hydrogen bonds of the double helix break. The single strands are then available to re-anneal with complementary-sequence DNA. The second step takes place at a reduced temperature. Two short DNA primers are annealed to complementary sequences on opposite strands of a template, thereby flanking the region to be amplified. They act as a starting point for DNA polymerase and consequently define the DNA region to be amplified. In the third step synthesis of complementary new DNA occurs through the extension of each annealed primer by the action of *Taq* polymerase in the presence of excess deoxyribonuclease triphosphates. The new strand formed consists of the primer at its 5′ end trailed by a string of linked nucleotides that are complementary to those of the corresponding template. An essential feature of the amplification process is that all previously synthesized products in the previous cycle act as templates for the subsequent cycle, resulting in geometric amplification. Twenty cycles, which take as little as 2 h, result in a millionfold increase in the amount of target DNA, which can be visualized by electrophoresis and staining with ethidium bromide, or by using a DNA probe.

The success of molecular instrumentation will be largely dependent on the development of appropriate sample extraction and purification technologies. Successful systems will need to integrate automated, high-throughput, extraction platforms with downstream detection systems. A number of systems are showing much promise. The QIAmp 96 DNA blood kit and BioRobot 9604 (Qiagen, Valencia, USA) are able to automate the isolation of DNA from blood, plasma, serum, bone marrow and body fluids. The process uses silica gel membrane technology for isolation of nucleic acids and a buffer system that allows for selective binding of nucleic acids to the membrane, washing and elution of the nucleic acids.

The BioRobot is reported to process up to 96 samples in microwell plate format within 2 h, and uses an automated tip change system to avoid cross-contamination and a barcode reader for sample identification.

The Biomek 2000 Laboratory Automation Workstation from Beckman Coulter (Fullerton, CA, USA) and the Dynal (Oslo, Norway) DNA DIRECT Auto 96 method based on magnetic-bead capture of DNA from whole blood is another promising high-throughput system. The Beckman Coulter system allows the simultaneous processing of up to 96 samples in 10 min.

Other manufacturers are marketing walk-away DNA extraction systems. Organon Teknika (Durhan, NC, USA) uses solid-phase technology in the form of silicon dioxide particles to bind nucleic acids. Autogen (Framingham, MA, USA) produces various models of automated DNA and RNA extraction systems, which are based on traditional nucleic acid extraction chemistry.

There are, however, many challenges to nucleic acid extraction. Automated instruments will need to consider quality control requirements before routine implementation of this technology can be achieved, and the ability to interface with high-throughput PCR systems will be an important factor. Roche Molecular Biochemicals (Indianapolis, IN, USA) and Applied Biosystems, Inc. (ABI; Foster City, CA, USA) have announced plans to launch automated nucleic acid extraction systems that interface with their corresponding systems for PCR and PCR product detection. The ABI PRISM 6700 instrument promises integration with the ABI PRISM 5700, and the Roche MagNApure will integrate with their LightCycler PCR instrument. Systems such as these can be programmed to automate both extraction and PCR set-up, providing a benchmark for high-throughput extraction technology.

Polymerase chain reaction technology is continually improving and now uses much shorter cycle times and improved formats for detecting PCR products with reduced incidence of cross-contamination.

These features are incorporated in some of the commercially available analytical systems. The semiautomated COBAS AMPLICOR system (Roche Diagnostics Corp., USA) uses microwell plates containing appropriate reagents and probes, and depends on a biotin-labelled, enzyme-linked method to detect amplified nucleic acids following multiple thermocycling steps. This system completes a run in about 4 h and is being used for qualitative testing of hepatitis C virus (HCV), *Mycobacterium tuberculosis*, *M. avium*, *M. intracellulare*, *Chlamydia trachomatis* and *Neisseria gonorrhoeae*, and quantitative testing of cytomegalovirus (CMV), hepatitis B virus (HBV), HCV and human immunodeficiency virus type 1 (HIV-1). Livengood and Wrenn (2001) have shown the system to be an accurate, rapid, and cost- and labour-efficient method for the detection of *Chlamydia trachomatis* and *Neisseria gonorrhoeae*.

'Real-time' PCR assays

'Rapid-cycle real-time PCR' is now available in several commercially available instruments, including the LightCycler from Roche Applied Science, the Smart Cycler from Cepheid in Sunnyvale (CA, USA) and the GeneAmp 5700 and its updated version, Prism 7700, from Applied Biosystems. Other rapid-cycle PCR instruments are available, but at this time there are few published clinical data with regard to their application for microbiology testing. The principal advantage of real-time PCR is that the formation of amplification products can be quantified and monitored in real time. The high sensitivity and short turnaround time of these systems could soon have a significant impact on patient care. Accuracy is improved and results are available rapidly, providing opportunities to diagnose infections sooner, improve patient care and decrease the overall costs of providing that care.

Microarray technology

Microarray technology will eventually be applicable to clinical microbiology laboratories and is likely to allow testing that may include detection of multiple gene targets for organism identification or drug-resistance patterns (Marshall and Hodgson, 1998; Tomb, 1998; Kozian and Kirschbaum, 1999; Diehn *et al.*, 2000). The basic technology revolves around DNA complementary to the genes of interest being generated and laid out in microscopic quantities on a solid surface at predetermined positions. DNA from samples is eluted over the surface, and complementary DNA binds and is generally detected by fluorescence after excitation.

Chips from several manufacturers are presently available. The GeneChip array (Affymetrix Santa Clara, CA, USA) has oligonucleotide probe sequences attached to a 1 cm^2 glass chip substrate. Each characterized probe is located in a specific and identifiable area on the probe array. The GeneChip instrument system can be purchased with a fluidics station, a hybridization oven, an Agilent GeneArray to measure emitted light, a PC workstation and GeneChip Data Analysis Suite software. Oligonucleotides can be placed on chips by a variety of technologies.

Current limitations of microarray technology include the high costs of instrumentation and disposables and our current inability for easy analysis of the enormous amount of information available through this technology. Production costs will decline over the coming years, although it is unlikely that this technology will play a significant role in diagnostic clinical microbiology in the immediate future. For a thorough review of suppliers of microarrays, see Cummings and Rehman (2000).

REFERENCES

Cantón, P-V. M., Pérez-Vázquez, M., Oliver, A. *et al.* (2001) Validation of the VITEK2 and the Advance Expert System with a collection of Enterobacteriaceae harboring extended spectrum or inhibitor resistant β-lactamases. *Diagnostic Microbiology and Infectious Disease* **41**: 65–70.

Cummings, C. A. and Relman, D. A. (2000) Using DNA microarrays to study host–microbe interactions. *Emerging Infectious Diseases* **6**: 513–525.

Diehn, M., Alizadeh A. A. and Brown P. O. (2000) Examining the living genome in health and disease with DNA microarrays. *Journal of the American Medical Association* **283**: 2298–2299.

Eden, R. and Eden, G. (1984) *Impedance Microbiology*. Herts: Research Studies Press Ltd.

Freeman, R., Goodfellow, M., Gould, F. K. *et al.* (1990) Pyrolysis mass spectrometry (Py-MS) for the rapid epidemiological typing of clinically significant bacterial pathogens. *Journal of Medical Microbiology* **23**: 283–286.

Hansen, D. S., Jensen, A. G., Nørskov-Lauritsen, N. *et al.* (2002) Direct identification and susceptibility testing of enteric bacilli from positive blood cultures using VITEK (GNI+/GNS-GA). *European Society of Clinical Microbiology and Infectious Diseases* **8**: 38–44.

Jorgensen J. H., Barry A. L., Traczewski M. M. *et al.* (2000). Rapid automated antimicrobial susceptibility testing of *Streptococcus pneumoniae* by use of the BioMerieux VITEK 2. *Journal of Clinical Microbiology* **38**: 2814–2818.

Joyanes P., Del Carmen Conejo M., Martinez-Martinez L. *et al.* (2001) Evaluation of the VITEK 2 System for the identification and susceptibility testing of three species of nonfermenting Gram-negative rods frequently isolated from clinical samples. *Journal of Clinical Microbiology* **39**: 3247–3253.

Kell, D. B. and Davey, C. L. (1990) Conductimetric and impedimetric devices. In: Cass, A. E. G. (ed.), *Biosensors: A practical approach*. Oxford: Oxford University Press, pp. 125–154.

Kelly, M. T., Matsen, J. M., Morello, J. A. *et al.* (1984) Collaborative clinical evaluation of the Autobac IDX system for identification of Gram-negative bacilli. *Journal of Clinical Microbiology* **19**: 529–533.

Kozian, D. H. and Kirschbaum, B. J. (1999) Comparative gene-expression analysis. *Trends in Biotechnology* **17**: 77.

Limond, A. and Griffiths, M. W. (1991) The use of the Bactofoss instrument to determine the microbial quality of raw milks and pasteurized products. *International Dairy Journal* **1**: 167–182.

Livengood III, C. H. and Wrenn, J. W. (2001) Evaluation of COBAS AMPLICOR (Roche): accuracy in detection of *Chlamydia trachomatis* and *Neisseria gonorrhoeae* by coamplification of endocervical specimens. *Journal of Clinical Microbiology* **39**: 2928–2932.

Livermore D. M., Struelens M., Amorim J. *et al.* (2002) Multicentre evaluation of the VITEK2 Advanced Expert System for interpretive reading of antimicrobial resistance tests. *Journal of Antimicrobial Chemotherapy* **49**: 289–300.

McAllister, J. M., Master, R. and Poupard, J. A. (1991) Comparison of the Microbial Identification System and the Rapid ANA II system for the identification of anaerobic bacteria. The 91st General Meeting of the American Society for Microbiology 1991 [abstract]. Washington DC: American Society for Microbiology.

Magee, J. T., Hindmarsh, J. M., Bennett, K. W. *et al.* (1989) A pyrolysis mass spectrometry study of fusobacteria. *Journal of Medical Microbiology* **28**: 227–236.

Marshall, A. and Hodgson, J. (1998) DNA chips: an array of possibilities. *National. Biotechnology* **16**: 27–31.

National Committee for Clinical Laboratory Standards (2000) *Performance Standards for Antimicrobial Disc Sensitivity Tests*. Document M2-A7, Vol. 17, No. 1. Villanova, PA: NCCLS.

Owens, J. D., Thompson, D. S. and Timmerman, A. W. (1989) Indirect conductimetry; a novel approach to the conductimetric enumeration of microbial populations. *Letters in Applied Microbiology* 9: 245–249.

Pettipher, G. L., Mansell, R., McKinnon, C. H. *et al.* (1980) Rapid membrane filtration epifluorescent microscopy technique for direct enumeration of bacteria in raw milk. *Applied and Environmental Microbiology* 39: 423–429.

Pettipher, G. L., Kroll, R. G., Farr, L. J. and Betts, R. P. (1989) DEFT: Recent developments for food and beverages. In: Stannard, C. J., Pettit, S. B. and Skinner, F. A. (eds), *Rapid Microbiological Methods for Foods, Beverages and Pharmaceuticals* Society for Applied Bacteriology Technical Series No. 25. Oxford: Blackwell, pp. 33–46.

Phillips, I., Andrews, J., Bint, P. *et al.* (1999) A guide to sensitivity testing. Report of a Working Party on Antibiotic Sensitivity Testing of the British Society for Antimicrobial Chemotherapy. *Journal of Antimicrobial Chemotherapy* 22(suppl D): 1–50.

Sisson, P. R., Freeman, R., Magee, J. G. and Lightfoot, N. F. (1992) Rapid differentiation of *Mycobacterium xenopi* from mycobacteria of the *Mycobacterium avium-intracellulare* complex by pyrolysis mass spectrometry. *Journal of Clinical Pathology* 45: 355–370.

Stager, C. E. and Davis, J. R. (1992) Automated systems for identification of microorganisms. *Clinical Microbiology Reviews* 5: 302–327.

Stannard, C. J. and Wood, J. M. (1983) The rapid estimation of microbial contamination of raw beef meat by measurement of adenosine triphosphate (ATP). *Journal of Applied Bacteriology* 55: 429–438.

Stevens, M., Feltham, R. K. A., Schneider, F. *et al.* (1984) A collaborative evaluation of a rapid automated bacterial identification system: the Autobac IDX. *European Journal of Clinical Microbiology* 3: 419–423.

Stewart, G. S. A. B. and Williams, P. (1992) *Lux* genes and the applications of bacterial bioluminescence. *Journal of General Microbiology* 138: 1289–1300.

Thorpe, T. C., Wilson, M. L., Turner, J. E. *et al.* (1990) BacT/Alert: an automated colorimetric microbial detection system. *Journal of Clinical Microbiology* 28: 1608–1612.

Tomb, J-F. (1998) A panoramic view of bacterial transcription. *National Biotechnology* 16: 23.

Truant, A. L., Starr, E., Nevel, C. A., Tsolakis, M. and Fiss, E. F. (1989) Comparison of AMS-Vitek, MicroScan, and Autobac Series II for the identification of Gram-negative bacilli. *Diagnostic Microbiology and Infectious Diseases* 12: 211–215.

Wilson, M. L., Weinstein, M. P., Reimer, L. G. *et al.* (1992) Controlled comparison of the BacT/Alert and Bactec 660/730. Nonradiometric blood culture systems. *Journal of Clinical Microbiology* 39: 323–329.

Zaman, Z., Roggeman, S. and Verhaegen, J. (2001) Unsatisfactory performance of flow cytometer UF-100 and urine strips in predicting outcome of urine cultures. *Journal of Clinical Microbiology* 39: 4169–4171.

9

Mycological methods

DIRECT EXAMINATION

Microscopic examination is a crucial part of the diagnostic process for most fungal infections and observation of fungal elements in clinical material can greatly enhance the significance of a subsequent isolation of a yeast or mould. Moreover, such examination can lead to a rapid presumptive diagnosis, often days before a positive culture is obtained. In tissue specimens such as biopsies, and nail and skin scrapings, the presence of hyphae is diagnostic, but it is important to culture the material to establish the identity of the pathogen. In some cases the pathogen may fail to grow in culture and the microscopy result will be the only evidence of a fungal aetiology.

Arrange hair, skin or nail fragments on a slide in a drop of mounting fluid (see below). Apply a cover glass and then leave for a few minutes or, in the case of nail specimens, up to 30 minutes, until the preparation softens and 'clears'. Then apply gentle pressure to the top of the cover glass to produce a monolayer of cells for microscopic examination.

Use the same procedure for small fragments of tissue from biopsy samples and the residue from centrifuged Sputasol-treated sputum, brochoalveolar lavage (BAL), peritoneal dialysis or other body fluids. Make wet films of pus or other exudates in equal volumes of potassium hydroxide (KOH) and glycerol if a more permanent preparation is required. Mix cerebrospinal fluid (CSF) deposit with an equal volume of 2.5% nigrosin in 50% glycerol or Indian ink. Glycerol prevents mould growth in the solution, delineates capsules more sharply and prevents preparations, which should be thin, from drying out. Smear swabs from cases of mucosal candidiasis on a glass slide and stain by Gram's method. As the small budding yeast cells of *Histoplasma capsulatum* are difficult to detect in wet mounts, add Calcofluor white, 0.1% in distilled water or examine Giemsa-stained preparations.

Fluorescence microscopy

An optical brightener, such as Calcofluor white, may be used to enhance the natural fluorescence of fungal cell walls. Mixed in equal volumes with KOH, this will facilitate the microscopic examination of specimens because it will bind to the chitin in the fungal cell walls. Examine preparations under a fluorescence microscope fitted with an ultraviolet (UV) light source, and appropriate excitation and barrier filters.

Mounting fluid and stains

Potassium hydroxide

Use a 20% solution of KOH in water to 'clear' specimens so that fungal mycelia and yeast cells may be more easily seen.

Lactophenol–cotton blue

This is commercially available or may prepared as follows.

Lactophenol

Phenol crystals	20 g
Lactic acid	20 ml
Glycerol	40 ml
Distilled water	20 ml

Heat gently to dissolve.

Lactophenol–cotton blue

Cotton blue	0.075 g
Lactophenol	100 ml

Store away from direct sunlight.

Lactofuchsin

Acid fuchsin	0.1 g
Lactic acid	100 ml

Store away from direct sunlight.

Mount small portions of moulds from foods in water or, if the material is water repellent, in lactophenol with or without cotton blue.

Polyvinyl alcohol-mounting fluid

Polyvinyl alcohol granules	17 g
Lactic acid 85%	80 ml
Distilled water	20 ml

Heat the polyvinyl alcohol (PVA) in the distilled water in a 55°C water bath with frequent stirring. Add the lactic acid and transfer to a boiling water bath, stirring occasionally. The resulting clear viscous fluid may be kept for several years at room temperature.

Direct methods for *Pneumocystis carinii* (*Pneumocystis jiroveci*)

Cysts of *P. carinii* in induced sputum or BAL samples may be stained with Grocott's methenamine silver (see original paper for method: Grocott, 1955), toluidine blue-O or cresyl violet. The trophozoites stain less readily with these agents and Giemsa, Diff-Quik or Gram–Weigart stains are used instead. Currently, there are several kits available for indirect immunofluorescence staining using monoclonal antibodies raised against *P.*

carinii, which work well although there may be some false positives as a result of cross-reactions.

Molecular methods are under development for the detection of this organism.

Histological examination

Histological examination of biopsy material can help to distinguish hyalohyphomycosis (infections with moulds with hyaline or colourless hyphae), phaeohyphomycosis (infections with moulds with pigmented hyphae), zygomycosis in which the hyphae are broad and aseptate, and yeast infections. There are several stains that can be used to enhance the microscopic appearance. Grocott's methenamine silver stain (Grocott, 1955), a specific stain for fungal cell walls, is the most useful. This stains fungal cell walls dark brown or black whereas the host tissue is counterstained green. Haematoxylin and eosin (H and E) produces less distinction between fungus and host cells, but may be useful for Zygomycetes and in the staining of mycetoma granules. Mayer's mucicarmine specifically stains the polysaccharide capsular material of *Cryptococcus neoformans*. Periodic acid–Schiff (PAS) is also used to stain cryptococci. Zygomycete hyphae, unlike the hyphae of other moulds, are specifically stained by cresyl fast violet. Giemsa stain is useful for *Histoplasma capsulatum* and may also be used on fresh specimens. Fungal immunostaining with antibodies raised to cell wall antigens can also be employed using immunofluorescence (bright green) or immunoperoxidase (dark). Such staining can be useful for diagnostic purposes when the organism has not been isolated or when only formalin-fixed tissue is available.

Direct examination of mould growths on foods and other perishable materials

Examine with a lens or low-power microscope *in situ* to see the arrangement of spores, etc. before disturbing the growth. Remove a small piece of mycelium with a wire or needle. It may help if the end of the needle has been bent at right angles to give a short, sharp hook. Use the point of the needle to cut out a piece of the growth near the edge where sporulation is just beginning. Transfer it to a drop of lactophenol–cotton blue, lactofuchsin or PVA mountant on a slide, apply a cover slip and then examine after 30 min when the stain has penetrated the hyphae. If bubbles are present warm gently to expel them.

ISOLATION OF MOULDS AND YEASTS

Primary isolation

Most fungi are not particularly fastidious in their nutritional requirements and will grow quite readily on many bacteriological agars. However, the medium most commonly employed for the isolation of fungi is glucose peptone agar (Sabouraud's agar) containing 0.5 g/100 ml chloramphenicol. If the medium is to be used to isolate dermatophytes add cycloheximide to inhibit overgrowth by contaminating moulds. Use glucose peptone agar, malt agar or one of the proprietary media. Note, however, that the culture media has a marked influence on the colonial growth of most fungi, particularly the dermatophytes, and most of the identification manuals describe or illustrate the colonial growth on glucose peptone agar. Place about 20 pieces of hair or skin into the surface of the medium. Do not push them into the agar. Cut nail clippings into small pieces with a scalpel before culturing. Pieces of skin, hair or nail powder may be picked up more easily if the heated inoculating needle is first pushed into the medium to moisten it.

Spread pus, CSF, sputum or BAL deposits, or biopsy material from suspected cases of the invasive mycoses, on slopes of glucose peptone agar with chloramphenicol. Medical flats containing this agar may be used because they provide a larger surface area to increase the chance of isolation.

Plates may also be used but they are more susceptible to contamination by environmental moulds, and offer less protection in the case of isolation of an unexpected Hazard Group 3 pathogen. It is also useful to set up a purity culture at the same time as the sample to alert one to the presence of contaminating moulds in the culture environment. Cut tissues into small pieces. Do not homogenize them because this would break fungal mycelium into non-viable fragments. This is particularly true for the Zygomycetes, which can be difficult to culture from necrotic tissue even when the hyphae have been seen on microscopic examination.

Wash mycetoma grains several times with sterile saline to remove most contaminants before culturing.

Incubate cultures at 28°C but if the material is from a deep body site incubate also at 37°C to hasten growth of the pathogen. Most yeasts grow in 2–3 days, although *Cryptococcus* spp. may require incubation for up to 6 weeks. Common moulds, too, will grow in a few days. Cultures for dermatophytes should be kept for 2 weeks although most are identifiable after 1 week. It is good practice to examine all mould cultures once or twice a week and ensure adequate aeration.

Dimorphic Hazard Group 3 pathogens

Specimens suspected of containing Hazard Group 3 pathogens should be handled in the appropriate containment Level 3 Laboratory. The above methods may be used but slopes or medical flats rather than plates should be used for isolation purposes. If the suspected pathogen is one of the group of dimorphic fungi, an additional culture on a slope of brain–heart infusion agar incubated at 37°C may yield the yeast form of the fungus and help to confirm its identity.

Isolation from food, soil, plant material, etc.

Inoculate plates of chloramphenicol-containing glucose peptone agar or other selective media or use dichloran glycerol (DG18) medium.

Dip slides, coated with appropriate media, are useful for liquid samples or homogenates and are available commercially. Zygomycetes, if present, will outgrow everything else but are markedly restrained when grown on Czapek-Dox agar. To grow moulds from physiologically dry materials such as jam, increase the sucrose content of the Czapek-Dox agar to 20%. Some yeasts can tolerate low pH levels and may be isolated on malt agar to which 1% lactic acid has been added after melting and cooling to 50°C. Incubate cultures at room temperature and at 30°C.

Enumeration of moulds in foods, etc.

Make a 10% suspension of the material in sterile water, Ringer's or peptone water diluent. Process in a Stomacher (other blenders may raise the temperature and injure the moulds). Make tenfold dilutions and surface plate on selective media (e.g. Aspergillus Flavus Parasiticus agar [AFPA], oxytetracycline yeast extract agar [OGYE] or, for xerophilic moulds, DG18). Incubate and count colonies on plates that have between 50 and 100. Calculate the numbers of mould propagules per gram.

Methods for individual foods are given in Chapter 14. Useful sources of information are Pitt and Hocking (1985), King et al. (1986), Krogh (1987), and Sampson and van Reenan-Hoekstra (1988).

Sampling air for mould spores

Various air-sampling instruments are described in Chapter 19. A cheap, simple device is the Porton impinger in which particles from a measured volume of air are trapped in a liquid medium on which viable counts are done. More information is gained by using the Andersen sampler, which captures the airborne particles using the principle that the higher the air speed the smaller the particle that will escape the airstream and be impacted on an agar surface. Plastic Petri dishes give lower counts than glass ones but they make the whole machine much easier to handle.

The medium and time of incubation will depend on the organism sought. For general use one run with Czapek-Dox agar and one with chloramphenicol glucose peptone or malt agar will serve; for thermophilic actinomycetes use glucose peptone agar without antibiotics; incubate one set at 40°C, the other at 50°C.

As invasive aspergillosis is often a nosocomial infection and high-risk patients may be subjected to the inhalation of potentially lethal fungal spores within the hospital environment, it is helpful to monitor for the presence of fungal spores in some patient areas. The air-sampling techniques outlined above may be employed but will provide only a snapshot of the spores that were airborne at the time of sampling and lack sensitivity unless a large volume of air is sampled. A longitudinal approach is to sample horizontal surfaces for the presence of fungal spores. This can be made more quantitative by sampling a given area in a room. Moisten a swab in sterile distilled water, wipe it over an area of 0.5 m² and on the surface of a Sabouraud agar plate containing chloramphenicol. Incubation of the plate at 45°C for 48 h will select for Aspergillus fumigatus. Incubation at 28°C will also allow the growth of other environmental moulds. A modification of the settle plate method may also be employed, in which a high-sided, screw-capped receptacle with at least a 5-cm aperture ('honey pot') is left open at a particular point in the area to be sampled for 7 days. After this time any spores that it contains are harvested by adding a volume of 4 ml sterile saline with 0.01% Tween 80, closing the lid, shaking and then plating on to Sabouraud's agar with chloramphenicol and incubating as before.

IDENTIFICATION OF MOULDS

Most molds may be identified by examining microscopic and macroscopic morphology. Colony form, site, surface colour and pigmentation are useful features but microscopic examination is essential. Well-sporing mould colonies may usually be readily identified. If spores are not seen, re-incubate and/or subculture on to other media (potato sucrose agar, malt agar, Borelli's lactrimel agar and half cornmeal agar are all useful for stimulating sporulation). Sporulation may sometimes be stimulated by incubation under near-ultraviolet ('black') light or by exposing the colonies to alternating cycles of light and dark (diurnal cycles).

The usual methods of examining sporing structures are needle mounts and/or tape mounts. To prepare a needle mount take a small fragment of mycelium from the colony with a sharp needle and tease it out in a drop of mounting fluid. Place a cover glass over the specimen and tap it a few times to help to spread the mycelial fragments within the mounting fluid. The spores are usually most visible at the edges of the mycelial mass.

Tape mounts have the advantage that spores can be viewed without disruption of their normal configuration. Cut a strip about 1.5 cm long from a roll of sticky tape preferably using double-sided tape and a plastic dispenser. Attach one end to a mounted needle, forming a flag. Touch the surface of the colony with this near the growing edge. To improve the optics mount the tape, sticky side up, in a drop of mountant, add another drop of mountant and overlay this with a cover glass, so that the structures are not being examined through a layer of tape. Examine immediately, because tape mounts are not suitable for long-term preservation: the mounting fluid renders the sticky tape opaque.

The centre of a well-sporing colony may show nothing but spores, in which case it may be desirable to make a second preparation nearer to the edge of the colony or to try a different preparation technique such as a needle mount. Conversely, a poorly sporing colony may show spores only in its centre and it may be necessary to dig down into the agar with the tip of a needle. It will frequently be necessary to make several mounts to be able to discern the sporing structures and precise mechanism of spore production.

Slide culture

This enables the spores and sporing structures to be examined *in situ* and permits a more critical exam-

ination. Cut small blocks of medium (0.5 × 0.5 cm) from an appropriate agar-poured plate and transfer them to the centres of sterile slides. Inoculate the edges of each block with material from a well-growing colony and apply a sterile cover glass on top of the block. Incubate in a Petri dish containing moist filter paper and glass rods to support the slide (wet chamber). Once visible growth has developed, monitor its progress with the low-power objective of the microscope. When typical structures are visible prepare a slide and cover slip with a drop of mountant. Lift the cover glass carefully off the slide culture and mount it on the prepared slide. Lift off and discard the agar block. Place a drop of mounting fluid on the original slide to produce a second mount. Examine both preparations after 30 min.

Exoantigen identification methods and Gene–Probe

The exoantigen test (Immuno-Mycologics, Inc.) is designed to permit the identification of the dimorphic fungal pathogens *Histoplasma capsulatum*, *Coccidioides immitis* and *Blastomyces dermatitidis*. The method consists of a double-diffusion test in which an antiserum specific to each of these three fungi is run against an extract of the isolate to be identified. The production of precipitin lines with one of the control extracts of the three pathogens confirms the identification. Another commercial method, Gen-Probe, has also been developed for these Hazard Group 3 pathogens.

Molecular identification methods

Molecular identification methods have the advantage of speed and may allow identification of non-sporing moulds and the differentiation of morphologically similar groups of yeasts and moulds. They may also prove useful in the identification of yeasts and moulds in tissue sections when culture of the organism has failed or when only formalin-fixed tissue sections are available. Amplification of the fungal DNA is carried out by polymerase chain reaction (PCR) and then subjected to a series of species-specific probes or DNA sequencing, e.g. various primers have been developed for the PCR amplification of fungal genomic ribosomal DNA sequences. Species-specific probes can then be used to detect matching sections or the precise order of bases may be sequenced and compared with stored database sequences for different fungal species. These methods of mould and yeast identification are in various stages of development and a careful morphological examination, and in the case of yeasts analysis of biochemical profiles, is an essential adjunct to confirm the identification.

Molecular strain typing

Examination of the degree of similarity of fungi of the same species may be a useful tool for tracing the source of infection in an outbreak. It can also help to track the development of drug-resistant strains over time in a given population or help to establish whether a current infection is re-infection or relapse. Most methods based on phenotypic characteristics have proved disappointing and recent developments have concentrated on an assessment of genetic relatedness. Various methods have been examined, including restriction fragment length polymorphism (RFLP), random amplified polymorphic DNA (RAPD), multilocus enzyme electrophoresis (MEE), probes based on repetitive elements, analysis of microsatellite regions and electrophoretic karyotyping. Numerous publications attest to the relative merits of each approach.

EXAMINATION OF YEAST CULTURES

Microscopic morphology

To study yeast morphology use a Dalmau plate comprising cornmeal agar with 1% Tween 80. Streak the organism on a plate of the medium, cover with a sterile cover glass and incubate at 30°C. Examine after 3–4 days through the cover glass with × 10 and × 40 objectives. Look for colony colour, cell shape, and the presence or absence of mycelia, pseudomycelia, arthrospores, chlamydospores and capsules.

To encourage ascospore formation inoculate sodium acetate agar made by adjusting the pH of 0.5% sodium acetate to 6.5 with acetic acid, then solidifying it with agar. Stain by the malachite green–safranin method. Prepare a smear in a drop of water and dry in air. Fix by heat and cover with 1% malachite green for 2 min. Heat until steam rises, wash with tap water and counterstain with safranin for 1 min. Wash, blot and dry. Ascospores are stained green and vegetative cells red.

The germ tube test

A test for the production of a true mycelium can differentiate *Candida albicans* and the recently described *Candida dubliniensis* from other yeasts. Add a small portion of a colony (over inoculation can lead to autoinhibition of germ tube production) to a tube containing 0.5 ml horse serum and incubate at 37°C for 3 h. Then remove a drop to view under a microscope. The production of a short germ tube initially without constriction at the point of emergence from the blastospore is diagnostic for these two species.

Fermentation tests

These differ from those used in bacteriology in that the sugar concentration is 3%. Durham tubes are always used to detect gas formation and a 10-ml volume of medium is preferable to encourage fermentation rather than oxidation.

Inoculate a set of the following sugar media: glucose, maltose, lactose, raffinose and galactose. Incubate cultures at 25–30°C for at least 7 days. Gas production indicates fermentation.

Auxanograms

Fermentation tests alone are insufficient to identify species of yeasts and assimilation tests should also be performed. Culture the test organism on a slope of pre-assimilation medium. Prepare a heavy suspension by adding about 5 ml sterile water to the slope and gently suspend the growth; it is not necessary to wash the organisms. Melt tubes of auxanogram carbon-free base (for carbon assimilation) and nitrogen-free base (for nitrogen assimilation) and cool to 45°C. Add about 0.25 ml of the yeast suspension to each 20 ml auxanogram medium and immediately prepare pour plates in 9-cm Petri dishes. When set, place up to five discs of carbon or nitrogen sources well apart on the surface of the appropriate medium. Incubate at 28°C and examine the plates daily for up to 7 days for a halo of growth around the substrate.

Carbon sources commonly used are glucose (control), maltose, sucrose, lactose, inositol, galactose, raffinose, mannitol and cellobiose. Nitrogen sources include sodium nitrate, asparagine and ethylamine hydrochloride. Alternatively prepare slopes (or tubes of liquid) of a basal medium containing only the test sugar and an indicator. Inoculate with a suspension of the test organism taken from pre-assimilation medium and note the presence or absence of growth for a 3-week period. Much more extensive series of test sugars may be employed to characterize the physiological properties of the test yeast and these may be analysed by reference to specialist texts, which, however, are beyond the scope of most routine microbiology laboratories.

Commercial 'kit' tests for yeast identification

There are several commercial kits that are satisfactory for the identification of most strains of yeast encountered in clinical laboratories. Some use methods based on traditional criteria, whereas others rely on novel approaches. In all of them, however, the databases are from a restricted list of species and occasional misidentifications may occur. When using the kits, therefore, it is crucial to examine the yeast morphology on a cornmeal agar plate; for many of them this forms an important part of the numerical identification profile. Examination of the morphology can help to minimize the danger of misidentification.

Chromogenic agars

There are commercial agars incorporating chromogenic substrates which lead to the production of different colony colours for different yeast species. Such agars may be used as the basis for identification of a limited range of yeast species and are particularly useful for detecting mixed cultures.

Mycotoxins and seed-borne moulds

Mycotoxins are metabolites of fungi, which may be produced during mould growth on foods and animal feeds. The most heavily contaminated commodities are usually cereals and oilseeds. The mycotoxigenic moulds most frequently encountered belong to the genera *Penicillium*, *Aspergillus* and *Fusarium* (Moss, 1989). Repeated ingestion of aflatoxin, produced by *Aspergillus flavus*, probably has the most serious long-term consequences, because it is a powerful carcinogen.

Surface sterilize seeds by immersion for 2 min in sodium hypochlorite (0.4% available chlorine). Place up to 10 small or 5 large seeds on DG-18 agar in a Petri dish and incubate at room temperature. Subculture from colonies to oatmeal agar (*Aspergillus*), Czapek yeast autolysate agar (*Penicillium*) or synthetic nutrient-poor agar (*Fusarium*).

SEROLOGICAL METHODS

Antigen production

There is now little call for laboratories to produce their own antigen preparations because commercially produced, standardized and quality controlled antigens are now available to test for antibody production to a wide range of fungal pathogens.

Antibody detection

Antibody detection is useful in immunocompetent patients but is of less benefit for patients who have an invasive disease and often have an underlying immunological deficiency. Detection of antibody to *Aspergillus* spp. can help to confirm aspergilloma, allergic aspergillosis, acute bronchopulmonary aspergillosis and aspergillus endocarditis. Detection of antibodies to *Candida* spp. is the single most consistent diagnostic finding in patients with candida endocarditis and can also be useful in patients with invasive candida infections after surgery where an antibody titre of $\geq 1:8$ is usually indicative of active infection. Complement fixation and immunodiffusion antibody tests are useful in the diagnosis of blastomycosis, histoplasmosis and coccidioidomycosis. Both yeast and mycelial phase antigens of *Histoplasma* are used. All the reagents required for these tests are commercially available.

Immunodiffusion

In the double diffusion (DD) test, serum and antigen are allowed to diffuse towards each other from wells cut in an agar gel. Lines of precipitation form after several days where individual antibodies and antigen meet at optimum concentration. Counterimmunoelectrophoresis (CIE) provides a more rapid and sensitive method for the detection of the precipitins. It differs from DD in that an electric current is passed through the agar gel for 90 min to hasten the migration process. In CIE tests, conditions are chosen in which antigens migrate towards the anode because their isoelectric points are lower than the pH of the buffer, and antibodies migrate towards the cathode as a consequence of electroendosmosis. Non-specific reactions and other arte-

facts resembling precipitin lines are more common and more obvious with CIE than DD.

Agar base

Dissolve 2 g agar (Oxoid No. 1) in 100 ml water by autoclaving. Add 100 ml buffer heated to 50°C and mix with the agar. Keep in a water bath at 50°C and discard if not used within 48 h of preparation.

Buffer

Boric acid (H_2BO_3)	10 g
Powdered borax ($Na_2B_4O_7 \cdot 10H_2O$)	20 g
EDTA, disodium salt	10 g
Water to	1000 ml

The pH should be 8.2. This buffer is double strength. Dilute it with agar for plates and slides or with an equal volume of water for use in electrophoresis tanks.

Plates

Place 30 ml agar into a plastic Petri dish and fit a Perspex jig with metal pegs in place of the lid to produce the pattern of wells shown in Figure 9.1. The large wells are 6 mm in diameter and the smaller wells 2 mm. The distance between the central and peripheral wells is also 6 mm.

The test serum occupies the central well, antigens

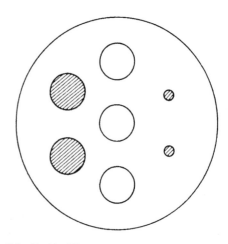

Figure 9.1 Double diffusion tests

the pairs of large and small wells (shown shaded), and appropriate control sera the top and bottom wells. Two antigens can thus be tested and a single solution suffices because the relative volumes (60 and 6 µl in the two holes) give high- and low-concentration gradients in the agar. This arrangement also allows reactions of identity to be obtained between test and control sera, thus eliminating some anomalous reactions, particularly among *Aspergillus* spp.

Counterimmunoelectrophoresis

Buffered gel

Veronal buffer	4 l
Purified agar	20 g
Agarose	20 g
Sodium azide	2 g

Add agarose and purified agar to buffer and steam at 100°C for 1 h. Add 2 g sodium azide, cool to 56°C and adjust pH to 8.2. Dispense in 200-ml amounts.

Veronal buffer 0.05 mol/l

Barbitone (*caution*)	13.76 g
Barbitone sodium (*caution*)	30.28 g
Distilled water	4.0 l

Adjust pH to 8.2 before use.

Saline–trisodium citrate

Sodium chloride	40 g
Trisodium citrate	100 g
Sodium azide	1 g
Distilled water	2.0 l

Buffalo black staining solution

Buffalo black	0.5 g
Distilled water	500 ml
Ethanol	400 ml
Acetic acid	100 ml

Gel destaining solution

Methanol	900 ml
Glacial acetic acid	200 ml
Distilled water	900 ml

Method

Two gels are required for *Aspergillus fumigatus* precipitins (●): one for the somatic antigen at low and high concentrations, and the second for the culture filtrate antigen at low and high concentrations (Figure 9.2). These should be labelled 1 and 2, 3 and 4, etc. depending on the number of tests. The top right-hand corner of each gel should be labelled with alcian blue.

- All wells marked ● must be filled with 10 µl of the appropriate antigen: 2 mg/ml (low concentration) or 20 mg/ml (high concentration).
- All wells marked ○ must be filled with 10-µl volumes of the patient's serum.
- The first row of each gel must contain the positive control serum.

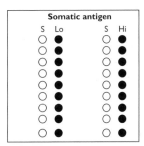

Figure 9.2 CIE template

Place the slide in an electrophoresis tank containing 0.05 mol/l veronal buffer. Connect the slide to the buffer with lint wicks, ensuring that good contact is made along the full width of the gel. The antigen wells should be adjacent to the cathode and the serum wells adjacent to the anode. Set the equipment to deliver a constant voltage of 30 V per slide and run for 90 min.

Once the gel has been run, immerse in a Petri dish containing saline–trisodium citrate for 18 h (overnight). Rinse the gel with tap water, cover it with a moistened filter paper square, and dry with a hair dryer or inside a heated cabinet for about 1 h. Once the gel has dried, moisten the filter paper with tap water, remove it and rinse the gel. Dry once more for about 10 min.

To stain the precipitin lines immerse the slide for

10 min in Buffalo black solution followed by two changes of destaining solution until the background is colourless.

If rapid results are required, wash the wet gel for 1 h and examine by dark-ground illumination. This gives a preliminary result only and must be confirmed after washing and staining.

Care must be taken in interpretation of the tests. Clearly defined lines between the serum and antigen wells are an indication of a positive result and the number of lines seen is an indication of the strength of the reaction. If necessary, the serum sample can then be tested at increasing dilutions in a doubling dilution series to determine the titre, which is the highest dilution at which an antibody–antigen precipitin line is still visible. Lines that appear as blurred smudges are unlikely to be significant, and also those that follow around the edges of the well.

Enzyme-linked immunosorbent assay test for antibody detection

The enzyme-linked immunosorbent assay (ELISA) test is a sensitive method and can be used to detect either antibody or antigen. To detect antibody the wells of a 96-well microtitre plate are coated with the appropriate antigen. Antigens are available from commercial sources and include both mycelial (somatic) and culture filtrate preparations. After coating, plates are washed three times in phosphate-buffered saline (PBS)–Tween, then 0.1-ml volumes of test and reference sera diluted 1 : 100 in PBS–Tween are added to designated wells. After incubation at room temperature for 2 hours the washing steps are repeated. Various detection methods can be used at this stage, including goat or rabbit anti-human IgG coupled to an enzyme, usually alkaline phosphatase or peroxidase, added in 0.1-ml volumes in PBS–Tween with 1% bovine serum albumin and incubated at room temperature. After 2 h, the washing steps are repeated and 0.1 ml of the appropriate enzyme substrate is added and the reaction is allowed to continue for

30 min, then stopped by the addition of 50 ml of 1 mol/l sulphuric acid. The absorbance value of each well is determined and results can be calculated by reference to the control sera.

There have been many modifications to this basic method including incubation at 37°C to speed up reaction times and the introduction of a further step known as sandwich ELISA to enhance sensitivity.

Commercial antibody detection methods

In addition to the commercial availability of a large array of fungal antigens and antibodies, some companies have developed methods for the detection of antibodies to *Candida* and *Aspergillus* spp. Commercial ELISA kits are available as well as others using the principle of haemagglutination.

Antigen detection

Commercial antigen detection kits are available for the early diagnosis of infections by *Cryptococcus, Candida* and *Aspergillus* spp. Latex agglutination kits have been available for the detection of cryptococcal antigen for more than 20 years and have sensitivity and specificity that are > 90%. The kit can be used with serum or CSF samples. It is advisable to test neat serum or CSF at a 1 : 10 dilution in an initial test to overcome potential problems with the prozone effect caused by an excess of antigen. In cases of a positive reaction, doubling dilutions should be tested until an endpoint is reached. High or rising titres are indicative of progressive infection and in some patient groups monitoring of antigen titres is a useful prognostic indicator. Antibody detection is less useful in the initial diagnosis of cryptococcosis, but can be indicative of a good prognosis because it is often detectable only once an infection has almost resolved.

Latex agglutination tests have proved less successful for the detection of infections with *Aspergillus* and *Candida* spp. because they have suffered from a lack of sensitivity and specificity. However, there are more sensitive commercial sandwich ELISA tests for the detection of aspergillus galactomannan and candida oligomannosides. These tests have proved useful in certain patient groups and show promise in the diagnosis or confirmation of invasive infection in some immunocompromised patients.

For further information on the serodiagnosis of fungal diseases, see MacKenzie *et al.* (1980).

MOLECULAR METHODS IN DIAGNOSIS

Currently molecular methods for the detection of circulating fungal genomic sequences are under evaluation for the diagnosis of invasive fungal infection. Panfungal, as well as species-specific, primers are employed with various detection methods including nested PCR, PCR–ELISA, PCR–blot, real-time PCR with light-cycler analysis and PCR sequencing. Test specimens include whole blood, serum, CSF, BAL fluid and tissue samples. Published DNA extraction methods vary from physical freeze–thaw techniques to enzymatic extraction. Development is still required to produce standardized tests with acceptable levels of sensitivity, specificity and reproducibility.

ANTIFUNGAL DRUGS

There are several different techniques for assessing the susceptibility of an organism to an antifungal agent. The simplest method, which is suitable for those agents that are water soluble, is based on agar diffusion. There are disc diffusion methods for flucytosine (5-fluorocytosine), fluconazole and voriconazole. This approach has been further refined in the commercially available E test.

Recent years have seen much effort expended on the development of reproducible, standardized methods for the susceptibility testing of yeasts and moulds. The National Committee of Clinical Laboratory Standards (NCCLS, 2002a, 2002b) has produced two approved methods: M27-A2 for yeasts and M38-A for filamentous fungi. Both methods are based on a broth dilution format that can be miniaturized and performed in microtitre plates. The test conditions are as follows.

Test conditions for M27-A2 and M38-A

Medium	RPMI 1640 with 0.165 mol/l MOPS pH 7.0 (Sigma)
Drug solvents	Dimethylsulphoxide (DMSO) or water
Inoculum (cells/ml)	M27-A2 for yeasts 0.5–2.5 × 10³ M38-A for moulds 0.4–5 × 10⁴
Incubation temperature	35°C
Incubation duration	48 h (rarely 72 h)
Endpoint readings	Complete inhibition, 80% inhibition or 50% inhibition depending on drug–organism combination
QC and reference organisms	*Candida albicans* ATCC 90028 *C. albicans* ATCC 24433 *C. krusei* ATCC 6258 *C. parapsilosis* ATCC 22019 *C. tropicalis* ATCC 750 *Aspergillus flavus* ATCC 204304 *A. fumigatus* ATCC 204305

Drug dilutions are prepared and placed into either tubes (1-ml final volume) or the wells of microtitre plates (0.2-ml final volume). The inoculum (0.1 ml) is added, tubes or plates are incubated at 35°C and read after 48 h, although it is sometimes possible to read after 24 hours. The endpoint is the lowest concentration at which there is complete inhibition of growth (amphotericin B and all drugs except flucytosine, fluconazole and ketoconazole in M38-A), 80% (flucytosine and azoles in M27-A) or 50% reduction in turbidity compared with a drug-free control (flucytosine and azoles in M38-A). There are proposed breakpoints for flucytosine, fluconazole and itraconazole for yeasts based on *in vitro–in vivo* correlations (Table 9.1).

Table 9.1 Proposed breakpoints (milligrams per litre) for antifungal susceptibility tests

	Susceptible	SDD	Resistant
Flucytosine	< 4.0	8.0–16.0	< 32.0
Fluconazole	< 8.0	16.0–32.0	< 64.0
Itraconazole	< 0.125	0.25–0.5	< 1.0

SDD, susceptible dependent on dose (intermediate)

Disc test sensitivity methods

Disc diffusion methods indicate relative susceptibility depending on zone size and do not provide a minimum inhibitory concentration (MIC). These are still useful for flucytosine, fluconazole and voriconazole which are relatively water soluble, although care should be exercised in selecting the basic medium because components of some complex media can obfuscate the antifungal activity.

Yeast morphology agar (YMA) can be used for flucytosine and high-resolution medium (HR) or Mueller–Hinton agar for fluconazole and voriconazole. The appropriate control organisms are *Candida kefyr* National Collection of Pathogenic Fungi (NCPF) 3234 for flucytosine and *Candida albicans* American Type Culture Collection (ATCC) 90028 for fluconazole and voriconazole.

Flucytosine

Invert the plate and bisect it with a line drawn on the base. Use a sterile swab to inoculate half of the plate with the control organism and the other half with the test organism, both at an inoculum concentration of 1×10^6 cells/ml in sterile water. Once the agar has dried, use sterile forceps to place a drug disc (1 μg) in the centre of each plate. Incubate the plate at 30°C for 48 h, but examine after 24 h for obvious resistance. If the inhibition zone for test strain is 80% or more of that for the control strain, the organism may be considered sensitive to flucytosine. Resistant isolates should be tested by another method to determine the precise MIC of the drug. Truly resistant organisms will often grow right up to the disc; intermediate zones or discrete colonies forming within the zone suggest the possibility of emergent resistance during therapy.

Fluconazole

Inoculate Mueller–Hinton agar supplemented with 2% glucose and 0.5% methylene blue with a yeast inoculum adjusted to a 0.5 McFarland standard. Once the agar has dried place a 25-μg fluconazole disc in the centre. Incubate at 35°C for 18–24 h and measure the zone of inhibition. Zone sizes can be correlated with the NCCLS breakpoints: ≥ 19 mm is susceptible, 13–18 mm is susceptible-dose dependent and ≤ 12 mm is resistant (Meis *et al.*, 2000). The QC strain should produce a zone of 32–43 mm. A similar method can be employed for voriconazole but there are as yet no established breakpoints for this drug.

E test methods

There has been a further refinement of the principle of agar diffusion in the commercially available E test strips, in which a concentration gradient of the drug is produced in an agar plate by surface application of a test strip. The technology has evolved to allow the use of this method with the majority of systemically active antifungal agents. Even amphotericin B, which is a large molecule that demonstrates notoriously poor agar diffusion, has been complexed with a carrier to allow a concentration gradient to be produced. In the E test method, a strip containing a concentration gradient of the test antifungal agent is placed on the surface of an agar plate that has been pre-inoculated with the test organism. After incubation a zone of inhibition is visible and the MIC of the organism is read at the point at which the inhibition zone intersects the E test strip. This method has proved to be useful and to correlate well with other standardized methods for both yeasts and moulds.

Other commercial methods for antifungal susceptibility testing

There is a commercial test system that follows the NCCLS methodology (Sensititre or Yeast one, TREK) in a microtitre format but with the incorporation of a chromogenic substrate to facilitate endpoint readings. There are other tests that use a breakpoint system of growth or no growth in wells containing one low and one high concentration of each of a panel of drugs (Fungitest, ATB Fungus).

Methods of assay

Assay of antifungal drugs is necessary either to confirm the attainment of therapeutic levels or to ensure that toxic levels are not reached. There are currently two antifungal agents for which routine assay is recommended. These are flucytosine, to help manage the toxicity associated with high levels of this drug, and itraconazole to ensure adequate absorption following oral administration.

The preferred method for assay is high-performance liquid chromatography (HPLC), which can detect the absolute level of an antifungal agent even in the presence of another agent. This is an important advantage because antifungal agents are increasingly used in combination. Moreover, HPLC assay will detect pure drug and, unlike a plate assay, will not be subjected to the vagaries of an active metabolite.

Extraction of itraconazole

Itraconazole is extracted from 1-ml volumes of serum with heptane–isoamyl alcohol.

HPLC conditions for itraconazole

Column	Chromsep Hypersil 5 ODS (100 × 3 mm) (Chrompak cat no. 28260)
Guard column	Chromsep Guard Column Reversed Phase (Chrompak cat. no. 28141)
Mobile phase	60% acetonitrile + 40% water with 0.03% diethylamine (adjust pH to 7.8 with orthophosphoric acid)
Flow rate	0.5 ml/min
Injection volume	20 μl
Run time	10 min
Wavelength	263 nm
Sensitivity	0.04 absorbance units fixed sensitivity (AUF)

Bioassay of flucytosine (5-fluorocytosine)

Medium

Yeast Nitrogen Base Glucose Citrate (YNBGC) solution

Bacto Yeast Nitrogen Base	67 g
Glucose	100 g
Trisodium citrate	59 g
Deionized water	1000 ml

Adjust the pH to 7.0. Filter sterilize and store in 20-ml amounts.

Agar

Purified agar	40 g
Cold deionized water	1800 ml

Heat to boiling to dissolve the agar and then autoclave at 121°C for 15 min. Dispense in 180-ml amounts.

Stock solutions

Dissolve 50 mg flucytosine in 5 ml sterile water. Filter sterilize and add 1 ml of this solution to 9 ml sterile water to obtain a 1000 mg/l stock solution. This stock solution may be stored at −20°C for at least 12 months. Use the 1000 mg/l stock solution to make up the following standards: 12.5, 25, 50, 100 and 200 mg/l and internal control solutions of 35 and 70 mg/l. Dispense in 0.5-ml amounts and store at −20°C until required. These solutions are stable for at least 12 months.

Preparation of inoculum

Add 2 ml concentrated YNBGC solution to 18 ml sterile water in a 25-ml glass bottle. Inoculate with the sensitive *Candida kefyr* strain NCPF 3234, harvested from a Yeast Morphology Agar (YMA) plate (up to 1-week old), to obtain a concentration of about 1×10^7 cells/ml (absorbance reading of about 1.6–1.7 on the spectrophotometer at 540 nm).

Assay procedure

Melt 180 ml purified agar, cool to 80°C and add 20 ml concentrated YNBGC solution. Pour the molten medium into a large square plate placed on a level surface and leave to set. Flood the entire surface of the agar with 20 ml of cell suspension in YNBGC. Pour off the excess and place the plate in a 37°C incubator with the lid removed. Leave until the surface of the agar has dried (about 20 min). Lay the plate on a sheet of paper (template) with a pattern of 30 positions to which test or standard numbers are assigned at random. Punch 30 wells of 4 mm diameter in the agar with a cork borer (six rows of five wells) and remove the agar plugs. Place 20 μl of standard internal control, or patient specimen, in each well. Each specimen should be tested in triplicate and must be placed on the plate in a randomized distribution. Incubate the plate at 37°C overnight (about 18 h). Measure the diameters of the zone of inhibition around each well using callipers or a zone reader. If performing a manual calculation, plot the mean diameters of the five standards against drug concentrations on semi-logarithmic paper, and construct a standard curve from which the drug concentrations in the patient specimens and internal controls may be interpolated. The drug concentrations should be plotted on the logarithmic ordinate.

Bioassay of itraconazole

The procedure should be followed as above with the following modifications:

Preparation of stock solutions

Dissolve 10 mg itraconazole in 1 ml dimethylformamide. Add 9 ml methyl alcohol to obtain a 1000 mg/l stock solution. Take 1ml of this solution and add 9 ml methyl alcohol to obtain a 100 mg/l stock solution. Use the 100 mg/l stock solution to make up the following standards: 0.5, 1, 2, 4 and 8 mg/l and internal control solutions of 3.5 and 7 mg/l in horse serum. Dispense in 0.5-ml amounts and store at −20°C until required. These standards are stable for at least 1 month.

Inoculum preparation

The indicator organism is *Candida albicans* strain NCPF 3281.

REFERENCES

Grocott (1955) A stain for fungi in tissue and smears using Gomori's methenamine–silver nitrate technique. *American Journal of Clinical Pathology* **25**: 957–959.

King, A. D., Pitt, J. I., Beauchat, L. R. and Corry, J. E. L. (1986) *Methods for the Mycological Examination of Food*, Plenum, New York.

Krogh, P. (1987) *Mycotoxins in Food*. London: Academic Press.

MacKenzie, D. W. R., Philpot, C. M. and Proctor, A. G. J. (1980) *Basic Serodiagnosis Methods for Diseases caused by Fungi and Actinomyces*, Public Health Laboratory Service Monograph No. 12. London: HMSO,

Meis, J., Petrou, M., Bille, J *et al.* (2000) A global evaluation of the susceptibility of *Candida* species to fluconazole by disk diffusion. *Diagnostic Microbiology and Infectious Disease* **36**: 215–223.

Moss, M. O. (1989) Mycotoxins of *Aspergillus* and other filamentous fungi. *Journal of Applied Bacteriology* **87**(suppl): 69S–82S.

National Committee for Clinical Laboratory Standards (2002a) *Reference Method for Broth Dilution Antifungal Susceptibility Testing of Yeasts.* Approved Standard M27-A2, 2nd edn. Wayne, PA: NCCLS.

National Committee for Clinical Laboratory Standards (2002b) *Reference Method for Broth Dilution Antifungal Susceptibility Testing of Filamentous Fungi.* Approved Standard M38-A. Wayne, PA: NCCLS.

Pitt, J. J. and Hocking, A. D. (1985) *Fungi and Food Spoilage.* Sydney: Academic Press.

Samson, R. A. and van Reenan-Hoekstra, E. S. (1988) *Introduction to Food Borne Fungi.* Baarn: Centraalbureau voor Schimmelcultures.

It is often necessary to report on the size of the bacterial population in a sample. Unfortunately, industries and health authorities have been allowed to attach more importance to these 'bacterial counts' than is permitted by their technical or statistical accuracy (see also p. 198).

If a *total count* is required, which is usually done by physical methods, many of the organisms counted may be dead or indistinguishable from other particulate matter. A *viable count*, which is a cultural method, assumes that a visible colony will develop from each organism. Bacteria are, however, rarely separated entirely from their fellows and are often clumped together in large numbers, particularly if they are actively reproducing. A single colony may therefore develop from one organism or from hundreds or even thousands of organisms. Each colony develops from one *viable unit*. Because any agitation, as in the preparation of dilutions, will either break up or induce the formation of clumps, it is obviously difficult to obtain reproducible results. Bacteria are seldom distributed evenly throughout a sample and as only small samples are usually examined very large errors can be introduced. Viable counts are usually given as numbers of colony-forming units (cfu).

In addition, many of the bacteria present in a sample may not grow on the medium used, at the pH or incubation temperatures or gaseous atmosphere employed, or in the time allowed.

Accuracy is often demanded where it is not needed. If, for example, it is decided that a certain product should contain less than, say, 10 viable bacteria/gram this suggests that, on average, of ten tubes each inoculated with 0.1 g seven would show growth and three would not, and of ten tubes each

inoculated with 0.01 g only one or two would show growth. If all the 0.1-g tubes or five of the 0.01-g tubes showed growth, there would be more than 10 organisms/g. It does not matter whether there are 20 or 10 000: there are too many. There is no need to employ elaborate counting techniques.

In viable count methods, it is recognized that large errors are inevitable even if numbers of replicate plates are used. Some of these errors are, as indicated above, inherent in the material, others in the technique. Errors of ± 90% in counts of the order of 10 000–100 000/ml are not unusual even with the best possible technique. It is, therefore, necessary to combine the maximum of care in technique with a liberal interpretation of results. The figures obtained from a single test are valueless. They can be interpreted only if the product is regularly tested and the normal range is known.

Physical methods are used to estimate total populations, i.e. dead and living organisms. They include direct counting, usually by instrumentation (see Chapter 8) and measurements of turbidity. Biological methods are used for estimating the numbers of viable units. These include the plate count, roll-tube count, drop count, surface colony count, dip-slide count, contact plate, membrane-filter count and most probable number estimations.

DIRECT COUNTS

Counting chamber method

The Helber counting chamber is a slide 2–3 mm thick with an area in the centre called the platform and surrounded by a ditch. The platform is 0.02 mm lower than the remainder of the slide. The top of the slide is ground so that, when an optically plane cover glass is placed over the centre depression, the depth is uniform. On the platform an area of 1 mm^2 is ruled so that there are 400 small squares each 0.0025 mm^2 in area. The volume over each small square is 0.02×0.0025 mm^3, i.e. 0.00005 ml.

Add a few drops of formalin to the well-mixed suspension to be counted. Dilute the suspension so that when the counting chamber is filled there will be about five or ten organisms per small square. This requires initial trial and error. The best diluent is 0.1% peptone water containing 0.1% lauryl sulphate and (unless phase contrast or dark field is used for counting) 0.1% methylene blue. Always filter before use.

Place a loopful of suspension on the ruled area and apply the cover glass, which must be clean and polished. The amount of suspension must be such that the space between the platform and the cover glass is just filled, and no fluid runs into the ditch; this requires practice. If the cover glass is applied properly, Newton's rings will be seen. Allow 5 min for the bacteria to settle.

Examine with a 4-mm lens with reduced light, dark-field or phase contrast if available. Count the bacteria in 50–100 squares selected at random so that the total count is about 500. Divide the count by number of squares counted. Multiply by 20 000 and by the original dilution factor to obtain the total number of bacteria per millilitre. Repeat twice more and take the average of the three counts. Clumps of bacteria, streptococci, etc. can be counted as units or each cell counted as one organism. It is difficult to count micro-organisms with a normal mode of growth that is in large aggregates (e.g. myco-bacteria).

With experience, reasonably reproducible counts can be obtained but the chamber and cover glass must be scrupulously cleaned and examined microscopically to make sure that bacteria are not left adhering to either.

Counts of viable cells

Environmental microbiologists employ a variety of methods for counting and estimating the number of viable cells. These include direct epifluorescence using stains such as acridine orange (see pp. 123 and 150). A Live/Dead stain is available from Molecular Probes Inc. (Eugene, OR, USA).

Opacity tube method

International reference opacity tubes, containing glass powder or barium sulphate, are available and are known as Brown's or McFarland opacity tubes. These are numbered narrow glass tubes of increasing opacity and a table is provided equating the opacity of each tube with the number of organisms per millilitre. The unknown suspension is matched against the standards in a glass tube of the same bore. It may be necessary to dilute the unknown suspension. These physical methods should not be used with Hazard Group 3 organisms.

Rapid automated methods

A number of rapid, automated methods have are now in general use. These use: electronic particle counting; bioluminescence (as measured by bacterial ATP); changes in pH and Eh by bacterial growth; changes in optical properties; detection of ^{14}C in CO_2 evolved from a substrate; microcalorimetry; changes in impedance or conductivity; and flow cytometry (see Chapter 8).

VIABLE COUNTS

In these techniques the material containing the bacteria is serially diluted and some of each dilution is placed in or on suitable culture media. Each colony developing is assumed to have grown from one viable unit, which, as indicated earlier, may be one organism or a group of many.

Diluents

Some diluents, e.g. saline or distilled water, may be lethal for some organisms. Diluents must not be used direct from a refrigerator because cold shock may prevent organisms from reproducing. Peptone–water diluent (see p. 76), generally known as 'maximum recovery diluent' (MRD), is most commonly used. If residues of disinfectants are present add suitable quenching agents, e.g. Tween 80 or lecithin for quaternary ammonium compounds and sodium thiosulphate for chlorine and iodine (see Chapter 4, Bloomfield, 1991 and Clesceri et al., 1998).

Pipettes

The 10-ml and 1-ml straight-sided, blow-out pipettes are commonly used and are specified for some statutory tests. Disposable pipettes are labour saving. Mouth pipetting must be expressly forbidden, regardless of the nature of the material under test. A method of controlling teats is given on p. 38. Automatic pipettors or pipette pumps, described on p. 39, should be used. Some of these can be pre-calibrated and used with disposable polypropylene pipette tips.

Pipettes used in making dilutions must be very clean, otherwise bacteria will adhere to their inner surfaces and may be washed out into another dilution. Hydrophobic micro-organisms (e.g. mycobacteria and corynebacteria) adhere to glass and plastic quite readily. Siliconed pipettes may be useful. Fast-running pipettes and vigorous blowing out should be avoided: they generate aerosols. As much as 0.1 ml may remain in pipettes if they are improperly used.

Preparation of dilutions

Dispense 9-ml volumes of diluent in screw-capped bottles. These are the dilution blanks. If tubes with loose caps are used, sterilize them first and dispense the diluent aseptically, otherwise a significant amount of diluent may be lost during autoclaving.

When diluting liquids, e.g. milk for bacterial counts, proceed as follows. Mix the sample by shaking. With a straight-sided pipette dipped in half an inch (1–1.5 cm) only, remove 1 ml. Deliver into the first dilution blank, about half an inch above the level of the liquid. Wait 3 s, then blow out carefully to avoid aerosol formation. Discard the pipette. With a fresh pipette, dip half an inch (1–1.5 cm) into the liquid, suck up and down 10 times to mix, but do not blow bubbles. Raise the pipette and blow out. Remove 1 ml and transfer to the next dilution blank. Discard the pipette. Continue for the required number of dilutions, and remember to discard the pipette after delivering its contents, otherwise the liquid on the outside will contribute to a cumulative error. The dilutions will be:

Tube no.	1	2	3	4	5
Dilution	10^{-1}	10^{-2}	10^{-3}	10^{-4}	10^{-5}

The dilution represents the volume of the original fluid in 1 ml

When counting bacteria in solid or semisolid material, weigh 10 g and place in a Stomacher or blender. Add 90 ml of diluent and homogenize.

Alternatively, cut into small pieces with a sterile scalpel, mix 10 g with 90 ml of the diluent and shake well. Allow to settle. Assume that the bacteria are now evenly distributed between the solid and liquid.

Both of these represent dilutions of 1 : 10, i.e. 1 ml contains or represents 0.1 g, and further dilutions are prepared as above. The dilutions will be:

Tube no.	1	2	3	4
Dilution	10^{-2}	10^{-3}	10^{-4}	10^{-5}

The dilution represents the weight of the original material in 1 g

Mechanical aids for preparing the initial 1 : 10 dilutions are available commercially.

Plate count

Melt nutrient agar or other suitable media in bottles or tubes of appropriate volumes (usually 15 or 20 ml) amounts. Cool to 45°C in a water bath.

Set out Petri dishes, two or more per dilution to be tested, and label with the dilution number. Pipette 1 ml of each dilution into the centre of the appropriate dishes, using a fresh pipette for each dilution. Do not leave the dish uncovered for longer than is absolutely necessary. Add the contents of one agar tube to each dish in turn and mix as follows: move the dish gently six times in a clockwise circle of diameter about 150 mm; repeat counterclockwise; move the dish back and forth six times with an excursion of about 150 mm; repeat with to-and-fro movements. Allow the medium to set, invert and incubate for 24–48 h.

Economy in pipettes

In practice, the pipette used to transfer 1 ml of a dilution to the next tube may be used to pipette 1 ml of that dilution into the Petri dish. Alternatively, after the dilutions have been prepared a single pipette may be used to plate them out, starting at the highest dilution to avoid significant carry-over.

Surface count method

This is often used instead of the pour-plate method in the examination of foods, particularly because the growth of some organisms (e.g. pseudomonads) may be impaired if agar is used at too high a temperature. Also, it is easier to subculture colonies from surface plates than from pour plates.

Place 0.1 ml or another suitable amount of the sample measured with a standard (commercial) loop or a micropipette in the centre of a well-dried plate of suitable medium and spread it with a loop or spreader all over the surface (see p. 81). Incubate and count colonies. Before incubating media spread with liquid, it is important to ensure that the plate is completely dry. If liquid still remains on the surface, micro-organisms can grow in the liquid before it is absorbed into the medium. This results in the appearance of many clustered colonies, especially for rapidly growing micro-organisms. If freshly poured plates are used, incubation brings liquid to the surface and confluent growth can occur, even with small numbers of colony-forming units in the original volume spread. Plates should be dried at least overnight before spreading.

Whether pour plates or surface spreading is employed, aggregates of cells cause problems in obtaining accurate and reproducible colony counts, because an aggregate, if undisturbed by pipetting or spreading, will yield a single colony. If an aggregate is composed of phenotypically different cells (lactose fermenting and non-lactose fermenting), the resulting colony on a fermentation indicator medium will be mixed (e.g. sectored). If suspensions of aggregates are spread, the extent of spreading (e.g. to dryness) and the age of the medium (e.g. fresh or old and dry) will influence the number of colonies. Lowest counts will be obtained by spreading the liquid to dryness on old and dry plates to the point where the liquid disappears and it becomes difficult to move the glass spreader over the surface.

When counting colonies, difficulties arise with large 'smears' of small colonies caused when a large viable unit is broken up but not dispersed during manipulation. Treat these as single units. Usually, other colonies can be seen and counted through 'smears'. There may be problems with swarming organisms, such as some *Proteus* spp. In this case pour melted agar over the inoculated plates.

Counting colonies on pour or surface plates

To count on a simple colony counter select plates with between 30 and 300 colonies for micro-organisms forming colonies of 2–5 mm diameter. Higher numbers can be counted for organisms that form small colonies (e.g. < 1 mm). Place the open dish, glass side up, over the illuminated screen. Count the colonies using a 75-mm magnifier and a hand-held counter. Mark the glass above each colony with a felt-tip pen. Calculate the colony or viable count per millilitre by multiplying the average number of colonies per countable plate by the reciprocal of the dilution. Preferably count the colonies from at least two dilutions and calculate the colony forming units in the original sample by the method of weighted means (BSI, 1991). Report as 'colony forming units/millilitre' (cfu/ml) or as 'viable count/millilitre' not as 'bacteria/gram' or 'bacteria/millilitre'.

If all plates contain more than 300 colonies rule sectors, e.g. of one-quarter or one-eighth of the plate, count the colonies in these and include the sector value in the calculations.

For large workloads semi-automatic or fully automatic counters are essential. In the former the pen used to mark the glass above the colonies is connected to an electronic counter, which displays the numbers counted on a small screen. In the fully automatic models a TV camera or laser beam scans the plate and the results are displayed or recorded on a screen or read-out device.

Roll-tube count

Instead of using plates, tubes or bottles containing media are inoculated with diluted material and rotated horizontally until the medium sets. After incubation, colonies are counted.

Dispense the medium in 2- to 4-ml amounts in 25-ml screw-capped bottles. The medium should contain 0.5–1.0% *more* agar than is usual. Melt and cool to 45°C in a water bath. Add 0.1 ml of each dilution and rotate horizontally in cold water until the agar is set in a uniform film around the walls of the bottle. This requires some practice. An alternative method employs a slab of ice taken from the ice tray of a refrigerator. Turn it upside down on a cloth and make a groove in it by rotating horizontally on it a bottle similar to that used for the counts but containing warm water. The count tubes are then rolled in this groove. If the roll-tube method is to be used often a commercial roll-tube apparatus, which includes a water bath, saves much time and labour.

Incubate roll-tube cultures inverted so that condensation water collects in the neck and does not smear colonies growing on the agar surface. To count, draw a line parallel to the long axis of the bottle and rotate the bottle, counting colonies under a low-power magnifier.

The roll-tube method is popular in the dairy and food industries. Machines for speeding up and taking the tedium out of this method are available.

A roll-tube method for counting organisms in bottles (container sanitation) is given on p. 249.

Drop count method

In this method, introduced by Miles and Misra (1938) and usually referred to by their names, small drops of the material or dilutions are placed on agar plates. Colonies are counted in the inoculated areas after incubation.

Pipettes or standard loops that deliver known volumes are required. Disposable cottonwool-plugged Pasteur pipettes are available commercially and will deliver drops of 0.02 ml. These are sufficiently accurate for the purpose, provided that the tips are checked immediately before use to ensure that they are undamaged. It is advisable, however, to check several pipettes in each batch by counting the number of drops of distilled water that yield 1 g.

Dry plates of suitable medium very well before use. Drop at least five drops from each dilution of sample from a height of not more than 2 cm (to avoid splashing) on each plate. Replace the lid but do not invert until the drops have dried. After incubation, select plates showing discrete colonies in drop areas, preferably one that gives less than 40 colonies per drop (10–20 is ideal). Count the colonies in each drop, using a hand magnifier. Divide the total count by the number of drops counted, multiply by 50 to convert to 1 ml and by the dilution used.

This method lends itself to arbitrary standards, e.g. if there are fewer than 10 colonies among five drops of sample, this represents fewer than 100 colonies of that material. If these drops are uncountable (> 40 colonies), then the colony count per millilitre is > 2000.

Spread-drop plate count

This is a modification of the Miles and Misra method, often used in the examination of foods.

Spread standard volume drops (e.g. 0.02 ml) on quadrants marked out on the bottom of the dish. Inoculate two dishes, each with drops from different decimal dilutions. Do counts on plates (usually two) from dilutions that yield ≤ 50 colonies.

Spiral plate method

The apparatus is described on p. 82. The results are said to compare favourably with those obtained by the surface spread plate method. A commercial imaged-based colony counting system is available for the rapid and accurate counting of colonies on spiral plates.

Membrane filter counts

These offer a more accurate alternative to the most probable number (MPN) estimation (see p. 150 and Clesceri et al., 1998). Liquid containing bacteria is passed through a filter that will retain the organisms. The filter is then laid on a sterile absorbent pad containing liquid culture medium or on the surface of an agar medium and incubated, when the colonies that develop may be counted.

The filter-carrying apparatus is made of metal, glass or plastic, and consists of a lower funnel, which carries a fritted glass platform surrounded by a silicone rubber ring. The filter disc rests on the platform and is clamped by its periphery between the rubber ring and the flange of the upper funnel. The upper and lower funnels are held together by a clamp.

The filters are thin, porous cellulose ester discs, varying in diameter and about 120 μm thick. The pores in the upper layers are 0.5–1.0 μm diameter enlarging to 3–5 μm diameter at the bottom. Bacteria are thus held back on top, but culture medium can easily rise to them by capillary action.

Filters are stored interleaved between absorbent pads in metal containers. A grid to facilitate counting is ruled on the upper surface of each filter.

Membrane filter apparatus may be reusable or sold as kits by the companies that make membranes. Shallow metal incubating boxes with close-fitting lids, containing absorbent pads to hold the culture media, are also required.

The most convenient size of apparatus for counting colonies takes filters 47 mm in diameter, but various sizes larger or smaller are available.

Sterilization

Assemble the filter carrier with a membrane filter resting on the fritted glass platform, which in turn should be flush on top of the rubber gasket. Screw up the clamping ring, but not to its full extent. Wrap in oven-resistant plastic film or foil and autoclave for 15 min at 121°C. Alternatively, loosely assemble the filter carrier, wrap and autoclave; sterilize the filters separately interleaved between absorbent pads in their metal container also by autoclaving. This is the most convenient method when a number of consecutive samples are to be tested; in this case, sterilize the appropriate number of spare upper funnels in the same way.

Sterilize the incubation boxes, each containing its absorbent pad, by autoclaving.

Culture media

Ordinary or standard culture media do not give optimum results with membrane filters. It has been found necessary to vary the proportions of some of the ingredients. Membrane filter versions of standard media may be obtained from most culture media suppliers. A 'resuscitation' medium to revive injured bacteria (e.g. damaged by chlorine) is also desirable for some purposes.

Method

Erect the filter carrier over a filter flask connected by a non-return valve to a pump giving a suction of 25–50 mmHg. Check that the filter is in place and tighten the clamping ring. Pour a known volume, neat or diluted, of the fluid to be examined into the upper funnel and apply suction.

Pipette about 2.5–3 ml of medium on a Whatman 5 cm, No. 17, absorbent pad in an incubation container or Petri dish. This should wet the pad to the edges but not overflow into the container.

When filtration is complete, carefully restore the pressure. Unscrew the clamping ring and remove the filter with sterile forceps. Apply it to the surface of the wet pad in the incubation box so that no air bubbles are trapped. Put the lid on the container and incubate. A fresh filter and upper funnel can be placed on the filter apparatus for the next sample.

For total aerobic counts, use tryptone soya membrane medium; for counting coliform bacilli, either 'presumptive' or *E. coli*, see 'Water examination' (Chapter 20); for counting clostridia in water, see p. 255.

For anaerobic counts, use the same medium incubated anaerobically, or roll the filter and submerge in a 25-mm diameter tube of melted thioglycollate agar at 45°C and allow to set. To count anaerobes causing sulphide spoilage, roll the filter and submerge in a tube of melted iron sulphite medium at 42°C and allow to set. For yeast and mould counts, use the appropriate Sabouraud-type or Czapek-type media.

Counting

Count in oblique light under low-power magnification. For some methods (e.g. counts of *E. coli* in drinking water), indicator media are available and only colonies with a specific colour are counted. If it is necessary to stain colonies to see them, remove the filter from the pad and dip in a 0.01% aqueous solution of methylene blue for half a minute and then apply to a pad saturated with water. Colonies are stained deeper than the filter. The grid facilitates counting. Report as membrane colony count per standard volume (100 ml, 1 ml, etc.).

Count anaerobic colonies in thioglycollate tubes by rotating the tube under strong illumination. Count black colonies in iron sulphite medium.

Micro-organisms in air

Membrane filters may be used with an impinger or cascade sampler. Methods are given on p. 247.

Dip slides

This useful method, commonly used in clinical laboratories for the examination of urine, now has a more general application, especially in food and drinks industries for viable coliform and yeast counts. The slides are made of plastic and are attached to the caps of screw-capped bottles. There are two kinds: one is a single- or double-sided tray containing agar culture media; the other consists of a membrane filter bonded to an absorbent pad containing dehydrated culture media. Both have ruled grids, on either the plastic or the filter, which facilitate counting.

The slides are dipped into the samples, drained, replaced in their containers and incubated. Colonies are then counted and the bacterial load estimated.

Direct epifluorescence filtration technique

This is a rapid, sensitive and economical method for estimating the bacterial content of milk and beverages. It may also be used for foods if they are first treated with proteolytic enzymes and/or surfactants.

A small volume (e.g. 2 ml) of the product is passed through a 24-mm polycarbonate membrane, which is then stained by acridine orange and examined with an epifluorescence microscope. Kits are available from several companies. For more information, see Chapter 8.

Most probable number estimates

Most probable number (MPN) estimates are described in *Standard Methods for the Examination of Water and Wastewater* (Clesceri *et al.*, 1998). These are based on the assumption that bacteria are *normally* distributed in liquid media, i.e. repeated samples of the same size from one source are expected to contain the same number of organisms *on average*; some samples will obviously contain a few more, some a few less. The average number is the *most probable number*. If the number of organisms is large, the differences between samples will be small; all the individual results will be nearer to the average. If the number is small the difference will be relatively larger.

If a liquid contains 100 organisms/100 ml, then 10-ml samples will contain, on average, 10 organisms each. Some will contain more – perhaps one or two samples will contain as many as 20; some will contain less, but a sample containing none is most unlikely. If a number of such samples are inoculated into suitable medium, every sample would be expected to show growth.

Similarly, 1-ml samples will contain, on average, one organism each. Some may contain two or three, and others will contain none. A number of tubes of culture media inoculated with 1-ml samples would therefore yield a proportion showing no growth.

Samples of 0.1 ml, however, could be expected to contain only one organism per 10 samples and most tubes inoculated would be negative.

It is possible to calculate the most probable number of organisms/100 ml for any combination of results from such sample series. Tables have been prepared (Tables 10.1–10.4[1]) for samples of 10 ml, 1 ml and 0.1 ml using five tubes or three tubes of

[1]Tables 10.1–10.3 indicate the estimated number of bacteria of the coliform group present in 100 ml of water, corresponding to various combinations of positive and negative results in the amounts used for the tests. The tables are basically those originally computed by McGrady (1918) with certain amendments due to more precise calculations by Swaroop (1951). A few values have also been added to the tables from other sources, corresponding to further combinations of positive and negative results which are likely to occur in practice. Swaroop has tabulated limits within which the real density of coliform organisms is likely to fall, and his paper should be consulted by those who need to know the precision of these estimates.

These tables are reproduced by permission of the Environmental Agency from *The Microbiology of Drinking Water Supplies. Part 4. Methods for the isolation and enumeration of coliform bacteria and Escherichia coli (including E. coli O 157: H7)*. London, 2002.

Table 10.1 MPN/100 ml, using one tube of 50 ml, and five tubes of 10 ml

50-ml tubes positive	10-ml tubes positive	MPN/100 ml
0	0	0
0	1	1
0	2	2
0	3	4
0	4	5
0	5	7
1	0	2
1	1	3
1	2	6
1	3	9
1	4	16
1	5	18 +

each sample size and, for water testing, using one 50-ml, five 10-ml and five 1-ml samples.

This technique is used mainly for estimating coliform bacilli, but it can be used for almost any organisms in liquid samples if growth can be easily observed, e.g. by turbidity, acid production. Examples are yeasts and moulds in fruit juices and beverages, clostridia in food emulsions and rope-spores in flour suspensions. 'Black tube' MPN counts for anaerobes are described on p. 377.

Mix the sample by shaking and inverting vigorously. Pipette 10-ml amounts into each of five tubes (or three) of 10 ml double-strength medium, 1-ml amounts into each of five (or three) tubes of 5 ml single-strength medium and 0.1-ml amounts (or 1 ml of a 1 : 10 dilution) into each of five (or three) tubes of 5 ml single-strength medium. For testing water, also add 50 ml water to 50 ml double-strength broth.

Use double-strength broth for the larger volumes because the medium would otherwise be too dilute. Incubate for 24–48 h and observe growth, or acid and gas, etc. Tabulate the numbers of positive tubes in each set of five (or three) and consult the appropriate table.

Table 10.2 MPN/100 ml, using one tube of 50 ml, five tubes of 10 ml and five tubes of 1 ml

50-ml tubes positive	10-ml tubes positive	1-ml tubes positive	MPN/100 ml
0	0	0	0
0	0	1	1
0	0	2	2
0	1	0	1
0	1	1	2
0	1	2	3
0	2	0	2
0	2	1	3
0	2	2	4
0	3	0	3
0	3	1	5
0	4	0	5
1	0	0	1
1	0	1	3
1	0	2	4
1	0	3	6
1	1	0	3
1	1	1	5
1	1	2	7
1	1	3	9
1	2	0	5
1	2	1	7
1	2	2	10
1	2	3	12
1	3	0	8
1	3	1	11
1	3	2	14
1	3	3	18
1	3	4	20
1	4	0	13
1	4	1	17
1	4	2	20
1	4	3	30
1	4	4	35
1	4	5	40
1	5	0	25
1	5	1	35
1	5	2	50
1	5	3	90
1	5	4	160
1	5	5	180 +

Table 10.3 MPN/100 ml, using five tubes of 10 ml, five tubes of 1 ml and five tubes of 0.1 ml

10-ml tubes positive	1-ml tubes positive	0.1-ml tubes positive	MPN/ 100 ml	10-ml tubes positive	1-ml tubes positive	0.1-ml tubes positive	MPN/ 100 ml
0	0	0	0	4	0	2	20
0	0	1	2	4	0	3	25
0	0	2	4	4	1	0	17
0	1	0	2	4	1	1	20
0	1	1	4	4	1	2	25
0	1	2	6	4	2	0	20
0	2	0	4	4	2	1	25
0	2	1	6	4	2	2	30
0	3	0	6	4	3	0	25
1	0	0	2	4	3	1	35
1	0	1	4	4	3	2	40
1	0	2	6	4	4	0	35
1	0	3	8	4	4	1	40
1	1	0	4	4	4	2	45
1	1	1	6	4	5	0	40
1	1	2	8	4	5	1	50
1	2	0	6	4	5	2	55
1	2	1	8	5	0	0	25
1	2	2	10	5	0	1	30
1	3	0	8	5	0	2	45
1	3	1	10	5	0	3	60
1	4	0	11	5	0	4	75
2	0	0	5	5	1	0	35
2	0	1	7	5	1	1	45
2	0	2	9	5	1	2	65
2	0	3	12	5	1	3	85
2	1	0	7	5	1	4	115
2	1	1	9	5	2	0	50
2	1	2	12	5	2	1	70
2	2	0	9	5	2	2	95
2	2	1	12	5	2	3	120
2	2	2	14	5	2	4	150
2	3	0	12	5	2	5	175
2	3	1	14	5	3	0	80
2	4	0	15	5	3	1	110
3	0	0	8	5	3	2	140
3	0	1	11	5	3	3	175
3	0	2	13	5	3	4	200
3	1	0	11	5	3	5	250
3	1	1	14	5	4	0	130
3	1	2	17	5	4	1	170
3	1	3	20	5	4	2	225
3	2	0	14	5	4	3	275
3	2	1	17	5	4	4	350
3	2	2	20	5	4	5	425
3	3	0	17	5	5	0	250
3	3	1	20	5	5	1	350
3	4	0	20	5	5	2	550
3	4	1	25	5	5	3	900
3	5	0	25	5	5	4	1600
4	0	0	13	5	5	5	1800+
4	0	1	17				

Table 10.4 MPN/100 ml, using three tubes each inoculated with 10, 1.0 and 0.1 ml of sample

Tubes positive 10 ml	1.0 ml	0.1 ml	MPN	Tubes positive 10 ml	1.0 ml	0.1 ml	MPN	Tubes positive 10 ml	1.0 ml	0.1 ml	MPN
0	0	1	3	1	2	0	11	2	3	3	53
0	0	2	6	1	2	1	15	3	0	0	23
0	0	3	9	1	2	2	20	3	0	1	39
0	1	0	3	1	2	3	24	3	0	2	64
0	1	1	6	1	3	0	16	3	0	3	95
0	1	2	9	1	3	1	20	3	1	0	43
0	1	3	12	1	3	2	24	3	1	1	75
0	2	0	6	1	3	3	29	3	1	2	120
0	2	1	9	2	0	0	9	3	1	3	160
0	2	2	12	2	0	1	14	3	2	0	93
0	2	3	16	2	0	2	20	3	2	1	150
0	3	0	9	2	0	3	26	3	2	2	210
0	3	1	13	2	1	0	15	3	2	3	290
0	3	2	16	2	1	1	20	3	3	0	240
0	3	3	19	2	1	2	27	3	3	1	460
1	0	0	4	2	1	3	34	3	3	2	1100
1	0	1	7	2	2	0	21	3	3	3	1100 +
1	0	2	11	2	2	1	28				
1	0	3	15	2	2	2	35				
1	1	0	7	2	2	3	42				
1	1	1	11	2	3	0	29				
1	1	2	15	2	3	1	36				
1	1	3	19	2	3	2	44				

From Jacobs and Gerstein's *Handbook of Microbiology*, D. Van Nostrand Company, Inc., Princeton, NJ (1960). Reproduced by permission of the authors and publisher

REFERENCES

Bloomfield, S. F. (1991) Method for assessing antimicrobial activity. In: Denyer, S. P. and Hugo, W. B. (eds) *Mechanism of Action of Chemical Biocides*. Society for Applied Bacteriology Technical Series No. 20. Oxford: Blackwells, pp. 1–22.

BSI (1991) BS 5763: Part I: Enumeration of microorganisms: Colony count technique at 30°C. London: British Standards Institution.

Clesceri, L. S., Greenberg, A. E. and Eaton A. D. (1998) *Standard Methods for the Examination of Water and Wastewater*, 20th edn. Washington DC: American Public Health Association.

Jacobs, M. B. and Gerstein, M. J. (1960) *Handbook of Microbiology*. Princeton, NJ: D Van Mostrand Co., Inc.

McCrady, M. H. (1918) Tables for rapid interpretation of fermentation test results. *Public Health Journal, Toronto* 9: 201–210.

Miles, A. A. and Misra, S. S. (1938) The estimation of

the bactericidal power of the blood. *Journal of Hygiene (Cambridge)* 38: 732–749.

Swaroop, S. (1951) Range of variation of Most Probable Numbers of organisms estimated by dilution methods. *Indian Journal of Medical Research* 39: 107–131.

11

Clinical material

All clinical material should be regarded as potentially infectious. Blood specimens may contain hepatitis B and/or human immunodeficiency viruses. For details of special precautions, see National Institutes of Health (NIH, 1988), World Health Organization (WHO, 1991), Advisory Committee on Dangerous Pathogens (ACDP, 1995, 2001), and Collins and Kennedy (1997, 1999). Material from the central nervous system should also be treated with care because it may contain prions. For details of special precautions see ACDP (1998).

The laboratory investigation of clinical material for microbial pathogens is subject to a variety of external hazards. Fundamental errors and omissions may occur in any laboratory. Vigorous performance control testing programmes and the routine use of controls will reduce the incidence of error but are unlikely to eliminate it entirely (see Chapter 2).

Most samples for microbiological examination are not collected by laboratory staff and it must be stressed that, however carefully a specimen is processed in the laboratory, the result can only be as reliable as the sample will allow. The laboratory should provide specimen containers that are suitable, i.e. leak proof, stable, easily opened, readily identifiable, aesthetically satisfactory, suitable for easy processing and preferably capable of being incinerated. It will be apparent that some of these qualities are mutually exclusive and the best compromise to suit particular circumstances must be made. Containers are discussed on p. 41. The specimen itself may be unsatisfactory because of faulty collection procedures, and the laboratory must be ready to state its requirements and give a reasoned explanation if cooperation is to be obtained from ward staff and doctors.

Specimens should be:

- Collected without extraneous contamination.
- Collected, if possible, before starting antibiotic therapy.
- Representative: pus rather than a swab with a minute blob on the end; faeces rather than a rectal swab. If this is not possible, then a clear note to this effect should be made on the request form, which must accompany each sample.
- Ideally collected and sent to the laboratory in a clear, sealed, plastic bag to avoid leakage during transit, or placed in a transport medium. The request form accompanying each specimen should not be in the same bag (bags with separate compartments for specimens and request forms are available) and should be clearly displayed to minimize sorting errors.
- Delivered to the laboratory without delay. If this is not possible, appropriate specimens must be delivered in a commercial transport medium. For other specimens, storage at 4°C will hinder bacterial overgrowth. If urine samples cannot be delivered promptly and refrigeration is not possible, boric acid preservative may be considered as a last resort. Dip-slide culture is to be preferred. Sputum and urine specimens probably account for the greater part of the laboratory workload and the problems associated with their collection are particularly difficult to overcome because the trachea, mouth and urethra have a normal flora.

It must be stressed that, in any microbiological examination, only that which is sought will be found and it is essential to examine the specimen in the light of the available clinical information (which, regrettably, is often minimal). This is even more important with modern techniques that rely on the direct detection of bacterial antigens or nucleic acids in clinical specimens. The examination of pathological material for viruses and parasites is not within the scope of this book, but the possibility of infection by these agents should be considered and material referred to the appropriate laboratory.

The laboratory has an obligation to ensure that its rules are well publicized so that the periods of acceptance of specimens are known to all. Any tests that are processed in batches should be arranged in advance with the laboratory. This may be particularly important for certain tests, e.g. microbiological assays of antibiotics for which the laboratory may need to subculture indicator organisms in advance. In the UK, it is now a requirement for clinical laboratory accreditation that the laboratory produces a user manual to enable the clinicians to use its services in an appropriate way.

BLOOD CULTURES

Blood must be collected with scrupulous care to avoid extravenous contamination. Blood culture is an important diagnostic procedure in cases of recurrent fever, and the highest success rate is associated with the collection of cultures just as the patient's temperature begins to rise, rather than at the peak of the rigor. In cases of septicaemia, the timing is less important, and a rapid answer and a susceptibility test result are essential. Culturing a larger volume of blood may increase the yield from blood cultures.

Most busy clinical laboratories now use automated systems for blood culture (see Chapter 8) or a kit such the Oxoid Signal or the Difco Sentinel, but there is still a place for the Castenada system. In this system, the culture bottles are biphasic, i.e. they have a slope of solid medium and a broth that may also contain Liquoid (sodium polyethanol sulphonate), which neutralizes antibacterial factors including complement. Large amounts of blood (up to 50% of the total volume of the medium) may be examined in Liquoid cultures; otherwise, it is necessary to dilute the patient's blood at least 1 : 20 with broth. The bottles are tipped every 72 h so that the blood–broth mixture washes over the agar slope. Colonies will appear first at the interphase and then all over the slope. The bottles should be retained for at least 4 and preferably 6 weeks. The Castenada system avoids repeated subculture and lessens the risk of laboratory contamination. It is important to avoid contamination because patients receiving immunosuppressive drugs are likely to be infected with organisms not usually regarded as pathogens. No organism can therefore be automatically excluded as a contaminant.

Vacuum blood collection systems also help to reduce contamination. Some of these ensure a 10% carbon dioxide atmosphere in the container. This is important when organisms such as *Brucella* spp. are sought. If other media are used they should be incubated in an atmosphere containing at least 10% carbon dioxide.

Evacuated containers or those containing additional gas may need 'venting' with a cottonwool-plugged hypodermic needle (*caution*) or a commercial device. This is important, especially for the recovery of *Candida* spp. and *Pseudomonas* spp. from neonatal samples, where only a small volume of blood is available for culture.

Blood cultures should be checked for growth at least daily and subcultured to appropriate media when growth is evident, either by visual inspection or by one of the current automated detection systems (see Chapter 8).

Likely pathogens

These are *Staphylococcus aureus*, *Escherichia coli* and other enterobacteria, including salmonellas, streptococci and enterococci, *Neisseria meningitidis*, *Haemophilus influenzae*, *Brucella* spp., clostridia and pseudomonads. In ill or immuno-compromised patients, any organisms, especially if recovered more than once, may be significant (e.g. coagulase-negative staphylococci or coryneforms). In some circumstances special media may have to be used to look for specific organisms (e.g. mycobacteria in AIDS patients). Blood from patients who have had surgery may yield a mixed growth of other organisms.

Commensals

Note that common contaminants (from skin) include coagulase-negative staphylococci and coryneforms. Transient bacteraemia with other organisms may occur occasionally.

All manipulations with open blood culture bottles should be done in a microbiological safety cabinet.

CEREBROSPINAL FLUID

It is important that these samples should be examined with the minimum of delay. As with other fluids, a description is important. It is essential to note the colour of the fluid; if it is turbid, then the colour of the supernatant fluid after centrifugation may be of diagnostic significance. A distinctive yellow tinge (xanthochromia) usually confirms an earlier bleed into the cerebrospinal fluid (CSF), e.g. in a sub-arachnoid haemorrhage as opposed to traumatic bleeding during the lumbar puncture. (It is necessary, of course, to compare the colour of the supernatant fluid with a water blank in a similar tube.)

Before any manipulation examine the specimen for clots, either gross or in the form of a delicate spider's web, which is usually associated with a significantly raised protein level and traditionally is taken as an indication that one should look particularly for *Mycobacterium tuberculosis*. The presence of a clot will invalidate any attempt to make a reliable cell count.

If the cells and/or protein concentrations are raised (> 3×10^6/l and/or > 40 mg/100 ml), a CSF sugar determination should be compared with the blood sugar level collected at the same time. This and the type of cell may indicate the type of infection.

Raised polymorphs with very low sugar usually indicate bacterial infection or a cerebral abscess. Raised lymphocytes with normal sugar usually indicate viral infection. Raised lymphocytes with lowered sugar usually indicate tuberculosis. Raised polymorphs or mixed cells with lowered sugar usually indicate early tuberculosis. It is important, however, to note that there are many exceptions to these patterns.

Make three films from the centrifuged deposit, spreading as little as possible. Stain one by the Gram method and one with a cytological stain, and retain the third for examination if indicated for acid-fast bacilli. A prolonged search may be necessary if tuberculosis is suspected.

Plate on blood agar and incubate aerobically and anaerobically and on a chocolate agar plate for incubation in a 5–10% CO_2 atmosphere. Examine after 18–24 h. If organisms are found, set up direct sensitivity tests according to local protocol. Culture for *M. tuberculosis* as indicated.

Commercial systems are now available for the detection (e.g. by latex agglutination) of the common bacterial causes of meningitis (*Streptococcus pneumoniae, N. meningitidis, H. influenzae*). They should be used in all cases of suspected bacterial meningitis, but they are particularly useful when the Gram stain is unhelpful or when the patient has already received antibiotics.

Likely pathogens

These include *H. influenzae, N. meningitidis, S. pneumoniae, Listeria monocytogenes, M. tuberculosis* and, in very young babies, coliform bacilli, group B streptococci and *Pseudomonas aeruginosa*. *Staphylococcal epidermidis* and some micrococci are often associated with infections in patients with devices (shunts) inserted to relieve excess intracranial pressure.

Opportunists

Any organism introduced during surgical manipulation involving the spinal canal may be involved in meningitis. Infections with other agents may occur in the immunocompromised patient (e.g. *Cryptococcus neoformans* in AIDS).

Commensals

There are none.

DENTAL SPECIMENS

These may be whole teeth, when there may have been an apical abscess, or scrapings of plaque or carious material. Plate on selective media for staphylococci, streptococci and lactobacilli.

Likely organisms

These include *S. aureus*, streptococci, especially *S. mutans* and *S. milleri*, and lactobacilli.

EAR DISCHARGES

Swabs should be small enough to pass easily through the external meatus. Many commercially available swabs are too plump and may cause pain to patients with inflamed auditory canals.

A direct film stained by the Gram method is helpful, because florid overgrowths of faecal organisms are not uncommon and it may be relatively difficult to recover more delicate pathogens. Plate on blood agar and incubate aerobically and anaerobically overnight at 37°C. Tellurite medium may be advisable if the patient is of school age. A medium that inhibits spreading organisms is frequently of value; CLED or MacConkey, chloral hydrate or phenethyl alcohol agars are the most useful.

Likely pathogens

These include *S. pyogenes*, *S. pneumoniae*, *S. aureus*, *Haemophilus* spp., *Corynebacterium diphtheriae*, *P. aeruginosa*, coliform bacilli, anaerobic Gram-negative rods, and fungi.

Commensals

These include micrococci, diphtheroids, *S. epidermidis*, moulds and yeasts.

EYE DISCHARGES

It is preferable to plate material from the eye directly on culture media rather than to collect it on a swab. If this is not practicable, it is essential that the swab be placed in transport medium.

Examine a direct Gram-stained film before the patient is allowed to leave. Look particularly for intracellular Gram-negative diplococci and issue a tentative report if they are seen, so that treatment can be started without delay; this is of particular importance in neonates. Infections occurring a few days after birth may be caused by *Chlamydia trachomatis*, which should be suspected if films have excess numbers of monocytes or the condition does not resolve rapidly. Detection of chlamydial infections may be attempted by microscopy (with Giemsa stain), cell culture or immunological methods, e.g. immunofluorescence (see p. 102).

Plate on blood agar plates and incubate aerobically and anaerobically and on a chocolate agar or 10% horse blood Columbia agar plate for incubation in a 5% carbon dioxide atmosphere.

Likely pathogens

These include *S. aureus*, *S. pneumoniae*, viridans streptococci, *N. gonorrhoeae*, and *Haemophilus*, *Chlamydia* and *Moraxella* spp., but rarely coliform organisms, C. *diphtheriae*, P. *aeruginosa*, and *Candida*, *Aspergillus* and *Fusarium* spp.

Commensals

These include S. *epidermidis*, micrococci and diphtheroids.

FAECES AND RECTAL SWABS

It is always better to examine faeces than rectal swabs. Swabs that are not even stained with faeces are useless. If swabs must be used, they should be moistened with sterile saline or water before use and care must be taken to avoid contamination with perianal flora. It may be useful to collect material from babies by dipping the swabs into a recently soiled napkin (diaper). Swabs should be sent to the laboratory in transport medium unless they can be delivered within 1 h.

As a result of the random distribution of pathogens in faeces, it is customary to examine three sequential specimens. Culture as below according to clinical information:

- Plate on desoxycholate citrate agar (DCA), xylose–lactose desoxycholate citrate agar (XLD) and MacConkey agar for salmonellas, shigellas and enteropathogenic *Escherichia coli*. Incubate at 37°C overnight.
- Inoculate selenite and/or tetrathionate broths, incubate overnight and subculture to DCA, etc.
- Plate stools from children, young adults and diarrhoea cases on campylobacter medium plus antibiotic supplements. Incubate for 48 h at 42°C and at 37°C in 5% oxygen and 10% carbon dioxide.
- In cases of haemorrhagic colitis or haemolytic–uraemic syndrome, culture for E. coli O157 on sorbitol MacConkey agar. Culture also for *Aeromonas* and *Plesiomonas* spp.
- Plate on TCBS and inoculate alkaline peptone

water if cholera is suspected, if the patient has recently returned by air from an area where cholera is endemic or if food poisoning caused by *Vibrio parahaemolyticus* is suspected.

- Plate on Yersinia Selective Agar and incubate at 32°C for 18–24 h (see also p. 318).
- Plate on cycloserine cefoxitin egg yolk fructose agar (CCFA) for *Clostridium difficile* and test the stool for C. *difficile* toxins either in cell culture or with a commercial kit.
- In cases of food poisoning caused by S. *aureus*, plate faeces on blood agar and on one of the selective staphylococcal media (see p. 329), and inoculate salt meat broth. Incubate all cultures overnight at 37°C and subculture the salt meat broth to blood agar and selective staphylococcal medium. It should be noted that here the symptoms relate to a toxin so the organisms may not be recovered.
- In food poisoning where C. *perfringens* may be involved, make a 10^{-1} suspension of faeces in nutrient broth and add 1 ml to each of two tubes of Robertson's cooked meat medium. Heat one tube at 80°C for 10 min and then cool. Incubate both tubes at 37°C overnight and plate on blood agar with and without neomycin and on egg yolk agar. Place a metronidazole disc on the streaked-out inoculum. Incubate at 37°C anaerobically overnight.

Likely pathogens

These include salmonellas, shigellas, *Vibrio* sp., enteropathogenic E. *coli*, C. *perfringens* (both heat resistant and non-heat resistant), S. *aureus*, *Bacillus cereus*, *Campylobacter*, *Yersinia* and *Aeromonas* spp., and C. *botulinum*.

Commensals

These include coliform bacilli, and *Proteus*, *Clostridium*, *Bacteroides* and *Pseudomonas* spp.

NASAL SWABS (SEE ALSO 'THROAT SWABS')

The area sampled will influence the recovery of particular organisms, i.e. the anterior part of the nose must be sampled if the highest carriage rate of staphylococci is to be found. For *N. meningitidis*, it may be preferable to sample further in and for the recovery of *Bordetella pertussis* in cases of whooping cough a pernasal swab is essential.

Inoculate as soon as possible on charcoal agar containing 10% horse blood and 40 mg/l cephalexin and incubate at 35°C in a humid atmosphere.

Direct films are of no value. If *M. leprae* is sought, smears from a scraping of the nasal septum may be examined after staining with Ziehl–Neelsen stain or a fluorescence stain.

Plate on blood agar and tellurite media and incubate as for throat swabs.

Likely pathogens

These include *S. aureus, S. pyogenes, N. meningitidis, B. pertussis, C. diphtheriae* and other pathogens of uncertain significance.

Commensals

These include diphtheroids, *S. epidermidis, S. aureus, Branhamella* spp., aerobic spore bearers and small numbers of Gram-negative rods (*Proteus* spp. and coliform bacilli).

PUS

As a general rule, pus rather than swabs of pus should be sent to the laboratory. Swabs are variably lethal to bacteria within a few hours because of a combination of drying and toxic components of the cotton wool released by various sterilization methods. Consequently, speed in processing is important; a delay of several hours may allow robust organisms to be recovered while the more delicate pathogens may not survive. Swabs should be moistened with broth before use on dry areas.

Examine Gram-stained films and films either direct or concentrated material (see p. 381) for acid-fast rods. If actinomycosis is suspected wash the pus with sterile water and look for 'sulphur granules' (see Chapter 47). Plate granules on blood agar and incubate aerobically and anaerobically. Include a chloral hydrate plate in case *Proteus* sp. is present. Prolonged anaerobic incubation of blood agar cultures, in a 90% hydrogen–10% carbon dioxide atmosphere, possibly with neomycin or nalidixic acid, may allow the recovery of strictly anaerobic *Bacteroides* spp. These organisms may be recovered frequently if care is taken and the number of 'sterile' collections of pus consequently reduced. Culture in thioglycollate broth may be helpful. The dilution effect may overcome specific and non-specific inhibitory agents. Culture pus from post-injection abscesses or infected surgical wounds for mycobacteria (see p. 382).

The recovery of anaerobes will be enhanced by the addition of menadione and vitamin K to the medium. Wilkins and Chalgren's medium is also of value.

If the equipment is available, gas–liquid chromatography is useful for the detection of volatile fatty acids, which may indicate anaerobes (but the bacteriologist's nose is cheaper, even if less sensitive).

Likely pathogens

These include *S. aureus, S. pyogenes*, anaerobic cocci, *Mycobacterium, Actinomyces, Pasteurella, Yersinia, Clostridium, Neisseria* and *Bacteroides* spp., *B. anthracis, Listeria, Proteus, Pseudomonas* and *Nocardia*, spp., fungi, and other organisms in pure culture.

Commensals

There are none.

SEROUS FLUIDS

These fluids include pleural, peritoneal synovial, pericardial, hydrocele, ascitic and bursa fluids.

Record the colour, volume and viscosity of the fluid. Centrifuge at 3000 rev/min for at least 15 min, preferably in a sealed bucket. Transfer the supernatant fluid to another bottle and record its colour and appearance. Make Gram-stained films of the deposit and also stain films for cytological evaluation. Plate on blood agar and incubate aerobically and anaerobically at 37°C for 18–24 h. Culture on blood agar anaerobically plus 10% CO_2 atmosphere and incubate for 7–10 days for *Bacteroides* spp. Use additional material as outlined for pus. Prolonged incubation may be necessary for these strictly anaerobic organisms.

Gonococcal and meningococcal arthritis must not be overlooked; inoculate one of the GC media and incubate in a 10% CO_2 atmosphere for 48 h at 22 and 37°C. Culture, if appropriate, for mycobacteria.

Likely pathogens

These include *S. pyogenes*, *S. pneumoniae*, anaerobic cocci, *S. aureus*, N. *gonorrhoeae*, *M. tuberculosis* and *Bacteroides* spp.

Commensals

There are none.

SPUTUM

Sputum is a difficult specimen. Ideally, it should represent the discharge of the bronchial tree expectorated quickly with the minimum of contamination from the pharynx and the mouth, and delivered without delay to the laboratory.

As a result of the irregular distribution of bacteria in sputum, it may be advisable to wash portions of purulent material to free them from contaminating mouth organisms before examining films and making cultures. Gram-stained films may be useful, suggesting a predominant organism.

Homogenization of the specimen with Sputasol, Sputolysin, N-acetylcysteine or pancreatin, for example, followed by dilution enables the significant flora to be assessed. In sputum cultures not so treated initially, small numbers of contaminating organisms may overgrow pathogens.

The following method is recommended. Add about 2 ml sputum to an equal volume of Sputolysin or Sputasol in a screw-capped bottle and allow to digest for 20–30 min with occasional shaking, e.g. on a Vortex Mixer. Inoculate a blood agar plate with 0.01 ml of the homogenate, using a standard loop (e.g. a 0.01-ml plastic loop) and make a Gram-stained film. Add 0.01 ml with a similar loop to 10 ml peptone broth. Mix this thoroughly, preferably using a Vortex Mixer. Inoculate blood agar and MacConkey agar with 0.01 ml of this dilution. Incubate the plates overnight at 35°C; incubate the blood agar plates in a 10% CO_2 atmosphere. Chocolate agar may be used but should not be necessary for growing *H. influenzae* if a good blood agar base is used.

Five to 50 colonies of a particular organism on the dilution plate are equivalent to 10^6–10^7 of these organisms/millilitre of sputum and this level is significant, except for viridans streptococci when one to five colonies are equivalent to 10^5–10^6 organisms/millilitre and this level is equivocal.

Set up direct sensitivity tests if the Gram film shows large numbers of any particular organisms.

Methods of examining sputum for mycobacteria are described in Chapter 45.

For the isolation of *Legionella* spp. and similar organisms from lung biopsies or bronchial secretions use blood agar or one of the commercial legionella media plus supplement.

Likely pathogens

These include *S. aureus*, *S. pneumoniae*, *H. influenzae*, coliform bacilli, *Klebsiella pneumoniae*, *Pasteurella*, *Mycobacterium* and *Candida* spp., *M. catarrhalis*, *Mycoplasma*, *Pasteurella*, *Yersinia* spp., *Y. pestis*, *L. pneumophila*, *Aspergillus*, *Histoplasma*, *Cryptococcus*, *Blastomyces* and *Pseudomonas* spp., particularly from patients with cystic fibrosis (also *Chlamydia pneumoniae*, *C. psittaci* and *Coxiella burnettii*).

Commensals

These include *S. epidermidis*, micrococci, coliforms, *Candida* spp. in small numbers, and viridans streptococci.

THROAT SWABS

Collect throat swabs carefully with the patient in a clear light. It is customary to attempt to avoid contamination with mouth organisms. Evidence has been advanced that the recovery of *S. pyogenes* from the saliva may be higher than from a 'throat' or pharyngeal swab. Make a direct film, stain with dilute carbol–fuchsin and examine for Vincent's organisms, yeasts and mycelia.

Plate on blood agar and incubate aerobically and anaerobically for a minimum of 18 but preferably 48 h in 7–10% CO_2 at 35°C. Plate on tellurite medium and incubate for 48 h in 7–10% CO_2 at 35°C.

For details of the logistics of mass swabbing, e.g. in a diphtheria outbreak, see Collins and Dulake (1983).

Likely pathogens

These include *S. pyogenes*, *C. diphtheriae*, *C. ulcerans*, *S. aureus*, *C. albicans*, *N. meningitides* and *B. vincenti* (*H. influenzae* Pittman's type b in the epiglottis is certainly a pathogen).

Commensals

These include viridans streptococci, *S. epidermidis*, diphtheroids, *S. pneumoniae*, and probably *H. influenzae*.

TISSUES, BIOPSY SPECIMENS, POSTMORTEM MATERIAL

As a general rule, retain all materials submitted from postmortem examination until the forensic pathologists agree to their disposal. If it is essential to grind up tissue, obtain approval.

Grind tissue specimens in a blender or, if the specimen is very small, in a Griffith's tube or with glass beads on a Vortex Mixer. The Stomacher Lab-Blender is best for larger specimens because it saves time and eliminates the hazard of aerosol formation. Examine the homogenate as described under 'Pus'.

URETHRAL DISCHARGES

Collect swabs of discharges into a transport medium or plate directly at the bedside or in the clinic. Examine direct Gram-stained films. Look for intracellular Gram-negative diplococci. In males a Gram-stained film of the urethral discharge is a sensitive and specific means of diagnosing gonorrhoea. In the female, the recovery of organisms is more difficult and the interpretation of films may occasionally present problems because over-decolourized or dying Gram-positive cocci may resemble Gram-negative diplococci.

Plate on chocolate agar and a selective medium for gonococci (e.g. Modified Thayer-Martin, Modified New York City medium) and incubate in a humid atmosphere enriched with 5% CO_2.

Likely pathogens

These include *N. gonorrhoeae*, and *Mycoplasma* and *Ureaplasma* spp. Occasionally other pathogens may be isolated, particularly coliforms when a urinary catheter is present.

Commensals

These include *S. epidermidis*, micrococci, diphtheroids and small numbers of coliform organisms.

URINE

The healthy urinary tract is free from organisms over the greater part of its length; there may, however, be a few transient organisms present, especially at the lower end of the female urethra.

Collection

Catheters may contribute to urinary infections and are no longer considered necessary for the collection of satisfactory samples from females.

A 'clean catch' mid-stream urine (MSU) is the most satisfactory specimen for most purposes if it is delivered promptly to the laboratory. Failing prompt delivery, the specimen may be kept in a refrigerator at 4°C for a few hours or, exceptionally, a few crystals of boric acid may be added.

Dip slides are now widely used for assessing urinary tract infections. They do, however, preclude examination for cells and other elements unless a fresh specimen is also submitted.

Examination

This includes counting leucocytes, erythrocytes and casts, estimating the number of bacteria/millilitre and identifying them. Normally, there will be none or fewer than 20×10^6/l of leucocytes or other cells and a colony count of less than 10^4 organisms/ml.

Mix the urine by rotation, never by inversion. Count the cells with a Fuchs Rosenthal or similar counting chamber or use an inverted microscope. Inoculate a blood agar plate and a MacConkey, CLED or EMB plate with 0.001 ml using a standard loop (disposable loops or micro-pipettes with disposable tips are satisfactory alternatives). Also 0.1 ml of a 1 : 100 dilution of urine may be used. Spread with a glass spreader, incubate overnight at 37°C and count the colonies:

< 10 colonies = 10^4 organisms/ml	not significant
10–100 colonies = 10^4–10^5 organisms/ml	doubtful significance
> 100 colonies = > 10^5 organisms/ml	significant bacteriuria

If dip slides are submitted, incubate them overnight and follow the manufacturer's directions for counting.

Note that the accuracy and usefulness of bacterial counts on urines are limited by the age of the specimen. Only fresh specimens are worth examining. Other limitations are the volume of fluid recently drunk and the amount of urine in the bladder.

Various other techniques, e.g. a filter paper 'foot', multipoint inoculator, dip sticks for leucocyte enzymes or simple visual inspection, have been advocated for screening urines. These save media and may be appropriate in laboratories where most samples tested are 'negative'.

For methods of examining urine for tubercle bacilli, see p. 382.

Automated urine examination

The Rapid Automatic Microbiological Screening system (RAMUS, Orbec), is a particle volume analyser, capable of counting leucocytes and bacteria. Counts of bacteria > 10^5/ml are regarded as significant.

Multipoint inoculation methods are very useful in the examination of large batches of urines (see Faiers *et al.*, 1991, and Chapter 8). The technique can allow for tests for the presence of antibacterial substances, direct inoculation of identification media and media for those fastidious organisms that are now considered to be of importance in urinary tract infections.

Likely pathogens

These include *E. coli*, other enterobacteria, *Proteus* spp., staphylococci including *S. epidermidis*, *S. saprophyticus*, enterococci, salmonellas (rarely), *Leptospira* spp. and mycobacteria.

Commensals

There can be small numbers ($< 10^4$/ml) of almost any organisms. It is unwise to be too dogmatic about small numbers of organisms isolated from a single specimen; it is better to repeat the sample.

VAGINAL DISCHARGES

Bacterial vaginosis is a condition in which there is a disturbance in the vaginal flora. Lactobacilli decrease in numbers whereas those of a variety of others, e.g. *Gardnerella vaginalis*, and *Mobiluncus* and *Prevotella* spp. increase. This condition should be differentiated from vaginitis caused by *C. albicans*. A Gram-stained film may reveal 'clue cells' – epithelial cells with margins obscured by large numbers of adherent Gram-variable rods. Culture for *G. vaginalis* is less helpful.

PUERPERAL INFECTIONS

A high vaginal swab is usually taken. Swabs that cannot be processed very rapidly after collection should be put into transport medium. Examine direct Gram-stained films. If non-sporing, square-ended, Gram-positive rods, not in pairs, and which appear to be capsulated, are seen they should be reported without delay, because they may be *C. perfringens*, which can be an extremely invasive organism.

Plate on blood agar and incubate aerobically and anaerobically overnight at 37°C and also plate on MacConkey or other suitable selective medium. Plate any swabs suspected of harbouring *C. per-fringens* on a neomycin half-antitoxin Nagler plate for anaerobic incubation.

Likely pathogens

These include *S. pyogenes*, *S. agalactiae*, *L. mono-cytogenes*, *C. perfringens*, *Bacteroides* spp., *G. vaginalis*, anaerobic cocci, *Candida* spp., and excess numbers of enteric organisms.

Commensals

These include lactobacilli, diphtheroids, micro-cocci, *S. epidermidis*, small numbers of coliform bacilli, and yeasts.

NON-PUERPERAL INFECTIONS

Examine a Gram-stained smear and also a wet preparation for the presence of *Trichomonas vaginalis*. If the examination is to be immediate, a saline suspension of vaginal discharge will be satisfactory; otherwise a swab in transport medium is essential. A wet preparation may still be examined for actively motile flagellates but culture in trichomonas medium usually yields more reliable results, especially in cases of light infection. An alternative to the wet preparation is the examination under the fluorescence micro-scope of an air-dried smear stained by acridine orange.

Examine a Gram-stained smear for *Mobiluncus* spp. If *N. gonorrhoeae* infection is suspected in a female, swabs from the urethra, cervix and poste-rior fornix should be collected, placed into trans-port medium or plated immediately on selective medium. Many workers do not make films from these swabs because the difficulties of interpreta-tion are compounded by the presence of charcoal from the buffered swabs; they rely on cultivation using VCN (vancomycin, colistin, nystatin) or VCNT (VCN + trimethoprim) medium as a selective inhibitor of organisms other than *N. gonor-rhoeae*.

Plate on blood agar and incubate, aerobically and anaerobically, on Sabouraud's agar (when looking for *Candida* spp.) and chocolate agar or another suitable gonococcal medium for incubation in a 5% CO_2 atmosphere to encourage isolation of *N. gonorrhoeae*.

If *L. monocytogenes* infection is suspected, plate on serum agar with 0.4 g/l nalidixic acid added.

Likely pathogens

These include *N. gonorrhoeae*, *C. albicans* and related yeasts in moderate numbers, *S. pyogenes*, *Gardnerella* spp., *L. monocytogenes*, *Haemophilus* and *Bacteroides* spp., anaerobic cocci, *Mobiluncus* spp., possibly *Mycoplasma* spp. and *T. vaginalis*.

Commensals

These include lactobacilli, diphtheroids, entero-cocci and small numbers of coliform organisms.

WOUNDS (SUPERFICIAL) AND ULCERS

Examine Gram-stained films and culture on blood agar and proceed as for pus, etc. Examine Albert-stained smears for corynebacteria. Culture also on Löwenstein–Jensen medium and incubate at 30°C because some mycobacteria that cause superficial infections do not grow on primary isolation at 35–37°C.

Likely pathogens

These include *S. aureus*, streptococci, *C. diphthe-riae*, *C. ulcerans* (both in ulcers), *M. marinum*, *M. ulcerans* and *M. chelonei*.

Contaminants

These include staphylococci, *Pseudomonas* spp., and a wide variety of bacteria and fungi.

WOUNDS (DEEP) AND BURNS

The collection of these specimens and their conse-quent rapid processing are two vital factors in the recovery of the significant organisms. Burn cases frequently become infected with staphylococci and streptococci as well as with various Gram-negative rods, especially *Pseudomonas* spp. It may be diffi-cult to recover Gram-positive cocci because of the overgrowing Gram-negative organisms.

Plate on blood agar and incubate aerobically and anaerobically. Plate also on MacConkey agar and CLED medium (to inhibit the spreading of *Proteus* spp.). Robertson's cooked meat medium may be useful.

Plate swabs, etc. from burns on media selective for *S. pyogenes* and *Pseudomonas* spp. Anaerobes are rarely a problem in burns.

Likely pathogens

These include *Clostridium* spp., *S. pyogenes*, *S. aureus*, anaerobic cocci, *Bacteroides* spp. and Gram-negative rods.

Contaminants

These include small numbers of a wide variety of organisms.

REFERENCES

Advisory Committee on Dangerous Pathogens (ACDP) (1995) *Protection against Blood-borne Infections in the Workplace: HIV and hepatitis*. London: HMSO.

ACDP (1998) *Transmissible Spongiform Encephalopathies: Safe working and the prevention of infection.* London: HMSO.

ACDP (2001) Supplement to *ACDP Guidance on Protection against Blood-borne Infections in the Workplace: HIV and hepatitis.* London: HMSO.

Collins, C. H. and Dulake, C. (1983) Diphtheria: the logistics of mass swabbing. *Journal of Infection* 6: 277–280

Collins, C. H. and Kennedy, D. A. (eds) (1997) *Occupational Blood-borne Infections.* Wallingford: CAB International.

Collins, C. H. and Kennedy, D. A. (1999) *Laboratory-acquired Infections*, 4th edn. Oxford: Butterworth–Heinemann, pp. 240–244.

Faiers, M., George, R., Jolly, J. and Wheat, P. (1991) *Multipoint Methods in the Clinical Laboratory.* London: Public Health Laboratory Service.

National Institutes of Health (1988) Working Safely with HIV in the Research Laboratory. Bethesda, MD: NIH.

World Health Organization (1991) *Biosafety Guidelines for Diagnostic and Research Laboratories Working with HIV*, WHO AIDS Series 9. Geneva: WHO.

The primary purpose of antimicrobial susceptibility testing is to guide the clinician in the choice of appropriate agents for therapy. Agents are often used empirically and routine testing provides up-to-date information on the most suitable agents for empirical use. Apart from routine clinical work, antimicrobial susceptibility tests are used to evaluate the *in vitro* activity of new agents.

Antimicrobial susceptibility may be reported qualitatively as: sensitive (S), which indicates that the standard dose of the agent should be appropriate for treating the patient infected with the strain tested; resistant (R), which indicates that an infection caused by the isolate tested is unlikely to respond to treatment with that antimicrobial agent; and intermediate (I), where the strains are moderately resistant or moderately susceptible and indicate that the strain may be inhibited by larger doses of the agent or may be inhibited in sites where the agent is concentrated (e.g. urine). It may also be regarded as a 'buffer-zone', which reduces the likelihood of major errors, i.e. reporting a resistant isolate as susceptible or vice versa. Caution is necessary when the basis of resistance is the production of inactivating enzymes because zone sizes or minimum inhibitory concentrations (MICs) (see below) may indicate intermediate susceptibility rather than resistance. The routine use of the intermediate category on patient reports should be avoided because clinicians rarely find it helpful

Quantitative testing determines the MIC, which is the lowest concentration of the agent that will prevent a microbial population from growing. The minimum bactericidal concentration (MBC) is the lowest concentration that kills 99.9% of the population.

The concentrations of agents defining the 'breakpoints' between sensitive, intermediate and resistant are based on clinical, pharmacological and microbiological considerations. These breakpoint concentrations may be used to interpret MIC results, as test concentrations in breakpoint methods, or as the basis for interpreting zone sizes in disc diffusion tests (Andrews, 2001a, 2001b; King, 2001).

TECHNICAL METHODS

The three principal methods used for susceptibility testing, Disc Diffusion, MIC/MBC determination and Breakpoint, are outlined briefly below, as are the technical factors that influence the results.

Disc diffusion tests

Diffusion tests with agents in paper discs are still the most widely used methods in routine clinical laboratories. They are technically straightforward, reliable if correctly standardized and also relatively cheap.

Minimum inhibitory concentration

In minimum inhibitory concentration methods, the susceptibility of organisms to serial twofold dilutions of agents in agar or broth is determined. The MIC is defined as the lowest concentration of the agent that inhibits visible growth. The methods are not widely used for routine testing in clinical laboratories in most countries, although commercial systems with restricted-range MICs in a variety of formats are available from a number of companies. In clinical laboratories MIC methods are used to test the susceptibility of organisms that give equivocal results in diffusion tests, for tests on organisms where disc tests may be unreliable, as with slow-growing organisms, or when a more accurate result is required for clinical management, e.g. infective endocarditis. MIC methods are accepted as the standard against which other methods are assessed and for testing new agents. MICs may also be derived from the size of the zone of inhibition in disc diffusion tests using a regression curve produced by plotting the diameter of the zone of inhibition against the MIC of a range of strains of a given species. In some cases it is necessary to know the MBC of an organism. This is determined by subculturing the broth of tubes showing no turbidity in an MIC test on to antimicrobial agent-free media and observing for growth after further incubation.

Breakpoint methods

Breakpoint methods are essentially MIC methods in which only one or two concentrations are tested. The concentrations chosen for testing are those equivalent to the breakpoint concentrations separating different categories of susceptibility. In recent years such methods have been used routinely as an alternative to disc diffusion methods. Breakpoint tests are suitable for automated inoculation and reading of tests, and may be combined with identification tests.

Factors affecting susceptibility tests

Many of the effects of technical variation on susceptibility testing are widely known (Barry, 1976, 1991; Brown and Blowers, 1978; Wheat, 2001).

Medium

The medium should support the growth of organisms normally tested and should not contain antagonists of antimicrobial activity or substances (e.g. albumin) that can bind to and inactivate microbial agents. Thymidine antagonizes the activity of sulphonamides and trimethoprim, leading to false reports of resistance to these agents. If the medium is not low in thymidine, lysed horse blood or thymidine phosphorylase should be added to remove free thymidine. Monovalent cations such as Na^+ increase the activity of bacitracin, fusidic acid and novobiocin against staphylococci, and of penicillins against *Proteus* spp. Divalent cations such as Mg^{2+} and Ca^{2+} reduce the activity of aminoglycosides and polymyxins against *Pseudomonas* spp., and of tetracyclines against a range of organisms.

Media specially formulated for susceptibility testing should be used. Iso-Sensitest agar (Oxoid) and similar media are generally used in the UK; Mueller–Hinton agar (various suppliers) is used in several other countries. There have been problems with consistency of batches of Mueller–Hinton agar which has led to the publication of a performance standard (National Committee for Clinical Laboratory Standards or NCCLS, 1996, 1997, 1999). Additional supplements are necessary for growth of some organisms (see also Chapter 2).

pH

The activity of aminoglycosides, macrolides, lincosamides and nitrofurantoin increases with increasing pH. Conversely, the activity of tetracycline, fusidic acid and novobiocin increases as the pH is lowered. The microbial growth rate is also influenced by pH, thus indirectly affecting susceptibility.

Depth of agar medium

With diffusion tests the size of the zone varies with the depth of the medium. The agar in the plate should be level and of a consistent depth – approximately 4 ± 0.5 mm.

Inoculum size

Increasing the inoculum size reduces the susceptibility to agents in both diffusion and dilution tests.

Tests of susceptibility to sulphonamides, trimethoprim and β-lactam agents, where resistance is based on β-lactamase production in staphylococci, *Haemophilus* spp. and *Neisseria* spp., and methicillin resistance in staphylococci, are particularly affected by variation in inoculum size. Among Gram-negative organisms, the effect of inoculum size on MICs of those producing β-lactamase depends on the amount of any β-lactamase produced, and the activity of the enzyme against the particular penicillin or cephalosporin under test. Standardization of the inoculum size is therefore necessary.

In most diffusion methods an inoculum size that gives a semi-confluent growth is used. This has the advantage that incorrect inocula can be readily identified and the test rejected. There is no evidence, however, that semi-confluent growth gives more reliable results than just confluent growth, as used in the NCCLS (2003) diffusion method.

In agar dilution methods (MIC or breakpoint) an inoculum of 10^4 colony-forming units (cfu)/inoculum spot is the accepted standard. This should be increased to 10^6 cfu/inoculum spot for methicillin susceptibility tests on staphylococci. With organisms producing extracellular β-lactamases, the standard inoculum may result in MICs only slightly higher than those obtained with susceptible strains. This can be avoided by increasing the inoculum to 10^6 cfu/inoculum spot on agar dilution plates. For broth dilution methods the standard inoculum should be 10^5 cfu/ml of test broth at the beginning of the incubation period.

Use four or five colonies of a pure culture to avoid selecting an atypical variant. Prepare the inoculum by emulsifying overnight colonies from an agar medium, by diluting an overnight broth culture or by diluting a late log-phase broth culture. The broth should not be antagonistic to the agent tested (e.g. media containing thymidine should not be used to prepare inocula for tests on sulphonamides and trimethoprim). Organisms should be diluted in distilled water to a concentration of 10^7 cfu/ml. A 0.5 McFarland standard gives a density equivalent to approximately 10^8 cfu/ml. Alternatively, inocula may be adjusted photometrically or by standard dilution of broth cultures if the density of such cul

tures is reasonably constant. If there is any doubt about whether a particular inoculum preparation procedure yields the requisite density of organisms, it should be checked by performing viable counts on a small number of representative isolates before embarking on a series of MIC tests. Inocula should be used promptly to avoid subsequent changes in their density. Plates should be inoculated within 15 min of standardizing the inoculum.

Preincubation and prediffusion

In diffusion tests preincubation and prediffusion of the inoculum decrease and increase the sizes of zones respectively. In practice these effects are avoidable by applying discs soon after plates are inoculated and incubating plates soon after discs are applied.

Antimicrobial agents/discs

Modern commercial discs are generally produced to a high standard and problems caused by incorrect content or incorrect agents in discs are rare. Humidity will cause deterioration of labile agents and many of the problems with discs are related to incorrect handling in the laboratory. Store discs in sealed containers in the dark (some agents, e.g. metronidazole, chloramphenicol and quinilones, are sensitive to light); the containers should include an indicating desiccant. Discs may be kept at < 8°C within the expiry date given by the maker. To minimize condensation of water on the discs allow containers to warm to room temperature before opening them. For convenience, leave closed containers at room temperature during the day and refrigerate them overnight only.

Problems with humidity also affect pure antimicrobial powders used in MIC testing and similar precautions to those described for discs should be taken. If powders absorb water the amount of active substance per weighed amount will be reduced even if the agent is not labile because of the weight of the water absorbed.

Incubation

Incubate plates at 35–37°C in air. Use alternative conditions only if essential for growth of the organisms. Incubation in an atmosphere containing additional carbon dioxide may reduce the pH, which may affect the activity of some agents. Incubation at 30°C is usually necessary for methicillin/oxacillin susceptibility tests against staphylococci, but should not be used for other agents because zone sizes will be increased. Do not stack plates more than four high in the incubator in order to avoid uneven heating during incubation.

Reading results

Unless tests are read by an automated system the reading is somewhat subjective without clear endpoints. Criteria for reading and interpreting susceptibility tests – whatever basic method is used – should be written down and available to staff who do these tests.

Definitions of endpoints for agar dilution tests differ, although up to four colonies or a thin haze in the inoculated spot are commonly disregarded. Endpoints trailing over several dilutions may indicate contamination or a minority resistant population.

For broth dilution tests the MIC is the lowest concentration of the agent that completely inhibits growth.

In reading diffusion tests take the point of abrupt diminution of growth as the zone edge. If there is doubt read the point of 80% inhibition of growth. Measure zones at points unaffected by interactions with other agents or by proximity to the edge of the plate. Ignore tiny colonies at the zone edge, swarming of *Proteus* spp., and traces of growth within sulphonamide and trimethoprim zones. Large colonies within zones of inhibition may indicate contamination or a minority resistant population.

Diffusion methods

The basic technique for disc diffusion tests has not changed for many years. Paper discs containing known amounts of antibiotic are applied to plates inoculated with the test and/or control organism. After incubation the diameter of the zone of inhibition of growth is used to decide whether the strain is susceptible to the agent being tested. The methods of control of variation and the basis of interpretation have fallen into two groups: the comparative methods and the standardized methods.

In the comparative methods the zone size of the test strain is compared with that of a control strain set up at the same time. It is assumed that variation in the test affects the test and control strains equally, and thereby cancels them out. The comparison may be made with a control strain on a separate plate or both control and test may be inoculated on the same plate. However, the lack of standardization and the difficulty in selecting suitable controls make these methods less reliable than standardized methods. This is particularly so for many of the newer antibiotics. The lack of standardization also makes the analysis of trends in antibiotic resistance by inter- and intra-laboratory comparison difficult, and the epidemiological value of results is low.

In the standardized methods all details of the tests are recommended in the report of an international collaborative study of susceptibility testing (Ericsson and Sherris, 1971). Interpretation of zone diameters is from a table of zone size breakpoints. The zone size breakpoints equivalent to defined MIC breakpoints have been established by regression analysis of the relationship of zone size to MIC, by error minimization, by study of the distribution of susceptibility of different species, and by clinical experience relating the results of tests to the outcome of treatment. These different approaches have been combined in some methods. An important feature of the standardized methods is that the zone size breakpoints are valid only if all aspects of the defined method are followed precisely. Hence there are several standardized methods that differ in detail and in the breakpoints used. Both the British Society for Antimicrobial Chemotherapy (BSAC) method (Andrews, 2001b) and, in the USA, the NCCLS (2003) method are examples.

STANDARDIZED METHODS

British Society for Antimicrobial Chemotherapy method

For full details of this method, the BSAC Guidelines (Andrews, 2001b) should be consulted. These are also available on the BSAC website (www.bsac.org.uk) which is updated regularly. A summary of the method is given below.

Media

Iso-Sensitest agar (ISTA, Oxoid) is used for testing Enterobacteriaceae, *Pseudomonas* spp., enterococci and staphylococci other than for the detection of methicillin/oxacillin resistance. Mueller–Hinton or Columbia agar with 2% sodium chloride is used for the detection of methicillin/oxacillin resistance in staphylococci. Iso-Sensitest agar (ISTA) with 5% defibrinated horse blood (ISTHBA) is used for *Streptococcus pneumoniae*, haemolytic streptococci, *Moraxella catarrhalis*, *Neisseria meningitidis*, and *N. gonorrhoeae*. ISTHBA supplemented with 20 mg/l β-nicotinamide adenine dinucleotide (NAD) is used for *Haemophilus* spp. and may also be used for organisms tested on ISTHBA. The medium should be approximately 4 mm deep (25 ml in a 90-mm Petri dish). Plates should be dried to remove excess surface moisture but it is important not to over-dry them.

Controls

Control organisms should be used to monitor the performance of the test by ensuring that the zone sizes obtained fall within the recommended limits. These are published by the BSAC. Examples are given in Table 12.1. They are also used to confirm that the method will detect resistance. They are not used for the interpretation of susceptibility. The following are the control organisms: *Escherichia coli* NCTC 12241 (ATCC 25922) or NCTC 10418; *E. coli* β-lactamase positive strain NCTC 11560; *Staphylococcus aureus* NCTC 12981 (ATCC 25923) or NCTC 6571; *S. aureus* methi-

cillin/oxacillin-resistant strain NCTC 12493; *Pseudomonas aeruginosa* NCTC 12934 (ATCC 27853) or NCTC 10662; *Enterobacter faecalis* NCTC 12697 (ATCC 29212); *H. influenzae* NCTC 12699 (ATCC 49247 or NCTC 11931); *S. pneumoniae* NCTC 12977 (ATCC 49619); *N. gonorrhoeae* NCTC 12700 (ATCC 49226). Control strains should be stored so as to minimize mutation. Storage at −70°C on beads in glycerol broth is ideal. Two vials should be used for each control, one 'in use' which is subcultured weekly on to a non-selective medium. If this subculture is pure it is used to prepare one subculture for each of the next 7 days. A bead from the second vial is subcultured when it is necessary to replenish the 'in use' vial.

Inoculum

The aim is to produce a semi-confluent growth. This has the advantage that an incorrect inoculum can easily be seen. A suspension of organisms in a suitable diluent (broth or peptone water) with a density equivalent to a 0.5 McFarland standard is prepared and then diluted appropriately. To prepare a 0.5 McFarland standard add 0.5 ml 0.048 mol/l $BaCl_2$ (1.17% w/v $BaCl_2 \cdot 2H_2O$) to 99.5 ml 0.18 mol/l H_2SO_4 (1% w/v) with constant stirring. This should be distributed into screw-capped bottles of the same volume and size as those used for the preparation of the test suspensions. These standards should be stored, thoroughly mixed before use and replaced every 6 months. Alternatively, pre-prepared standards can be obtained from commercial suppliers. There are also pre-calibrated photometers available from a number of suppliers. Organisms may be suspended directly in Iso-Sensitest broth, or non-fastidious organisms inoculated into Iso-Sensitest broth and incubated at 37°C until the appropriate density has been achieved. The density of the suspension is adjusted to match the 0.5 McFarland standard and then further diluted, within 10 min of preparation, in sterile distilled water. Enterobacteriaceae, enterococci, haemolytic streptococci, *Pseudomonas* spp. *Acinetobacter* spp. and *Haemophilus* spp. are diluted 1:100, staphylococci, *Serratia* spp.,

Table 12.1 The zone diameters, in millimetres, of control strains used in the BSAC disc diffusion method growing on Iso-Sensitest agar incubated at 35–37°C in air[a]

Antibiotic	Disc content (μg)	*Escherichia coli*			*Pseudomonas aeruginosa*		*Staphylococcus aureus*		*Enterobacter faecalis*
		NCTC 10418	ATCC 25922	NCTC 11560	NCTC 10662	ATCC 27853	NCTC 6571	ATCC 25923	ATCC 29212
Amikacin	30	24–27	23–27		21–30	26–32			
Ampicillin	10	21–26	16–22						26–35
	25	24–30							
Cefotaxime	30	36–45	34–44		20–29	20–24			
	5						26–32[b]		
Ceftazidime	30	32–40	31–39		29–37	27–35			
Cefuroxime	30	25–32	24–29						
Ciprofloxacin	1	31–40	31–37		21–28	24–30	25–32	17–22	
Co-amoxiclav	3							27–32	
	30	18–31	20–26	18–23					
Erythromycin	5						22–31	22–29	
	5						25–29[b]		
	5						22–29[c]	23–29[c]	
Fusidic acid	10						32–40	30–37	
Gentamicin	10	21–27	21–27		20–26	22–28	24–30	20–26	
	200								22–27
Imipenem	10	32–37	33–37		20–27	23–28			
Methicillin[d]	5						18–30	18–28	
Oxacillin[d]	1						19–30	19–29	
Penicillin	1 unit						32–40	29–36	
	1 unit						30–41[b]	27–35[b]	
	1 unit						39–43[b]		
Piperacillin/ Tazobactam	85				28–35	29–37			
Teicoplanin	30						17–23	16–20	19–25
Trimethoprim	2.5	28–34	20–26				25–30	20–28	
	5						24–34		
Vancomycin	5						14–20	13–17	13–19

		H. influenzae (with **NAD**)[b]		*S. pneumoniae*[b]
		NCTC 11931	ATCC 49247	ATCC 49619
Ampicillin	2	22–30	6–13	
Cefotaxime	5	33–45	27–38	
Co-amoxiclav	3	20–27	10–20	
Erythromycin	5	12–23	9–16	23–36
Oxacillin	1			8–16

[a]After BSAC guidelines (Andrews, 2001b) with permission
[b]Using Iso-Sensitest agar with 5% horse blood with or without NAD incubated at 35–37°C in 5% CO_2
[c]Using Iso-Sensitest agar with 5% horse blood with or without NAD incubated at 35–37°C in air
[d]Using Mueller–Hinton with 2% NaCl incubated at 30°C in air

S. pneumoniae, N. meningitides and *M. catarrhalis* are diluted 1 : 10. *N. gonorrhoeae* is not diluted any further. This suspension is used to inoculate plates within 15 min of preparation. A sterile cottonwool-tipped swab is dipped into the suspension, and the excess fluid removed and used to spread the inoculum over the entire surface of the plate.

Discs

Discs should be stored as detailed above. Up to six may be placed on the surface of a 90-mm Petri dish.

Incubation

With most organisms plates should be incubated within 15 min of disc application at 35–37°C in air for 18–24 h. For *Neisseria* spp., *Haemophilus* spp. and *S. pneumoniae,* however, 5% CO_2 at 35–37°C is required. Also 24-h incubation is essential before reporting enterococci as sensitive to vancomycin or teicoplanin. Plates for methicillin/oxacillin susceptibility testing are incubated at 30°C in air before inoculation.

Reading

The diameter of the zone of inhibition is measured to the nearest millimetre with a ruler, callipers or an automated zone reader. A template may also be used for interpreting susceptibility (a programme for constructing templates may be downloaded from the BSAC website: www.bsac.org.uk). The susceptibility of a strain is determined by comparing the zone size obtained with breakpoint sizes given in the table published in conjunction with the method. The zone size for a given antibiotic will vary according to the species being tested and the concentration of the disc. Examples of disc concentrations, MIC and zone size breakpoints and their interpretation are shown in Tables 12.2 and 12.3.

Anaerobes and fastidious organisms

Anaerobes and fastidious organisms require special conditions (King, 2001). For those that will grow well in 24 h the BSAC recommend Wilkins–Chaldren agar (Oxoid) with 5% horse blood. A suspension equivalent to a 0.5 McFarland Standard is prepared and diluted 1 : 100 for Gram-negative organisms and 1 : 10 for Gram-positive organisms. Incubation is at 35–37°C under strict anaerobic conditions for 18–24 h.

For slow growing anaerobes and *Helicobacter pylori* the E test (see below) is recommended. When testing *Campylobacter* spp. an undiluted 0.5 McFarland suspension is inoculated on to Iso-Sensitest agar with 5% horse blood, incubated at 42°C or 37°C under microaerophilic conditions for 18–24 h. Disc testing is not suitable for *Legionella* spp. MICs should be determined on Buffered Yeast Extract agar with 5% horse blood to which different concentrations of the antimicrobials have been added (at 45°C).

National Committee for Clinical Laboratory Standards method

This is the standard method used in the USA and is summarized below. For full details including control organisms, disc concentrations and the interpretation of zone sizes consult relevant NCCLS (2003) publications.

Media

Mueller–Hinton agar is used for most organisms. For *Streptococcus* spp. 5% defibrinated sheep blood is added. For *Haemophilus* spp. use Mueller–Hinton agar supplemented with 15 mg/l β-NAD, 15 mg/l bovine haemin and 5 g/l yeast extract. GC agar with a defined growth supplement is used for *N. gonorrhoeae*.

Inoculum

A suspension in broth or other suitable diluent, with a turbidity equivalent to a 0.5 McFarland Standard, is used as the inoculum.

Incubation

Plates are normally incubated at 35°C for 16–18 h in air. For streptococci, *Neisseria* spp. and *Haemophilus* spp. they are incubated in 5% CO_2, and for anaerobes in an anaerobic cabinet.

Table 12.2 Examples of MIC breakpoints, disc contents and zone diameters for the BSAC susceptibility method[a]

Organisms and antibiotic	MIC breakpoint (mg/l)			Disc content (μg)	Interpretation of zone diameters (mm)		
	R ≥	I	S ≤		R ≥	I	S ≤
Enterobacteriaceae and *Acinetobacter*							
Ampicillin	16	–	8	10	17	–	18
Cefotaxime	2	–	1	30	29	–	30
Cefuroxime	16	–	8	30	19	–	20
Ciprofloxacin	2	–	1	2	17	–	18
Co-amoxiclav	16	–	8	20/10	17	–	18
Gentamicin	2	–	1	10	19	–	20
Trimethoprim	4	1–2	0.5	2.5	14	15–19	20
Pseudomonas spp.							
Amikacin	32	8–16	4	30	17	18–21	22
Ceftazidime	16	–	8	30	23	–	24
Ciprofloxacin	8	2–4	1	1	9	–	10
Gentamicin	8	2–4	1	10	14	15–21	22
Imipenem	8	–	4	10	21	–	22
Piperacillin/Tazobactam	32	–	16	75/10	23	–	24
Staphylococci							
Ciprofloxacin	2	–	1	1	17	–	18
Erythromycin	1	–	0.5	5	19	–	20
Fusidic acid	2	–	1	10	29	–	30
Gentamicin	2	–	1	10	19	–	20
Linezolid	8	–	4	10	19	–	20
Methicillin	8	–	4	5	14	–	15
Oxacillin	4	–	2	1	14	–	15
Penicillin	0.25	–	0.12	1 unit	24	–	25
Teicoplanin	8	–	4	30	14	–	15
Vancomycin	8	–	4	5	11	–	12
S. pneumoniae							
Cefotaxime	2	–	1	5	29	–	30
Erythromycin	1	–	0.5	5	19	–	20
Penicillin	2	0.12–1	0.06	Oxacillin 1	19	–	20
Enterococci							
Ampicillin	16	–	8	10	19	–	20
Gentamicin	1024	–	512	200	9	–	10
Linezolid	8	–	4	10	19	–	20
Teicoplanin	8	–	4	30	19	–	20
Vancomycin	8	–	4	5	12	–	13
Haemolytic streptococci							
Penicillin	0.25	–	0.12	1 unit	19	–	20
Erythromycin	1	–	0.5	5	19	–	20
Cefotaxime	2	–	1	5	27	–	28

(continued)

Table 12.2 (continued)

Organisms and antibiotic	MIC breakpoint (mg/l)			Disc content (μg)	Interpretation of zone diameters (mm)		
	R ≥	I	S ≤		R ≥	I	S ≤
M. catarrhalis							
Ampicillin	2	–	I	2	29	–	30
Co-amoxiclav	2	–	I	2/1	18	–	19
Erythromycin	I	–	0.5	5	27	–	28
Cefuroxime	2	–	I	5	19	–	20
N. gonorrhoeae							
Penicillin	2	0.12–1	0.06	I unit	17	18–25	26
Cefuroxime	2	–	I	5	19	–	20
Nalidixic acid[b]				30			
N. meningitidis							
Penicillin	0.12	–	0.06	I unit	24	–	25
Cefotaxime	2	–	I	5	29	–	30
Rifampicin	2	–	I	2	29	–	30
Ciprofloxacin	2	–	I	I	31	–	32
H. influenzae							
Ampicillin	2	–	I	2	19	–	20
Co-amoxiclav	2	–	I	2/1	19	–	20
Cefotaxime	2	–	I	5	24	–	25

[a]After BSAC guidelines (Andrews, 2001b) with permission
[b]Quinolone resistance is most reliably detected with nalidixic acid. Strains with reduced susceptibility to fluoroquinilones, e.g. ciprofloxacin, have no zone of inhibition with nalidixic acid
R, resistant; S, sensitive; I, intermediate

Table 12.3 Examples of MIC breakpoints, disc contents and zone diameters for the BSAC susceptibility method for urinary tract infections (Gram-negative rods)[a]

Antibiotic	MIC BP (mg/l)			Disc content (μg)	Interpretation of zone diameters (mm)								
					Coliforms			*Escherichia coli*			*Proteus mirabilis*		
	R ≥	I	S ≤		R ≥	I	S ≤	R ≥	I	S ≤	R ≥	I	S ≤
Ampicillin	64	–	32	25				15	–	16	24	–	25
Cephalexin	64	–	32	30				15	–	16	11	–	12
Ciprofloxacin	8	–	4	I	19	–	20	19	–	20	19	–	20
Co-amoxiclav	64	–	32	20/10				17	–	18	17	–	18
Trimethoprim	4	–	2	2.5	16	–	17	16	–	17	16	–	17
Nitrofurantoin	64	–	32	200				19	–	20			
Mecillinam	16	–	8	10				13	–	14	13	–	14
Gentamicin	2	–	I	10	19	–	20	19	–	20	19	–	20

[a]After BSAC guidelines (Andrews, 2001b) with permission
R, resistant; S, sensitive; I, intermediate

COMPARATIVE METHODS

Stokes' method

This is a commonly used disc method in the UK. The diameter of the zone of inhibition of test organism is compared with that of an appropriate control growing on the same plate (Phillips *et al.*, 1991).

Media

Any medium designed specifically for susceptibility testing may be used. It must be supplemented with 5% defibrinated horse blood for tests on fastidious organisms. If the medium is not free from thymidine, add lysed horse blood for tests on sulphonamides and trimethoprim.

Controls

Use *E. coli* NCTC 10418 and NCTC 11560 to control discs containing both a β-lactam agent and a β-lactamase inhibitor, e.g. a mixture of amoxicillin and clavulanic acid (co-amoxiclav), *P. aeruginosa* NCTC 10662, *H. influenzae* NCTC 11931, *S. aureus* NCTC 6571, *Cl. perfringens* NCTC 11229, *Bacteroides fragilis* NCTC 9343.

Inoculum

In general, an acceptable inoculum can be prepared from a fully grown, nutrient-broth culture or from a suspension of several colonies emulsified in broth to give a density similar to a broth culture. It should give semi-confluent growth of colonies on the plates after overnight incubation. Reject tests with a confluent growth or clearly separated colonies and repeat.

Inoculation of plates

With the band plating method the control culture is applied in two bands on either side of the plate leaving a central area uninoculated. Use pre-impregnated control swabs or apply a loopful of an inoculum prepared as described above to both sides of the plate and spread it evenly in the two bands with a dry sterile swab. Seed the test organism

evenly in the band across the centre of the plate by transferring a loopful of the broth culture or suspension to the centre of the plate, and spreading it with a dry sterile swab. This method usually allows up to four discs to be applied to the plate.

If a rotary plating method is used, apply the control to the centre of the plate leaving an uninoculated 1.5-cm band around the edge of the plate. Apply the test organism to the 1.5-cm band. This method usually allows up to six discs to be applied to the plate.

Application of discs

After the inoculum has dried apply discs to the inoculated medium with forceps, a sharp needle or a dispenser, and press down gently to ensure even contact. Suitable concentrations of agents in discs are listed in Table 12.2.

Place discs on the line between the test and control organisms Four discs can be accommodated on a 9-cm circular plate. If a rotary plating method is used apply six discs on a 9-cm circular plate.

Incubation

Incubate tests and controls overnight at 35–37°C (30°C is recommended for methicillin/oxacillin susceptibility tests of staphylococci).

Reading of zones of inhibition

Measures zones of inhibition from the edge of the disc to the edge of the zone. As the control and test organism are adjacent on the same plate the difference between the respective zone sizes may be easily seen. If the test zones are obviously larger than the control or give no zone at all, it is not necessary to make any measurement. If there is any doubt, measure zones with callipers (preferably), dividers or a millimetre rule.

Interpretation

Interpret each zone size as follows:

- Sensitive: zone diameter equal to, wider than, or not more than 3 mm smaller than the control.
- Intermediate: zone diameter > 2 mm but smaller than the control by > 3 mm.

- Resistant: zone diameter ≤ 2 mm.

Exceptions

- Penicillinase-producing staphylococci show heaped-up, clearly defined zone edges, and should be reported as resistant, irrespective of zone size.
- Polymyxins diffuse poorly in agar so that zones are small and the above criteria cannot be applied. Tests with polymyxins should therefore be interpreted thus:
  - sensitive: zone diameter equal to, wider than, or not more than 3 mm smaller than the control
  - resistant: zone diameter > 3 mm smaller than the control.
- Zones around ciprofloxacin discs are large with some of the sensitive control strains so tests should be interpreted thus (1) if *S. aureus* or *P. aeruginosa* is used as control; or (2) if *H. influenzae* or *E. coli* is used as control:
  - sensitive: zone diameter equal to, wider than or not more than 7 mm, or 10 mm smaller than the control
  - intermediate: zone diameter > 2 mm but smaller than the control by > 7 mm or 10 mm
  - resistant: zone diameter ≤ 2 mm.

Agar dilution methods

The agar dilution method for determining MICs has advantages over broth dilution methods in that contamination is more easily seen and re-isolation of the required organism is usually not a problem.

Medium

As for diffusion tests a medium optimized for susceptibility testing should be used.

Antimicrobial agents

Use generic rather than proprietary names of agents. Obtain antimicrobial powders directly from the manufacturer or from commercial sources. The agent should be supplied with a certificate of analysis stating the potency (μg or international unit [IU]/mg powder, or as percentage potency), an expiry date, details of recommended storage conditions and data on solubility. Ideally, hygroscopic agents should be dispensed in small amounts and one used on each test occasion. Allow containers to warm to room temperature before opening them to avoid condensation of water on the powder.

Range of concentrations tested

The range of concentrations will depend on the particular organisms and antimicrobial agent being tested. Recommended ranges for major groups of organisms are given by the BSAC (Andrews, 2001b).

Preparation of stock solutions

To ensure accurate weighing of agents, use an analytical balance and, whenever possible, weigh at least 100 mg of the powder. With a balance weighing to five decimal places the amount of powder weighed may be reduced to 10 mg. Use the following formula to make allowance for the potency of the powder:

$$\text{Weight of powder (mg)} = \frac{\text{Volume (ml)} \times \text{Concentration (mg/l)}}{\text{Potency of powder (μg/mg)}}$$

Alternatively, given a weighed amount of antimicrobial powder, calculate the volume of diluent needed from the formula:

$$\text{Volume of diluent (ml)} = \frac{\text{Weight (mg)} \times \text{Potency (μg/ml)}}{\text{Concentration (mg/l)}}$$

Concentrations of stock solutions should be made up to ≥ 1000 mg/l although the insolubility of some agents will prevent this. The actual concentrations of stock solutions will depend on the method of preparing working solutions (see below). Agents should be dissolved and diluted in sterile water where possible and some agents may require other solvents or diluents (Andrews, 2001b). If required, sterilize by membrane filtration and compare samples before and after sterilization to ensure that adsorption has not occurred. Unless otherwise instructed by the manufacturer, store stock solutions frozen in small amounts at –20°C or below. Most agents will keep at –60°C for at least 6 months. Use stock solutions promptly on defrosting and discard unused solutions.

Preparation of plates

Volumes of 20 ml agar are commonly used in 9-cm Petri dishes for agar dilution MICs. Prepare a series twofold dilutions of the agent at 20 times the final required concentration. Add 1 ml of dilution to 19 ml cooled molten agar, mix and pour into pre-labelled plates immediately. Allow the plates to set at room temperature and dry them so that no drops of moisture remain on the surface of the agar. Do not over-dry the plates.

It is best to use plates immediately, particularly if they contain labile agents such as β-lactams. If the plates are not used immediately, store them in a refrigerator (4–8°C) in sealed plastic bags and use within 1 week.

Preparation of inoculum

Standardization of the density of inoculum is essential if variation in results is to be avoided. Follow the procedures described above.

Inoculation of plates

Mark the plates so that the orientation is obvious. Transfer diluted bacterial suspensions to the inoculum wells of an inoculum-replicating apparatus. Use the apparatus to transfer the inocula to the series of agar plates, including control plates without antimicrobial agent at the beginning and end of the series. Inoculum pins 2.5 mm in diameter will transfer close to 1 μl, i.e. an inoculum of 10^4 cfu/spot if the bacterial suspension contains 10^7 cfu/ml. If in doubt as to the volume delivered, check with the supplier and adjust the inoculum preparation procedure to yield a final inoculum of 10^4 cfu/spot. Allow the inoculum spots to dry at room temperature before inverting the plates for incubation.

Incubation of plates

Incubate plates at 35–37°C in air for 18 h. To avoid uneven heating do not stack them more than four high. If the incubation period is extended for slow-growing organisms, assess the stability of the agent over the incubation period. Avoid incubation in an atmosphere containing 5% CO_2 unless absolutely necessary for growth of the organisms (e.g. *Neisseria* spp.). Incubation at 30°C should be used for methicillin/oxacillin susceptibility tests on staphylococci.

Reading the results

The MIC is the lowest concentration of the agent that completely inhibits growth, disregarding up to four colonies or a thin haze in the inoculated spot. A trailing endpoint, with a small number of colonies growing on concentrations that are several dilutions above that which inhibits most organisms should be investigated. Subculture and re-testing may be necessary because this may be caused by contamination. It may also be the result of resistant variants, β-lactamase-producing organisms or, if incubation is prolonged, regrowth of susceptible organisms after deterioration of the agent. With sulphonamides and trimethoprim. endpoints may be seen as only a reduction in growth if the inoculum is too heavy or antagonists are present.

Broth dilution methods

This macrodilution method involves the use of test tubes and culture volumes of 1–2 ml, and is most suitable for small numbers of tests because preparation of many sets of tubes is tedious. Microdilution methods involve the use of plastic microdilution trays with culture volumes of 0.1–0.2 ml. In the microdilution method, multiple plates may be conveniently produced by the use of apparatus for automatically diluting and dispensing solutions.

Medium

The requirements for broth media are similar to

those for agar media. There are fewer broth media than agar media designed specifically for antimicrobial susceptibility testing. Oxoid Iso-Sensitest and Mueller–Hinton media are available in both agar and broth formulations. Supplements, other than blood, may be added to broth for fastidious organisms as for agar media.

Preparation of dilutions

The total volume of each dilution prepared will depend on the number of tests to be done. The volumes should be adjusted according to the number of tests to be done, but should not be less than 10 ml. Dispense the solutions in 1-ml volumes in tubes for the macrodilution method and 0.1-ml volumes into the wells of the microdilution plate for the microdilution method. Include control tubes or wells containing broth without antimicrobial agent to act as growth and sterility controls.

Macrodilution tubes are commonly used immediately after preparation (preferable for labile agents such as β-lactam agents), but may be stored at 4°C for up to 1 week. Microdilution trays may be kept in sealed plastic bags stored at –60°C or below for up to 1 month. Do not refreeze trays once they have been thawed.

Preparation of inoculum

Prepare the inoculum as for the agar dilution method but adjust the density so that, after inoculation, the final volume of test broth contains $10^5 – 5 \times 10^5$ cfu/ml. The density of the inoculum required depends on the volume of broth in the tubes or wells and the inoculation system used.

Inoculation of tubes/plates

Inoculate macrodilution tubes containing 1 ml broth with 0.05 ml inoculum containing 5×10^6 cfu/ml or add 1 ml inoculum of 5×10^5 cfu/ml in broth to 1-ml volumes of broth. The latter inoculum requires the preparation of a double-strength dilution series of the agent to compensate for the 1 in 2 dilution on inoculation.

Inoculate microdilution wells containing 0.1-ml broth with 0.005 ml inoculum containing 5×10^6 cfu/ml. Manual inoculation with a pipette is easier if 0.05 ml inoculum is added to 0.05-ml volumes of broth in the dilution series. As with the macrodilution method, inoculum density and antimicrobial concentrations are adjusted to compensate for the 1 in 2 dilution of agents on inoculation.

Incubation of tubes/plates

Seal microdilution plates in a plastic bag, with plastic tape or some other method to prevent the plates from drying. To avoid uneven heating, do not stack microdilution plates more than four high. Incubate tubes or trays at 35–37°C as with agar dilution plates.

Reading results

The MIC is the lowest concentration of the agent that completely inhibits growth. The control containing no agent should be turbid.

MINIMAL BACTERICIDAL CONCENTRATION

The MBC is the lowest concentration of an agent that kills a defined proportion (usually 99.9%) of the population after incubation for a set time. The method is usually an extension of the broth dilution MIC. After reading the MIC, organisms are quantitatively subcultured from tubes or wells showing no growth to antimicrobial-free agar medium. After incubating the plates, the proportion of non-viable organisms, compared with the original inoculum, is assessed.

The MBC is used to guide therapy in situations such as endocarditis, where the bactericidal activity of an agent is considered essential.

TIME–KILL CURVES

These are an extension of the MBC and give information on the rates at which organisms are killed.

The rate is measured by counting the number of viable cells at different time intervals after exposure to various concentrations of an agent.

TOLERANCE

This is the ability of strains to survive but not grow in the presence of an agent and relates particularly to Gram-positive organisms tested against agents acting on the terminal stages of peptidoglycan synthesis, i.e. β-lactam and glycopeptide agents. Strains that have an MIC : MBC ratio of $\geq 1 : 32$ after 24-h incubation are usually termed 'tolerant'. However, MICs and MBCs are determined after fixed periods of incubation and tolerance may be a reflection of a lower rate of killing. Technical variation has a marked effect on the detection of tolerance (Sherris, 1986; Thrupp, 1986).

BREAKPOINT METHODS

Agar dilution breakpoint susceptibility testing methods are essentially highly abbreviated agar dilution MIC tests, in which only one or two critical antimicrobial concentrations are tested. With a single breakpoint concentration growth = *resistant* and no growth = *susceptible*.

Using a lower and a higher breakpoint concentration no growth at either concentration = *susceptible*, growth at the lower concentration only = *intermediate*, and growth at both concentrations = *resistant*.

The general considerations detailed in the section on dilution methods apply equally to agar incorporation breakpoint susceptibility testing methods. It is essential to inoculate growth control plates of the same basal medium as the antimicrobial-containing plates at the beginning and end of the inoculation run and to incubate them under the same conditions as the test. Adequate growth of the test isolates on these control plates allows a confident interpretation of susceptibility based on the absence of growth on antimicrobial containing plates because the antimicrobial content is the only variable.

Purity plates

Purity plates (which may be half, third or quarter plates) are essential and should be of a suitable non-selective growth medium, e.g. blood agar. Prepare these by subculturing a standard 1-µl loopful from the inoculum 'pots' used for the inoculation 'run', after all antimicrobial test and growth control plates have been inoculated. In this way mixed cultures should be detected and the use of a standard loop allows an approximate check on the actual inoculum density used for each test organism.

Preparation of breakpoint plates

Preparing sets of plates for breakpoint testing from pure antimicrobial agent powders is a complex task. Accurate weighing of powders, dilution of the solutions and quality control of the plates are essential. Pre-prepared antimicrobials at specific concentrations for addition to molten agar are available from a number of manufacturers (e.g. Mast Adatabs).

Breakpoint concentrations to be tested

Some examples of appropriate concentrations are shown in Tables 12.2 and 12.3; for a full list consult the BSAC or NCCLS publications (Andrews, 2001b; NCCLS, 1997, 1999).

Reading and interpretation of breakpoint susceptibility tests

The 'interpretation' of breakpoint sensitivity testing has two elements: the reading of the laboratory tests and the assessment of their clinical relevance. In practice, the selected breakpoints are those considered to provide clinically relevant results and are usually based on one or two concentrations of a given agent. The results can then be interpreted as 'susceptible' or 'resistant' when one concentration is used or as 'susceptible', 'intermediate' and 'resistant' when two are used.

The reading of breakpoint plates is the final stage of a susceptibility testing process, which should adhere to standard methods for preparation and storage of media and antimicrobial agents, for preparation of the inoculum and for conditions of incubation.

The results are usually unambiguous and free from subjective interpretation. Automated computerized equipment is available for reading plates and recording the results, e.g. Mastascan Elite.

Quality assessment breakpoint susceptibility tests

Check that the inoculum is pure, of the expected species and of approximately the correct density – the use of a standard loop to inoculate the purity plate provides for both of these requirements.

Check that the control has grown adequately. It is useful to have at least one antimicrobial agent in the test set to which isolates are likely to be resistant, e.g. aztreonam for Gram-positive and vancomycin for Gram-negative isolates. This acts as an additional control of adequate growth and can aid the recognition of mixtures of Gram-positive and Gram-negative organisms. Computer printouts of data or manual lists are useful for this purpose. These lists should also be checked for unexpected sensitivity patterns that could be caused by mixed cultures, incorrect subculture and inversion of the plates, or simply reading the wrong plate. They are also useful for alerting the control of infection and medical teams to potential epidemiological or treatment problems.

Organization of breakpoint susceptibility testing schemes

The range of antimicrobial agents selected for inclu-

sion within a breakpoint testing scheme must be related to the species of organisms to be tested and the clinical site of infection from which they are isolated. Some laboratories use breakpoint methods only for infections for which the usual causative organisms can be predicted with some accuracy, and the range of agents to be tested is limited. The most common example is urinary tract infections. Others adopt a broader approach and test all aerobic Gram-negative bacilli whatever their clinical source against the same set of antimicrobial agents. Similarly a set of antimicrobial agents may be selected for breakpoint testing against staphylococci and enterococci. If the susceptibility testing medium used is supplemented with growth additives, e.g. 5% lysed horse blood and nicotinamide adenine dinucleotide (NAD), the set of antimicrobial agents may be extended to cover susceptibility testing of more fastidious species such as pneumococci, other streptococci, *H. influenzae* and *Branhamella catarrhalis*.

One system described by Franklin (1990) uses a susceptibility testing medium capable of supporting the growth of virtually all bacteria of medical importance. All bacterial isolates are tested against the same wide range of agents, usually at more than one breakpoint concentration. The results produced by such a scheme will not be relevant to a particular species/genus or site of infection. Selectivity in reporting is necessary.

THE E TEST

This system for testing MICs represents a different approach from quantitative testing of antimicrobial activity and has some advantages over conventional methods. The commercial E test strips are 50 mm × 5 mm plastic carriers with an exponential antimicrobial gradient dried on one side and a graduated MIC scale on the other. To set up an E test, inoculate agar medium by flooding or with a swab in the same way that plates are inoculated for disc diffusion susceptibility tests. Up to six E test strips can be placed in a radial pattern on a 15-cm plate, or a single strip may be placed on a 9-cm plate. On incubation, elliptical zones of inhibition are produced and the MIC is read directly from the graduated E test strip at the point of intersection of the zone of inhibition with the strip. Include a control strain with each E test by inoculating one half of a plate with the test organism and the other half of the plate with the control organism and apply the E test strip to the line between the two organisms (ABBiodisk, Sweden).

β-LACTAMASE TESTING

Rapid biochemical tests may be used to detect β-lactamase production (Livermore and Brown, 2001). There are three main methods: those based on chromogenic cephalosporins are the most sensitive but are not suitable for staphylococci. Staphylococcal β-lactamase may require induction and can be detected with iodometric or acidometric methods. The later methods are also suitable for detecting production by *N. gonorrhoeae* and *H. infuenzae*. Extended spectrum β-lactamases (ESBLs) are plasmid-mediated enzymes produced by a number of enterobacteria and *P. aeruginosa*. In *E. coli* and *Klebsiella* spp., these can be detected by a combined disc method (Livermore and Brown, 2001). An Iso-Sensitest plate is inoculated with the test strain to give a semi-confluent growth. A ceftazidime 30 μg disc and an amoxicillin/clavulanic acid 20 + 10 μg disc are placed 25–30 mm apart. After overnight incubation at 37°C, if the zone of inhibition round ceftazidime disc is expanded ESBL production is inferred. ESBL production may also be detected by an E-test ESBL strip and by inference from cephalosporin resistance. Methicillin (5 μg) or oxacillin (1 μg) discs may be used to detect intrinsic resistance to penicillin of pneumococci.

DIRECT (PRIMARY) SUSCEPTIBILITY TESTS

Tests in which the specimen is the inoculum can be useful because the results are available a day earlier than tests on pure cultures; isolation from mixed cultures may be easier if there are differences in susceptibility and small numbers of resistant variants may be seen within zones of inhibition in diffusion tests. There are some potential disadvantages, in that the inoculum cannot be controlled, care must be taken to avoid reporting results on commensal organisms, and the specimens on which tests are likely to be useful are limited to those from sites that are normally sterile and, in the case of urine, to specimens shown microscopically to contain organisms. Direct tests should not be undertaken with specimens from patients on antimicrobial therapy (Waterworth and Del Piano, 1976)

For direct tests by the Stokes' method, pus swabs may be streaked in place of the test organisms or well-mixed urines used in place of broth cultures or suspensions of pure cultures of organisms. Use *E. coli* NCTC 10418 as a control for urines and *S. aureus* NCTC 6571 for other specimens. With breakpoint methods only urine specimens can be tested. With the important and common exception of direct susceptibility testing on urines, repeat all tests on pure cultures and only report direct tests as 'provisional'.

QUALITY CONTROL

Departure from the usual range of routine susceptibility may indicate gross errors in the method. Routine quality control is necessary, however, to maintain high standards (Franklin, 1990). External quality control schemes provide independent assessment of the performance of a test, and allows performance to be compared with that of other laboratories, although internal quality control is necessary to detect day-to-day variation.

Standard strains must be used to control the performance of the method. Test colonies of control cultures in the same manner as test cultures. Include control cultures in each batch of tests and use them to evaluate new lots of agar or broth before they are put into routine use.

Diffusion tests

A rapid examination of control zone sizes will indicate major problems. Recording of control zone sizes on a chart allows problems to be more easily detected. Greater control may be exercised by establishing zone-size limits. If control tests indicate problems the source of error should be investigated. Common problems are:

- Inactivation of labile agents in discs as a result of inadequate storage or handling in the laboratory may be indicated by a gradual decrease in zone sizes. New batches should be tested before being put into routine use.
- Too heavy or too light an inoculum may be indicated by a general decrease or increase in zone sizes.
- Errors in transcription or measuring zone sizes. In particular, different observers reading zone edges differently will result in fluctuating zone sizes.
- Problems with medium, e.g. high pH, will produce larger zones with aminoglycoside antimicrobials and erythromycin, and smaller zones with tetracycline, methicillin and fusidic acid. The reverse may occur if the pH is too low. New batches of medium should be tested before being put into routine use. Fluctuating zone sizes may indicate variation in the depth of medium.
- Contamination or mutational changes in the control strain.

Minimum inhibitory concentrations

In general, the MICs of control organisms should be within plus or minus one dilution step of the target values. In addition:

- For broth dilution tests, incubate one set of tubes or one microdilution plate uninoculated to check for sterility (media contaminants on agar dilution plates are usually obvious).
- The control without antimicrobial agents should show adequate growth of test and control strains.
- Plate a sample of inoculum prepared for each strain on a suitable agar medium to ensure that the inoculum is pure.
- Occasionally check that the inoculum density is correct by counting organisms in inocula (taken from growth controls with broth dilution methods).
- Check that endpoints are read consistently by all staff independently reading a selection of tests.

Breakpoint methods

In addition to the foregoing comments on quality control, the following points should be borne in mind by users of breakpoint methods. Monitor the concentrations of antimicrobial agents in the plates by using control strains with known MICs for the agents. To achieve suitable control these strains should have MICs only one or two dilutions above (for the resistant control) or below (for the sensitive control) the concentration in the breakpoint plates.

When breakpoint control strains of known MIC are used, the breakpoint testing method must be identical to that used for the initial selection of the control strains in order to ensure reliability of performance. However, a batch of 'breakpoint' plates with, say, six different antimicrobial agents with two breakpoints each would require up to 24 control organisms with appropriate MICs to check the levels of agents adequately in all plates. If performed every day, this extensive control procedure would leave little room for testing the clinical isolates! As plates can be stored without deterioration for up to 1 week, a sample of each batch of media may be subjected to this degree of control. In addition, new batches of media may be checked by initially running them in parallel with the previously controlled plates. For certain agents, where there is some evidence that degradation of antimicrobial activity may occur during storage of plates (e.g. co-amoxiclav and imipenem), daily microbiological quality control may be advisable. Similarly it is advisable to include a 'difficult' control strain of methicillin-resistant *S. aureus* (MRSA) on a daily basis to ensure reliable detection of methicillin-heteroresistant staphylococci. (Control strains for some breakpoints are available from the Antibiotic Reference Laboratory, Public Health Laboratory Service, Colindale, London.)

Franklin (1990) described a method to overcome the practical difficulties of using numerous control strains. By cutting plugs of agar from a batch of the prepared plates and placing them on the surface of a seeded, non-inhibitory culture plate, the antimicrobial content can be monitored by the size of the inhibition zone around the agar plug. As deviation from the expected zone size indicates error, an acceptable range has first to be determined. Establishing and validating this method of quality control may be at least as complex as searching for a small number of strains with appropriate MICs.

Inoculum size is critical; therefore, the careful cleaning, maintenance and calibration of inoculating pins are necessary.

Transcription errors are always possible and appropriate staff training can reduce their likelihood. Automated reading of the plates reduces this risk. If plates are read visually then plates containing antimicrobial agents with similar sounding names (e.g. the various cephalosporins) should be separated in the order of plate reading to avoid confusion. A legibly labelled template showing pin positions should be available to the plate reader. Worksheets should be simple and easy to fill in. All isolates with unusual susceptibility patterns should be checked and, if necessary, investigated further.

REFERENCES

Andrews, J. M. (2001a) Determination of minimum inhibitory concentrations. *Journal of Antimicrobial Chemotherapy* 48(suppl 1): 5–16.

Andrews, J. M. (2001b) BSAC standardized disc susceptibility testing method. *Journal of Antimicrobial Chemotherapy* 48(suppl 1): 43–57.

Barry, A. L. (1976) *Antimicrobial Susceptibility Test: Principles and practice.* Philadelphia, PA: Lea & Febiger, pp. 163–179.

Barry, A. L. (1991) Procedures and theoretical considerations for testing antimicrobial agents in agar medium. In: Lorian, V. (ed.), *Antibiotics in Laboratory Medicine* 3rd edn. Baltimore, MA: Williams & Wilkins, pp. 1–16.

Brown, D. F. J. and Blowers, R. (1978) Disc methods of sensitivity testing and other semiquantitative methods. In: Reeves, D. S., Phillips, I., Williams, J. D. *et al.* (eds), *Laboratory Methods in Antimicrobial Chemotherapy.* Edinburgh: Churchill Livingstone, pp. 8–30.

Ericsson, M. M. and Sherris, J. C. (1971) Antibiotic sensitivity testing. Report of an international collaborative study. *Acta Pathologica et Microbiologica Scandinavica* (Section B) Supplement 217.

Franklin, J. C. (1990) Quality control in agar dilution sensitivity testing by direct assay of the antibiotic in the solid medium. *Journal of Clinical Pathology* 33: 93–95

King, A. (2001) Recommendations for susceptibility tests on fastidious organisms and those requiring special handling. *Journal of Antimicrobial Chemotherapy* 48(suppl 1): 77–80.

Livermore, D. M. and Brown, D. F. J. (2001) Detection of β-lactamase mediated resistance. *Journal of Antimicrobial Chemotherapy* 48(suppl 1): 59–64

National Committee for Clinical Laboratory Standards (NCCLS) (1996) *Evaluating Production Lots of Dehydrated Mueller–Hinton Agar.* Approved standard M6-A. Wayne, PA: NCCLS.

NCCLS (2003) *Performance Standards for Antimicrobial Disc Susceptibility Tests.* Approved standard M2-A8. Wayne, PA: NCCLS.

Phillips, I., Andrews, J., Bint, *et al.* (1991) A guide to sensitivity testing. Report of the Working Party on Antibiotic Sensitivity Testing of the British Society for Antimicrobial Chemotherapy. *Journal of Antimicrobial Chemotherapy* 27(suppl D): 1–50.

Sherris, J. C. (1986) Problems of *in vitro* determination of antibiotic tolerance in clinical isolates. *Antimicrobial Agents and Chemotherapy* 30: 633–637.

Thrupp, L. D. (1986) Susceptibility testing of antibiotics in liquid media. In: Lorian, V. (ed.), *Antibiotics in Laboratory Medicine*. Baltimore, MA: Williams & Wilkins, pp. 93–150.

Waterworth, P. M. and Del Piano, M. (1976) Dependability of sensitivity tests in primary culture. *Journal of Clinical Pathology* **29**: 179–184.

Wheat, P. (2001) History and development of antimicrobial susceptibility testing methodology. *Journal of Antimicrobial Chemotherapy* **48**(suppl 1): 1–4.

13
Food poisoning and food-borne disease

In 1992, the Advisory Committee on the Microbiological Safety of Food defined food poisoning as 'any disease of an infectious or toxic nature caused by or thought to be caused by the consumption of food or water'. The definition had been previously adopted by the World Health Organization (WHO, 1998). Although the words 'food poisoning' are used in public health legislation, 'food-borne disease' is now the generally preferred term for these conditions (Department of Health, 1994). Some predominantly food-borne infections, e.g. salmonellosis, can also be transferred by the faecal–oral route, and some infections predominantly spread by the faecal–oral route, e.g. hepatitis A and *Shigella* spp., can occasionally be food borne.

An outbreak of food-borne disease is an incident in which two or more people, thought to have a common exposure, experience a similar illness or proven infection with at least one of them having been ill. A general outbreak is one that affects members of more than one household, or residents of an institution. Food-borne disease encompasses illness caused by a variety of different agents, including:

- bacterial infections (e.g. with salmonellas, campylobacters)
- pre-formed bacterial toxins (e.g. botulism)
- other biological toxins (e.g. paralytic shellfish, scombrotoxin, poisoning)
- viral infections (e.g. noroviruses)
- parasitic infections (e.g. protozoa)
- toxic chemicals (e.g. heavy metals).

As a result of the diversity of the agents involved, the clinical features of food-borne disease vary considerably. Typical symptoms and incubation times provide an indication of the likely cause, but microbiological tests are required to confirm any such diagnosis. Most food-borne disease organisms exert their influence on the gastrointestinal tract, giving rise to 'classic' food-borne disease symptoms such as nausea, vomiting, diarrhoea, abdominal pains, fever, etc.

Some toxins, however, act on other systems (e.g. nervous, immune), producing a wide variety of symptoms. Some characteristics of the more important food-borne disease bacteria are summarized in Table 13.1, which gives details of main sources, the types of food they are most commonly associated with and their preferred growth conditions.

The true incidence of food-borne infection is difficult to determine because asymptomatic infection is common, only a minority of people with symptomatic infection will seek medical treatment and only a minority of patients will be investigated microbiologically. There is a statutory requirement to notify suspected food-borne disease in most countries, and many countries also conduct surveillance by collating laboratory results. Food-borne infection appears to be one of the most common infectious diseases, and the incidence is increasing. For a review of the factors that have contributed to the changing patterns of food-borne disease, see WHO (1998).

Table 13.1 Food-borne disease: agents, symptoms and sources

Bacteria	Symptoms and possible consequences	Incubation period	Duration of illness	Infective dose (cells/g)	Microbiology/Growth characteristics	Sources
Salmonella spp.	Diarrhoea, vomiting, abdominal pain, high temperature. Blood poisoning (septicaemia) and inflammation of the abdominal wall (peritonitis in severe cases)	Up to 4 days[a]	Up to 3 weeks[b]	Risk varies with nature of food and susceptibility of individual	Non-sporing, temperature range 10–50°C but some grow down to 6–8°C, acid resistant (grow at pH > 4.0)	Foods not cooked at an adequate temperature or contaminated by raw foods. Mainly meat (especially poultry), eggs and dairy products. Also infected food handlers and pets
Clostridium perfringens	Diarrhoea, abdominal pain	8–22 h[c]	~24 hours	~1 000 000 cells in total	Heat-resistant spores. Temperature 15–50°C. Optimum pH 6–7, anaerobe	Cooked meat dishes stored at the wrong temperature
Escherichia coli O157	Abdominal pain, vomiting, diarrhoea that may contain blood, kidney failure. Can be fatal. Young children and elderly people particularly vulnerable	12–60 h	Varies	Very high infectivity (e.g. tens of cells)	Non-sporing. Temperature 10–45°C. Acid resistant (grows at pH 4.5)	Beef and beef products, unpasteurized dairy products. Unpasteurized fruit juice, raw vegetables. Direct contact with animals, infected people and bathing in contaminated water
Bacillus cereus	Nausea, vomiting, diarrhoea, abdominal pain	10 min–16 h	24–36 h	~100 000 or preformed toxin	Heat resistant spores. Temperature 10–50°C. pH > 4.3	Cooked food stored at the wrong temperature – mainly rice dishes and pastry products containing meat or vegetables. Occasionally dairy products and bakery products

Table 13.1 (continued)

Bacteria	Symptoms and possible consequences	Incubation period	Duration of illness	Infective dose (cells/g)	Microbiology/Growth characteristics	Sources
Campylobacter spp.	Fever, headache and dizziness followed by abdominal pain and diarrhoea	Up to 10 days[d]	Varies, usually a week	~100	Non-sporing, only grows > 30°C, pH 6.5–7.5, prefers low oxygen, does not generally grow in foods	Raw or undercooked meat, especially poultry, unpasteurized milk, milk from bird-pecked bottles, untreated water Also pets
Staphylococcus aureus	Vomiting, abdominal pain, diarrhoea	1–7 h[e]	Up to 2 days	100 000 to 1 000 000 cells will produce enough toxin to cause illness	Non-sporing, some strains grow down to 7°C, pH range 4.5–9.3 (optimum ~7)	Human contamination of food Mainly fish, prawns and cream cakes handled and stored at the wrong temperature
Listeria monocytogenes	Mild flu to septicaemia and meningitis Can cause miscarriage, stillbirth or premature labour in pregnant women	3–70 days	Varies	Relatively low infectivity	Non-sporing, temperature optimum 30–37°C but grows down to –1°C, acid resistant (grows at pH > 4.3)	Occurs naturally in the environment Mainly raw vegetables and meat, pâtés and soft cheeses

[a] Usually 12–48 hours
[b] But may be a carrier for 12 weeks or more after symptoms have gone
[c] Usually 12–18 hours
[d] Usually 2–5 days
[e] Usually 2–4 hours

Source: Health Protection Agency, Communicable Disease Surveillance Centre

INFECTIVITY

The infectivity of the different organisms depends on the following:

- Type of bacteria: relatively large numbers (e.g. ≥ 100 000 cells) of most types of food-borne disease bacteria are generally required to cause illness, but verocytotoxin-producing *Escherichia coli* (VTEC) and *Campylobacter* spp., and some species of *Shigella* are notable exceptions, where illness can result from just a few tens or hundreds of cells.
- Type of food: some types of food (e.g. high fat content) protect bacteria from the harsh acidic conditions in the human stomach, so that food-borne disease may result from very much lower doses of bacteria than would normally be the case (e.g. *Salmonella* in chocolate).
- Susceptibility of the individual: in general, very young, old or immunocompromised individuals are most at risk.

DEALING WITH OUTBREAKS OF FOOD-BORNE DISEASE

In the UK, responsibility and management of outbreaks of food-borne disease fall jointly to the local authority (LA) and the Primary Health Care Trust (PHA). All LAs and PHAs have contingency plans drawn up in consultation with other experts such as the Health Protection Agency (HPA), local consultants in communicable disease control (CCDCs), chief environmental health officers, etc. Outbreaks are identified via surveillance systems, public complaints, etc. Physicians have a statutory duty to notify the appropriate local officer of all cases (and suspected cases) of food-borne disease. Identification may also arise by other routes, such as through diagnostic laboratories alerting CCDCs or LAs if they encounter pathogenic bacteria in specimens that they have tested. Initial investigations will lead to a preliminary hypothesis as to the likely cause of the outbreak, in terms of the

food and the food-borne disease agent involved, the number of people exposed to it, etc. This, in turn, should lead to an indication of the control measures needed to protect public health.

In the US, physicians report outbreaks of food-borne disease to the local Health Departments, which inform the State Health Departments. They, in turn, report to the Centers for Disease Control and Prevention.

TYPES OF FOOD-BORNE AGENTS

Salmonellas

The two major groups are those causing enteric fevers (typhoid and paratyphoid) and those causing food-borne disease (non-typhoid salmonellas). The enteric fever group consists of *Salmonella typhi* and *Sal. paratyphi* A and B. With rare exceptions these organisms infect only humans. The non-typhoid group consists of 2200 serotypes, all of which have their primary reservoirs in animals. Salmonellas can grow in many foods at temperatures from 10°C to 50°C. Pasteurization temperatures destroy them. Salmonellas survive well outside a person or an animal in faeces, on vegetables, in animal feeds and in many other foods for long periods. Fewer than 10 serotypes are responsible for most of the cases of human salmonellosis in the UK – predominantly *S. enteritidis* and *S. typhimurium* (PHLS, 2000).

Campylobacters

The species causing food-borne disease are *Campylobacter jejuni* and *C. coli*. They are present as part of the normal gut flora of many animals including mammals, birds and reptiles. Their growth temperatures are more exacting than for salmonellas, and the organisms do not grow in food under normal circumstances. Growth occurs up to 42°C and the organisms survive well in water and raw milk. Poultry appear to be responsible for some sporadic outbreaks. The organisms are heat sensitive and destroyed by pasteurization, and do not survive in acid foods. In common with salmonellas, control measures such as adequate cooking and prevention of recontamination are successful.

Clostridium perfringens

This organism is found in animal faeces, soil, dust, vegetation and elsewhere in the environment. It will grow only anaerobically, and produces a variety of enterotoxins, some of which act on the gastrointestinal tract. Spores may be formed in adverse conditions and enable the organism to survive circumstances that would normally kill vegetative bacteria. Five types are recognized on the basis of toxins and enzymes produced. Type A strains are responsible for most food-borne disease outbreaks. Young and long-stay patients and hospital staff tend to become colonized with C. perfringens. Counts of 10^5/gram in the faeces of sufferers are not uncommon.

As the heat resistance of C. perfringens spores varies, the organisms survive some cooking processes. Much depends on the amount of heat that reaches the cells and the period of exposure. If the spores survive and are given suitable conditions, they will germinate and multiply. It is common practice to cook a large joint of meat, allow it to cool, slice it when cold and re-heat it before serving. This is hazardous unless cooling is rapid and re-heating is thorough and above 80°C. Growth of heat-sensitive strains in cooked food may be the result of too low a cooking temperature or recontamination after cooking.

Staphylococcus aureus

Many people carry S. aureus in their noses and many carry it on their skin. Enterotoxin production is common and there are eight serologically distinct types (A, B, C_1, C_2, C_3, D, E and F), all of which cause an intoxication type of disease. Growth occurs on many foods, particularly those with a high protein content, and the toxin is readily produced at ambient temperatures. The organisms are killed at 60°C in 30 min, but the toxin will withstand heating at 100°C for 30 min. Infection occurs usually in the kitchen from human contamination (infected cuts, boils and other lesions or from nasal carriers).

Bacillus cereus

This organism is widespread in the environment and is particularly, but not exclusively, associated with foods involving grains and cereals. Spores are formed that resist boiling and frying for short periods, and will germinate in ideal conditions for growth. Subsequent heating does not destroy any toxins that may have formed. B. cereus strains produce two kinds of toxin: diarrhoeal and emetic. The emetic type is more common. Occasionally other Bacillus spp. may be responsible for similar symptoms (see Gilbert et al., 1981).

Clostridium botulinum

The pre-formed toxin of C. botulinum affecting the central nervous system (botulism) is rare in the UK, but more prevalent in Europe and the USA where home canning of meat and vegetables is common practice. This anaerobic, spore-forming organism is extensively distributed in the environment. The organism will grow particularly well on low-acid foods (i.e. with a pH value higher than 4.5) such as fish, fruit and vegetables. There are seven antigenic types, A–G, each with its own distinct toxin. Human intoxications are usually with types A, B and E, although type G has caused one outbreak. The toxins, powerful as they are, are readily destroyed by heat. The spores of type A resist boiling for several hours; those of other types are slightly less resistant. Conditions must be optimum for toxin formation: complete anaerobiosis, neutral pH and absence of competing organisms. Between 12 and 36 h after eating food containing the toxin, patients develop symptoms including thirst, vomiting, double vision and pharyngeal paralysis, lasting 2–6 days and usually fatal. Vehicles are usually home-preserved meat or vegetables, contaminated with soil or animal faeces and insufficiently heated. Correct commercial canning processes destroy the organisms.

Listeria monocytogenes

This organism is widely distributed in the gastrointestinal tract of animals and therefore the environment generally. *L. monocytogenes* can grow over a wide range of temperatures from as low as –1°C, are comparatively resistant to disinfectants and changes in pH, and can survive extreme environmental conditions (it is hardly surprising that they are recovered from some ice-creams). Strains may be distinguished by serotyping, types 1/2a and 4b being most commonly found in strains isolated from pathological sources. Listeriosis affects mainly young children, pregnant women and elderly people. The organisms are present in small numbers in many foods, but are unlikely to cause disease unless they have the opportunity to multiply. This may happen at quite low temperatures, e.g. in improperly maintained cold storage cabinets and refrigerators.

Yersinia enterocolitica

This is an uncommon cause of food poisoning but its ability to grow at 4°C makes it unusual and a potential hazard. The organism is widely distributed in animals and the environment. Foods implicated include milk, ice-cream and seafoods. The symptoms are diarrhoea, fever, abdominal pain and vomiting.

Vibrio parahaemolyticus

This marine organism is found in coastal and brackish waters. It is halophilic. Infection is rare in the UK but it is one of the most common causes of food-borne disease in Japan and has been incriminated in outbreaks in the USA and Australia.

Infection is associated with eating raw and processed fish products. The organism has been frequently isolated from seafoods, particularly those harvested from warm coastal waters. Symptoms of infection may appear from 2 to 96 h after inges-

tion, depending on the size of the infecting dose, type of food and acidity of the stomach. They vary from acute gastroenteritis with severe abdominal pain to mild diarrhoea.

Escherichia coli

Large numbers of *E. coli* are present in the human and animal intestines. They are often found in raw foods where they may indicate poor quality. At least four different types (see p. 287) may be involved in food poisoning. *E. coli* may cause travellers' diarrhoea when the source may be contaminated food or water. Verocytotoxin-producing *E. coli* O157 is by far the most frequently reported strain in human cases and appears to be largely of beef or milk origin.

Shigellas

The dysentery group of organisms are pathogens of humans and are usually transmitted by the faecal–oral route in circumstances of poor general hygiene. Food contamination readily occurs, although *Shigella* spp. are rarely associated with food-borne disease in developed countries.

Mycotoxins and aflatoxins

Certain fungi, mainly *Aspergillus* spp., form toxins in nuts and grains during storage. These toxins can cause serious disease in humans and animals. The toxins are detected chemically and commercial kits are available. For information about these toxins, see Moss *et al.* (1989).

Other organisms

Other agents of food-borne disease include *Brucella* spp. and *Mycobacterium bovis* (both milk borne), *Aeromonas* spp. and enterobacteria other than those mentioned above. Some viruses, e.g. hepatitis A, and parasites, not considered in this book are also known to be associated with food-borne disease.

EXAMINATION OF PATHOLOGICAL MATERIAL AND FOOD

Do a total viable count (methods are described in Chapter 10) on all foods suspected of causing food poisoning. In most cases this will be high. Relate the findings to conditions of storage after serving. If possible count the numbers of causative organisms.

Table 13.1 suggests the causative organism. Isolation and identification methods are given in the relevant chapters:

- Salmonellas: Chapter 28
- Campylobacters: Chapter 33
- *Staphylococcus*: Chapter 36
- *Y. enterocolitica*: Chapter 34
- *B. cereus*: Chapter 42
- *V. parahaemolyticus*: Chapter 24
- Botulism: Chapter 44
- *E. coli*: Chapter 26
- *C. perfringens*: Chapter 44
- *Listeria*: Chapter 40.

For further information on food poisoning, see Eley (1996), Hobbs and Roberts (1987), Parliamentary Office of Science and Technology (POST, 1997) and the US Centers for Disease Control.

REFERENCES

Centers for Disease Control and Infection. http://www.cdc.gov.health/foodill.htm. http://www.cdc.gov/ncidod/eid/vol5no5/pdf/mead/pdf.

Eley, A. R. (ed.) (1996) *Microbial Food Poisoning*. London: Chapman & Hall.

Gilbert, R. J., Turnbull, P. C. B., Parry, J. M. and Kramer, J. M. (1981) *Bacillus cereus* and other *Bacillus* species: their part in food poisoning and other clinical infections. In: Berkeley, R.C.M and Goodfellow, M. (ed.), *The Aerobic Endosporing Bacteria: Classification and Identification*. London: Academic Press, pp. 297–314.

Hobbs, B. C. and Roberts, D. (eds) (1987) *Food Poisoning and Food Hygiene*, 5th edn. London: Edward Arnold.

Moss, M. O., Jarvis, B. and Skinner, F. A. (1989) Filamentous Fungi in Foods and Feeds. *Journal of Applied Bacteriology* [Symposium Supplement] **67**: 1S–144S.

Health Protection Agency (2001) Cases in Humans, England and Wales, by Serotype, 1981–2001 (www.hpa.org.uk/infections/topics_az/salmonella/data_human.htm).

Parliamentary Office of Science and Technology (1997) *Safer Eating. Microbiological Food Poisoning and its Prevention*. London: HMSO.

World Health Organization (1998) Surveillance Programme for Control of Foodborne Infections and Intoxicants in Europe. Newsletter 57 (October 1998). Copenhagen: WHO.

14

Food microbiology: general principles

There are many textbooks on the microbiological safety and the keeping quality of food, e.g. those of Adams and Moss (2000), Jay (2000), and Lund *et al.* (2000), as well as the publications of the International Commission for Microbiological Standards for Foods (ICMSF, 1980, 1986, 1988, 2001) (website: www.dfst.csiro.au/icmsf.htm).

Food microbiologists often employ media and methods first developed for clinical microbiology. The International Standards Organization (ISO), European Standards Committee (CEN) and International Dairy Federation (IDF) have published various procedures that include general methods for the preparation of dilutions, total colony counts, and various plate counts and MPN (most probable number) counting methods for coliforms, members of the Enterobacteriaceae and *Escherichia coli*. There are also standard methods for the detection of pathogens, including salmonellas, thermophilic campylobacters, *Listeria monocytogenes, E. coli* O157:H7, *Staphylococcus aureus* and others, as well as methods for enumeration of spoilage organisms, including lactic acid bacteria, pseudomonads, *Brochothrix thermosphacta*, and yeasts and moulds. Almost all these methods are also published by the British Standards Institution (BSI). In North America the AOAC International also publishes standard methods, as does the Food and Drug Administration (FDA, 1998) and the American Public Health Association (APHA) (Dowens and Ito, 2001). The UK Public Health Laboratory Service (PHLS) has also published a book of methods (Roberts and Greenwood, 2002). Many laboratories use these methods, which are mostly traditionally cultural, but others use simplified or more rapid versions

involving molecular (polymerase chain reaction or PCR) or immunological methods (e.g. enzyme-linked immunosorbent assay or ELISA).

Food microbiology laboratories doing official testing are now required to be accredited to a quality assurance scheme. In the UK this is the UK Accreditation Service (UKAS) scheme. Microbiology laboratories attached to food factories in the UK are normally also accredited to UKAS or to similar organizations (e.g. 'LabCred' run by Law Laboratories or 'CLAS' [Campden Laboratory Accreditation Scheme], run by Campden and Chorleywood Food Research Association]) approved by the major supermarkets, such as Tesco, J. Sainsbury and Marks and Spencer.

Many UKAS-accredited contract laboratories have opened up in the UK in recent years, taking over the testing previously done in laboratories owned by supermarkets and food manufacturing companies. Quality assurance systems required in accredited laboratories include: traceability of samples and results, recording of results, standard operating procedures, training of analysts, regular calibration and checking of equipment, documented methods (and validated, if these differ from standard methods) and monitoring of the quality of media. Laboratories may have to validate non-standard methods themselves, or they may be able to use methods validated by organizations such as AOAC International or MicroVal. In addition, laboratories have to take part in an external proficiency testing scheme, such as those in the UK run by the UK Central Public Health Laboratory (see Snell *et al.*, 1991) or the UK Central Science Laboratory (FEPAS – Food Examination Proficiency Scheme) as well as several commercial schemes. Participating laboratories receive shelf-stable, simulated food samples with instructions to examine them for particular groups of microbes (presence/absence or enumeration). Alternatively, information about the type of food and the illness associated with it are given, and the laboratory has to decide which organisms to look for.

SCOPE OF MICROBIOLOGICAL INVESTIGATIONS IN FOOD MICROBIOLOGY

Four kinds of investigations are usually carried out:

1. Estimation of the viable (colony) count (colony-forming units [cfu]/g, cfu/cm^2 or cfu/ml).
2. Estimation of the numbers of *E. coli*, coliforms (more often Enterobacteriaceae) and/or *Enterococcus* spp. ('faecal streptococci'). These may indicate the standard of hygiene during the preparation of the foods, especially where processing (e.g. heating) would be expected to kill these organisms (see 'Indicator organisms', below).
3. Detection of specific organisms known to be associated with spoilage. This may be more important than the viable count in determining the potential shelf-life.
4. Detection of pathogens.

No single method is completely satisfactory for the examination of all foodstuffs. The methods selected will depend on local conditions, e.g. availability of staff, space and materials, numbers of samples to be examined and the time allowed to obtain a result. To these ends, rapid, automated (validated) methods have much to offer (see Chapter 8). It may be more desirable to examine a large number of samples by a slightly suboptimal method than a smaller number by an exhaustive method.

Many of the micro-organisms in foods may be sublethally damaged, e.g. by heat, cold or adverse conditions such as low pH, low water activity, or high salt or sugar content. On the other hand, these conditions may favour the growth of yeasts and moulds, when mycotoxins may be formed.

An important problem in the microbiological examination of foods is that micro-organisms are usually not evenly distributed throughout a product. This affects any sampling procedure; there is no easy answer.

INDICATOR ORGANISMS

These are organisms that may indicate evidence of contamination or pollution, especially of a faecal nature, e.g. *E. coli* and coliform bacilli. The family name Enterobacteriaceae is sometimes used because 'coliforms' are poorly defined taxonomically. Enterobacteriaceae includes other bacteria, e.g. important pathogens such as salmonellas and various non-lactose fermenters that may be present in human and animal faeces. It should be noted, however, that the presence of these latter organisms is not evidence, at first sight, of faecal contamination: many of them are normally present in soil and on plant material.

MICROBIOLOGICAL CRITERIA (STANDARDS)

Limits for numbers of organisms (or their toxins) are given various names. The terms are defined by the ICMSF (2001) and summarized here.

Standard

This is a criterion set out in a law or regulation controlling foods produced, processed or stored in the area of jurisdiction, or imported into that area. Foods not meeting the standard would be removed from the market.

Guidelines

These are criteria used by manufacturers or regulatory agencies in monitoring a food ingredient, process or system (can be applied at points during manufacture). Guidelines are advisory, and need not lead to rejection of the food – but could be used to help bring about changes to improve the safety or quality of the food.

Microbiological (purchasing) specifications

These are criteria that determine the acceptance of a specific food or food ingredient by a food manufacturer or other private or public purchasing agency.

QUALITY ASSURANCE SYSTEMS

It is widely appreciated that satisfactory results of microbiological testing of the endproducts of food manufacture cannot guarantee that the whole consignment is safe, although it may detect an unsatisfactory batch. It is enough here to state that endproduct testing alone is not sufficient to check, with a reasonable degree of certainty, that a given batch of food is satisfactory unless an impracticable number of samples is examined. Microbiological testing of the endproduct is only one of many methods of checking that the system is providing safe food and is useful only as a check on the whole process. The food manufacturing and catering industry is obliged by law, in the European Union and many other countries, to use a hazard analysis critical control point (HACCP) system, which attempts to anticipate possible hazards in the manufacturing process and devise methods of controlling or minimizing them, as well as methods of checking and correcting deviations. An example is monitoring the temperature and time of pasteurization. Details can be found in one of the many publications on this subject (see, for example, ICMSF, 1988).

If a sample of a perishable food is taken at a point after manufacture and after an unknown period on the shelf, the total colony count result is often difficult to interpret and is of little use in assessing the quality at the point of manufacture. It is more appropriate to test for pathogens.

SAMPLING PLANS

Limits are normally set in terms of numbers of organisms (cfu/g or ml) or presence/absence of a pathogen in a given amount of the food. The exact number of organisms can never be determined precisely and it is therefore inappropriate to set criteria with exact limits beyond which a food is unacceptable. For this reason it is customary to use three-class sampling plans for the numbers of organisms. These classes are: 'acceptable' – under a specified number (m); 'marginally acceptable' – under a higher specified number (M); and 'unacceptable' when numbers exceed M. M is usually at least 10 times higher than m, e.g. the total count limit for liquid, pasteurized, whole-egg products suggested by ICMSF is:

$$n = 5, c = 2, m = 5 \times 10^4, M = 10^6$$

where n is the total number of samples tested and c the number that may exceed m. None should exceed M.

Two-class plans are usual for salmonella testing, e.g. samples of 5, 10, 20 or > 25 g might be tested for the presence/absence of salmonellas and the food would be unacceptable if any sample was positive ($n = 5, 10, 20$ or more, $c = 0, m = 0$). The number of samples examined would depend on the perceived risk if the food was contaminated: foods with the highest risk would be those eaten without prior heating and by a very susceptible class of consumer, e.g. dried milk for babies.

Limits in microbiological standards that are applied by enforcement authorities normally include those for pathogens and indictor organisms, not total colony counts.

SAMPLING

The usefulness of laboratory tests depends largely on correct sampling procedures. The sampling plan (see above) must be applied to each situation so that the samples submitted to the laboratory are fully representative of the batch. In factory sampling it is more useful to take small numbers of samples at different times during the day than larger numbers at any one time. Where resources are limited sampling may concentrate on the finished product, but if there is an increase in counts it is necessary to go back through the process and take 'in-line' samples, especially of raw (starting) materials because they may be the source of contamination. It is also important to sample the food-processing instruments.

Containers and samplers

Use screw-capped aluminium or polypropylene jars or plastic bags; never use glass containers where any breakages could contaminate the product or cause injury to personnel. Instruments for sampling from bulk material vary: for frozen samples use a wrapped sterilized brace and bit, a cork borer or a chopper that has been washed in 70% ethanol or methylated spirit and flamed. Take care to avoid cuts from ice splinters. Use individually wrapped spoons, spatulas or wooden tongue depressors for soft materials. With pre-packed food take several packages as offered for sale.

When sampling carcass meat and fish, where contamination occurs only on the surface, use a surface swab over a defined area or excise a surface layer 2–3 mm thick (see p. 205).

Transport and storage of samples

Carry samples (other than canned and bottled foods) in insulated containers, cooled with ice packs ('coolboxes'). The ice should not have melted by the time the samples arrive at the laboratory.

Small refrigerated containers that are run off a car battery can also be used. The samples should be transferred to a laboratory refrigerator and examined as soon as possible. There is a code of practice for sampling by authorized officers under the Food Safety Act 1990 (see also Harrigan and Park, 1991; Roberts and Greenwood, 2002).

PRE-TESTING CONSIDERATIONS

Before examining any particular food check the type of spoilage and/or pathogens that might be expected. In general, products with low pH and/or low water activity (a_w) tend to be spoiled by yeasts and moulds. At intermediate pH levels and low a_w lactic acid bacteria and other Gram-positive organisms such as micrococci and enterococci may predominate. At neutral pH and/or high a_w spoilage is mostly the result of Gram-negative bacteria such as *Pseudomonas* spp. In some heat-processed foods, such as meat pies, spoilage is the result of outgrowth of surviving spores of *Bacillus* and/or *Clostridium* spp., especially in the absence of competing flora.

GENERAL METHODS

If the organisms are expected to be distributed throughout the material, homogenize 10 g of the food (weighed aseptically) in 90 ml maximum recovery diluent (MRD, see p. 76) in a homogenizer (e.g. a Stomacher). Make serial dilutions, e.g. 10^{-1}–10^{-4}, and do colony counts on appropriate agar media and inoculate enrichment media for relevant pathogens (see below).

If the organisms are likely to be on the surface only (e.g. of pieces of vegetable or dried fruit) weigh 50 or 100 g into a sterile jar and add 100 ml diluent. Shake for 10 s, stand on the bench for 30 min and shake again. The organisms are assumed to have been washed into the diluent, i.e. 100 ml now contains the organisms from the original weight of sample. Make dilutions and proceed as above.

Total and viable counts

Methods are given in Chapter 10. For total counts, counting-chamber and DEFT (direct epifluorescence technique) counts are useful and suggest appropriate dilutions for viable counts. For counting mould hyphae and yeasts the Neubauer chamber, used in haematology, is better than the Helber chamber because it is deeper.

The methods and incubation temperatures for viable counts depend on the nature and storage conditions of the food. For most foods use pour plates and incubate at 30°C for 24–48 h.

For chilled and frozen meat and frozen fish use surface plating because the psychrophiles and psychrotrophs that predominate in such foods may be damaged if exposed to the temperature of melted agar used in pour plating. Incubate at 20–25°C. To count psychrophiles incubate at 1°C for 14 days. A maximum of 200 μl can normally be spread on a 9-cm agar plate. Two plates of each medium are normally inoculated from each dilution. Surface plating can be more economic than pour plating if more than one dilution is plated on each Petri dish. Commonly four dilutions (10–50 μl of each dilution) per dish can be used if spread, or six if not spread.

Use the 'black tube' method (see p. 377) for counting anaerobes. Report counts as colony-forming units per gram or millilitre. The tables on p. 146 show the amount of food in grams contained in the various dilutions.

Presence/absence tests

It may not be necessary to set up full viable counts. If a standard is set on the basis of experience the technique can be modified to give a present or absent ('pass' or 'fail') response. If, for example, an upper limit of 100 000 cfu/g is set, then a plate count method that uses 1 ml of a 1 : 1000 dilution will suffice. More than 100 colonies, easily observed without laborious counting, will fail the sample. Similarly, more than 100 colonies in roll tubes containing 0.1 ml of a 1 : 100 dilution, or more than 40 per drop in Miles and Misra counts from a 1 : 50 dilution, will suggest rejection.

Enterobacteriaceae, coliforms and *E. coli*

There is a wide choice of media for these tests. Most 'standard' tests specify EE broth (which is brilliant green glucose broth), violet red bile glucose agar (VRBG) and violet red bile lactose agar (VRBL or VRB). VRBG and VRB should not be autoclaved – see manufacturers' instructions. If these are not available MacConkey broth and agar may be used. Lauryl sulphate tryptose broth is also specified for some tests. Some workers prefer minerals modified glutamate medium (MMGM) instead of EE broth.

If small numbers are anticipated use the MPN or presence/absence (P/A) methods, but for heavily contaminated material plate or surface counts are better.

Enterobacteriaceae

Presence/absence

If resuscitation of sublethally damaged organisms is indicated add dilutions of the original suspension (see above) to MRD, incubate at room temperature for 6 h.

Add 10 ml suspension to 10 ml double-strength EE broth and 1 ml suspension to 10 ml single-strength EE broth. Incubate at 30°C for 18–24 h and then plate on VRBG. Incubate plates at 30°C and examine for typical colonies.

MPN method

For this method, see p. 150.

Pour plates

If resuscitation is necessary incubate the dilutions in MRD (see p. 76) for 90 min. Use freshly prepared VRBG for the counts.

Surface counts (with resuscitation)

Spread 0.1-ml volumes of the dilutions on non-selective medium (e.g. plate count agar). Incubate for 6 h at room temperature to resuscitate damaged cells, overlay with 15 ml VRBG and incubate at 30°C overnight.

Coliforms

Use the methods outlined above but with lauryl sulphate tryptose broth. Subculture positive tubes to brilliant green lactose broth to check gas production. For colony counts use VRB agar (with lactose).

Incubation temperatures

Incubate coliform tests on dairy products, etc. at 30°C. For other products use 37°C.

Escherichia coli ('faecal coli')

Subculture suspect colonies or broth tubes to brilliant green broth and peptone water and incubate at 44 ± 0.2°C overnight. Only E. coli produces gas and indole at 44°C. E. coli can be confirmed in the membrane method (above) by immersing the membrane in indole reagent and exposing it to ultraviolet (UV) light for 30 min. E. coli colonies are pink. For other confirmation tests, see Chapter 26.

Membrane method

Place the membranes on MMGM or other non-selective medium. Spread 1-ml volumes of the diluted suspension on the membranes (do not invert plates). Incubate for 4 h at 37°C. Transfer membranes to tryptose bile agar and incubate at 44°C for 18–24 h. Count colonies and subculture if necessary.

Counting enterococci

Do surface counts as described on p. 147 with kanamycin azide aesculin agar, azide blood agar, MacConkey or Slanetz and Bartley medium. Consult manufacturers' manuals for colonial appearance.

Counting clostridia and bacilli

As counting these organisms usually requires identifying them as well, the techniques are described under Clostridium, in Chapter 44. Direct plating for C. perfringens can be done on tryptone sulphite cycloserine (TSC) or oleandomycin polymyxin sulphadiazine perfringens (OPSP) agar. Bacillus cereus is counted on polymyxin egg yolk mannitol bromothymol blue agar (PEMBA) or mannitol egg yolk polymyxin agar (MEYP).

Spoilage organisms

Table 14.1 shows organisms associated with non-sterile food. Methods for enumerating spoilage organisms are given under the headings of the foods concerned. Table 14.2 shows the media used for examining foods.

Food poisoning bacteria

'Routine' examination of all foods for salmonellas and staphylococci, or other food-poisoning organisms, is hardly worth while on account of the relative infrequency of isolation. Only those foods and ingredients known to be vehicles need be tested. Methods are given in Chapter 13 and under the appropriate headings in later chapters.

Yeasts and moulds

Sampling

Circumstances vary. Obtain as much 'background' information as possible before deciding on the method of examination. The number of samples should be as large as is practicable. Sample whole packages where possible and examine the casing for damage and water staining.

Visual inspection

Open carefully and look for evidence of moulds. Microscopy is rarely helpful.

Table 14.2 Plating media commonly used to examine foods

Medium	Group selected	Selective agents	Indicator agents	Inoculation	Incubation: time, temperature, atmosphere
	Spoilage organisms				
CFC	*Pseudomonas* spp.	Cephaloridine, fucidin, cetrimide	None	Surface	48 h, 25°C, aerobic
MRS	Lactic acid bacteria	None	None	Surface	5 days, 25°C, anaerobic
STAA	*Brochothrix thermosphacta*	Streptomycin, thallium acetate, cycloheximide (actidione)	None	Surface	25°C, 48 h, aerobic
Iron agar	*Shewanella putrefaciens* and other H$_2$S producers	None	Sulphite, cysteine and thiosulphate	Pour plate (with overlayer)	20 or 25°C for 2–5 days
DG18	Yeasts and moulds	Chloramphenicol, dichloran, glycerol	None	Surface	25°C, 7 days, in dark, lids up
RBC	Yeasts and moulds	Chloramphenicol, Rose Bengal	None	Surface	25°C, 5 days in dark, lids up
DRBC	Yeasts and moulds	Chloramphenicol, Rose Bengal dichloran	None	Surface	25°C, 5 days in dark, lids up
OGY	Yeasts and moulds	Oxytetracycline	None	Surface	25°C, 5 days in dark, lids up
	Indicator organisms				
VRBG	Enterobacteriaceae	Bile salts, crystal violet	Glucose, neutral red	Pour or surface with or without overlayer	37 or 30°C for 24 h
VRBL	Coliforms (*E. coli*)	Bile salts, crystal violet	Lactose, neutral red	Pour or surface with or without overlayer	37 or 30°C for 24 h (44°C 24 h for *E. coli*)
KAA	*Enterococcus* spp.	Sodium azide, kanamycin	Aesculin, ferric ammonium citrate	Surface	37 or 42°C 18–24 h, aerobic
M-E	*Enterococcus* spp.	Sodium azide	Triphenyltetrazolium chloride	Surface	37°C for 4 h, then 44°C for 44 h, air
TBA	*Escherichia coli*	Bile salts	None	On membrane	44°C 18–24 h, aerobic
TBX	*Escherichia coli*	Bile salts	5-bromo-4-chloro-3-indolyl β-D glucoronic acid (BCIG)	Pour or surface or membrane	

Table 14.2 (continued)

Medium	Group selected	Selective agents	Indicator agents	Inoculation	Incubation: time, temperature, atmosphere
	Pathogens				
BP	*Staphylococcus aureus*	Lithium chloride, tellurite	Tellurite, egg yolk	Surface	37°C, 24–48 h aerobic
mCCD	*Campylobacter jejuni* and *C. coli*	Deoxycholate, cefoperazone, amphotericin	None	Surface	37 or 42°C, 48 h microaerobic (5% O_2, 10% CO_2, 85% N_2 or H_2)
PEMBA	*Bacillus cereus* and other *Bacillus* spp.	Polymyxin, cycloheximide	Mannitol, bromothymol blue, egg yolk	Surface	30°C, 18–24 h, and, if needed, overnight at room temperature, aerobic
MEYP	*Bacillus cereus* and other *Bacillus* spp.	Polymyxin B	Mannitol/phenol red, egg yolk	Surface	30°C, 24–30 h, aerobic
Oxford	*Listeria* spp.	Lithium chloride, acriflavine, colistin, fosfomycin, cefotetan, cycloheximide	Aesculin, ferric ammonium citrate	Surface	30 or 37°C, 48 h aerobic
PALCAM	*Listeria* spp.	Lithium chloride, acriflavine, polymyxin, ceftazidime	Aesculin, ferric ammonium citrate	Surface	30°C, 24–48 h, microaerobic (5% O_2, 10% CO_2, 85% N_2 or H_2)

	Pre-enrichment	Enrichment	Plating	Details
Salmonella spp.	25 g food in 225 ml buffered peptone water (37°C, 18–24 h)	0.1 ml in 10 ml Rappaport Vassiliadis broth (42.5°C 24 h) or 1 ml in 10 ml selenite broth (37°C 24 and 48 h)	Streak on to brilliant green agar, XLD agar, Rambach agar, 37°C 24 h	Confirm biochemically and/or with salmonella antisera
Thermophilic *Campylobacter* spp.		25 g food in 225 ml Bolton broth 37°C 48 h or 4 h 37°C and 42°C 44 h.	Streak on mCCD agar, 37 or 42°C 48 h, microaerobic (see Table 25.3)	Presumptive confirmation of oxidase-positive Gram-negative curved rods not growing aerobically at 37°C
E. coli O157:H7		25 g food in 225 ml of tryptone soy broth with novobiocin. Examine after 6–7 h and 41.5°C using immuno-magnetic beads (with anti O157 antibodies)	Plate on CT-SMAC agar; incubate at 37°C for 24 h	Send sorbitol-negative, indole positive colonies agglutinating anti-O157 polystyrene beads to a reference laboratory

Different manufacturers use alternative names for some media

BP, Baird-Parker medium; CFC, cephaloridine fucidin cetrimide medium; CT-SMAC, cefoxine tellurite sorbitol MacConkey agar; DG18, dichloran glycerol medium; DRBC, dichloran Rose Bengal chloramphenicol agar; KAA, kanamycin aesculin azide medium; mCCD, membrane campylobacter-selective medium; M-E, membrane enterococcus medium; MEYP, mannitol egg yolk polymyxin agar; MRS, deMan, Rogosa and Sharp agar; Oxford, Oxford listeria medium; PALCAM, Listeria selective medium; PEMBA, polymyxin egg yolk mannitol bromothymol blue agar; RBC, Rose Bengal chloramphenicol agar; STAA, Gardner's (1966) medium; VRBL, violet red bile lactose agar

Direct examination

Pre-incubation may be useful (Jarvis *et al.*, 1983). Place a filter paper soaked with glycerol in a large glass Petri dish and autoclave it. Suspend a sample above the filter paper, e.g. on glass rods, replace the lid and incubate for up to 10 days. Examine daily for moulds.

To estimate shelf-life relative to mould growth under adverse conditions, incubate unopened packages in a controlled humidity cabinet.

Culture

Prepare a 10^{-1} suspension in MRD containing 0.1% Tween 80. Use a Stomacher if possible. Make serial dilutions in the same diluent and surface plate (in duplicate) on Rose Bengal chloramphenicol agar. Incubate at 22°C for 5 days. For low a_w foods use DG18 agar.

REFERENCES

Adams, M. R. and Moss, M. O. (2000) *Food Microbiology*, 2nd edn. London: Royal Society of Chemistry.

Dowens, F. P. and Ito, K. (eds) (2001) *Compendium of methods for the Microbiological Examination of Foods*, 4th edn. Washington DC: American Public Health Association (APHA).

Food and Drug Administration (1998) *Bacteriological Analytical Manual*, 8th edn. Arlington, VA: AOAC International.

Gardner, G. A. (1966) A selective medium for the enumeration of *Microbacterium thermosphactum* in meat and meat products. *Journal of Applied Bacteriology* 29: 455–460.

Harrigan, W. F. and Park, R. W. A. (1991) *Making Food Safe*. London: Academic Press.

International Commission of Microbiological Standards for Foods (ICMSF) (1980) Cereal and Cereal Products. In: Microorganisms in Foods. 3. *Microbial Ecology of Foods*, Vol. 2. New York: Academic Press.

ICMSF (1986) *Microorganisms in Foods. 2. Sampling for Microbiological Analysis. Principles and Specific Applications*. Toronto: Toronto University Press.

ICMSF (1988) *Microorganisms in Foods. 4. Application of the HACCP system to ensure Microbiological Safety and Quality*. Oxford: Blackwell Scientific.

ICMSF (1998) *Microorganisms in Foods. 6. Microbial Ecology of Food Commodities*. London: Blackie Academic & Professional. Available from Kluwer Publishers.

ICMSF (2001) *Microorganisms in Foods. 7. Microbiological Testing in Food Safety Management*. New York: Kluwer Academic/ Plenum Publishers.

Jarvis, B., Seiler, D. A. L., Ould, S. J. L and Williams, A. P. (1983) Observations on the enumeration of moulds in food and feedingstuffs. *Journal of Applied Bacteriology* 55: 325–326.

Jay, M. (2000) *Modern Food Microbiology*, 6th edn. MA: Aspen Publishers.

Lund, B. M., Baird-Parker, T. C. and Gould, G. W. (eds) (2000) *The Microbiological Safety and Quality of Food*. New York: Aspen Publishers.

Roberts, D. and Greenwood, M. (eds) (2002) *Practical Food Microbiology*, 2nd edn. Oxford: Blackwell.

Snell, J. J. S., Farrell, I. D. and Roberts, C. (eds) (1991) *Quality Control. Principles and practice in the microbiology laboratory*. London: Public Health Laboratory Service.

15

Meat and poultry

FRESH AND FROZEN RAW RED MEAT

Table 14.1 (see p. 200) summarizes the types of bacteria that cause spoilage. Many different species of bacteria are present on carcasses immediately after slaughtering, but Gram-negative organisms multiply and predominate during chill storage. The deep muscle tissue is usually sterile but occasionally may be contaminated. 'Bone taint' is revealed by an unpleasant odour when the carcass is boned, and is most common in the hip and shoulder joints, which cool most slowly. Microbiological investigation most often reveals clostridia or enterococci. The source of contamination is not certain, but could be from contaminated slaughter instruments, via infected lymph nodes or by spread from the intestine at the time of death. Meat from stressed animals may be of poor keeping quality because the pH is higher than normal (> 6.0).

Carcass sampling

Microbial contamination is usually confined to the skin or surface of carcasses. The most reliable methods are destructive (which can affect the value of the carcass), involving excising the top 2–3 mm of a measured area (e.g. 25 or 50 cm²). The most practical non-destructive method is swabbing. Impression plates or slides may also be used. In general, excision methods recover more organisms than swab methods, which in turn recover more than impression methods. Microbiological testing of five to ten carcasses on one day per week, after dressing and before chilling (aerobic colony count

at 30°C and Enterobacteriaceae at 37°C) from four sites on each carcass and from the abattoir environment is now required in the European Union (EC, 2001; SI, 2002). Excision sampling is recommended, with swabbing as an alternative.

Excision sampling

Use aluminium templates with, for example, 25 or 50 cm² apertures. Remove the top 2–3 mm of the meat defined by the template area with a scalpel or, for frozen meat, a cork borer.

Homogenize an appropriate volume (e.g. 25 ml) in 50 ml of maximum recovery diluent (MRD – see p. 76) and do counts as described in Chapter 10 and above.

Swab sampling

Moisten a large plain cottonwool swab with a wooden shaft with MRD and then rub it vigorously over the area to be sampled, e.g. as defined by a template. Repeat with a dry swab. Break off the ends of both swabs into 10 ml MRD in the same container and do total counts at 30°C as described in Chapter 10 and above. This method cannot be used for frozen meat.

Whole carcass sampling

To sample the whole carcass by a non-destructive technique, use the following method (Kitchell *et al.*, 1973). Sterilize large cottonwool pads, wrapped in cotton gauze in bulk, and transfer them to individual plastic bags. Moisten a pad with diluent, e.g. MRD, and wipe the carcass with it, using the bag as a glove. Take a dry pad and wipe the carcass again. Place both pads in the same bag, then seal and label it. Add 250 ml diluent to the pads and knead, e.g. in a Stomacher, to extract the organisms. Prepare suitable dilutions from the extract and do counts as above. Inoculate other media as desired. This method is not usually practicable for beef carcasses.

The areas on beef carcasses that are most likely to be contaminated after dressing are the neck, rump, brisket and flank – and similar parts of other red meat carcasses. For excised samples, the daily log mean limits of aerobic colony counts at 30°C and of enterobacteria at 37°C for cattle, sheep, goats and horses are shown in Table 15.1. Values for pig carcasses are in most cases 0.5 log cycles higher. Values for swab samples are about 20% of the excision sample values.

In dry (about 80% relative humidity) chill conditions, mould spoilage may occur on chilled (−1°C) carcasses stored for several weeks. Carcasses are now, however, seldom stored chilled for long periods. Mould and yeast colonies can also be seen on primals (see below) in frozen storage when the correct temperature (≤ −12°C) has not been maintained. 'Green spots' are usually caused by *Penicillium* spp., 'white spots' by *Sporotrichum* spp., 'black spots' by *Cladosporium* spp. and 'whiskers' by *Mucor* and *Thamnidium* spp.

Boned-out meat

Most carcasses are split into primals a few days after slaughter. These are large pieces of meat that are sometimes stored at 0 ± 1°C in vacuum-packs for several weeks before being divided up into smaller cuts for retail sale. Much meat is now retailed in modified atmosphere packs (about 80% O_2, 20% CO_2) in supermarkets. The oxygen preserves the red colour of oxymyoglobin for as long as possible, whereas the carbon dioxide inhibits the pseudomonads, which predominate in meat stored in aerobic packs.

As with carcass sampling, primals or smaller cuts of red meat are usually sampled either by excision or by swabbing. It may also be possible to sample the liquid exudate ('purge') that accumulates in vacuum packs. Two areas of 5 or 10 cm^2 are usually sampled per piece of meat. These could include a sample from a cut edge and another from the original exterior surface. The samples may be pooled before microbiological examination. Results of excision or swabbing are expressed as colony-forming units per square centimetre (cfu/cm^2). In a meat production facility, where the hygiene is good, more than 70% of samples may be expected to have counts of less than 1000 cfu/cm^2 (or $\log_{10}$ 3 cfu/ml).

Microbial flora of raw meat during storage

In moist, aerobic chill conditions the predominant spoilage organisms are pseudomonads, with lower

Table 15.1 Acceptable and unacceptable daily mean values of colony and enterobacteria counts for cattle, sheep, pig, goat and horse carcasses

	Daily mean values (log cfu/cm²) as			
	Acceptable range for		Marginal range (> *m* but < *M*) for cattle/pig/sheep/ goat/horse	Unacceptable range (> *M*) for cattle/pig/ sheep/goat/horse
	Cattle/sheep/ goat/horse	Pig		
Total viable count	< 3.5	< 4.0	3.5 (pig: 4.0) to 5.0	> 5.0
Enterobacteriaceae	< 1.5	< 2.0	1.5 (pig: 2.0) to 2.5 (pig: 3.0)	> 2.5 (pig: > 3.0)

See Chapter 14 for definitions of sampling plans
From EC (2001) and SI (2002)

numbers of *Acinetobacter* spp., psychrotrophic Enterobacteriaceae, including *Enterobacter* and *Hafnia* spp., and *Brochothrix thermosphacta*. Spoilage is imminent when numbers of pseudomonads reach about $10^7/cm^2$.

In meat stored in a modified atmosphere or vacuum packs, lactic acid bacteria (LAB) normally outgrow the pseudomonads; *Lactobacillus sake* and *L. curvatus* are the most common LAB in vacuum-packed raw meat. *Carnobacterium*, *Leuconostoc* and *Weissella* spp. occur in lower numbers. Psychrotrophic enterobacteria and *B. thermosphacta* are sometimes present. *B. thermosphacta* is more often found in lamb and pork. Numbers of LAB on vacuum-packed meat frequently exceed 10^8 (log 8)/cm^2 for several weeks before the meat is judged spoiled (an initial rather cheesy smell often dissipates if the pack is left open for a few minutes). This is because the LAB produce much less obnoxious products than pseudomonads in aerobically stored meat. There are a number of possible reasons why vacuum-packed meat has apparently spoiled before the end of its normal shelf-life. Meat with pH > 6 will spoil rapidly as a result of the multiplication of Enterobacteriaceae and/or *Shewanella putrefaciens*. Meat with apparently normal pH sometimes has elevated numbers of enterobacteria or *B. thermosphacta* (> $10^5/cm^2$), which may be the result of temperature abuse during storage, or possibly poor hygiene. In addition, psychrotolerant clostridia, most commonly the psychrophilic species *C. estertheticum*, have emerged during the last 15 years as a very important cause of spoilage in vacuum-packed beef and, to a lesser extent, lamb, stored at proper chill temperature and with normal pH. Packs sometimes 'blow' (become grossly distended with gas, see Chapter 16) within 2 weeks of packing and economic losses can be extremely high. The gas consists largely of a mixture of CO_2 and H_2. The meat usually appears normal in colour, but has a very unpleasant vomit-like odour. Gram stains of the drip fluid sometimes reveal large Gram-positive rods among the normal small rod-shaped LAB and the occasional yeast. Isolation of *C. estertheticum* is difficult because there is no selective medium, but spores may be selected by use of mild heat (65°C) or ethanol for a few minutes before plating on to blood agar and incubation at 10°C anaerobically for 2–3 weeks. Colonies of *C. estertheticum* are normally β-haemolytic. A more reliable and rapid method of detection is to use a polymerase chain reaction (PCR)-type method (e.g. Helps *et al.*, 1999). Fortunately, there is no evidence that *C. estertheticum* is pathogenic or produces toxins.

Many of the well-known pathogens may be found on red meat in low numbers, as a result of cross-contamination from hides and intestinal contents. Salmonellas are not often present, however, except in pork and minced products. Campylobacters are most often found on offal, such as liver, and survive poorly during their shelf-life. *C. perfringens*, and (less often) *C. botulinum* may be isolated from raw meat, but pose no threat unless the meat is mishandled during processing and/or storage. Enterohaemorrhagic *E. coli* O157:H7, which causes severe disease and has a low infectious dose, colonizes cattle and sheep, and so can be found on raw meat. Careful attention to hygiene during slaughter will minimize its presence on meat, but cannot guarantee its absence.

The microflora detected on meat sampled during chilled shelf-life will depend on the method of storage (aerobic, gas or vacuum pack) and the point in shelf-life. Thus, aerobic plate counts alone are unlikely to be meaningful. Enumeration of pseudomonads on aerobically stored meat might be useful, because > log 7 cfu/cm^2 would indicate imminent spoilage. The finding of high numbers of LAB (> log 8 cfu/cm^2) on vacuum- or gas-packed meat would only confirm the presence of normal flora. Detection of numbers of Enterobacteriaceae or *B. thermosphacta* at 30°C (> log 5 cfu/cm^2) would be cause for concern that the meat might have been stored at elevated temperature and/or that hygiene had been lax during slaughter and cutting.

COMMINUTED MEAT

This is fresh meat, minced or chopped and with no added preservative, sometimes prepared from trimmings. Instead of being confined to the surface, organisms are therefore distributed throughout and further contamination may occur during mincing. In addition, the normal structure of the meat is disrupted, making the nutrients more readily accessible to the microflora. Raw minced meat tends to be spoiled by the same micro-organisms as larger pieces of raw meat, with the exception that it often contains significant numbers of yeasts.

Pathogens

As with whole red meat, salmonellas, clostridia and staphylococci may be present. Tests for *E. coli* O157:H7 in 25 g minced beef are indicated because minced beef is often used to make beefburgers (hamburgers in the USA), which have been implicated in outbreaks of disease. This organism is more of a hazard in beefburgers than on whole pieces of meat because the microflora occur throughout the meat, and are more likely to survive under-cooking.

The EU (EC, 1994) has set limits for freshly prepared minced meat (Table 15.2).

This standard is likely to be revised soon to include a requirement for the absence of *E. coli*

O157:H7. This standard is most appropriate for minced meat intended for consumption without further cooking, e.g. in steak tartare or filet americain.

Numbers of microbes in stored minced meat can reach considerably higher figures than those indicated above. As the product may not be homogeneous, it is useful to take several samples. In a study of retail minced beef in the UK (Nychas *et al.*, 1991) the microflora were similar whether the product had been stored in air or with CO_2. Counts of *B. thermosphacta*, pseudomonads, LAB, Enterobacteriaceae and yeasts were all in the region of 10^4–10^7 cfu/g and total counts were around 10^8 cfu/g. Aerobic colony counts at 30°C and Enterobacteriaceae at 30 or 37°C should be checked. Numbers of *E. coli* may be useful as an indicator of hygiene during preparation, and *E. coli* O157:H7 could also be sought. This organism is almost certain to be present occasionally, however, because it can be found in the faeces of many cattle. The action to be taken if it is detected should therefore be decided in advance

BRITISH FRESH SAUSAGES

Fresh sausages contain comminuted raw meat (usually pork) and fat, and water mixed with other ingredients such as phosphates, rind, rusk, seasoning, spices and herbs. This is filled into a natural or

Table 15.2 European Union limits of bacterial counts for freshly prepared minced meat

Microbes	Limits	Sampling plan			
		n	*c*	*m* (/g)	*M* (/g)
Aerobic mesophilic bacteria		5	2	5×10^5	5×10^6
Escherichia coli		5	2	50	500
Salmonella spp.	Absence in 10 g	5	0		
Staphylococcus aureus		5	2	100	5000

See Chapter 14 for definitions of sampling plans; *m*, specified number; *M*, higher specified number; *n*, total number of samples tested; *c*, the number that may exceed *m*
From EC (1994)

artificial casing, with sodium metabisulphite or sulphite as preservative (not more than 450 ppm measured as SO_2 at the point of sale). This product should not be confused with smoked sausages, which are popular in Europe and America

Microbial content

Without the preservative and spices, the microflora that would develop in this product would be similar to that in raw minced meat. The preservative selectively inhibits the Gram-negative organisms, allowing yeasts and *B. thermosphacta* to predominate, with lower numbers of LAB. Spoilage occurs as a result of slow souring. Shelf-life is longer than for minced meat, and is extended by gas or vacuum packing, which slows yeast growth. STAA medium (Gardner, 1982) is useful for the isolation and enumeration of *B. thermosphacta*. Freshly prepared sausages often contain about log 4 cfu/g yeasts and log 5 cfu/g *B. thermosphacta*. After 7 days of chill storage the numbers of yeasts can reach log 7 cfu/g and numbers of *B. thermosphacta* log 8 cfu/g, whereas numbers of pseudomonads, Enterobacteriaceae and LAB remain static at around log 4 cfu/g (Banks *et al*., 1985).

Pathogens

Sulphite may have played a role in the relatively good safety record of sausages as a cause of salmonella food poisoning. Although salmonellas can be isolated from sausages, sulphite will prevent them multiplying if the product is stored at elevated temperature. *E. coli* O157:H7 should be a hazard only in sausages containing beef or lamb.

Viable counts

The Institute of Food Science and Technology (IFST, 1999) suggests limits for aerobic plate counts (cfu/g) on sausages immediately after pro-

duction of < log 5 (and an upper limit of log 7, and of log 4 and log 6 respectively for yeasts).

Remove the casing, if present, and do total viable counts using dilutions from 10^{-4} to 10^{-8}. Incubate at 30°C for 48 h and at 22°C for 3 days. The use of surfactants, e.g. Tween 80, does not seem particularly advantageous. Sometimes they are inhibitory.

Counts of several million colony-forming units per gram may be expected. Skinless sausages may have lower counts than those with casings because the process involves blanching with hot water to remove the casings in which they are moulded.

Sausage casings, whether natural (stripped small intestine of pig or sheep) or artificial, usually do not present any microbiological problems.

MEAT PIES

Meat pies may be 'hot eating', e.g. steak and chicken pies, or 'cold eating', e.g. pork, veal and ham pies, and sausage rolls. Consumers, however, cannot, be relied on to reheat 'hot-eating' pies thoroughly before consumption. In both cases, the pastry acts as a physical barrier against post-cooking contamination of the filling. The filling in most cold-eating pies includes preservatives, e.g. nitrite in pork pies and sulphite in sausage rolls. Temperature fluctuations during storage of these products in films can result in condensation, soggy pastry and subsequent spoilage.

'Hot eating' pies

The meat filling is pre-cooked, added to the pastry casing and the whole is cooked again.

Viable counts

Examine the meat content only. Open the pie with a sterile knife to remove the meat. Prepare a 1 : 10 suspension in MRD. Do counts on dilutions of 10^{-1} and 10^{-2} and incubate aerobically and anaerobically on blood agar at 37°C for 48 h. A reasonable level is not more than 100 cfu/g. Numbers in excess of 1000 cfu/g may indicate temperature abuse after cooking (*Bacillus* or *Clostridium* spp.), or under-cooking.

Little other than heat-resistant spores should survive in these products.

'Cold eating' pies

These contain cured meat, cereal and spices which are placed in uncooked pastry cases. The pies are then baked and jelly, made from gelatin, spices, seasoning, flavouring and water, is added. Provided that the jelly is heated to, and maintained at, a sufficiently high temperature until it reaches the pie, there should be no problem.

Bad handling, insufficient heating and re-contamination of cool jelly can cause gross contamination by both spoilage organisms and pathogens.

The pies are cooled in a pie tunnel through which cold air is blown. At this stage mould contamination is likely if the air filters are not properly cleaned. Air in pie tunnels may be monitored by exposing plates of malt agar or similar medium. As with hot-eating pies, bad storage may result in outgrowth of spores of *Bacillus* and *Clostridium* spp.

Viable counts

Examine only the filling. Do counts on homogenized material using dilutions of 10^{-1} and 10^{-2}. Incubate at 30°C for 48 h. A reasonable level is less than 1000 cfu/g.

CURED MEAT

Cured raw meat

This is pork or beef that has been treated with salt and nitrite, but could be applied to any meat. The haemoglobin is altered so that the characteristic pink colour is produced when the meat is cooked. This is sold as bacon, gammon, cured shoulder, salt beef or brisket, and has a lower water activity (a_w) and a higher salt content than raw meat.

Bacon has been studied most, but other cured raw meats are likely to have similar characteristics. Numbers and species of microbes on sides of bacon can be very variable, as a result of the variety of ecological niches present – rind, fat, lean, variations in pH, a_w, curing salts, etc. Surface spoilage of refrigerated bacon is mainly caused by *Micrococcus*, non-pathogenic *Vibrio* spp. and *Acinetobacter*. Yeasts occur more commonly on smoked and dried rather than 'green' (unsmoked) bacon. Their growth may sometimes be visible to the naked eye, and they can be physically removed without any problem. The generally lower a_w and higher salt result in much lower total numbers of microbes on the surface than on green bacon. Gram-negative bacteria are rarely found. As smoking and drying are sometimes carried out at elevated temperatures, and curing salts are often injected, 'bone taint' can sometimes be a problem (see p. 205).

Vacuum-packed sliced bacon

Most bacon is now sold sliced and packed in vacuum or CO_2, which prevents the oxidation of the red-coloured nitrosohaemoglobin to brown products such as metmyoglobin. This type of packaging also extends the microbiological shelf-life to several weeks at chill temperatures. The predominant flora consist of LAB, with lower numbers of micrococci. Numbers of organisms on fat tend to be higher than on lean. As with vacuum-packed raw red meats, numbers of these bacteria can reach high levels (up to 10^8 cfu/g) during chill storage, producing gradual souring before spoilage is evident. Table 14.1 (see p. 200) indicates the types of spoilage organisms. Storage at temperatures above ambient can result in putrefactive spoilage as a result of multiplication of *Vibrio* spp., and enterobacteria such as *Proteus*, *Serratia* and *Hafnia* spp., *Staphylococcus aureus*, *Clostridium perfringens* and *C. botulinum* could theoretically multiply and produce toxins if bacon were stored at elevated temperatures, but in practice this product has not been associated with food-borne illness. This is most probably because the product is usually cooked before consumption and the toxins produced are heat labile. In addition, these organisms do not compete well with the majority microflora. Spoilage also intervenes and the product becomes inedible. Salmonellas might also be found on bacon, but this food has not been associated with disease caused by this organism.

Microbial content

The main groups of organisms found immediately after packaging are micrococci and coagulase-negative staphylococci. During storage at 20°C, lactobacilli, enterococci and pediococci become dominant. Yeasts may also be found. If large numbers of Gram-negative rods are present it is usually an indication that the initial salt content was low.

Vacuum-packed whole and sliced cooked meats and cooked cured meats

These meats include cured products, such as ham, as well as uncured ones, such as cooked beef, boned chicken or turkey, and pork. The meat is usually cooked inside a plastic film, which is left on during distribution and storage, or removed after cooking. In the former case the product normally has very low counts; in the latter case some contamination is inevitable. The meat may be used in catering, or sliced and sold in vacuum or gas packs. Spore formers will survive cooking, and so pathogens such as *C. perfringens* or *C. botulinum* could germinate and multiply if the products were temperature abused. Spoilage caused by psychrotolerant clostridia has occasionally been reported in cooked uncured meat after storage at slightly elevated temperatures (Lawson *et al.*, 1994; Broda *et al.*, 1997; Kalinowski and Tompkin, 1999).

Other organisms that sometimes survive cooking include enterococci and some lactobacilli. As with vacuum- or gas-packed raw meats, LAB normally predominate in packed sliced cooked meats, reaching high numbers before gradual spoilage as a result of souring. Spoilage organisms in aerobically packed cooked sliced meat are similar to those in aerobically stored raw meats (predominantly pseudomonads). Multiplication of small numbers of *S. aureus*, introduced during slicing, may occur, especially in cured meats, such as ham, if the temperature is raised about 10°C. Outbreaks of poisoning caused by staphylococcal toxins in ham are not uncommon, but more often occur as a result of poor practices in the catering industry (e.g. leaving meat for extended periods at ambient temperature) than in industrially produced sliced packed ham. Multiplication of *L. monocytogenes*, which can grow at refrigeration temperature, is possibly a more important hazard, especially with the current trend towards longer shelf-life. Levels in excess of 100 cfu/g at the end of the shelf-life are considered potentially hazardous (EC, 1999).

Laboratory examinations

Whole joints in cooking bags

Do surface counts on blood agar and on MacConkey agar for enterococci. These are useful organisms for assessing the efficiency of cooking. Total counts should be less than 1000 cfu/g.

Sliced meats

Do total (surface) counts immediately after slicing and packing. These should not exceed 10^4 cfu/g, but much higher numbers can be found during shelf-life. Numbers of *L. monocytogenes* should not exceed 100 cfu/g by the end of the shelf-life (EC, 1999), and numbers of *S. aureus* should be < 1000 cfu/g.

Cooked sausages and paté

As with joints cooked in sealed film, whole cooked sausages are protected by their casing and, provided cooking is sufficient, only low numbers of spore formers will survive. Sometimes, however, a few enterococci or lactobacilli may survive and grow. The spore formers cause no problem if the product is stored at low temperature. Hydrogen peroxide produced by lactobacilli can sometimes cause grey–green, yellow or white discoloration caused by oxidation of the nitrosyl myoglobin.

Belgian-type paté has similar characteristics to cooked sausages, except that it is more prone to post-cooking contamination. It is sometimes decorated with gelatin and pieces of vegetable. LAB may found at levels > 10^5 cfu/g. *L. monocytogenes* should not be present at levels > 100 cfu/g at end of the shelf-life.

POULTRY

Intensive rearing produces birds with very tender flesh, but the close confinement necessary in this method of production often results in cross-infection. Poultry meat is thought to be the source of a large proportion of cases of salmonellosis, and most cases of campylobacter diarrhoea in the UK and the USA. Salmonella infections occur either as a result of undercooking or of raw-to-cooked food cross-contamination. Campylobacter infections probably occur by a similar mechanism, as well as by the hand-to-mouth route after handling the raw product. Table 14.1 (see p. 200) indicates the type of spoilage organisms.

Sampling

The neck skin is a good sampling site because (1) it is the lowest point during processing and all fluids pass over it, and (2) it is easy to sample and may be removed without affecting the value of the remainder of the carcass. A good alternative is to rinse the whole carcass.

Place the carcass in a large sterile plastic bag. Add 500 ml diluent (MRD) and (holding the bag tightly closed) shake thoroughly for approximately 1 min. Carefully decant the fluid back into the bottle that held the sterile diluent. Do viable plate counts on tenfold dilutions of this fluid. Normal counts

on carcass rinsed immediately after chill are listed below.

Viable counts

Discard the subcutaneous fat from the neck skin, homogenize a known weight of the remaining material, prepare dilutions as described in Chapter 10 and use plate count agar. Incubate at 30°C or 37°C for 24–48 h. Express the counts as colony-forming units per gram of neck skin. These counts may be about 10 times higher than those obtained from the skin on other parts of the carcass. Normal ranges at the abattoir immediately after chill ($\log_{10}$ cfu/g neck skin) are: viable count, 4.5–5.3; coliforms, 2.7–3.8; pseudomonads, 2.9–3.9 (Mead *et al.*, 1993). Counts on carcass rinse are expressed as $\log_{10}$ cfu or per millilitre rinse fluid: viable count 3.8–6.0; Enterobacteriaceae 2.6–4.6; pseudomonads 3.2–5.9 (Corry, unpublished, 2002). As chlorine in wash water at levels above about 5 ppm is no longer permitted for washing poultry in the EU, numbers of coliforms/Enterobacteriaceae tend to be a little higher than pre-2001.

Pathogens

Examine the carcass rinse or neck skin. Culture for salmonellas and campylobacters as described in Chapters 28, 33 and 36 (see also Chapter 14). High *S. aureus* counts are sometimes found; these result from colonization of plucking machines. They are rarely toxin-producing, food-poisoning strains.

REFERENCES

Banks, J. G., Dalton, H. K., Nychas, G. J. *et al.* (1985) Review: sulphite, an elective agent in the microbiological and chemical changes occurring in uncooked comminuted meat products. *Journal of Applied Biochemistry* 7: 161–179.

Broda, D. M., DeLacey, K. M., Bell, R. G. and Penney, N. (1997) Prevalence of cold tolerant clostridia associated with vacuum-packed beef and lamb stored at abusive and chill temperatures. *New Zealand Journal of Agricultural Research* **40**: 93–98.

European Commission (EC) (1994) Minced Meat Directive. Council Directive 94/65/EEC of December 1994 laying down the requirements for the production and placing on the market of minced meat and meat preparations. Brussels: EU.

EC (1999) Opinion of the Scientific Committee on Veterinary Measures relating to Public Health on *Listeria monocytogenes*. (europa.eu.int/comm/food/fs/sc/scv/out25_en.html. Brussels, EU).

EC (2001) Commission Decision 2001/471/EC: of 8 June 2001 laying down rules for the regular checks on the general hygiene carried out by the operators in establishments according to Directive 1964/433/EEC on health conditions for the production and marketing of fresh meat and Directive 1971/118/EEC on health problems affecting the production and placing on the market of fresh poultry meat. *Official Journal of the European Communities* L 165, 21/06/2001 P. 0048–0053

Gardner, G. A. (1982) Microbiology of processing: bacon and ham. In: Brown, M.H. (ed.), *Meat Microbiology*. Oxford: Elsevier Applied Science, pp. 129–178.

Helps, C. R., Harbour, D. A. and Corry, J. E. L. (1999) PCR-based 16S ribosomal DNA detection technique for *Clostridium estertheticum* causing spoilage in vacuum-packed chill-stored beef. *International Journal of Food Microbiology* **52**: 57–65.

Institute of Food Science and Technology (IFST) (1999) *Development and Use of Microbiological Criteria for Foods*. London: IFST.

Kalinowski, R. M. and Tompkin, R. B. (1999) Psychrotrophic clostridia causing spoilage in cooked meat and poultry products. *Journal of Food Protection* **62**: 766–772.

Kitchell, A. G., Ingram, G. C. and Hudson, W. R. (1973) Microbiological sampling in abattoirs. In: Board, R.G and Lovelock, D.W. (eds), *Sampling – Microbiological Monitoring of the Environment*. Society for Applied Bacteriology Technical Series No. 7. London: Academic Press, pp. 43–54.

Lawson, P., Dainty, R. H., Kristiansen, N., *et al.* (1994) Characterization of a psychrotrophic *Clostridium* causing spoilage in vacuum-packed cooked pork: description of *Clostridium algidicarnis* sp. nov. *Letters in Applied Microbiology* **19**: 153–157.

Mead, G. C., Hudson, W. R. and Hinton, M. H. (1993) Microbiological survey of five poultry processing plants in the UK. *British Poultry Science* **17**: 71–82.

Nychas, G. J., Robinson, A. and Board, R. G. (1991) Microbiological and physico-chemical evaluation of ground beef from retail shops. *Fleischwirtschaft* **71**: 1057–1059.

Statutory Instrument (SI) (2002) Statutory Instrument No. 889. *The Meat (Hazard Analysis and Critical Control Point) (England) Regulations 2002*. London: HMSO.

16

Fresh, preserved and extended shelf-life foods

FRESH FRUIT AND VEGETABLES

Spoilage of fresh fruit is mostly caused by moulds. Mould and yeast counts may be indicated in fruit intended for jam making or preserving. Orange serum agar is useful for culturing fungi from citrus fruit (Lund and Snowdon, 2000).

Washed, peeled and chopped vegetables, in packs where the atmosphere has been modified either deliberately, or as a result of the metabolic activity of the vegetables, to give a raised level of CO_2 and reduced O_2 content, are sold in most supermarkets. They can include mixtures of vegetables for 'stir-fry' dishes or stews, as well as various types of salad and lettuce. High viable counts – up to 10^7–10^8 – are normal; products such as bean sprouts, cress, watercress and spring onions have the highest counts (Roberts et al., 1981; Lund, 2000; Nyugen-the and Carlin, 2000). Spoilage is caused by pseudomonads, psychrotrophic non-pathogenic Enterobacteriaceae (mostly Entero-bacter, Erwinia and Rahnella spp.), various pectinolytic bacteria and lactic acid bacteria, especially leuconostocs, as well as lower numbers of yeasts and moulds. With the exception of the lactic acid bacteria, which seem to be encouraged by the raised level of CO_2, similar numbers and types of microflora can be found on loose vegetables.

Although the incidence of contamination with pathogenic micro-organisms in plant foods is lower than that on foods of animal origin, they can be found from time to time. Contamination sometimes occurs as a result of the use of sewage-polluted water for irrigation, or contamination with animal or human faeces during harvest. Seeds used to produce sprouts (e.g. mung beans for bean sprouts) can sometimes contain low numbers of salmonellas that multiply during germination, and can be extremely difficult to eliminate from the seeds. For this reason consumers are often advised to blanch bean sprouts before eating them in salads. For products that will be cooked before consumption, the risk posed by pathogens is clearly less. There have been outbreaks of various types of food-borne disease from raw salad vegetables and from fruit (mostly soft fruit, pre-sliced fruit or unpasteurized juice), including Shigella sonnei, salmonellas, pathogenic E. coli (including E. coli O157:H7), hepatitis A virus, and the protozoa Cryptosporidium, Giardia and Cyclospora spp. A very large outbreak of E. coli O157:H7 infection, caused by contaminated radish sprouts, occurred in Japan (Michino et al., 1999). Listeria monocytogenes can sometimes be isolated from packed vegetable products, but rarely in numbers sufficient to cause concern (i.e. > 100 colony-forming units (cfu)/g).

Routine examination of these products for all the pathogens listed above is not practicable (or even possible). E. coli type 1 can be sought as an indicator of faecal contamination, but coliform, 'faecal coliform' or Enterobacteriaceae counts are not appropriate because normal spoilage flora will be included. Limits at use-by date for vegetables in France are: salmonellas: absent in 25 g; E. coli per gram: $m = 100$, $M = 1000$, $n = 5$, $c = 2$; L. monocytogenes: $M = 100$/g (Nyugen-the and Carlin, 2000). (For this m is the threshold value for bacterial count, result satisfactory if not exceeded; M is the maximum value for bacterial count, result unsatisfactory if exceeded in any sample; n is the number of sample units comprising the sample; and c is the number of sample units that may have bacterial counts between *m* and *M* if the bacterial count of the other samples is *m* or less.)

DEHYDRATED FRUIT AND VEGETABLES

These rarely cause outbreaks of disease, although they too may be contaminated in the same way as fresh fruit and vegetables. Total counts on dried vegetables are very variable, and can be as high as 10^8 cfu/g. Yeasts and moulds are also often present at levels up to 10^4 cfu/g. Numbers of bacterial spores can sometimes be raised, especially if the vegetables were blanched before drying. In products that may be consumed with minimal heating (e.g. 'instant' soups and dried herbs and spices) salmonellas should be sought in 25 g, and counts of *E. coli* type 1 (not more than 5/g) and *Bacillus cereus* and *Clostridium perfringens* ($m = 10^2$, $M = 10^4$) should be done.

Total counts on dried fruit are dominated by moulds and yeasts. Hazards are most likely to come from mycotoxins – particularly aflatoxin, in dried figs or nuts (e.g. peanuts, pistachios, cashews, Brazil nuts). Level should be < 4 ppb. Aflatoxin-producing moulds can be enumerated on Aspergillus Flavus Parasiticus Agar (AFPA), although direct detection of aflatoxin is preferable, because the moulds may no longer be viable. Desiccated coconut has sometimes been found to be contaminated with salmonellas.

FROZEN VEGETABLES AND FRUIT

Freezing affects the microbial flora little, and the microbiological profile should be similar to that of the fresh product – similar tests should be carried out (see above). Some vegetable products are blanched before freezing, lowering the total count by about 2 log cycles, but contamination frequently occurs from the processing environment before freezing. Frozen peas sometimes carry large numbers of *Leuconostoc* spp. and other lactic acid bacteria (Lund, 2000). A particular hazard to frozen vegetables from the processing environment could be *L. monocytogenes*, although outbreaks from this source have not been reported. Freezing cannot be relied on to eliminate pathogens such as salmonellas or hepatitis A virus.

PICKLES

Vegetables are first pickled in brine. The salt is then leached out with water and they are immersed in vinegar (for sour pickles) or vinegar and sugar (for sweet pickles). Some products are pasteurized. Spoilage is the result of low salt content of brine, poor quality vinegar, underprocessing and poor closures.

Counts

Use glucose tryptone agar at pH 6.8 for total counts and at pH 4.5 for counts of acid-producing colonies (these have a yellow halo caused by a colour change of indicator). Count lactic acid bacteria on Rogosa, MRS (de Man, Rogosa and Sharp) or other suitable medium. Estimate yeasts either by the counting chamber method (stain with 1 : 5000 erythrosin) or on Rose Bengal chloramphenicol agar.

Culture

Use glucose tryptone and Rogosa or MRS agar for lactic acid bacteria. Grow suspected film yeasts in a liquid mycological medium, containing 5 and 10% sodium chloride for 3 days at 30°C. For obligate halophiles, use a broth medium containing 15% sodium chloride.

Microbial content

Pickles are high-acid foods. Counts are usually low, e.g. 1000 cfu/g. Yeasts are a frequent source of spoilage and may be either gas producing or film producing. In the former, enough gas may be generated to burst the container. Bacterial spoilage may be caused by acid-producing or acid-tolerant bacteria such as acetic acid bacteria, lactic acid bacteria and aerobic spore bearers. Infected pickles are often soft and slimy.

Pickles of the sauerkraut type are fermented and therefore contain large numbers of lactic acid bacteria (*Lactobacillus* and *Leuconostoc* spp.) which are responsible for their texture and flavour.

Acid-forming bacteria are active at salt concentrations below 15%. Above this concentration, obligate halophiles are found.

KETCHUPS AND SAUCES

The most common cause of spoilage is *Zygosaccharomyces* (*Saccharomyces*) *baillii*. This yeast grows at pH 2, and from < 5°C up to 37°C, in 50–60% glucose; it is relatively heat resistant.

SUGAR AND CONFECTIONERY

Sugars, molasses and syrups

The importance of micro-organisms is related to the intended use of the products, e.g. flat-sour bacteria (see pp. 221 and 365) are more important in baking than in canning. Do total counts, examine for spore-forming thermophiles (*B. stearothermophilus*, *Desulfotomaculum* (formerly *Clostridium*) *nigrificans*, *C. thermosaccharolyticum*), and for yeasts and moulds. Spoilage is often caused by osmotolerant yeasts; culture for these on a medium of low water activity (DG18 agar).

Chocolate and cocoa powder

Chocolate has been shown to be the vehicle of salmonella infection in a number of cases (see the review by D'Aoust, 1977).

The standard method of examination for salmonellas needs to be modified in order to neutralize the natural inhibitory effect of cocoa products. Add 25 g cocoa powder, or shave 25 g chocolate into 225 ml buffered peptone water with 10% non-fat dried milk and then proceed according to the standard method (enrichment in selenite or tetrathionate and Rappaport Vassiliadis broth, etc. – see Chapter 14). Do not place large lumps of chocolate into liquid medium. Automated methods, especially those employing impedimetric principles, are useful for salmonella screening in industry. Examine spoiled soft-centred chocolates for yeasts and moulds by direct microscopy and culture on DG18 medium.

CAKE MIXES AND INSTANT DESSERTS

Spores are likely to be present, e.g. of *C. perfringens* and *B. cereus*. Salmonellas and staphylococci may survive processing or be post-processing contaminants. Any of these organisms may multiply if the product is reconstituted and then kept under unsuitable conditions. The powdered product may also cross-contaminate other products that could provide suitable conditions for growth.

Do total viable counts; test for *C. perfringens* and *B. cereus* ($m = 10^2$, $M = 10^4$), *E. coli* ($m = 10$, $M = 100$), *Staphylococcus aureus* ($M = 10^3$) and salmonellas (absent in 25 g).

Salmonellas and *E. coli* are potentially much more hazardous in instant desserts than in cake mixes, which will be cooked before consumption.

CEREALS

Contamination occurs from many sources – soil, manure, insects, animals, birds and fungal infection before or after harvest (see Legan, 2000).

Flour

White flour, freshly milled, contains relatively high numbers of micro-organisms – total count 30°C:

$5 \times 10^3 - 5 \times 10^5$; yeasts and moulds a factor of 10 lower, with higher numbers in whole-grain flour. Numbers of vegetative bacteria, yeasts and moulds decline during storage. Low numbers of coliforms can be found in most samples of freshly milled wheat flour, with *E. coli* in about 50% and salmonellas in 0.3–2.3% of 1-g samples. Low numbers of the food poisoning bacteria *S. aureus* and *B. cereus* are commonly found ($< 10^3$/g) and could be a source of illness if allowed to multiply to high levels. Mycotoxins are probably the most serious potential hazards in flour. Trichothecenes (deoxynivalenol and to a lesser extent nivalenol) and ochratoxin, produced by moulds of the genus *Fusarium*, are relatively common in wheat, especially wheat imported into the UK. Aflatoxin and deoxynivalenol can sometimes be found in North American maize. These toxins are formed in the field, not during storage. If flour gets damp it will spoil as a result of the growth of moulds.

If the flour will be used to make bread without preservatives (e.g. propionic or acetic acid), the concentration of spore formers implicated in 'ropey bread' (*B. subtilis* and *B. lichenformis*) will be of interest. If it is to be used to make raw chilled pastry, the numbers of lactic acid bacteria present in the flour will affect its shelf-life – this should be < 1000/g. Numbers of spores of thermophilic spore formers, such as *B. stearothermophilus* and *B. coagulans*, are relevant if the flour is to be used as an ingredient of a canned food.

Pastry

Uncooked pastry, prepared for factory use or for sale to the public, may suffer spoilage caused by lactobacilli. Do counts of lactic acid bacteria on Rogosa or MRS agar, and incubate at 30°C for 4 days. Counts of up to 10 000/g are not unreasonable.

Pasta products

These are made without a heating step from wheat flour, semolina, farina and water. Egg (powdered or frozen), spinach, vitamins and minerals may be added. The egg in particular may contain salmonellas and, although these may be destroyed in the subsequent cooking, there is the possibility of cross-contamination from uncooked to cooked products.

High levels of *S. aureus* and/or preformed staphylococcal enterotoxin have been reported in dried pasta with or without filling (International Commission on the Microbiological Specifications for Foods, ICMSF, 1998).

Examination

As contamination is likely to occur during manufacture, it is necessary to liberate the organisms from within the pasta.

Add 25 g pasta to 225 ml MRD and allow it to soften at room temperature for about 1 h. Macerate, e.g. in a Stomacher, and do total, *E. coli* and *S. aureus*, *B. cereus*, and yeast and mould counts.

Inoculate salt meat broth. Incubate at 37°C for 24 h and plate on Baird-Parker medium for staphylococci. If staphylococcal enterotoxin is suspected, in the absence of viable *S. aureus* (e.g. if microscopy reveals large numbers of non-viable Gram-positive cocci), use one of the commercial kits see (p. 332). Examine 25 g for salmonellas.

Extrusion-cooked products

These include breakfast cereals and crispbreads, and may be textured with vegetable protein and bread crumbs. The liquid and solid ingredients are blended, shaped and cooked within 1–2 min. Examine as for pasta.

GELATIN (DRY PRODUCT)

Gelatin is often used to top up pastry cases of cold-eating pies, in canned ham production and in ice-cream manufacture. It should be free from spores and coliforms.

If the process involves low-temperature reconstitution examine for *S. aureus*, salmonellas and clostridia.

Examination

Weigh 5 g gelatin into a bottle containing 100 ml sterile water and allow to stand at 0–4°C for 2 h. Place the bottle in a water-bath at 50°C for 15 min and then shake well. Mix 20 ml of this solution with 80 ml sterile water. This gives a 1 : 100 dilution. Use 1.0 and 0.1 ml for total counts by the pour-plate method. Incubate at 37°C for 48 h. Examine 25 g of dry product for salmonellas.

Gelatin for ice-cream manufacture

Do a semi-quantitative coliform estimation using MacConkey or similar broth. Add 10 ml 1 : 100 gelatin to 10 ml double-strength broth; add 1 ml 1 : 100 gelatin to 5 ml single-strength broth and 0.1 ml 1 : 100 gelatin to 5 ml single-strength broth. Incubate at 37°C for up to 48 h and do confirmatory tests where indicated (see p. 253). Thus, the presence or absence of coliforms and *E. coli* in 0.1, 0.01 or 0.001 g of the original material can be determined. It is desirable that coliforms should be absent from 0.01 g and *E. coli* absent from 0.1 g. The total count in gelatin to be used for ice-cream manufacture should not exceed 10 000 cfu/g.

Gelatin for canned ham production

This should have a low spore count. After doing the total counts, heat the remaining 1 : 100 solution of gelatin at 80°C for 10 min. Plate 4 × 1 ml of this solution on standard plate count medium. Incubate two plates at 37°C and two plates at 55°C for 48 h. There should be no more than one colony per plate. The total count should not exceed 10 000 cfu/g.

SPICES AND ONION POWDER

A plastic bag inverted over the hand is a satisfactory way of sampling spices. Some of them are toxic to bacteria and the initial dilution should be 1 : 100 in broth (except for cloves which should be 1 : 1000). Do total viable counts and, if indicated by intended usage, consider inoculating *B. cereus*-selective agar (e.g. PEMBA, see Table 14.1) and one of the media for staphylococci and salmonellas.

The total counts on these materials vary widely. Total viable counts of 10^8/g are generally acceptable, but low spore counts are important if these materials are to be used for canned foods: an acceptable level at 30 and 55°C is < 100/g.

COCONUT (DESICCATED)

In the 1950s and 1960s this product was often contaminated with salmonellas. In spite of improved processing it is still a potential hazard in the confectionery trade. Examine samples for salmonellas by the method described on p. 295.

OILY MATERIAL

Mix 1.5 g tragacanth with 3 ml ethanol and add 10 g glucose, 1 ml 10% sodium tauroglycocholate and 96 ml distilled water. Autoclave at 115°C for 10 min. Add 25 g of the oily material under test and shake well. Make dilutions for examination in warm diluent.

Salad creams

These contain edible oils and spoilage may be caused by the lipolytic bacteria. Test for these with tributyrin agar. Examine for coliform bacilli and thermophiles. All these should be absent. There should not be more than five yeasts or moulds per gram.

Mayonnaise-based salads

Examine for lactobacilli. Yeasts will also grow so colonies should be examined by Gram staining. Alternatively, do parallel counts on both MRS medium (lactobacilli plus yeasts) and MRS containing 100 mg/ml chloramphenicol (yeasts only) (Rose, 1985).

CANNED, PREPACKED AND FROZEN FOODS

The practice of prolonging the shelf-life of foods by sealing them in metal or glass containers and heat processing them has long been established, but modern packaging materials have given the consumer a wider choice. Some products are sterilized and then aseptically packed into sterile containers – usually flexible. Excellent quality assurance during these processes is of paramount importance, particularly for low acid products (Codex Alimentarius Commission, CAC, 1993a), but the principles for microbiological examination are similar to those for products sterilized in pack.

Eating habits have changed. Single portions that can be taken from the freezer or cupboard and heated in a microwave oven are popular. With these, packaging is important. It should:

- prevent contamination by micro-organisms
- preserve quality and nutritional value
- be inert and offer no hazard in use
- be economic to manufacture and distribute
- be easily labelled.

Routine control of the product in the factory is the responsibility of the quality assurance and laboratory departments of the manufacturer, but other laboratories may be asked to help when defects or spoilage has developed after the products have left the factory or where the product is suspected of having caused food poisoning. Occasionally, arbitration between suppliers of ingredients and manufacturers of finished goods is necessary.

Sealed containers intended to be stored at ambient temperature for long periods pose the highest risk if errors occur during or after processing. Canned foods are divided into two categories, depending on whether they can support germination, multiplication and toxin production by *C. botulinum* spores. This organism cannot grow in a product with a pH < 4.5. Hence only products with a pH > 4.5 ('low acid') need to be heated sufficiently to kill *C. botulinum*. This type of product is usually given at least a '12D' heat treatment, i.e. the time at a particular temperature sufficient to kill 10^{12} spores of *C. botulinum* (3 min at 121°C, or equivalent treatment at another temperature). Low acid foods include meat and fish, as well as many vegetables. Acid canned foods are mostly fruits, and can be preserved by a more gentle heat treatment (< 100°C), although they are still occasionally spoiled by spore formers if the raw material is heavily contaminated. These can be *Bacillus* or *Clostridium* spp. or ascospores of *Byssochlamys fulva* and *B. nivea* or spores of *Alicyclobacillus*.

Microbiological examination of canned foods should be carried out by specially trained personnel. Detailed instructions can be found in various publications, e.g. CAC (1979b, 1993b), Food and Drug Administration (FDA, 1998). Figures 16.1 and 16.2 are flow charts of procedures adapted from CAC publications.

Investigations normally include detailed examination of the can seams. The following sections provide information about the defects that can occur during or after canning. Faults are usually uncommon (< 1 per 1000 cans) and cannot be detected by post-production quality control sampling – the canning process must be run using a *quality assurance* system such as the Hazard Analysis Critical Control Point system (HACCP) (ICMSF, 1988a; Shapton and Shapton, 1991). It is customary, however, to hold all cans at 37°C for about 14 days after production, while routine quality control checks are carried out (e.g. examination of can seams) and to examine the cans visually for signs of spoilage (i.e. bulging – Shapton and Shapton, 1991).

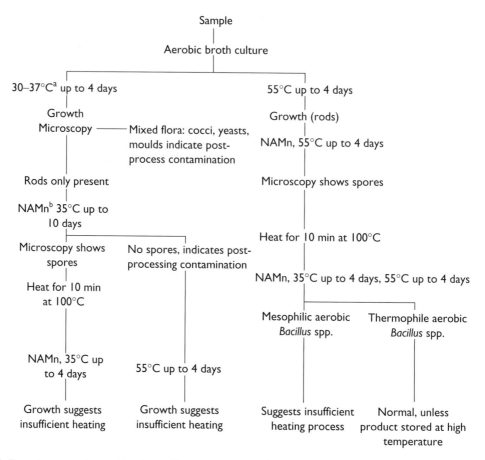

Figure 16.1 Flow diagram for the aerobic culture of low-acid canned foods.
[a] Incubation temperature may be varied according to environmental conditions. [b] Nutrient agar plus 0.4 mg/l manganese chloride to encourage sporulation of *Bacillus subtilis* (adapted from Codex Alimentarius Commission, 1979b)

Types of container

Cans

Tin plate and aluminium are widely used. Lacquering gives resistance to corrosion. Welded three-piece cans have largely superseded the soldered type. Base metal thickness has been reduced, giving lighter and cheaper cans, but these have to be 'beaded' to withstand processing. Two-piece cans with 'easy-open ends' are popular.

Jars and bottles

These are made of glass and closed with metal cap and resilient seals, sometimes under vacuum. Jars and bottles should be tamper proof.

Trays

These are made of aluminium or aluminium/polypropylene laminated, with lids.

Semi-rigid flexible plastic and laminate containers

These are easy to stack, light and allow a shelf-life of up to 2 years. Plastic-sided, metal-ended containers are used for fluids. For a review of containers, see Dennis (1987).

Types of blown can

Gas-producing organisms cause the can to swell. The first stage is the 'flipper' when the end of the

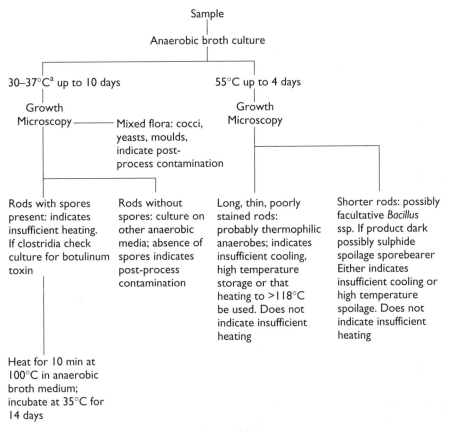

Figure 16.2 Flow diagram for the anaerobic culture of low-acid canned foods.
Note that spores of thermophilic anaerobes are rarely seen or found in microscopy or first subculture, hence no heating is suggested. It may be used, however, followed by anaerobic subculture at 55°C. [a] Incubation temperature may be varied according to environmental conditions (adapted from Codex Alimentarius Commission, CAC, 1979b)

can flips outward if the can is struck sharply. A 'springer' is caused by more gas formation. Pressing the end of the can causes the other end to spring out. The next stage is the 'swell' or 'blower' when both ends bulge. A 'soft swell' can be pressed back but bulges again when the pressure is released. 'Hard swells' cannot be compressed.

Spoilage may not result in gas formation and may be apparent only when the can is opened. This kind of spoilage includes the 'flat-sour' defect. In addition, hydrogen gas is sometimes produced in cans where microbial spoilage has not occurred. This is usually the result of a reaction between the contents of the can and the metal.

Spoilage caused by micro-organisms

Microbiological examination of cans can be done with or without prior incubation. If a can already shows signs of spoilage, e.g. swelling, it may be sufficient to open it with suitable precautions and examine the contents. Cans that appear normal may be incubated for up to 14 days at 37°C, examining at intervals for progressive swelling. Swollen cans are removed, and examined. Cans that do not swell are examined after 14 days. Initial microscopic examination is followed by cultural tests. Cans must be opened in a way that the seams are not damaged and so that the possibility of contamination from outside the can is minimized. This involves the use of a laminar flow or safety cabinet, decontamination of the outside of the can, e.g. with 4% iodine in 70% ethanol, and sterile instruments to open the can and sample the contents. Samples of the contents should be taken from below the surface. Blown cans (grossly distended) should be refrigerated before opening, and precautions taken to prevent spillage of contents during opening (e.g. put a cloth or funnel over the can).

Leaker spoilage

Contrary to the usual understanding of this word, leakage in canning usually refers to entry inwards rather than exit outwards. After autoclaving and during cooling there is negative pressure in the can, and bacteria may be drawn in through minute holes. These holes can occur because of faults in can seams, particularly where the end-seams cross the body seams, but leakage can occur in cans with normal seams if hygienic conditions have not been maintained during cooling and drying of the cans. It is important therefore that water used to cool cans is treated, e.g. with chlorine, to minimize numbers of micro-organisms in the water. Even dry surfaces in contact with wet cans need to be cleaned and disinfected frequently in order to minimize contamination. Leakage can also be caused by rough treatment of cans during processing or cooling. When the seams are dry, the chances of contamination are slight. Wetting cans during shelf-life can also result in leakage. Cans contaminated by leakage usually contain a mixture of bacteria, which can include pathogens such as *S. aureus* or even *C. botulinum*, as well as (more commonly) spoilage organisms. The cans may or may not appear 'blown'. Presence of holes can be checked on the empty and thoroughly cleaned can by use of vacuum or pressure.

Underprocessing or no processing

Gross underprocessing or no processing will usually have been found by the manufacturer's tests. Mixtures of micro-organisms (rods, cocci, yeasts and moulds) will be found. When the processing treatment is only slightly suboptimal, it is common to find only one type of (spore-forming) organism consistently from can to can, often in a similar position in each can.

Inadequate cooling

If cooling is too slow or inadequate or the cans are stored at very high temperatures, e.g. in tropical countries, there may be sufficient time for the highly resistant spores of *B. stearothermophilus* to grow and multiply. The optimum growth temperature for this organism is between 59 and 65°C. It will not grow at 28°C or at a pH of less than 5. Together with *B. coagulans*, it is the chief cause of 'flat-sour' spoilage in canned foods. Most strains of *B. coagulans* will grow at 50–55°C. They will also grow in more acid conditions.

Preprocessing spoilage

This can occur when the material to be canned is mishandled before processing, e.g. if precooked meats with a large number of surviving spores are kept for too long at a high ambient temperature, this may result in the outgrowth of spores and the production of gas. When canned, the organisms will be killed, leaving the gas to give the appearance of a blown can.

Types of food and organisms causing spoilage

Low-acid foods

If the pH is 4.5 or above, as in canned soups, meat, vegetables and fish (usually about pH 5.0), the cans are given a '12D' process. One of the following thermophiles is usually found in spoilage caused by underprocessing.

- *B. stearothermophilus*, causing 'flat-sour' spoilage

- *C. thermosaccharolyticum*, causing 'hard swell'
- *Desulfotomaculum* (previously *Clostridium*) *nigrificans*, causing 'sulphur stinkers'
- Mesophilic spore bearers, obligate or facultative anaerobes, causing putrefaction.

Leaker spoilage

This may be the result of a variety of organisms: aerobic and anaerobic spore bearers, Gram-negative non-sporing rods and various cocci, including *Leuconostoc* and *Micrococcus* spp., may be found. *S. aureus* (food poisoning type) has been isolated.

Acid foods

Canned foods with pH 3.7–4.5 include most canned fruits. As *C. botulinum* cannot grow at or below pH 4.5, these foods are usually given a much milder heat treatment than low acid foods. Aerobic and anaerobic sporebearers may cause spoilage but this is not common. Lactobacilli and *Leuconostoc* spp. have been reported. Osmophilic yeasts and the mould *Byssochlamys* spp., which produces heat resistant spores, are sometimes found. Acid canned foods are too acidic for the growth of most bacteria and leaker spoilage is uncommon.

High-acid foods

These foods are given a mild heat treatment, as with acid foods. Spoilage is rare in processed foods with a pH of 3.7 or less, e.g. pickles and citrus fruits. Spores of *Alicyclobacillus* spp. sometimes spoil these products. Yeasts may be found when there has been serious underprocessing.

Canned cured meats, e.g. ham, corned beef

As these are preserved with salt and nitrite, they can be given a less severe heat treatment than other low acid foods because, although *C. botulinum* spores can survive this treatment, they are unable to germinate and multiply.

Some cured meats are deliberately given a non-sterilizing process because they are rendered less palatable by autoclaving. Catering-size cans of ham and mixtures of ham and other meat are preserved with salt and nitrite, and given the minimum heat treatment. The manufacturers do not claim sterility and the label on the can invariably recommends cold or cool storage. The pH, salt, and nitrite content and storage temperature should be such that the bacteria in the can are prevented from multiplying. There is evidence that some species of enterococci produce antibacterial substances in canned hams, which act antagonistically on some species of clostridia, lactobacilli and members of the genus *Bacillus*. This is most likely a factor in the successful preservation of commercial products of this nature (Spencer, 1966; Kafel and Ayres, 1969; Gardner 1983).

Frozen foods

Some frozen convenience foods are mentioned here, others – frozen meat, fish and ice-cream – are included under their appropriate headings in Chapters 15, 17 and 18. For frozen vegetables, see p. 215 (see also CAC, 1999; Lund, 2000).

The minimum temperature for multiplication of microbes in foods is generally accepted to be about −10°C. Frozen foods are normally stored at −18 to −20°C, and so should be microbiologically stable. If the temperature is allowed to rise to around −10°C, spoilage is likely to occur as a result of growth of yeasts and moulds. Freezing cannot be relied on to destroy micro-organisms, however, although it can be used in inactivate parasites such as *Trichinella* spp. Frozen foods will therefore have a microflora similar to the fresh foods from which they were prepared. Their shelf-life is normally limited by biochemical deterioration.

Examination

This can be carried out in a similar way to examining fresh foods, except that it should be understood that a proportion of the population is likely to be sublethally damaged by freezing (Mackey, 2000). This means, for instance, that colony counts of Enterobacteriaceae or coliforms may be affected by bile salts in VRB or VRBG agar. To compensate for this, a resuscitation step can be inserted, such as inoculating plates of non-selective agar for a few hours before over-layering with selective medium, or by putting the inoculum on a membrane filter that can be transferred from a non-selective to a selective medium. Enrichment protocols for recovering pathogens such as salmonellas or listerias from foods usually include a resuscitation step, so special procedures are not usually needed for frozen foods.

Precooked chilled ('cook–chill') products

The safety and shelf-life of these foods depend on sublethal treatment and aseptic packaging. Major advances in transportation and handling of chilled and frozen products have increased the availability of these foods. Chilled foods account for over 40% of expenditure on foods in the UK. Temperatures should be controlled between +1°C and +4°C throughout handling and storage. As many of these foods are capable of supporting the growth of food-poisoning organisms, it is important that they are not mishandled.

The packaging and the food are sterilized separately and united under 'commercially sterile' conditions. Flexible pouches and semi-rigid pots and cartons are used for low acid foods, e.g. ice-cream, custards, soups and sauces.

Cook–chill catering

This system is used by hospitals, works canteens, prisons, schools, and banqueting and travel organizations. It demands a high level of technical control to ensure safety and quality. If handled and stored according to the guidelines (Department of Health, 1989) they should be safe.

Bacteriological examination

These products may be examined by the same method as other foods, and the same standards should apply. These foods should conform to the following standards:

- Viable count after 48 h at 37°C – < 100 000/g
- *E. coli* – < 10/g
- *S. aureus* – < 100/g
- *Salmonella* spp. not present in 25 g
- *Listeria monocytogenes* not present in 25 g
- *C. perfringens* not present in 100 g.

'Sous vide' products

These are cook–chilled foods that are vacuum packed and given a mild heat treatment in the range 65–75°C. They are used primarily in the catering industry. The packaging prevents post-cook recontamination, but the mild heat treatment allows spores to survive, so these products must be stored below 3°C in order to prevent growth of psychrotrophic spore formers – especially non-proteolytic *C. botulinum*. Shelf-life can be several weeks. Clearly, it is important that the storage temperature of these products is carefully controlled. It has been suggested (Gould, 1999) that products intended to have a long shelf-life should either be treated in order to give a 6-log cycle reduction of psychrotrophic *C. botulinum* spores, which is 90°C for 10 min, or equivalent, or have other preservative factors (pH < 5 or water activity (a_w) < 0.97 or NaCl > 3.5% [salt on water]). The Advisory Committee on the Microbiological Safety of Food (ACMSF, 1992) recommended that 'sous vide' products given a mild heat treatment should have a shelf-life not exceeding 10 days.

Nissen *et al.* (2002) found that all studied 'sous vide' meals heat treated at 85–100°C for 10–60 min had an initial total viable count (TVC) < 10^3/g, and 75% < 10/g. Similar numbers were found after 3–5 weeks' shelf-life, storing at 4°C for the first third and 7°C for the rest of the time. In another study (Nyati, 2000), meat and other dishes were heated to internal temperatures varying from 68 to 85°C. Reductions in total count varied from 4 to 6 log cycles and satisfactory results were obtained after storage at 3°C for 4 weeks, but after 2 weeks at 8°C 8 of the 18 products were spoiled – with high counts of lactic acid bacteria and sometimes pseudomonads. *B. cereus* (> 3×10^4 cfu/g) was found in one chicken product.

Apart from *B. cereus* and non-proteolytic *C. botulinum*, other psychrotrophic pathogens that might multiply in sous vide foods and hence could be tested for, are *Listeria* spp., *Yersinia* spp. and *Aeromonas* spp.

BABY FOODS

Direct breast-feeding allows little chance of infection of the infant with enteric pathogens. Testing of human milk is described on p. 243. With other feeds contamination may occur during the time lag between production and consumption.

There are many milk preparations and weaning formulae, as well as dried, bottled and canned foods designed for small children. Dehydrated milk may contain organisms that have survived processing and these may multiply if the product is not stored correctly after it is reconstituted. Dirty equipment may contribute bacteria to an otherwise sterile material. Unsuitable water may be used for reconstitution. Since an outbreak of salmonellosis in the UK, from a contaminated dried milk-based

formula, attention has been paid to sampling plans for salmonellas in dried milks. Sampling plans in the Codex Code of Hygienic Practice for Foods for infants and children (CAC, 1979a) are shown in Table 16.1. *Enterobacter sakazaki* has recently been found in infant food formulae and implicated as the cause of severe disease in babies. It may be sought using a sampling plan similar to that for salmonellas (Iverson *et al.*, 2004).

Central milk kitchens in maternity units should have very high standards of hygiene (see p. 245 for sampling surfaces, etc.). In-bottle terminal heating, with teat in place, is good practice.

Viable counts should be low and pathogens should be absent (see Robertson; 1974; Collins-Thompson *et al.*, 1980). Nasogastric feeds should conform to the same standards.

Heat-treated jars and cans should be examined as described for extended shelf-life foods on p. 219.

SOFT DRINKS

Total counts should be low and coliform bacilli absent, as in drinking water. Membrane filters can

Table 16.1 Advisory microbiological specifications for foods for infants and children

Product	Test	n	c	Limit per g	
				m	M
Dried biscuit type products:					
Plain	None	–	–	–	–
Coated	Coliforms	5	2	< 3 (by MPN)	20
	Salmonellas[a] (chocolate coatings only)	10	0	0	–
Dried and instant products (e.g. milk)	Mesophilic aerobic bacteria	5	2	10^3	10^4
	Coliforms	5	1	< 3 (by MPN)	20
	Salmonellas[a]	60	0	0	–
Dried products requiring heating before consumption	Mesophilic aerobic bacteria	5	3	10^4	10^5
	Coliforms	5	2	10	100
	Salmonellas[a]	5	0	0	–

[a] 25 g of sample examined
m is the threshold value for bacterial count, result satisfactory if not exceeded; *M* is the maximum value for bacterial count, result unsatisfactory if exceeded in any sample; *n* is the number of sample units comprising the sample; and *c* is the number of sample units that may have bacterial counts between *m* and *M* if the bacterial count of the other samples is *m* or less (from Codex Alimentarius Commission, (1979a)

be used to test water intended for soft drink production as well as the end product. Millipore publish a very useful booklet on the microbiological examination of soft drinks. Yeast and mould spoilages are not uncommon and they may raise the pH and allow other less acid-tolerant organisms to grow.

In non-carbonated fruit drinks, yeasts are not inhibited by the amounts of preservatives that are permitted by law. Spoilage is generally controlled by acidity (except *Zygosaccharomyces baillii*) (see 'Ketchups', p. 216). In both these cases and carbonated drinks, the microbial count diminishes with time.

FRUIT JUICES

Lactic acid and acetic acid bacteria may grow at pH 4.0 or less in some fruit juices.

Heat-resistant fungi can cause problems in concentrated juices. Screen by heating at 77°C for 30 min. Cool and pour plates with 2% agar. Incubate for up to 30 days. Dip slides, including those with medium for yeasts and fungi, are useful.

Automated counting methods (e.g. ATP assay) are useful. Fruit juices are incubated at 25°C for 24–48 h (72 h for tomato juice), the reagent is added and incubated for 45 min when the instrument gives the result.

BOTTLED WATERS

For information on bottled waters, see p. 257.

MILK-BASED DRINKS

For information on milk-based drinks, see p. 239.

VENDING MACHINES

Water and flavoured drinks in vending machines may be of poor quality, with high counts (> 1000 cfu/ml) and coliforms. This may be the result of inadequate cleaning (see Hunter and Burge, 1986).

USEFUL REFERENCES

Food microbiology is a very large subject and in addition to the references cited in this chapter and those preceding it the following are recommended: Hersom and Hulland (1980); Jowitt (1980); Harrigan and Park (1991); Roberts and Skinner (1983); ICMSF (1986, 1988, 1998a, 1998b, 2002); Shapton and Shapton (1991); Hayes (1992); Jay (2000); Lund *et al.* (2000); Dowens and Ito (2001).

REFERENCES

Advisory Committee on the Microbiological Safety of Food (ACMSF) (1992) *Report on Vacuum Packaging and Associated Processes.* London: HMSO.

Codex Alimentarius Commission (CAC) (1979a) *Recommended International Code of Hygienic Practice for Foods for Infants and Children.* CAC/RCP 21-1979.

CAC (1979b) *Recommended International Code of Hygienic Practice for Low and Acidified Low Acid Canned Foods.* CAC/RCP 23-1979, rev. 2 (1993) – see Appendix V.

CAC (1993a) *Code of Hygienic Practice for Aseptically Processed and Packaged Low Acid Canned Foods.* CAC/RCP 40-1993.

CAC (1993b) *Guideline Procedures for the Visual Inspection of Lots of Canned Foods.* CAC/GL 17.

CAC (1999) *Code of Hygienic Practice for Refrigerated Packaged Foods with Extended Shelf Life.* CAC/RCP 46-(1999).

Collins-Thompson, D. L., Weiss, K. F., Riedel, G. W. and Charbonneau, S. (1980) Microbiological guidelines and sampling plans for dried infant cereals and dried infant formulae. *Journal of Food Protection* **43**: 613–616.

D'Aoust, J. Y. (1977) Salmonella and the chocolate industry: a review. *Journal of Food Protection* **40**: 718–726.

Dennis, C. (ed.) (1987) *Symposium on the Microbiological and Environmental Health Problems in Relation to the Food and Catering Industries.* Chipping, Campden: Campden Food Preservation Association.

Department of Health (1989) *Chilled and Frozen Food. Guidelines on Cook-Chill and Cook-Freeze Catering Systems.* London: HMSO.

Dowens, F. P. and Ito, K. (eds) (2001) *Compendium of Methods for the Microbiological Examination of Foods,* 4th edn. Washington DC: American Public Health Association.

Food and Drug Administration (1998) *Bacteriological Analytical Manual,* 8th edn. Arlington, VA: AOAC International.

Gardner, G. A. (1983) Microbiological examination of curing brines. In: Board, R. G. and Lovelock, D. W. (eds), *Sampling – Microbiological Monitoring of the Environment.* Society for Applied Bacteriology Technical Series No. 7. London: Academic Press.

Gould, G. W. (1999) Sous vide foods: conclusions of an ECFF Botulinum Working Party. *Food Control* **10**: 47–51.

Harrigan W. F. and Park, R. W. A. (1991) *Making Food Safe.* London: Academic Press.

Hayes, R. (1992) *Food Microbiology and Hygiene,* 2nd edn. London: Elsevier.

Hersom, A. C. and Hulland, E. D. (1980) *Canned Foods. Thermal processing and microbiology.* London: Churchill-Livingstone.

Hunter, P. R. and Burge, S. H. (1986) Bacteriological quality of drinks from vending machines. *Journal of Hygiene (Cambridge)* **97**: 497–504.

International Commission on the Microbiological Specification for Food (ICMSF) (1986) *Microorganisms in Foods. 2. Sampling for Microbiological Analysis. Principles and specific applications.* 2nd edn. Toronto: University of Toronto Press (updated as ICMSF 2002).

ICMSF (1988) *Microorganisms in Foods. 4. Application of the HACCP system to ensure Microbiological Safety and Quality,* Oxford: Blackwell Scientific.

ICMSF (1998a) *Microorganisms in Foods. 6. Microbial Ecology of Food Commodities.* London: Blackie Academic & Professional. Available from Kluwer Publishers.

ICMSF (1998b) Cereal and Cereal Products. In: *Microorganisms in Foods. 6. Microbial Ecology of Food Commodities.* Blackie Academic & Professional. Available from Kluwer Publishers.

ICMSF (2002) *Microorganisms in Foods. 7. Microbiological Testing in Food Safety Management.* New York: Kluwer Academic/ Plenum Publishers.

Iverson, C., Druggan, P. and Forsythe, S. (2004) A selective medium for *Enterobacter sakazaki. International Journal of Food Microbiology,* in press.

Jay, M. (2000) *Modern Food Microbiology,* 6th edn. New York: Aspen Publishers.

Jowitt, R. (ed.) (1980) *Hygienic Design and Operation of Food Plants.* Chichester: Ellis Horwood.

Kafel, S. and Ayres, J. C. (1969) The antagonism of enterococci on other bacteria in canned hams. *Journal of Applied Bacteriology* **32**: 217–232.

Legan, J. D. (2000) Cereals and cereal products. In: Lund, B., Baird-Parker, A. C. and Gould, G. W. (eds), *The Microbiological Safety and Quality of Food,* Vol. 1. London: Chapman & Hall, pp. 759–783.

Lund, B. M. (2000) Freezing. In: Lund, B., Baird-Parker, A. C. and Gould, G. W. (eds), *The Microbiological Safety and Quality of Food,* Vol. 1. London: Chapman & Hall, pp. 122-145.

Lund, B. M. and Snowdon, A. L. (2000) Fresh and processed fruits. In: Lund, B., Baird-Parker, A. C. and Gould, G. W. (eds), *The Microbiological Safety and Quality of Food,* Vol. 1. London: Chapman & Hall, pp. 738-758.

Mackey, B. M. (2000) Injured bacteria. In: Lund, B., Baird-Parker, A. C. and Gould, G. W. (eds), *The Microbiological Safety and Quality of Food,* Vol. 1. London: Chapman & Hall, pp. 315-341.

Michino, H., Araki, K., Minami, S. *et al.* (1999) Massive outbreak of *Escherichia coli* O157:H7 infection in schoolchildren in Sakai City, Japan, associated with consumption of white radish sprouts. *American Journal of Epidemiology* **150**: 787–796.

Nissen, H., Rosnes, J. T., Brendehaug, J. and Kleiberg, G. H. (2002) Safety evaluation of sous vide-processed foods. *Letters in Applied Microbiology* **35**: 433–438.

Nyati, H. (2000) An evaluation of the effect of storage and processing temperatures on the microbiological status of sous vide extended shelf-life products. *Food Control* **11**: 471–476.

Nyugen-the, C. and Carlin, F. (2000) Fresh and processed vegetables. In: Lund, B., Baird-Parker, A. C. and Gould, G. W. (eds), *The Microbiological Safety and Quality of Food*, Vol. 1. London: Chapman & Hall, pp. 620–684.

Roberts, T. A. and Skinner, F. A. (eds) (1983) *Food Microbiology*. Society for Applied Bacteriology Symposium Series No. 11. London: Academic Press.

Roberts, T. A., Hobbs, G., Christian, J. H. B. and Skovgaard, N. (eds) (1981) *Psychrotrophic Microorganisms in Spoilage and Pathogenicity*. London: Academic Press.

Robertson, M. H. (1974) The provision of bacteriologically safe infant feeds in hospitals. *Journal of Hygiene (Cambridge)* **73**: 297–303.

Rose, S. A. (1985) A note on yeast growth in media used for the culture of lactobacilli. *Journal of Applied Bacteriology* **59**: 53–56.

Shapton, D. A. and Shapton, N. F. (eds) (1991) *Principles and Practices for the Safe Processing of Foods*. London: Butterworths.

Spencer, R. (1966) Non-sterile canned cured meats. *Food Manufacture* **41**: 39–41, 43.

17

Fresh fish, shellfish and crustaceans

Examples of these three broad groups are shown in Table 17.1. Quality is assessed best by appearance and odour. As with other foods, bacteriological examination involves assessing the presence and numbers of micro-organisms that affect the quality of the food (i.e. the spoilage flora that influence the freshness or shelf-life of the product) and those that may affect human health.

Table 17.1 Broad definition of fish species

Fish type	Examples of species
Finfish	Cod, plaice, mackerel, salmon, trout
Molluscan shellfish	Mussels, oysters, scallops, clams
Crustaceans	Shrimp, prawns, crab, lobster

FRESH FISH

Microbial content

Total bacterial counts are still used today, although usually in combination with other microbiological indices such as numbers of spoilage bacteria and the presence of potentially pathogenic bacteria. Fish muscle taken aseptically from a newly killed fish is sterile, but micro-organisms increase in numbers during storage. Numbers on the surface of fish differ depending on where the sample was taken. Liston *et al.* (1976) found that numbers on freshly caught fish varied between 10^2 and 10^6 colony-forming units (cfu)/cm² on the skin, 1 log less in the gills and up to 10^7 per g in the intestines of actively feeding fish. Large numbers of pseudomonads, flavobacteria, coryneforms, and *Acinetobacter*, *Aeromonas* and *Cytophaga* spp. are often associated with the slime that occurs naturally on the skin. Micrococci and photobacteria (luminescent in the dark on some media) are often present. Mycobacteria are also found in fish and cause problems, especially in farmed fish raised in large impoundments of water or in tanks in the southern USA.

Numbers of bacteria increase with time of storage but only a critical proportion of these affects the organoleptic quality of the food. The total count method should vary depending on the conditions under which or where the fish was caught. In the UK, local sea temperatures are generally low and the temperature of agar plate incubation should reflect this to estimate bacterial numbers as accurately as possible. The former fisheries research institute at Torry in Aberdeen, UK recommended incubation at 20°C, which is the optimum growth temperature for the majority of bacteria found in local waters. Some fish-associated micro-organisms grow rather slowly, however, and an incubation period of 5 days was also recommended. This incubation time produced maximum counts, although numbers at 3 days were sometimes not significantly lower.

Viable count

The method may vary depending on the type of fish under investigation. For whole fish, select several sites and aseptically peel back the skin and remove

25 g muscle for testing. In the case of small fish such as whitebait, the whole fish is often consumed and so the whole should be tested. For fillets, select from the cut surface. Counts will be influenced greatly by any inclusion of skin or intestines in the examination of filleted fish portions

Add 225 ml chilled phosphate-buffered saline to 25 g fish and homogenize for 2 min. Plate 0.1-ml volumes (in duplicate) of serial dilutions on nutrient agar containing 0.5% NaCl (some nutrient media do not normally contain NaCl), and incubate at 20°C for 3–5 days. Count the colonies on plates containing between 30 and 300, and calculate the numbers of colony-forming units per gram. Counts of $> 10^6$/g in most fish indicate little or no remaining shelf-life but if intestinal or skin tissue were assayed then this value could be one- to tenfold greater.

count because it can be used to estimate remaining shelf-life. Spoilage bacteria may be selectively assessed (Wood and Baird, 1943) by their production of hydrogen sulphide (black colonies) on cysteine-iron agar. Gibson and Ogden (1987) used numbers of H_2S-producing bacteria to predict the shelf-life of packaged fish, as did Gram and Huss (1999), who observed good correlation between numbers of H_2S-producing bacteria and remaining shelf-life of iced cod (Table 17.2).

Plate homogenates directly on Iron Agar (Oxoid) or add 1 : 10 dilutions to the melted medium (at 45°C), mix and allow to solidify. Then pour on an overlayer (1–2 ml) of melted medium. This minimizes oxidation of ferrous sulphide, which would cause instability of the black precipitate. Incubate the plates at 20–25°C for 3–4 days and count the numbers of black colonies.

Selective counts

A significant proportion of the viable count may not contribute to the spoilage of the fish during storage. Some, e.g. *Moraxella* spp. (now *Psychrobacter* spp.) and *Acinetobacter* spp. are biochemically inactive (relative to others found in fish) and large numbers of them do not significantly reduce shelf-life or cause food poisoning if consumed. Others, however, including *Shewanella* spp. and pseudomonads, are metabolically active and are mainly responsible for spoilage. An independent count of these is of more use than the total

Rapid methods

Conductance techniques (Richards *et al.*, 1978) may be particularly useful to estimate the microbiological status of fish because the metabolism of the spoilage flora gives rise to significant changes in conductance (see Chapter 8). Ogden (1986) showed acceptable correlation between conductance measurements and total viable counts in different fish species and improved correlation when plotted against numbers of spoilage bacteria (counted selectively using the Wood and Baird method).

Table 17.2 Relationship between numbers of H_2S-producing bacteria and remaining shelf-life for iced cod

Shelf-life remaining (days)	Log$_{10}$ cfu/g H_2S-producing bacteria
14	2
11	3
9.5	4
7	5
5	6
2	7

Adapted from Gram and Huss (1999)

SHELLFISH AND CRUSTACEANS

These are usually harvested from inshore waters or river estuaries that are prone to sewage pollution. The bivalve molluscs, in particular, are filter feeders and hence tend to concentrate bacteria and viruses from their environment. *Escherichia coli* is used as an index of pollution. The most probable number (MPN) technique is generally used but alternatives have been investigated (Ogden *et al.*, 1998) because the MPN method is time-consuming, labour intensive and of limited accuracy. Although the methods given below are primarily for shellfish, they may also be used for crustaceans.

MPN method for *E. coli* determination in shellfish

Use protective gloves to select 15–30 shellfish. Reject any that show signs of gaping or damage. Scrub them clean under running water of potable standard and open with a sterile shucking knife. Collect both meat and liquor in a weighed sterile container (beaker or bag) and transfer to a Stomacher for homogenization for 2-3 min in a filter bag.

Prepare 10^{-1} and 10^{-2} dilutions in sterile 0.1% peptone water. Use minerals modified glutamate medium (MMG) for the MPN test described on p. 150, adding 10 ml of the 10^{-1} dilution to five 10-ml tubes of double-strength MMG, 1 ml of the same dilution to five 9-ml tubes of single-strength MMG and 1 ml of the 10^{-2} dilution to five 9-ml tubes of single-strength MMG media. Thus, there are 1 g, 0.1 g and 0.01 g tissue per tube in each set.

Incubate all tubes at $37 \pm 1°C$ for 18–24 h in a water bath and note gas production. Re-incubate the tubes that show no gas for a further 18–24 h. Absence of gas confirms a negative result for *E. coli*.

Subculture tubes with gas into (1) 5 ml brilliant green bile broth (BGBB) and (2) 5 ml 1% tryptone water, and incubate both at $44 \pm 0.5°C$ for 24 h. Positive tubes for gas production in BGBB and indole production from tryptone (Kovacs reagent) confirm *E. coli*. Alternatively, the presence of β-glucuronidase confirms *E. coli* and is recognized by fluorescent colonies grown on MUG agar or blue colonies on BCIG agar, the latter being preferable because the colour remains within the bacterial colony (unlike fluorescence), which aids accurate counting. Calculate numbers from conventional MPN tables (see p. 152).

Crustaceans

Wash crabs, lobsters, etc. in running water as above. Break open, remove some meat and proceed as above. Wash prawns, shrimps, etc. in running water and proceed as above.

Standards

The European Union (EC, 1991) has set categories of shellfish for human consumption. These are shown in Table 17.3.

Those in category A are fit for human consumption whereas those in category B require further treatment. This can include repositioning in clean seawater and waiting for *E. coli* levels to fall naturally (relaying), or transferring the animals to a tank where the water is circulated through a ultraviolet (UV) source to kill bacteria (depuration). Category C shellfish are deemed unfit for human consumption.

Table 17.3 Microbiological definition of shellfish categories

Category	Numbers of *E. coli*/100 g shellfish
A	< 230
B	230–4600
C	> 4600

Pathogens

The pathogens that have been particularly associated with sea food poisoning include vibrios, notably *Vibrio parahaemolyticus* (see Chapter 24), *Staphylococcus aureus* (see Chapter 36), *Listeria monocytogenes* (see Chapter 40) and *Clostridium botulinum* type E (see Chapter 44). Aeromonads (see Chapter 24) and salmonellas (see Chapter 28) have also been incriminated. Methods for the isolation of these micro-organisms from foods are given in the appropriate chapters.

The possibility that fish might be contaminated by pathogenic viruses should be considered. Molluscs are filter feeders and depuration, which removes bacteria, may not remove viruses; they are not always killed by heat treatment either. Bivalves are therefore often associated with outbreaks of viral diseases such as hepatitis A and winter vomiting disease.

Molluscs are sometimes vehicles for algal toxins. Consumption may result in diarrhoeic and amnesic shellfish poisoning, the onset of which may be as short as 30 min. The origins of these toxins are the varied marine phytoplankton, which are the natural food of the shellfish. Some 40 of the 5000 species (mainly dinoflagellates and diatoms) are able to produce powerful toxins, which give a range of symptoms after consumption (Whittle and Gallacher, 2000).

In the last 25 years there has been a marked increase in the incidence of scombroid fish poisoning. The symptoms resemble those of food poisoning (i.e. diarrhoea and vomiting, headache, giddiness, rash on head and neck) 10 min to a few hours after ingestion of scombroid fish, e.g. mackerel, tuna, etc. Raised levels of histamine resulting from the decarboxylation of histidine in fish tissue by bacteria during poor storage of the fish is always involved. Levels of ≥ 5 mg/100 g fish are considered indicative of mishandling, and ≥ 50 mg/100 g hazardous (International Committee for the Microbiological Standards of Foods or ICMSF, 1998).

REFERENCES

EC (1991) Council Directive laying down the health standards for the production and placing on the market of bivalve molluscs. 91/492/EEC. *Official Journal of the European Communities* 24 September, 1991. No. L268/1.

Gibson, D. M. and Ogden, I. D. (1987) Estimating the shelf life of packaged fish. In: Kramer, D. E. and Liston, J. (eds), *Developments in Food Science: Seafood quality determination*. Amsterdam: Elsevier, pp. 437–451.

Gram, L. and Huss, H. H. (1999) Microbiology of fish and fish products. In: Lund B, Baird-Parker, A. C. and Gould, G. W. (eds), *The Microbiological Safety and Quality of Foods*. London: Chapman & Hall, pp. 472–506.

International Committee for the Microbiological Standards of Foods (ICMSF) (1998) *Micro-organisms in Foods 6. Microbial Ecology of Food Commodities*. Amsterdam: Kluwer, pp. 130–189.

Liston, J., Stansby, M. E. and Olcott, H. S (1976) Bacteriological and chemical basis for deteriorative changes. In: Stansby, M. E. (ed.), *Industrial Fishery Technology*. New York: Kreiger, pp. 345–358.

Ogden, I. D. (1986) The use of conductance methods to predict bacterial counts in fish. *Journal of Applied Bacteriology* 61: 263–268.

Ogden, I. D., Brown, G. C., Gallacher, S. *et al.* (1998) An interlaboratory study to find an alternative to the MPN technique for enumerating *Escherichia coli* in shellfish. *International Journal of Food Microbiology* 40: 57–64.

Richards, J. C. S., Jason, A. C., Hobbs, G. *et al.* (1978) Electronic measurement of bacterial growth. *Journal of Physics E. Scientific Instruments* 11: 560–568.

Whittle, K. J. and Gallacher, S. (2000) Marine toxins. *British Medical Journal* 56: 236–253.

Wood, A. J. and Baird, E. A. (1943) Reduction of trimethylamine oxide by bacteria, 1. The Enterobacteriaceae. *Journal of Fisheries Research Board of Canada* 6: 194–201.

18

Milk, dairy produce, eggs and ice-cream

These foods have a long history of spoilage and as vehicles for microbial diseases, including tuberculosis, scarlet fever, diphtheria, and various kinds of food poisoning and food-borne illnesses, all of which result from low standards of hygiene in production and distribution. This has resulted in legislation that impinges on bacteriological examinations.

MILK

In Europe, the European Commission (EC, 1992) lays down microbiological tests and standards for milk, etc. These are given under each test description. Similar tests and criteria are used in other countries such as the USA (Downs and Ito, 2001).

Raw milk

Raw milk, which includes that from cows, goats, ewes and buffaloes, is milk that is intended for:

1. The production of heat-treated drinking milk, fermented milk, junket, jellies and flavoured milk drinks
2. The manufacture of milk-based products other than those specified in (1), e.g. cows' milk, after packing for drinking in that state; and cows', ewes' or goats' milk sold directly to the ultimate consumer.

The bacteriological tests required to satisfy legal standards for these milks are:

- the plate count (total viable count, TVC) at 30°C

- enumeration of coliforms and *Staphylococcus aureus*
- detection of salmonellas.

Pasteurized milk

This is raw milk that meets the above requirements and has been heated at 71.7°C for 15 s or an equivalent time/temperature combination, and then immediately cooled to below 10°C. The tests required by legislation are:

- the plate count at 21°C
- enumeration of coliforms
- detection of pathogens (*Listeria monocytogenes* and salmonellas)
- the phosphatase test.

Ultra-heat-treated (UHT) milk

This is raw milk, as above, that has been heated at 135°C for 1 s. The statutory bacteriological test is the plate count at 30°C.

Sterilized milk

This is raw milk, as above, that has been heated at 100°C for such a time that it will pass the turbidity test. The statutory tests are:

- the plate count at 30°C
- the turbidity test.

Collection and transport of samples

Raw milk intended for the manufacture of milk drinks or milk-based products should be sampled at the point of production using sampling programmes drawn up on risk assessment principles. At least two samples a month should, however, be sampled for the plate count. Cows' milk packed for drinking should be sampled at the point of sale. Samples should be transported to the laboratory at 0–4°C.

Statutory testing

For legal purposes the reference methods specified in the EC (1992) Directive or other internationally accepted methods of analysis must be used, but for routine practice the methods described below are adequate. The targets, i.e. acceptable highest levels, usually corresponding to legal standards, are given below (see, however, the comments on the accuracy of viable counts in Chapter 10).

Total viable count

The method is that given on p. 145. Test raw milks on arrival, use standard milk plate count agar and for raw milks incubate at 30°C for 72 h. Pre-incubate pasteurized milks at 6°C for 5 days before testing and then incubate plate counts at 21°C. The targets levels (as total viable counts per millilitre or TVC/ml) are:

- Raw milk not intended for drinking in that state: < 100 000
- Raw milk packed for drinking: < 50 000
- Raw milk sold directly to the ultimate consumer: < 20 000
- Pasteurized milk: m = 50 000; M = 500 000; n = 5; c = 1; where m is the threshold value for bacterial count, result satisfactory if not exceeded; M is the maximum value for bacterial count, result unsatisfactory if exceeded in any sample; n is the number of sample units comprising the sample; and c is the number of sample units that may have bacterial counts between m and M if the bacterial count of the other samples is m or less. The test fails if any sample has a count exceeding M.

Total viable count, UHT and sterilized milk

Pre-incubate samples at 30°C for 15 days or 55°C for 7 days after sampling at point of production. Perform plate counts in duplicate using 0.1-ml volumes of milk and plate count agar. Incubate at 30°C for 72 h. The target level is:

- UHT and sterilized milk: < 100/ml.

Coliform count

Make serial tenfold dilutions and perform plate counts with violet red bile lactose agar. Incubate at 30°C for 24 h. Count only red colonies that are 0.5 mm in diameter or larger. Confirm if necessary by subculture in brilliant green bile lactose broth at 30°C for 24 h. It is useful to test confirmed coliforms by the 44°C and indole tests (see p. 254) to determine whether they are *Escherichia coli*. The target coliform level per millilitre is:

- Raw milk: < 100/ml
- Pasteurized milk: m, 0; M, 5; n, 5; c, 1.

Pathogens

Methods for these are given elsewhere: salmonellas, p. 295; *S. aureus*, p. 339; *Listeria*, p. 356; *E. coli* O157, p. 288; and campylobacters, p. 314. Pathogens should be absent in 25 ml in milk tested.

Phosphatase test

The enzyme phosphatase in milk is destroyed by the time/temperature conditions used for pasteurization.

Detection of the enzyme indicates inadequate pasteurization.

The fluorimetric method is an internationally recognized method and is more sensitive and reproducible than the non-fluorimetric technique (Greenwood and Rampling, 1997; BSI, 2000). The alkaline phosphatase activity of the sample is measured by a continuous fluorimetric direct kinetic assay. A non-fluorescent aromatic monophosphoric ester substrate, in the presence of any alkaline phosphatase (ALP) derived from the sample, undergoes hydrolysis of the phosphatase radical, producing a highly fluorescent product. Fluorimetric measurement of ALP activity is measured at 38°C for a 3-min period (BSI, 1997). The fluorimeter will perform the necessary calculations and display the sample number, the average increase in fluorescence and the ALP activity in milliunits per litre.

The legal level of phosphatase activity permitted in pasteurized milk is 4 µg phenol/ml. This is equivalent to 500 mU/l by the fluorimetric test and 10 µg p-nitrophenol/ml by the Aschaffenberg–Mullen (A–M) test.

The A–M test (SI, 1989) may be used but should be validated against the European reference method (EC, 1992a).

Turbidity test, sterilized milk

Weigh 4.0 g ammonium sulphate (AnalaR) into a 50-ml conical flask. Add 20 ml milk and shake for 1 min to ensure that the ammonium sulphate dissolves. Stand for 5 min and filter through a 12.5-cm Whatman No. 12 folded filter paper into a test tube. When 5 ml clear filtrate has collected, place the tube in a water-bath at 100°C for 5 min. Remove the tube and allow to cool. Examine the filtrate for turbidity. Properly sterilized milk gives no turbidity.

Sterilization alters the protein constituents and the ammonium sulphate precipitates all non-heat-coagulable protein. If heating is insufficient, some protein remains unaltered and is not precipitated by the ammonium sulphate. It coagulates, giving turbidity when the filtrate is boiled.

Microbial content of milk

Milk is at risk of faecal contamination from the cow or other producer species, and is subject to potential contamination from equipment and the environment during milking and handling. Refrigerated storage, which is normal practice after milking, retards the multiplication of bacteria present. Pasteurization, intended to kill pathogenic bacteria, does not necessarily reduce the count of other organisms. It may increase the numbers of thermophiles.

The 'normal microflora' depends on temperature: at 15–30°C S. lactis predominates and many streptococci and coryneform bacteria are present, but at 30–40°C they are replaced with lactobacilli and coliform bacilli. All of these organisms ferment lactose and increase the lactic acid contents, which causes souring. The increased acid content prevents the multiplication of putrefactive organisms. Spoilage during cold storage is caused by psychrophilic pseudomonads and Alcaligenes spp., psychrotrophic coliforms, e.g. Klebsiella aerogenes and Enterobacter liquefaciens, which are anaerogenic at 37°C; in pasteurized milk, thermoduric coryneform organisms (Micrococcus lactis) may be significant. These coryneforms probably come from the animal skin or intestine and from utensils. At temperatures above 45°C, thermophilic lactobacilli (Lactobacillus thermophilus) rapidly increase in numbers.

Gram-negative bacilli are rarely found in quarter samples collected aseptically. These enter from the animal skin and dairy equipment during milking and handling. Alcaligenes spp. are very common in milk. Pseudomonas fluorescens gels UHT milk.

Undesirable micro-organisms responsible for 'off flavours' and spoilage include psychrophilic Pseudomonas, Achromobacter, Alcaligenes and Flavobacterium spp., which degrade fats and proteins and give peculiar flavours. Coliform bacilli produce gas from lactose and cause 'gassy milk'. *S. cremoris*, *Alcaligenes viscosus* and certain *Aerobacter* species, all capsulated, cause 'ropy milk'. *Oospora lactis* and yeasts are present in stale milk. *P. aeruginosa* is responsible for 'blue milk' and *Serratia marcescens* for 'red milk' (differentiate from bloody milk).

Pathogenic organisms that may be present include *E. coli* O157, *S. aureus*, campylobacters, salmonellas, *Yersinia enterocolitica*, listerias, *Streptococcus pyogenes* and other streptococci from infected udders in mastitis. Tubercle bacilli may be found by culture of centrifuged milk deposits and gravity cream. *B. abortus* is excreted in milk and can be isolated by culture, or antibodies can be demonstrated by the ring test or whey agglutination test. *Rickettsia burnetti*, the agent of Q fever, may be found by animal inoculation of milk from suspected animals.

Mastitis

Looking for evidence of mastitis in bulk milk is obviously unrewarding in any but farm or other small samples. Heat the test sample in the water bath at 35°C. Mix the test sample carefully and cool to 20°C. Place 0.01 ml of the prepared test sample on a clean slide and make a film. Dry the film on a level hot plate. Dip the dried film in dye solution containing methylene blue for 10 min. Dip the film in tap water to wash away surplus dye, and dry (BSI, 1997). Examine for somatic cells and count the number. For bacteriological examination and identification of causative organisms, see p. 335.

EXAMINATION OF DAIRY EQUIPMENT

See Chapter 19 for methods, etc.

Microbial content

The types of organisms found will be similar to those in milk. Soil organisms, e.g. *Bacillus* spp., may be present. *B. cereus* reduces methylene blue very rapidly and may also cause food poisoning. Organisms originating from faeces, e.g. various coliforms and *Pseudomonas* spp. are found. Coliform bacilli also rapidly reduce methylene blue. Pseudomonads may produce oily droplets in the milk.

CREAM

Natural cream

The Dairy Products (Hygiene) Regulations (SI, 1995) specify the temperature requirements for production of all types of heat-treated cream except clotted cream in the UK. These are national rather than EC requirements which have been retained on public health grounds. They do not apply to cream from sheep and goats. There is a requirement to satisfy the phosphatase test. The bacteriological tests and targets required for cream are as follows.

Pasteurized cream

- Salmonellas: absent in 25 ml
- *L. monocytogenes*: absent in 1 ml
- Coliforms (guideline level): m, 0; M, 5; n, 5; c, 2 (see above).

Sterilized and UHT cream

Incubate in a closed container at 30°C for 15 days or 55°C for 7 days.

- Plate count at 30°C: < 100/ml.

Coliform test for pasteurized cream

Place 10 g of the test sample into a container containing glass beads. Add 90 ml buffered peptone diluent (see p. 76) and shake to disperse. A Stomacher may be used, but omit the glass beads. The temperature of the diluent should be approximately that of the test sample. Prepare further decimal dilutions if required.

The following method combines detection of coliforms and presumptive *E. coli* in a liquid medium containing 4-methylumbelliferyl-β-D-glucuronide (MUG) (from which fluorescent umbelliferone is released by glucuronidase activity) and calculation of the number of coliforms and/or presumptive *E. coli* by the most probable number (MPN) technique after incubation at 30°C. It is particularly suitable for use when levels are likely to be low (< 10/ml). The enumeration of coliforms and *E. coli* by the MPN technique involves six stages:

1. Inoculation of three tubes of lauryl tryptose broth with Durham tubes and containing 0.01% MUG per dilution of the test sample, using those dilutions appropriate to obtaining the required detection requirements for that product.
2. Incubation of those tubes at 30°C for 48 h.
3. Identification of tubes showing gas production as positive for presumptive coliforms.
4. Confirmation of coliform presence by subculture to brilliant green bile broth and detection of gas formation after incubation at 30°C for 24 h.
5. Identification of tubes showing fluorescence and formation of indole as positive for presumptive *E. coli*.
6. Determination of the MPN index from the number of positive tubes of selected dilutions using an MPN table and calculating the coliform count and/or *E. coli* count per gram or millilitre of sample.

Plate (total viable) count

Do plate counts or Miles and Misra counts (see p. 148) on serial dilutions ranging from 10^{-2} to 10^{-6} in a 0.1% peptone water solution, using milk plate count agar.

If the cream is solid or difficult to pipette, make the initial 10^{-1} dilution by weight as described on p. 197. The 10^{-1} dilution may be used, with appropriate media, for spiral plating for total viable, coliform and *S. aureus* counts.

Culture for pathogens

- Plate serial dilutions on Baird-Parker medium for *S. aureus*.
- For salmonellas pre-enrich a 1 in 10 mixture of cream (preferably 25 g) in buffered peptone water (BWP) containing glass beads and shake to disperse. Proceed as on p. 295.
- For campylobacters, use a 10-g sample and the method on p. 314.
- For *L. monocytogenes*, use a 25-g sample and the method on p. 356.

Phosphatase test

This is as for milk (see p. 234).

Microbial content

Fresh cream may contain large numbers of organisms, including coliforms, *B. cereus*, *E. coli*, *S. aureus*, other staphylococci, micrococci and streptococci. Most fresh cream has been pasteurized and therefore contamination by many of these organisms is the result of pasteurization failure or faulty dairy hygiene.

Cream should not be used in catering in circumstances where these organisms can multiply rapidly.

Imitation cream

Do plate counts as for milk, at 30°C, coliform counts, and also examine for *S. aureus* (see p. 330). Examine for staphylococci only, as for natural cream.

PROCESSED MILKS

Dried milk

This often has a high count and, if stored under damp conditions, is liable to mould spoilage. Spray-dried milk may contain *S. aureus*. The bacteriological tests and targets required for dried milk are:

- Salmonellas: absent in 25 g; *n*, 10; *c*, 0
- *L. monocytogenes*: absent in 1 g
- *S. aureus* (per g): *m*, 10; *M*, 100; *n*, 5; *c*, 2
- Coliforms 30°C (per g, guideline level): *m*, 0; *M*, 5; *n*, 5; *c*, 2.

TVC, coliform and *S. aureus* tests for dried milk

Warm a solution of dipotassium hydrogen phosphate pH 7.5 to 45°C in a water-bath. Thoroughly mix the contents of the closed sample container by repeatedly shaking and inverting. If the volume of the contents is too great to allow thorough mixing, transfer the contents to a larger container and mix. Remove 10 g sample and add to a bottle containing 90 ml dipotassium hydrogen phosphate. To aid dissolution, swirl slowly to wet the powder, then shake the bottle (e.g. 25 times in 7 s). The addition of sterile glass beads may help reconstitution, particularly with roller-dried milk. A Stomacher may be used, but omit the glass beads. Replace the bottle in the water bath for 5 min, shaking occasionally. Prepare further decimal dilutions if required:

- Do total counts as for fresh milk but prepare extra plates for incubation at 55°C for 48 h for thermophiles.
- Enumerate coliforms using the MPN method as described for coliform testing of cream (see p. 237).
- The 1 in 10 and further serial dilutions may be used to plate on Baird-Parker medium for *S. aureus* (see p. 330).

Salmonellas in dried milk

Thoroughly mix the contents of the sample container as described above. Place 225 ml sterile distilled water in an appropriate container and add 1 ml brilliant green solution. Weigh 25 g of the test sample and pour it over the surface of the liquid. Close the container but do not shake. Allow to stand undisturbed for 60 ± 10 min before incubation at 37°C. If after 3 h incubation the dried milk has not dissolved, shake the container to mix the contents. Proceed as on p. 295.

Yeasts, moulds and spores

Examine for yeasts and moulds as described for butter examination (see p. 239).

The number of spore bearers is important in milk and milk powder intended for cheese making. Inoculate 10-ml volumes of bromocresol purple milk in triplicate with 10-, 1.0- and 0.1-g amounts of milk. Heat at 80°C for 10 min, overlay with 3 ml 2% agar and incubate for 7 days. Read by noting gas formation and use Jacobs and Gerstein's MPN tables to estimate the counts (see p. 154).

Condensed milk

This contains about 40% of sugar and is sometimes attacked by (osmophilic) yeasts and moulds that form 'buttons' in the product. Examine as for butter (see p. 239).

Evaporated milk

This is liable to *underprocessing* and *leaker spoilage*. *Clostridium* spp. may cause *hard swell* and coagulation. Yeasts may cause swell. Open aseptically and examine as appropriate (see p. 219).

Fermented milk products

Yoghurt, leben, kefir, koumiss, etc., are made by fermentation (controlled in factory-made products) of milk with various lactic acid bacteria, and streptococci and/or yeasts. The pH is usually about 3.0–3.5 and only the intended bacteria are usually present, although other lactobacilli, moulds and yeasts may cause spoilage. The presence of large numbers of yeasts and moulds can be indicative of poor hygiene. Plate on potato glucose agar (pH adjusted to 3.5) and incubate at $23 \pm 2°C$ for 5 days.

Yoghurt is made from milk with *Lactobacillus bulgaricus* and *Streptococcus thermophilus*. In the finished product there should be more than 10^8/ml of each and they should be present in equal numbers. Lower counts and unequal numbers predispose to off-flavours and spoilage.

Make doubling dilutions of yoghurt in 0.1% peptone water and mix 5 ml of each with 5 ml melted LS Differential Medium. Pour into plates, allow to set and incubate at 43°C for 48 h. Both organisms produce red colonies, because the medium contains triphenyl tetrazolium chloride. Those of *L. bulgaricus* are irregular or rhizoid, surrounded by a white opaque zone, whereas those of *S. thermophilus* are small, round and surrounded by a clear zone. Count the colonies and calculate the relative numbers of each organism.

Post-pasteurization contamination with coliforms may occur. One indicator of contamination or spoilage is an increased pH, which is usually very acid. If this is pH > 5.6, examine for coliforms, *E. coli*, salmonellas and other enteric pathogens. The bacteriological tests and targets required for yoghurt are:

- Salmonellas: absent in 25 ml; n, 5; c, 0
- *L. monocytogenes*: absent in 1 ml
- Coliforms (per ml, guideline level): m, 0; M, 5; n, 5; c, 2.

The methods are as described for cream above.

MILK-BASED DRINKS

The bacteriological tests and targets required for liquid milk-based drinks are:

- Salmonellas: absent in 25 ml; n, 5; c, 0
- *L. monocytogenes*: absent in 1 ml
- Coliforms (per ml, guideline level): m, 0; M, 5; n, 5; c, 2.

The methods are as described for cream above.

Phosphatase test

This is as for milk (see p. 233).

BUTTERMILK

Test for coliforms only, using the MPN method as described for coliform testing of cream (see p. 237).

BUTTER

Soured cream is pasteurized and inoculated with the starter, usually *S. cremoris* or a *Leuconostoc* sp. Salted butter contains up to 2% of sodium chloride by weight.

Examination and culture

Emulsify 10 g butter in 90 ml warm (45°C) peptone/saline diluent and do plate counts on tryptose agar. Culture on a sugar-free nutrient medium and tributyrin agar.

The bacteriological tests and targets required for butter made from pasteurized milk or cream are:

- Salmonellas: absent in 25 g; n, 5; c, 0
- *L. monocytogenes*: absent in 1 g
- Coliforms (per gram, guideline level): m, 0; M, 10; n, 5; c, 2.

The methods are as described for cream above.

Microbial content

Correct flavours are caused by the starters, which produce volatile acids from the acids in the soured raw material.

Rancidity may be caused by *Pseudomonas* and *Alcaligenes* spp., which degrade butyric acids. Anaerobic organisms (*Clostridium* spp.) may produce butyric acid and cause gas pockets. Some lactic acid bacteria (*Leuconostoc* spp.) cause slimes and others give cheese-like flavours. Coliforms, lipolytic psychrophiles and casein-digesting proteolytic organisms, which give a bitter flavour, may be found. To detect these, plate on Standard Plate Count agar containing 10% sterile milk and incubate at $23 \pm 2°C$ for 48 h. Flood the plates with 1% hydrochloric acid or 10% acetic acid. Clear zones appear round the proteolytic colonies.

Yeasts and moulds may be used as an index of cleanliness in butter. Adjust the pH of potato glucose agar to 3.5 with 10% tartaric acid and inoculate with the sample. Incubate at $23 \pm 2°C$ for 5 days. Report the yeast and mould count per millilitre or gram. Various common moulds grow on the surface but often these grow only in water droplets or in pockets in the wrapper.

CHEESE

Many different cheeses are now available, including varieties that contain herbs which may contribute to the bacterial load. Some are traditionally made from raw milk and may contain pathogens such as *S. aureus*, brucellas, campylobacters and/or listerias.

Milk is inoculated with a culture of a starter, e.g. *S. lactis*, *S. cremoris*, *S. thermophilus*, and rennet added at pH 6.2–6.4. To make hard cheeses, the curd is cut and squeezed free from whey, incubated for a short period, salted and pressed. The streptococci are replaced by lactobacilli (e.g. *L. casei*) naturally at this stage and ripening begins. Some cheeses are inoculated with *Penicillium* spp. (e.g. *P. roquefortii*), which give blue veining and characteristic flavours as a result of the formation, under semianaerobic conditions, of caproic and other alcohols.

Soft cheeses are not compressed, have a higher moisture content and are inoculated with fungi, usually *Penicillium* spp., which are proteolytic and flavour the cheese.

Types and flavours of cheeses are caused by the use of different starters and ripeners and varying storage conditions.

Examination and culture

Take representative core samples aseptically, grate with a sterile food grater, thoroughly mix and subsample. Fill the holes, left as a result of boring, with Hansen's paraffin cheese wax to prevent aeration, contamination and texture deterioration.

Make films, and de-fat with xylol and stain. Bacteria may be present as colonies: examine thin slices under a lower-power microscope. Weigh 10 g and homogenize it in a blender with 90 ml 1 mol/l sodium citrate solution or dipotassium hydrogen phosphate pH 7.5. The methods for coliforms, *E. coli*, *S. aureus*, *L. monocytogenes* and salmonellas (pre-warm BPW to 45°C) are as described for cream above.

The bacteriological tests and targets required for cheese are as follows.

Hard cheese made from raw milk or thermised milk

- Salmonellas: absent in 25 g; *n*, 5; *c*, 0
- *L. monocytogenes*: absent in 1 g
- *S. aureus* (per gram): *m*, 1000; *M*, 10 000; *n*, 5; *c*, 2
- *E. coli* (per gram): m, 10 000; M, 100 000; *n*, 5; *c*, 2.

Hard cheese made from heat-treated milk

- Salmonellas: absent in 25 g; *n*, 5; *c*, 0
- *L. monocytogenes*: absent in 1 g.

Soft cheese made from raw or thermised milk

- Salmonellas: absent in 25 g; *n*, 5; *c*, 0
- *L. monocytogenes*: absent in 25 g; *n*, 5; *c*, 0
- *S. aureus* (per gram): *m*, 1000; *M*, 10 000; *n*, 5; *c*, 2
- *E. coli* (per gram): *m*, 10 000; *M*, 100 000; *n*, 5; *c*, 2.

Soft cheese made from heat-treated milk

- Salmonellas: absent in 25 g; *n*, 5; *c*, 0
- *L. monocytogenes*: absent in 25 g; *n*, 5; *c*, 0
- *S. aureus* (per gram): *m*, 100; *M*, 1000; *n*, 5; *c*, 2
- *E. coli* (per gram): *m*, 100; *M*, 1000; *n*, 5; *c*, 2
- Coliforms 30°C (per gram): *m*, 10 000; *M*, 100 000; *n*, 5; *c*, 2 (guideline level).

Fresh cheese

- Salmonellas: absent in 25 g; *n*, 5; *c*, 0
- *L. monocytogenes*: absent in 25 g; *n*, 5; *c*, 0
- *S. aureus* (per gram): *m*, 10; *M*, 100; *n*, 5; *c*, 2.

Microbial content

Apart from streptococci, lactobacilli and fungi that are deliberately inoculated or encouraged, other micro-organisms may be found. Contaminant moulds, and *Penicillium*, *Scopulariopsis*, *Oospora*, *Mucor* and *Geotrichum* spp., give colours and off-flavours. Putrefying anaerobes (*Clostridium* spp.) give undesirable flavours. *Rhodotorula* sp. gives pink slime and *Torulopsis* sp. yellow slime. Gassiness (unless deliberately encouraged by propionibacteria in Swiss cheeses) is usually caused by *Enterobacter* spp., but these are not found if the milk is properly pasteurized.

Gram-negative rods, some of which may hydrolyse tributyrin, may also be present. Psychrophilic spoilage is common and is caused by *Alcaligenes* and *Flavobacterium* spp. Counts may be very high. Bacteriophages that attack the starters and ripeners can lead to spoilage.

Butyric blowing

Clostridium butyricum and *C. tyrobutyricum* may cause gas pocket spoilage during the ripening of hard cheeses.

Homogenize cheese samples in 1 mol/l sodium citrate solution or dipotassium hydrogen phosphate solution pH 7.5 and add dilutions to melted sodium polyethanol sulphate (SPS) agar in tubes. Layer paraffin wax on the agar and incubate at 30°C. Butyric clostridia will produce gas bubbles in the medium. The organisms may be enumerated by the three- or five-tube MPN method with inocula of 10, 1 and 0.1 ml homogenate.

Propionibacteria

Although deliberately introduced into some cheeses, propionibacteria can cause spoilage in others. Accurate counts require microaerophilic conditions and media.

Prepare doubling dilutions of cheese homogenate in 1 mol/l sodium citrate solution or dipotassium hydrogen phosphate solution pH 7.5, and add 1 ml of each to tubes of yeast extract agar containing 2% sodium lactate or acetate. Mix and seal the surface with 2% agar. Incubate anaerobically or under carbon dioxide at 30°C for 7–10 days. Propionibacteria produce fissures and bubbles of gas in the medium. Choose a tube with a countable number of gas bubbles and calculate the numbers of presumptive propionibacteria (some other organisms may produce gas bubbles).

EGGS

Bacterial contamination of eggs can occur at different stages ranging from production to storage, processing, distribution and preparation. Contaminants on the shell are potential contaminants of the liquid egg. Washing eggs in water is an effective method of removing shell contaminants. Spoilage bacteria of eggs are mostly Gram-negative rods (*P. fluorescens*, *Alicaligenes bookeri*, *Paracolobacterium intermedium*, *P. melanovogenes*, *S. putrefaciens*, *P. vulgaris*, *Flavobacterium* spp.). Salmonella serotypes (*enteritidis*, *gallinarum*, *pullorum*) are associated with eggs and egg products. The EC Directive on egg products (EC, 1992b) contains microbiological standards for salmonellas, Enterobacteriaceae, *S. aureus*, and TVCs in the product from the treatment establishment.

Shell eggs

Shell eggs may be contaminated with various salmonellas during their passage down the oviduct or from the environment after laying. On the shell, *Salmonella* spp. may be present with other bacteria including coliforms, *Pseudomonas* spp., *Bacillus* spp., staphylococci and faecal streptococci. Shell eggs are normally examined for the presence of *Salmonella* spp. Batch examination of eggs can be carried out by testing pools of six eggs.

Disinfect the shells by swabbing or immersing with industrial alcohol. Break the eggs in batches of six, keeping the shells and contents in separate, sterilized containers, e.g. screw-capped jars. Add to each an equal volume of buffered peptone water (BPW) (pre-enrichment). Incubate at 37°C for 24 h and subculture 0.1 ml to 10 ml Rappaport–Vassiliadis Soya Broth (RVSB) and 1 ml to 10 ml selenite broth (SB). Incubate the RVSB at 42°C and SB at 37°C for up to 48 h. Subculture to xylose lysine deoxycholate (XLD) and brilliant green agar (BGA) or mannitol lysine crystal violet agar (MLCB), and proceed as on p. 295.

Pasteurized liquid egg

The American regulations require that liquid whole egg be heated to at least 60°C and held for 3.5 min, whereas the EC Directive specifies a heat treatment at 64°C for 2.5 min. The EC Directive bacteriological tests required are:

- TVC at 30°C
- enumeration of staphylococci (see p. 329) and coliforms (see p. 284), and detection of salmonellas (as above and see p. 295).

The targets levels are:

- Plate count: M, 100 000/g or ml
- Enterobacteriaceae: M, 100/g or ml
- Staphylococci: absent in 1 g or ml
- Salmonellas: absent in 25 g or ml.

Liquid egg albumen and crystalline egg albumen

Bacteriological tests and target levels required are as above for pasteurized liquid egg.

Powdered egg

Powdered egg products including whole egg, egg yolk and albumen are recognized vehicles of salmonellas and should be examined for them. Bacteriological tests and target levels required are as above for pasteurized liquid egg.

Frozen egg

Frozen egg must be pasteurized or otherwise treated to destroy salmonellas present. The α-amylase, bacteriological tests and target levels required are as above for pasteurized liquid egg.

ICE-CREAM AND EDIBLE ICES (ICE-LOLLIES)

These products vary in their composition and may be ice-cream, as defined in the various food regulations, mixtures of ice-cream and fruit or fruit juices, or contain no ice-cream. In general, viable (plate) counts, coliforms and tests for pathogens should be applied to ice-cream and commodities that are said to contain it whether as a core or as a covering. Otherwise the pH should be ascertained. If this is ≤ 4.5 there is no point in proceeding further. The EC (1992a) Directive specifies the temperature requirements for production of pasteurized and sterilized ice-cream. The bacteriological tests and targets required for ice-cream and other frozen milk-based products are:

- Salmonellas: absent in 25 g; n, 5; c, 0
- *L. monocytogenes*: absent in 1 g
- *S. aureus* (per gram): m, 10; M, 100; n, 5; c, 2

- Coliforms 30°C (per gram): *m*, 10; *M*, 100; *n*, 5; *c*, 2 (guideline level)
- Plate count 30°C (per gram): *m*, 100 000; *M*, 500 000; *n*, 5; *c*, 2 (guideline level).

Sampling

Send samples to the laboratory in a frozen condition. Remove the wrappers of small samples aseptically and transfer to a screw-capped jar. Sample loose ice-cream with the vendor's utensils (which can be examined separately if required, see Chapter 19). Keep larger samples in their original cartons or containers.

Hold edible ices by their sticks, remove the wrapper and place the lolly in a sterile screw-capped 500-ml jar. Break off the stick against the rim of the jar. Reject samples received in a melting condition in their original wrappers. Allow the samples to melt in the laboratory but do not allow their temperature to rise above 20°C.

Examination

Weigh 10 g of the test sample into a container. Place the container in a water-bath at 37°C until the whole test portion has just melted. Add 90 ml BPW diluent pre-warmed to 37°C. Mix in the stomacher until thoroughly dispersed (1 min).

The methods for coliforms, *S. aureus*, *L. monocytogenes* and salmonellas are as described for cream above.

Do plate counts using plate count agar and incubate at 30°C for 72 h. The spiral plating method (see p. 82) may be used for total counts.

Miles and Misra counts (see p. 148) are very convenient. Dilute the ice-cream 1 : 5 and drop three 0.02-ml amounts with a 50-dropper on blood agar plates. Allow the drops to dry, incubate at 37°C overnight and count the colonies with a hand lens. Multiply the average number of colonies per drop by 250 (i.e. dilution factor of 5 and volume factor of 50) and report as count per millilitre. If there are no colonies on any drops, the count can be reported as < 100/ml. It is difficult to count more than 40 colonies per drop; therefore, report such counts as > 10 000 cfu/ml.

In counts, note the presence of *B. cereus*. This sporebearer may be present in the raw materials. Its spores resist pasteurization.

Counts at 0–5°C for psychrophiles may be useful in soft ice-cream.

Ingredients and other products

Do plate counts at 5, 20 and 37°C. Although ice-cream is pasteurized and then stored at a low temperature, only high-quality ingredients with low counts will give a satisfactory product.

In cases of poor grading or high counts on the finished ice-cream, test the ingredients and also the mix at various stages of production. Swab rinses of the plant are also useful (see p. 248).

HUMAN MILK

There are effective bactericidal and bacteriostatic systems within raw breast milk that prohibit the further proliferation of micro-organisms in the milk. There is therefore no need for routine microbiological screening. Microbiological testing of mother's own milk should be carried out only when there is a clinical indication.

Bacteriological criteria for feeding raw human milk to babies on neonatal units have been recommended by Carroll *et al.* (1979). Samples containing 100 000 cfu/ml or less are classified as:

- class I – milk considered sterile (no growth after overnight incubation)
- class II – milk containing saprophytes
- class III – milk containing potential pathogens (*S. aureus*, enterococci, group B streptococci, Enterobacteriaceae and *P. aeruginosa*) – this is rejected.

Samples containing more than 100 000 cfu/ml are rejected. Pasteurization is recommended when the bacterial contamination is too high and when potential pathogens are present in the milk. The phosphatase test should be used to establish the efficiency of pasteurisation (as for cows' milk, p. 234). For more information see Carroll *et al.* (1979) and DHSS (1982).

For further information on the microbiology of milk and milk products see Greenwood (1995), Rampling (1998) and Robinson (1990a, 1990b).

REFERENCES

British Standards Institution (BSI) (1997) BS EN ISO 13366-1: *Milk – Enumeration of somatic cells. Part 1. Microscopic method.* London: British Standards Institution.

BSI (2000) BS EN ISO 11816-1: *Milk and milk-based drinks – Determination of alkaline phosphatase activity – Fluorimetric method.* London: British Standards Institution.

Carroll, L., Davies, D. P., Osman, M. *et al.* (1979). Bacteriological criteria for feeding raw breast-milk to babies on neonatal units. *Lancet* ii: 732–733.

Department of Health and Social Security (DHSS) (1982) *The Collection and Storage of Human Milk*, Reports on Health and Social Subjects No. 29. London: HMSO.

Downs, F. P. and Ito, K. (eds) (2001). *Compendium of Methods for the Microbiological Examination of Foods*, 4th edn. Washington: American Public Health Association.

EC (1989) European Council Directive 89/437/EEC of 20 June 1989 on hygiene and health problems affecting the production and placing on the market of egg products. *Official Journal of the European Communities* L212: 87–100.

EC (1992) European Council Directive 92/46/EEC of 16 June 1992 laying down the health rules for the production and placing on the market of raw milk, heat-treated milk and milk-based products. *Official Journal of the European Communities* L268: 1–32.

Greenwood, M. (1995) Microbiological methods for the examination of milk and dairy products in accordance with the Dairy Products (Hygiene) Regulations 1995. *PHLS Microbiological Digest* 12: 74–82.

Greenwood, M. H. and Rampling, A. M. (1997) Evaluation of a fluorimetric method for detection of phosphatase in milk. *PHLS Microbiology Digest* 14: 216–217.

Rampling, A. (1998) The microbiology of milk and milk products. In: *Topley & Wilson's Microbiology and Microbial Infections*, 9th edn, Vol. 2, *Systematic Bacteriology*. London: Arnold, pp. 367–393.

Robinson, R. K. (1990a) *Dairy Microbiology. The Microbiology of Milk*, Vol 1, 2nd edn. London: Elsevier Applied Science.

Robinson, R. K. (1990b) *Dairy Microbiology. The Microbiology of Milk Products*. Vol. 2, 2nd edn. London: Elsevier Applied Science.

Statutory Instrument (SI) (1989) SI 2383 *The Milk (Special Designation) Regulations*. London: HMSO.

SI (1995) SI 1986 *The Dairy Products (Hygiene) Regulations*. London: HMSO.

19

Environmental microbiology

It may be necessary to estimate the numbers of viable organisms on surfaces and in the air of buildings, in containers and on process equipment:

- to verify the standard of hygiene and efficiency of cleaning procedures
- to trace contamination from dirty to clean areas
- to test for the escape of 'process' organisms in industrial microbiology and biotechnology plants
- to assess the level of bacterial contamination in 'sterile' environments
- to investigate premises where humidifier fever, sick building syndrome, etc. are suspected.

SURFACE SAMPLING

Few standards or guidelines have been published on what is an acceptable level of micro-organism contamination on surfaces. The EC (2001) Decision calls for regular checks to be carried out on the general hygiene by meat establishment operators. It provides that cleaned and disinfected surfaces sampled using the agar contact plate or swab technique should have an acceptable range of 0–10 colony-forming units (cfu)/cm^2 for total viable counts (TVCs) and 0–1 cfu/cm^2 for Enterobacteriaceae. As a guideline, the US Public Health Service (Downs and Ito, 2001) recommends that adequately cleaned and disinfected food service equipment should have a total number of not more than 100 viable micro-organisms per utensil or surface area of equipment sampled.

Agar contact plates

Contact plates consist of an agar surface that is applied to the surface to be sampled. They are commercially available as Rodac (Replica Organism Direct Contact) plates. Non-selective or selective media may be used. If the surface to be sampled has been cleaned with a phenolic or quaternary ammonium agent, 0.5% Tween 80 and 0.7% soya lecithin should be added to the medium to neutralize them.

To sample a surface remove the lid of the Rodac plate and press the agar firmly on to the surface, applying a rolling uniform pressure on the back of the plate to effect contact. Replace the lid and incubate the plate at an appropriate temperature. When used on some surfaces, agar debris left on the surface during sampling will have to be removed, e.g. with an alcohol-impregnated wipe, to prevent microbiological growth associated with such nutrient debris.

Contact plates can be used only on flat non-porous surfaces that are relatively easy to clean and disinfect. They are suitable only to assess surface populations in a range that can be accurately counted on their surface, i.e. 10–100 colonies. The technique itself is less accurate in predicting surface populations than the surface swab technique (Holah, 1999). This is not a problem, however, when used for historical trend analysis because it is the deviation from normal internal standards in results over a period of time that is important, not the accuracy of the results themselves.

Surface swab counts

The swab method may be used to sample any surface up to 1 m². Wipes and sponges are preferred for large surface areas (floors, walls), and fabric-tipped swabs or sponges for equipment and small food contact surfaces (100 cm²). They are commercially available in a range of sizes as individually packaged units in a purpose-designed transport case, or individually wrapped in a disposable sleeve and packed in bulk. Sterile templates are required to define the swabbing area for flat surfaces, and are also commercially available in a range of sizes.

Place the sterile template on the surface that is to be swabbed. There should be one sterile pre-moistened swab, cloth or sponge per container. Open the package containing the sterile swab, and aseptically remove it. Reseal the package. Depending on the entire surface to be sampled, either sterile forceps may be used or a sterile glove worn to hold the swab. Rub swab firmly and thoroughly over the surface to be sampled. Rub five times from the bottom to the top and five times from the left to the right. Return the swab to its container and reseal. If fabric-tipped swabs have been used, ensure that the part handled during swabbing is broken off. For those surfaces that are difficult to swab because of their contour, it may not be possible to swab a defined area with this procedure.

Soak and squeeze swabs in nutrient broth or an appropriate selective medium to recover the organisms. If the surface has been washed or treated with a phenolic or quaternary ammonium (QAC) disinfectant, add 0.5% Tween 80 and 0.7% soya lecithin to the medium. Swabs are available commercially with neutralizing buffer. The count per 25 cm² (or the area swabbed) is given by the number of colony-forming units per millilitre of rinse or solvent multiplied by 10. With Miles and Misra counts (see p. 148), it is given by the number of colony-forming units per five drops multiplied by 100.

Rapid methods

Rapid monitoring methods generate results in a sufficiently short timeframe (usually within 20 min) to allow process control, e.g. by measuring ATP levels after cleaning to determine whether re-cleaning is necessary. Current methods allow the quantification of micro-organisms (DEFT, DEM, ATP), food soils (ATP, protein) or both (ATP). There are no commercially available methods that allow the detection of specific microbial types within this timeframe. For more details on rapid sampling methods, see Holah (1999).

AIR SAMPLING

Contamination of air in hospitals and other workplaces may be responsible for the spread of infection. In addition, in offices and factories, especially where there is 'air conditioning', it may be related to allergic illnesses such as humidifier fever and the 'sick building syndrome'. It may also lead to contamination of products (e.g. foods, pharmaceuticals).

Airborne contaminants may be released by people, industrial processes or poorly maintained air-conditioning plants. Bacteria and fungal spores may be suspended in the air singly, in large clumps, or on the surfaces of skin scales and dust from the materials being processed. Small particles (e.g. 5 μm) may remain suspended in the air for long periods and be moved around by air currents. Filtration, as in some air-conditioning plants, may not remove them. Larger particles settle rapidly and contaminate surfaces.

Settle plates

Settle plates are typically 90-mm Petri dishes containing appropriate agar. Non-selective or selective media may be used. Several plates are exposed for a given time, incubated at the appropriate temperature, and colonies counted. Settle plates do not give an assessment of the number of micro-organisms in the air, only those that settle on the surface. The accuracy of the technique is limited to the number of colonies that can be accurately counted on the surface of the plate, which in turn is related to exposure time. Therefore, prior trials are required to establish such exposure times. Those who need to monitor the air over long periods, e.g. in hospital cross-infection work, favour settle plates.

Mechanical air samplers

In principle, known volumes of air are drawn through the equipment and micro-organisms and other particles are deposited on the surface of agar media, on membrane filters or into broth media where they may be enumerated. Some samplers (e.g. the Andersen) and other cascade samplers not only allow colony counting, but also provide information about the sizes of the particles from which the colonies develop. Particles deposited on membranes may be analysed chemically, immunologically or by ATP detection kits.

Table 19.1 lists the characteristics of several devices. Choice depends on the volume of air to be sampled and the convenience of handling, e.g. portability and power supply. The SAS and RCS models are hand held and battery operated. Each of the monitoring methods has limitations of which the user should be aware. None of these samplers gives the total number of viable organisms per unit volume. Not all the bacteria or spores are trapped on to the medium; some adhere to other parts of the equipment. In addition, viability may be impaired by physical conditions or inactivated by impact. Nevertheless they are useful in the assessment of the microbiological quality of air.

For more detailed information on air sampling see Bennet *et al.* (1991), Ashton and Gill (1992), Hambleton *et al.* (1992), Holah (1999), and Downs and Ito (2001).

Table 19.1 Examples of commonly used air-sampling devices

Type of sampler	Advantages	Disadvantages
Cascade sieve impacter, e.g. Andersen 6-stage sampler	Reliable results, information on particle size distribution	Not practical in industry use
Single-stage sieve impacter, e.g. Andersen one-stage Surface Air System (SAS) MAS-100 and MicroBio MB2	MAS-100 and SAS are practical in industrial use	Inefficient at small particle sizes (< 5 μm)
Slit impacter, e.g. Casella MK II	Efficient, easy to use	Static, bulky
Centrifugal airstream, e.g. Reuter Centrifugal Sampler (RCS)	Practical in industrial use	Selectivity for larger particles
Impingement, e.g. All Glass Impinger (AGI-30) Multistage Liquid Impinger Biosampler three-nozzle impinger	Efficient for sampling of bacteria, yeast, and spores	Not practical in industry use
Filtration, e.g. MD8	Practical in air sampling of isolaters; efficient for sampling of spores	Desiccation of vegetative microbial cells

WASHED BOTTLES AND CONTAINERS

Swab-rinse method for utensils, cartons and containers

Dip a cotton-tip swab in sterile 0.1% peptone water and rub it over the surface to be tested, e.g. the whole of the inside of the carton. Use one swab for five such articles. Use one swab for a predetermined area of a worktop surface, chopping board, etc. Swab both sides of knives, ladles, etc. Return this swab to the tube and swab the same surfaces again with another, dry swab.

Add 10 ml 0.1% peptone water to the tube containing both swabs. Shake and stand for 20–30 min. Do plate counts with 1.0- and 0.1-ml amounts on yeast extract agar (YEA). Divide the count per millilitre by 5 to obtain the count per article. Inoculate three tubes of single-strength MacConkey broth with 0.1-ml amounts to obtain coliform counts.

Rinse method

For churns, bins and large utensils

Add 500 ml quarter-strength Ringer's solution containing 0.05% sodium thiosulphate to the vessel to neutralize chlorine, if present. Rotate the vessel to wash the whole of the inner surface and then tip the rinse into a sterile screw-capped jar.

Do counts with 1.0- and 0.1-ml amounts of the rinse in duplicate. Incubate one pair at 37°C for 48 h and one at 22°C for 72 h. Take the mean counts/ml at 37°C and 22°C and multiply by 500 to give the count per container.

For churns, cans and other large containers, counts of not more than 10 000 cfu/container are regarded as satisfactory, between 10 000 and 100 000 the cleaning procedure needs improvement, and over 100 000 as unsatisfactory (Harrigan, 1998).

For milk bottles, soft drink bottles and jars

If bottles are sampled after they have been through a washing plant in which a row of bottles travels abreast through the machine, test all those in one such row. This shows if one set of jets or carriers is out of alignment. In any event, examine not less than six bottles. Cap them immediately. To each bottle add 20 ml quarter-strength Ringer's solution containing 0.05% sodium thiosulphate and recap. Roll the bottles on their sides to rinse all of the internal surface. Leave the bottles for 15–30 min in a horizontal position. Roll the bottles again to achieve thorough wetting.

Pipette 5 ml of the rinse from each bottle into each of two Petri dishes. Add 15–20 ml molten YEA, mix and incubate one plate from each bottle at 37°C for 48 h and one at 22°C for 72 h. Incubation of both plates at 30°C or 25–30°C is also acceptable. Pipette 5 ml from each bottle into 10 ml double-strength MacConkey broth and incubate at 37°C for 48 h.

Take the mean of the count of the 37°C and 22°C plates and multiply by 4 to give the count per bottle. Find the average of the counts, omitting any figure that is 25 times greater than the others (indicating a possible fault in that particular line).

The figure obtained is the average colony count per container. Milk bottles giving average colony counts ≤ 200 cfu are regarded as satisfactory, from 200 to 1000 indicate that the cleaning procedure needs improvement, and over 1000 as unsatisfactory (Harrigan, 1998).

Coliforms should not be present in 5 ml of the rinse.

Membrane filter method

Examine clean bottles and jars by passing the rinse through a membrane filter. Place the membrane on a pad saturated with an appropriate medium, e.g. double-strength tryptone soya broth for total count. Incubate at 35°C for 18–20 h. Stain the membrane with methylene blue to assist in counting colonies under a low-power lens. Examine another membrane for coliforms using MacConkey membrane broth or membrane-enriched lauryl sulphate broth. Incubate at 35°C for 18–24 h. Staining to reveal colonies is unnecessary. Subculture suspect colonies into lactose peptone water and incubate at 37°C for 48 h to confirm gas production.

Roll-tube method for bottles

Use nutrient or similar agar for most bacteria, deMan, Rogosa and Sharp (MRS) medium for lactobacilli, reinforced clostridial medium (RCM) for clostridia and buffered yeast agar for yeasts. Increase the agar concentration by 0.5%. Use the roll-tube MacConkey agar for coliforms.

Melt the medium and cool it to 55°C. Add 100 ml medium to 1-litre bottles and proportionally less to smaller bottles. Stopper the bottles with sterile rubber bungs and roll them under a cold water tap to form a film of agar over the entire inner surface (see 'Roll-tube counts', p. 148). Incubate vertically and count the colonies. For lactobacilli and clostridia, replace the bung with a cottonwool plug and incubate anaerobically.

Vats, hoppers and pipework

Large pieces of equipment are usually cleaned in place (CIP). Test flat areas by agar contact or swabbing. Take swab samples of dead ends of pipework and crevices. CIP pipelines are examined by the rinse method where large quantities are pumped around the system. Samples of rinse fluid are collected from the discharge end of the system from the first, middle and final portions of rinse fluid. The first will contain any residual bacteria not killed by CIP treatment. If this has not been effective, the first sample will give a higher count than the middle and final samples. If all three samples are satisfactory then the cleaning was efficient. Rinse fluid samples may also be taken at various points throughout the system.

Examination of cloths, towels, etc.

To demonstrate unhygienic conditions and the necessity for frequent changes of cloths, examine as follows.

Spread a washcloth or drying cloth over the top of a screw-capped jar of known diameter. Pipette 10 ml 0.1% peptone water on to the cloth so that the area over the jar is rinsed into it. Do plate counts. If a destructive technique is possible, cut portions of cloths, sponges or brushes with sterile scissors and add to diluent. Where the cloth is of a small size add directly to diluent.

Mincers, grinders, etc.

After cleaning, rinse with 500 ml 0.1% peptone water. Treat removable parts separately by rinsing them in diluent in a plastic bag. Do colony count on rinsings as for milk bottles.

Sampling dried material (scrapings, sweepings, etc.)

Take a sample at each site selected, using appropriate sterile equipment, and insert into sterile containers. If sufficient sample is available, the quantity should approximate that used for determinations applied to the finished product.

Biotechnology plant

To detect the release of aerosols sample the air by one of the methods described on p. 246. To detect leakage around pipe joins and valves use the swab rinse method (see p. 248). Test surface contamination with Rodac plates (see p. 245). For more details on sampling and monitoring in process biotechnology, see Bennet *et al.* (1991) and Tuijnenburg-Muijs (1992).

Pharmaceutical, etc. 'clean rooms'

Use one of the mechanical air samplers (see p. 247), preferably placed outside the room and sampling through a probe. There are several different levels of clean rooms and the standards, i.e. the number of particles per unit volume, are set out in a British Standard (BSI, 1999). Settle plates are also useful but do not give comparable results.

Humidifier fever, etc.

Use one of the mechanical samplers (see p. 247) to test the air in premises where there are suspected cases of humidifier fever, sick building syndrome, extrinsic allergic alveolitis, etc. Do counts for both fungi and bacteria.

Sample the water and the biofilm in the humidifier reservoir for bacteria and fungi (see pp. 256 and 257). See also Collins and Grange (1990), Flannigan *et al.* (1991), and Ashton and Gill (1992).

For further information on effective environmental microbiological sampling, see Holah (1999) and Downs and Ito (2001).

REFERENCES

Ashton, I. and Gill, F. S. (eds) (1992) *Monitoring Health Hazards at Work*, 2nd edn. Oxford: Blackwells.

Bennet, A. M., Hill, S. E., Benbough, J. E. and Hambleton, P. (1991) Monitoring safety in process biotechnology. In: Grange, J. M., Fox, A. and Morgan, N. L. (eds), *Genetic Manipulation: Techniques and Applications*. Society for Applied Bacteriology Technical Series No. 28. London: Academic Press, pp. 361–376.

BSI (1999) BS EN ISO 14644-1: *Cleanliness and associated controlled environments. Part 1. Classification of air cleanliness.* London: British Standards Institution.

Collins, C. H. and Grange, J. M. (1990) *The Microbiological Hazards of Occupations*, Occupational Hygiene Monograph No. 17. Leeds: Science Reviews.

Downs, F. P. and Ito K. (eds) (2001) *Compendium of Methods for the Microbiological Examination of Foods,* 4th edn. Washington: American Public Health Association.

EC (2001) European Commission Decision 2001/471/EC laying down rules for the regular checks on the general hygiene carried out by operators in establishments according to Directive 64/433/EEC on health conditions for the production and marketing of fresh meat and Directive 71/118/EEC on health problems affecting the production and placing on the market of fresh poultry meat. *Official Journal of the European Communities* L165: 48–53.

Flannigan, B., McCabe, M. E. and McGarry, F. (1991) Allergenic and toxigenic micro- organisms in houses. In: Austin, B. (ed.), *Pathogens in the Environment*. Society for Applied Microbiology Symposium Series No. 20. London: Blackwells, pp. S61–S74.

Hambleton, P., Bennett, A. M., Leaver, J. and Benbough, J. E. (1992) Biosafety monitoring devices for biotechnology processes. *Trends in Biotechnology* 10: 192–199.

Harrigan, W. E. (ed.) (1998) *Laboratory Methods in Food Microbiology*. London: Academic Press.

Holah, J. (ed.) (1999) *Effective Microbiological Sampling of Food Processing Environment*. Guideline No. 20. Chipping Campden: Campden & Chorleywood Food Research Association.

Tuijnenburg-Muijs G. (1992) Monitoring and validation in biotechnological processes. In: Collins, C. H. and Beale, A. J. (eds), *Safety in Industrial Microbiology and Biotechnology*. Oxford: Butterworth–Heinemann, pp. 214–238.

20

Water microbiology

The microbiological examination of water is used world wide to monitor and control the quality and safety of various types of water. These include potable waters, treated recreational waters and untreated waters, such as the sea, used for recreational purposes. Microbiological examination of water samples is usually undertaken to ensure that water is safe to drink or to bathe in. The quality of water intended for human consumption in the EU is covered by a Directive (EC, 1998). This has been enacted into UK law by the Water Industry Act 1991, the Water Supply (Water Quality) Regulations 2000 (SI, 2000) and the Private Water Supplies Regulations 1991 (SI, 1991a). In the UK bottled water is subject to the Natural Minerals Water, Spring Water, and Bottled Drinking Water Regulations 1999 (SI, 1999). For a review of microbiological standards of water and their relationship to health risk, see Barrell *et al.* (2000).

Officially approved methods for the bacteriological examination of water are given by the UK Environment Agency (EA, 2002) and in the USA by the American Public Health Association (APHA, 1998). In relation to public health the principal tests applied to water are the viable plate or colony count and those for coliforms, *Escherichia coli*, faecal coliforms (includes species of *Klebsiella*, *Enterobacter* and *Citrobacter* spp.), enterococci and sulphite-reducing clostridia (*Clostridium perfringens*). In addition to the conventional microbiological parameters, UK legislation requires continuous monitoring of 'at-risk' water treatment works for cryptosporidial oocysts (the Water Supply [Water Quality] Regulations – SI, 2001). Table 20.1 provides the UK microbiological guidelines and standards for drinking water. In the UK, the Health

Protection Agency (HPA) provides a water microbiology external quality assurance scheme for laboratories to assess proficiency in testing.

SAMPLING

Water samples are usually collected by environmental health officers and water engineers, who use sterile 300-ml or 500-ml plastic or glass bottles supplied by the laboratory.

For samples of chlorinated water the bottles must contain sodium thiosulphate (0.1 ml of a 1.8% solution per 100 ml capacity) to neutralize chlorine.

COLONY COUNTS

These are required under UK regulations and EC Directives and are standard procedures in the USA. The technique is described in Chapter 10. Yeast extract agar (YEA) is used and tests are done in duplicate with undiluted and serially diluted samples. One set is incubated at 22°C for 72 h and the other at 37°C for 24 or 48 h. Permitted values are provided in Table 20.1.

TOTAL COLIFORMS AND *E. COLI* TEST

Most probable number method with minerals modified glutamate medium

The principle of the most probable number (MPN) method for estimating the numbers of micro-

Table 20.1 Microbiological guidelines and standards for drinking water

Type of water	Parameters	Permitted concentration or value	Guide level	Imperative level
Mains water (continuous sampling recommended)	Coliforms/*E. coli* Cryptosporidium oocysts[a]	0/100 ml		< 10/100 l
Private supplies Classes A–E and 1–4	Coliforms/*E. coli* Colony counts at 22°C and 37°C	0/100 ml	No significant increase over normal levels	
Class F	Coliforms/*E. coli*		No organism detrimental to public health	
Natural mineral waters Sampled any time up to sale	Coliforms/*E. coli* Enterococci *Pseudomonas aeruginosa* Sulphite-reducing clostridia Parasites/pathogens	0/250 ml 0/250 ml 0/250 ml 0/50 ml Absent		
Within 12 h of bottling	As above + colony count 22°C/72 h Colony count 37°C/48 h	100/ml 20/ml		
Drinking water in containers At any time	Coliforms/*E. coli* Enterococci Sulphite-reducing clostridia Colony counts at 22°C and 37°C	0/100 ml 0/100 ml ≤ 1/20 ml Should show no appreciable increase after bottling		
Within 12 h of bottling	As above +: Colony count 22°C/72 h Colony count 37°C/48 h	100/ml 20/ml		
Vending machines	Coliforms/*E. coli* Colony count 22°C/72 h Colony count 37°C/24 h	0/100 ml ≤ 10 000/ml ≤ 1000/ml[b]		
Meat premises Routine sampling	Coliforms/*E. coli* Colony count 22°C/72 h Colony count 37°C/48 h	0/100 ml ≤ 100/ml ≤ 10/ml		
If coliforms found	Resample as for routine plus test for: Enterococci Sulphite-reducing clostridia	0/100 ml 0/20 ml		

[a] Applies to treatment works shown by risk assessment to be potentially contaminated
[b] Provided that colony counts are no more than 10 times greater than those in water entering the machine
Reproduced from Barrell *et al.* (2000) with permission of the publishers, HPA Communicable Disease Surveillance Centre

organisms is given in Chapter 10. For water examination the test involves the addition of known volumes of the sample to a range of bottles or tubes containing a fluid medium, in this case minerals modified glutamate medium (MMGM), which has replaced the original MacConkey broth.

Select the range according to the expected purity of the water:

- Mains chlorinated water: *A* and *B*
- Piped water, not chlorinated: *A*, *B* and *C*
- Deep well or borehole: *A*, *B* and *C*
- Shallow well: *B*, *C* and *D*
- No information: *A*, *B*, *C* and *D*.

A: Add 50 ml water to 50 ml double-strength medium.

B: Add 10 ml water to each of five tubes of 10 ml double-strength medium.

C: Add 1 ml water to each of five tubes of 5 ml single-strength medium.

D: Add 0.1 ml water to each of five tubes of 5 ml single-strength medium.

Incubate the bottles or tubes at 37°C for 18–24 h and examine for acid production (as demonstrated by the presence of a yellow coloration) and again after a further 24 h. Retain all positive cultures because they are required for confirmatory testing for coliforms and *E. coli*.

The number of tubes or bottles for each series of volume of sample is recorded where a positive reaction is given, demonstrated by growth within the medium and the production of a yellow coloration.

When dilutions of sample have been used, a consecutive series of volumes is chosen whereby some of the tubes or bottles show a positive reaction and some show a negative reaction.

From the results determine the MPN of bacteria in the sample from the probability tables in Chapter 10 (Tables 10.1–10.3).

Confirmation of coliforms

The confirmation procedure outlined here is based on the demonstration that lactose fermentation is indicative of the possession of the β-galactosidase enzyme. Alternative procedures based on the direct detection of this enzyme, e.g. using the substrate *o*-nitrophenyl-β-D-galactopyranoside, may be more appropriate in some cases.

Subculture from each tube or bottle showing growth in the MMGM medium to MacConkey agar (MA) and nutrient agar (NA). Incubate at 37°C for 18–24 h. If a pure culture is obtained on NA, do the oxidase test. If the isolate is oxidase negative, test for lactose fermentation or possession of β-galactosidase (see p. 98). If there is any doubt about the purity of the culture, subculture at least one typical colony from MA to NA, incubate at 37°C for 18–24 h and carry out the oxidase test.

For each isolate to be tested, subculture to lactose peptone water (LPW), incubate at 37°C and examine for acid production after 24 h. If the results are negative, re-examine after a further 24 h. Confirmation of acid production is demonstrated by the change of colour from red to yellow. Further identification may be carried out using characteristic colonies on MA by means of appropriate biochemical and other tests. Commercial kits may be used.

Read the most probable number of coliforms from Tables 10.1–10.3.

Confirmation of *E. coli*

In addition to confirmatory tests for coliforms, subculture to MA each tube or bottle showing growth in the medium and incubate at 44°C for 18–24 h. Inoculate tryptone water (TW) with typical coliform colonies and incubate at 44°C for 24 h. Test for indole production by Kovac's method (see p. 97). *E. coli* gives a positive reaction. A test for β-glucuronidase (see p. 286) may assist in the early confirmation of *E. coli*. Suitable commercial test kits may be used.

Read the most probable numbers of *E. coli* from Tables 10.1–10.3. Permitted values for coliforms and *E. coli* are provided in Table 20.1.

Rapid methods

Rapid test procedures are now available commercially to detect coliforms and *E. coli* in water samples. The Colilert 18 medium (Indexx Inc., Portland, Maine) is one of these (EA, 2002).

Membrane filter method

Advantages of using membrane filter techniques for waters

- Speed with which results can be obtained as direct results
- Saving of labour, media, glassware and cost of materials if the filter is washed, sterilized and re-used
- Incubation conditions can be varied easily to encourage the growth of attenuated or slow-growing organisms
- False-positive reactions that occur with some media in the MPN method for coliforms as a result of the growth of aerobic and anaerobic spore-bearing organisms, or to mixtures of organisms, are unlikely to occur with membranes.

Disadvantages of using membrane filter techniques for waters

- There is no indication of gas production (some waters contain large numbers of non-gas-producing lactose fermenters capable of growth in the medium).
- Membrane filtration is unsuitable for waters with high turbidity because the filter will become blocked before sufficient water can pass through it, whereas the accumulated deposit on the membrane may also inhibit the growth of indicator organisms.
- Membrane filtration is unsuitable for waters with low numbers of coliforms in the presence of large numbers of non-coliform organisms capable of growth on the media used. These will overgrow or obscure colonies of coliforms.

If large numbers of water samples are to be examined and much fieldwork is involved, the membrane method is undoubtedly the most convenient. For each sample, pass two separate 100-ml

volumes of the water sample through 47-mm diameter sterile membrane filters of 0.45 μm pore size. If the supply is known or is expected to contain more than 100 coliforms/100 ml, either filter smaller volumes or dilute the sample before filtration, e.g. 10 ml water diluted with 90 ml quarter-strength Ringer's solution.

Place sterile absorbent pads of at least the same size as the membranes and approximately 1 mm thick in sterile Petri dishes and pipette 2.5–3 ml membrane lauryl sulphate broth (MLSB) over the surface. Place a membrane face up on each pad.

Incubate both membranes at 30°C for 4 h then transfer one membrane to 37°C for 14 h for total coliforms and the other to 44°C for 14 h for *E. coli*.

Counting

Count all yellow colonies within a few minutes of being removed from the incubator and report as *presumptive* coliform and *E. coli* count/100 ml of water. Membrane counts may be higher than MPN counts because they include all organisms producing acid, not only those producing acid and gas.

Confirmatory tests for coliforms

Subculture colonies to LPW and incubate at 37°C. After 6 h, subculture to MA and NA. (Large colonies may be subcultured to MA and NA directly from the membrane.) Examine the LPW cultures at 37°C after 24 h for acid production and, if the results are negative, after a further 24 h.

Incubate the MA and NA plates at 37°C for 24 h and do oxidase tests on colonies only from the NA. Coliforms are oxidase negative, possess β-galactosidase and produce acid from lactose within 48 h at 37°C. Further identification may be carried out using colonies on characteristic colonies on MA by means of appropriate biochemical and other tests (Barrow and Feltham, 1993). Commercial kits may be used.

Confirmatory tests for E. coli

Subculture yellow colonies from the membranes incubated at 37°C and 44°C to MA and NA, to LPW (containing a Durham tube if information regarding gas production is required) and to TW.

Incubate all media at 44°C for 24 h. *E. coli* is oxidase negative, produces acid (and usually gas) in LPW, and indole in TW. Tests for β-glucuronidase may assist in the early confirmation of *E. coli*. Commercial test kits may be used.

FAECAL STREPTOCOCCI IN WATER

These organisms are useful indicators when doubtful results are obtained in the coliform test. They are more resistant than *E. coli* to chlorine and are therefore useful when testing repaired mains. Only group D streptococci are significant.

Membrane method

Prepare and filter the sample as described above for coliforms and *E. coli*. Place the membrane on the surface of a well-dried plate of membrane enterococcus agar (MEA). For potable waters, incubate at 37°C for 48 h and for untreated waters incubate at 37°C for 4 h followed by 44°C for 44 h.

Count all red, maroon or pink colonies. These are presumptive faecal streptococci. Subculture a suitable number of red, maroon, pink or colourless colonies from the membrane to bile aesculin agar and incubate at 44°C for 18 h. Development of discrete colonies, surrounded by a brown or black halo from aesculin hydrolysis, within a few hours confirms the presence of faecal streptococci.

SULPHITE-REDUCING CLOSTRIDIA

Membrane method

Pass 100 ml of the sample (heated at 75°C for 10 min) through a 47-mm diameter sterile membrane filter of 0.45 µm pore size. Place the membrane face upwards on the tryptose suphite cycloserine agar with egg yolk (TSCA) and pour 15–20 ml TSCA,

pre-cooled to 50°C, over the membrane and allow to set. Incubate anaerobically at 37°C. Examine the plates after 24 and 48 h and count all black colonies. The specificity of TSCA is such that confirmation of isolates is not required (EA, 2002). If there is any doubt about the organisms isolated by this method, examine Gram-stained films from the colonies for spore-forming rods.

Permitted values for sulphite-reducing clostridia are provided in Table 20.1.

PATHOGENS IN WATER

Methods for the detection of certain bacterial pathogens are given below, but other pathogens, outside the scope of this book, such as viruses (e.g. norovirus, rotaviruses, hepatitis A) and parasites (e.g. *Cryptosporidium* and *Giardia* spp., and amoebae) should not be overlooked.

As a general rule the direct search for pathogenic bacteria has no place in routine microbiological examination of a water supplier. There may be occasions, for example, when the water is polluted by *E. coli*, when there is, however, a need to search for faecal pathogens.

Filter the sample using 47-mm diameter sterile membrane filter of 0.45 µm pore size (0.2 µm pore size for campylobacters).

Salmonellas

Transfer the membrane to 90 ml enrichment medium (e.g. BPW) in a wide-mouthed screw-capped container. Incubate at 37°C for 24 h and subculture 0.1–10 ml Rappaport–Vassiliadis Soya Broth (RVSB). Incubate the RVSB at 42°C for 24 h. Subculture to xylose lysine deoxycholate (XLD) and brilliant green agar (BGA) and incubate at 37°C for 24 h. Return the RVSB to the incubator at 42°C for a further 24 h. Again, subculture to XLD and BGA and incubate at 37°C for 24 h. Proceed as on p. 295.

Vibrio cholerae and other vibrios

Transfer the membrane to 100 ml alkaline peptone water. Incubate at 25°C for 2 h followed by 12–16 h at 37°C. Subculture to thiosulphate–citrate–bile salt–sucrose agar (TCBSA) and incubate at 37°C for 16–24 h. Proceed as on p. 275.

Legionellas

Legionellas may be present in the water in cooling towers, humidifier tanks, etc. and also in the biofilm that is usually attached to the walls of the containers. Examination of water alone may give a false impression.

Membrane filtration is the usual method of testing the water for legionellas (BSI, 1998). Examine several samples. If the water is cloudy, pre-filter it through glass fibre pads.

Transfer the membrane to 5–25 ml peptone saline diluent in a screw-capped bottle. Vortex mix for not less than 2 min to detach the micro-organisms. Decant the diluent and centrifuge it for 30 min at $3000 \times g$ or 10 min at $6000 \times g$.

Inoculate buffered charcoal yeast extract agar medium (BCYE) or other legionella-selective media with 100-μl amounts of resuspended deposit. Incubate under high humidity at 37°C. Include stock control cultures.

If the samples are highly contaminated use one of these decontamination methods:

- Heat the deposit at 50°C for 30 min and then culture 100-μl amounts.
- Suspend 0.1 ml of the deposit in 0.9 ml acid buffer
  - KCl (0.2 mol/l, 14.91 g/l) 25 ml
  - HCl (0.2 mol/l, 17.2 ml conc. acid/l) 3.9 ml
  - sterilize by autoclaving.

Plate out 100-μl amounts every min for 5 min. Identify legionellas as described on p. 322. Centrifuge biofilm suspensions and treat the deposit as described for contaminated samples (above).

Campylobacters

Transfer the membrane in a suitable screw-capped container filled almost completely with 90 ml campylobacter-enrichment broth. Seal tightly and incubate at 37°C for 24 h and then at 42°C for a further 24 h. Subculture to campylobacter-selective agar or the blood-free modification of this agar and incubate in a gas jar containing a microaerobic gas-generating kit at 37°C for 48 h. Proceed as on p. 314.

E. coli O157

Transfer the membrane into containers with 90 ml modified tryptone soya broth (MTSB) or BPW. MTSB is suitable for polluted water, whereas BPW is more appropriate for the recovery of stressed E. coli O157 from drinking water and relatively unpolluted waters (EA, 2002). Incubate the MTSB at 42°C for 24 h, and the BPW at 37°C for 24 h. Subculture on to cefoxime tellurite sorbitol MacConkey agar (CT-SMAC) and incubate at 37°C for 24 h, and proceed as on p. 287.

Recovery of E. coli O157 from enrichment broth is improved by using immunomagnetic separation with antibody-coated magnetic beads (Wright et al. 1994; EA, 2002).

Pseudomonas aeruginosa

Place the membranes on pseudomonas agar and incubate at 37°C for 48 h. Count all colonies that produce a green, blue or reddish–brown pigment and those that fluoresce under ultraviolet (UV) light. Regard those colonies that produce pyocyanin (green/blue pigmented) as P. aeruginosa.

Confirm fluorescent colonies by subculture on milk agar with cetyl trimethylammonium bromide incubated at 37°C for 24 h. Record as confirmed P. aeruginosa colonies that are 2–4 mm in diameter, show typical pigment production and possess a 'halo of clearing' around the colony where the casein has been hydrolysed. Where pigment is not produced, and *P. aeruginosa* is suspected, subculture colonies to a fresh plate of milk agar with cetyl trimethylammonium bromide; incubate at 37°C for 24 h. This provides a pure culture that can then be confirmed using commercially available test kits. For rapid identification of *P. aeruginosa*, which grows at 42°C, incubate subcultures at that temperature.

Microfungi and actinomycetes

These organisms cause odours and taints and often grow in scarcely used water pipes, particularly in warm situations, e.g. basements of large buildings where the drinking water pipes run near to the heating pipes.

Sample the water when it has stood in the pipes for several days. Regular running or occasional long running may clear the growth. Centrifuge 50 ml of the sample and plate on Rose Bengal agar and malt agar medium. Add 100 μg/ml kanamycin to the malt medium to suppress most eubacteria.

Alternatively, pass 100 ml or more of the sample through a membrane filter and apply this to a pad soaked in liquid media of similar composition.

BOTTLED NATURAL WATERS

These may be 'still' or carbonated. The tests used, with guidance levels at the point of sale and within 12 h of bottling, are given in Table 20.1. Environmental mycobacteria have been found in moderate numbers in some bottled waters.

SWIMMING POOLS

As with drinking water, microbiological examination of recreational water is aimed largely at detecting markers of faecal pollution. However, colony counts play a rather greater role as a marker of general water quality, and tests for *P. aeruginosa* are also often carried out. *P. aeruginosa* is associated with disease, usually folliculitis or otitis externa, in people who have used swimming pools or spa pools (Hunter, 1996).

Use a commercial kit to determine the chlorine content. This may be more useful than bacteriological examination but, if the latter is required, take 300-ml samples in thiosulphate bottles (see p. 251) from both the deep and the shallow ends. Do plate counts and test for coliforms, *E. coli* and pseudomonads. In satisfactory pools target levels are as follows:

	Guide level	Target level
Colony count, 37°C/24 h	< 100/ml	< 10/ml
Coliforms	< 10/100 ml	0/100 ml
E. coli	0/100 ml	
P. aeruginosa (optional)	0/100 ml	

Staphylococci are more resistant than coliforms to chlorination. They tend to accumulate on the surface of the water in the 'grease film'. To find staphylococci take 'skin samples'. Open the sample bottle so that the surface water flows into it. Filter the test volume and place each test membrane on the agar surface in a Petri dish containing modified lipovitellin–salt–mannitol agar (M-5LSMA). Incubate at 30°C for 48 h. Count and record presumptive staphylococci, and proceed as on p. 330.

SPA, JACUZZI AND HYDROTHERAPY POOLS

These have been associated with human infections with pseudomonads and legionellas. As they are maintained at 37°C the chlorine (or bromine) levels should be closely monitored.

Do plate counts on tryptone agar and test for coliforms, *E. coli* and pseudomonads. Target levels are:

	Guide level	Target level
Total viable count, 37°C/24 h	< 100/ml	< 50/ml
Coliforms and *E. coli*	0/100 ml	
P. aeruginosa	0/100 ml	

Pools with colony counts > 100 cfu/ml need further investigation (PHLS, 1994, 1999; HPA, 2003).

BATHING BEACHES AND OTHER RECREATIONAL WATERS

Test undiluted and serially diluted samples for coliforms, faecal coliforms and enterococci. Test for salmonellas and enteroviruses when there are grounds for suspecting deterioration in water quality or that the organisms are present. The EU guide levels (SI, 1991b) are as follows:

	Guide level	Imperative level
Total coliforms	500/100 ml	10 000/100 ml
Faecal coliforms	100/100 ml	2000/100 ml
Enterococci	100/100 ml	
Salmonella	0.1	
(Enteroviruses	0 plaque-forming units (PFU)/10 l)	

OTHER ORGANISMS OF PUBLIC HEALTH IMPORTANCE

These are not considered in this book although some references are cited.

Viruses

Viruses (especially rotaviruses) have been implicated in outbreaks associated with recreational and potable waters (see Sellwood, 1998).

Cyanobacteria

Blue-green algae produce a variety of dangerous toxins. Serious illnesses have been associated with their presence in recreational and potable waters and fish (see Codd, 1994).

Cryptosporidium spp.

Oocysts of this organism may be present in recreational and potable waters and offer a serious hazard to health (see Badenoch, 1995).

WATER TESTING IN REMOTE AREAS

Problems arise in the bacteriological examination of water in remote parts of tropical and subtropical countries. Some of the culture media described above are expensive or may not be available. It may be necessary to use simpler media such as MacConkey and litmus lactose broths and agars. These will affect results and standards. In addition, the 44°C tests may be unreliable because some indigenous, non-faecal organisms may grow at those temperatures.

Water samples collected in remote areas and transported over long distances to laboratories are unlikely to give reliable results. A portable kit, the Oxfam-Del Agua, devised for the WHO project for drinking water in rural areas, will give reliable faecal coli counts, turbidity and free chlorine levels. It is available from Del Agua (RCPEH, Guildford, UK; see Lloyd and Helmer, 1991) and Septi-Check, Indexx Inc. Portland, Maine, USA).

SEWAGE

Effluents and sludge are usually monitored microbiologically to ensure that treatment has reduced the loads of salmonellas, listerias and campylobacters, all of which can survive in water; if untreated sludge is spread on land.

For further information on water microbiology and technical directions on sampling see: APHA (1996); EA (2002); Barrell *et al.* (2000); HSE (2001); and HPA (2003).

REFERENCES

APHA (1986) *Standard Methods for the Examination of Water and Wastewater*, 20th edn. Washington: American Public Health Association,

Badenoch, J. (1995) *Cryptosporidium in Water Supplies*. Report of a group of Experts. London: HMSO.

Barrell, R. A. E., Hunter, P. R. and Nichols, G. (2000) Microbiological standards for water and their relationship to health risk. *Communicable Disease and Public Health* 3: 8–13. Corrections published in *Communicable Disease and Public Health* 3: 221.

Barrow, G. I. and Feltham, R. K. A. (eds) (1993) *Cowan and Steel's Manual for the Identification of Medical Bacteria,* 3rd edn. London: Cambridge University Press.

BSI (1998) British Standard BS ISO 11731: *Water quality – Part 4: Microbiological methods – Section 4.12: Detection and enumeration of Legionella.* London: British Standards Institution.

Codd, G. A. (1994) Blue-green algal toxins: water-borne hazards to health. In: Golding, A. M. B, Noah, N. and Stanwell-Smith, R. (eds), *Water and Public Health.* London: Smith Gordon, pp. 271–278.

EA (2002) *The Microbiology of Drinking Water, Parts 1–10: Methods for the Examination of waters and Associated Materials.* London: Environment Agency

EC (1998) European Council Directive 98/83/EC of 3 November 1998 on the quality of water intended for human consumption. *Official Journal of the European Communities* L330: 32–54.

HPA (2003) *Guidelines for water quality on board merchant ships including passenger vessels.* London: Health Protection Agency.

HSE (2001) *Legionnaire's Disease. The control of Legionella bacteria in water systems.* Approved Code of Practice and Guidance. London: Health and Safety Executive.

Hunter, P. R. (1996) Outbreaks of disease associated with treated recreational waters. *Microbiology Europe* 4: 10–12.

Lloyd, B. and Helmer, R. (eds) (1991). *Surveillance of Drinking Water in Rural Areas.* London: Longman.

Public Health Laboratory Service (1994) *Hygiene for Spa pools.* London: PHLS.

PHLS (1999) *Hygiene for Hydrotherapy pools,* 2nd edn. London: PHLS.

Sellwood, J. (1998) Viruses implicated as potential emerging pathogens. In: de Louvois, J. and Dadswell, J. (eds), *Emerging Pathogens and the Drinking Water Supply.* London: PHLS, pp. 55–1.

SI (1991a) *The Private Water Supply (Water Quality) Regulations.* SI 2790. London: HMSO.

SI (1991b) *The Bathing Waters (Classification) Regulations* SI 1597. London: HMSO.

SI (1999) *The Natural Minerals Water, Spring Water and Bottled Drinking Water Regulations.* SI 1540. London: HMSO.

SI (2000) *The Water Supply (Water Quality) Regulations.* SI 3184. London: HMSO.

SI (2001) *The Water Supply (Water Quality) (Amendment) Regulations.* SI 2885. London: HMSO.

Wright, D. J., Chapman, P. A. and Siddons, C. A. (1994) Immunomagnetic separation as a method for isolating *Escherichia coli* O157:H7 from food samples. *Epidemiology and Infection* 113: 31–39.

21

Key to common aerobic, non-sporing, Gram-negative bacilli

Aerobic Gram-negative non-sporing bacilli that grow on nutrient and usually on MacConkey agars are frequently isolated from human and animal material, from food and from environmental samples. Some of these bacilli are confirmed pathogens; others, formerly regarded as non-pathogenic, are now known to be capable of causing human disease under certain circumstances, e.g. in 'hospital infections', after chemotherapy or treatment with immunosuppressive drugs; others are commensals and many are of economic importance.

The organisms described in this and Chapters 22 and 23 have received much attention from taxonomists in recent years. Names of species and genera have been changed and new taxa created. The nomenclature used here is that in common use at the time of going to press, mostly as in the current edition of *Topley and Wilson's Microbiology and Microbial Infections* (Collier *et al.*, 1998). See also Bergey's manual (Holt *et al.*, 1994). Table 21.1 lists the general properties of aerobic Gram-negative rods that grow on nutrient agar.

Table 21.1 Properties of Gram-negative rods that grow on nutrient agar

Organism	O/F[a]	Oxidase	Arginine hydrolysis	Gelatin liquefaction	Motility
Pseudomonas spp.	OA	+	v	v	+
Acinetobacter spp.	ON	−	−	−	−
Alcaligenes spp.	N	−	−	−	+
Flavobacteria	O	+	−	+	−
Chromobacterium spp.	O	+	−	+	+
Janthinobacterium spp.	F	+	+	+	+
Aeromonas spp.	F	+	+	+	+
Plesiomonas spp.	F	+	+	−	+
Vibrio spp.	F	+	+	−	+
Enterobacteria	F	−	v	v	v
Pasteurella spp.	F	+	−	−	−
Yersinia spp.	F	−	−	−	v
Moraxella spp.	N	+	−	−	v
Bordetella spp.	N	v	−	−	v

[a] Oxidation/fermentation test (Hugh and Liefson)

O, oxidative; A, alkaline; N, no reaction; F, fermentative; v, variable

REFERENCES

Collier, L., Balows, A. and Sussman, M. (eds) (1998) *Topley and Wilson's Microbiology and Microbial Infections*, 9th edn. Vol. 2, *Systematic Bacteriology*. London: Arnold.

Holt, J. G., Kreig, N. R., Sneath, P. H. A. *et al.* (eds) (1994) *Bergey's of Determinative Bacteriology*, 9th edn. Baltimore, MA: Williams & Wilkins.

22

Pseudomonas and *Burkholderia* and other pseudomonads

Up to 1984, over 100 species were included in the genus *Pseudomonas*. These were mainly plant pathogens and only a minority was pathogenic for humans and animals. Today only members of the ribosomal RNA (rRNA) group I (of five rRNA groups) are included in the genus. Other rRNA groups have been reclassified in the genera *Burkholderia* (group II), *Comamonas* (now *Delftia*, group III), *Brevundimonas* (group IV) and *Stenotrophomonas* (group V). They are all aerobic, Gram-negative, non-sporing rods about 3 μm × 0.5 μm, they are motile by polar flagella and, with the exception of *Stenotrophomonas maltophilia*, they are usually oxidase positive, use glucose oxidatively (or produce no acid) and do not produce gas. Some produce a fluorescent pigment. In this book 'pseudomonad' is used as a general term for *Pseudomonas* spp. and related species

A recent phylogenetic analysis of the genus *Pseudomonas* which used combined *gyrB* and *rpoD* nucleotide sequences identified two intrageneric subclusters. Cluster I was further divided into two complexes and cluster II into three complexes (Yamamoto *et al.*, 2000).

Pseudomonads commonly occur in soil and water. Some species are recognized human and animal pathogens but some others, formerly regarded as saprophytes and commensals, have been incriminated as opportunist pathogens in hospital-acquired infections and susceptible patient groups such as those with cystic fibrosis (CF). In hospitals they are commonly isolated as contaminants of intravenous fluids, distilled water supplies, soaps, disinfectants and antiseptics, infusions and other pharmaceuticals. Species colonizing swimming pools, particularly *P. aeruginosa*, have been associated with skin and ear infections.

ISOLATION AND IDENTIFICATION

Inoculate nutrient agar, MacConkey agar, nutrient agar containing 0.03% cetrimide, pseudomonas agar with supplements and King's medium A. Incubate at 20–25°C (food, environment, etc.) or 35–37°C (human and animal material). Temperature and length of incubation are crucial for accurate identification.

Colonies on nutrient agar vary in appearance according to species (see below). A greenish-yellow or bluish-yellow fluorescent pigment may diffuse into the medium. Other less common pigments are pyomelanin (brown) and pyorubrin (port-wine). Mucoid strains are sometimes found in clinical material, particularly in sputum from patients with CF, and may be confused with klebsiellas.

Do oxidase test, reaction in Hugh and Leifson medium, and inoculate Thornley's arginine broth (Moeller's method is too anaerobic) and MacConkey agar. Inoculate nutrient broths and incubate at 4 and 42°C. Use a subculture from a smooth suspension for this test. Examine growth on King's B medium with a Wood's lamp for fluorescence. If this is seen, test for reduction of nitrate and acid production from glucose and mannitol in ammonium salt sugars and gelatin liquefaction (Tables 22.1 and 22.2).

The API 20NE kit, designed for non-fermenting Gram-negative rods, is useful for identifying pseudomonads. Other kits include Minitek NF and Oxi-ferm.

Table 22.1 Reactions of medically important pseudomonads in biochemical tests

Species	Pigment	Oxidase	O/F test[a]	Growth at 42°C	Growth at 4°C	Arginine	ONPG	NO$_3$ reduction	NO$_2$ reduction
Pseudomonas aeruginosa	Green	+[b]	O	+	-	+	-	+	+
P. fluorescens	-	+	O	-	+	+	-	v	-
P. putida	-	+	O	-	v	+	-	v	-
P. alcaligenes	-	+	A	+	-	v	-	v	-
P. pseudoalcaligenes	-	+	A	+	v	v	-	+	-
P. stutzeri	Yellow/-	+	O	+	v	v	-	+	+
Burkholderia cepacia	-/Yellow	v	v	v	v	-	v	v	-
B. pseudomallei	-	+	O	+	-	+	-	+	-
B. mallei	-	v	O	v	-	+	-	+	-
Delftia acidovorans	-	+	A/-	-	-	-	-	+	-
Brevundimonas diminuta	-	+	A	-	-	-	-	-	-
Brevundimonas vesicularis	Orange	+	O	v	-	-	v	-	-
Sphingomonas paucimobilis	Yellow	+	O	-	-	-	+	-	-
Stenotrophomonas maltophilia	-	v	O/A/-	v	-	-	v	v	-

[a] F, fermentative; O, oxidative; A, alkaline; -, no change; v, variable

[b] +, ≥ 90% positive; -, ≤ 10% positive; v, 11–89% positive

Table 22.2 Fluorescent pseudomonads

Species	Growth at		NO₃	Acid from	
	4°C	42°C		Mannitol	Maltose
P. aeruginosa	−	+	+	+	−
P. fluorescens	+	−ᵃ	+	+	+
P. putida	+ᵃ	−	−	−	−

ᵃ 40% of strains

Pseudomonads are motile (occasionally non-motile strains may be met), oxidative and non-reactive, or produce alkali in Hugh and Leifson medium. The oxidase test is usually positive except for *Stenotrophomonas* (formerly *Xanthomonas*) *maltophilia* and *Burkholderia cepacia*. Those pseudomonads that give a positive arginine test are common and give a positive arginine test more rapidly than other Gram-negative rods. This test is very useful for recognizing non-pigmented strains. *Alcaligenes* and *Vibrio* spp. do not hydrolyse arginine. Pseudomonads vary in their ability to liquefy gelatin and grow at 4 and 42°C (see Table 22.1).

Some of the strains associated with food spoilage are psychrophiles and may be lipolytic. Brine-tolerant strains and phosphorescent strains are not uncommon.

THE FLUORESCENT PSEUDOMONADS

Pseudomonas aeruginosa

The colonies of *P. aeruginosa* on nutrient agar after 24 h incubation in air are usually large (2–4 mm), flat and spreading Many strains produce a diffusible pigment. Some, however, form 'coliform-like' colonies with entire edges and a creamy consistency. Other colonial forms are rarely isolated from clinical material and include rough and dry 'pepper-corn' colonies, gelatinous and mucoid forms. The last two are most commonly found in cultures of sputum from CF patients and if not pigmented can be confused with klebsiellas, particularly on MacConkey agar. Small colony variant 'dwarf forms' may appear on extended incubation. *P. aeruginosa* characteristically forms a blue–green diffusible pigment on King's A agar which is caused by pyocyanin (blue) and fluorescein (yellow) pigments (see above).

Production of the blue–green pigment and a positive oxidase test reaction is sufficient to indicate that a strain is *P. aeruginosa*. Other confirmatory tests for the species are oxidation of glucose in Hugh and Leifson medium, deamination of arginine in Thornley's agar, and growth on three successive subcultures in nutrient broth at 42°C. Non-pigmented cultures should be examined on King's B medium for fluorescence and, if positive, should be tested further for reduction of nitrate to nitrogen gas and failure to grow at 4°C. Anaerobic growth is possible only in the presence of an alternative terminal electron acceptor such as nitrate. The demonstration of motility in Craigie tubes is therefore inappropriate. Sugar use is best demonstrated in ammonium salts-based medium because the weak acid formed by oxidation of the sugar is neutralized by alkali from the breakdown of peptone.

Pseudomonas aeruginosa is ubiquitous in nature and about 10–15% of humans will carry the organism transiently in the bowel with a rapid turnover of strain types. Infections in healthy individuals are rare but skin infections after prolonged exposure to contaminated recreational waters are not uncommon. Less frequent infections of the healthy include endophthalmitis and otitis externa. In hospital patients, *P. aeruginosa* is a common cause of infection ranging from relatively benign colonization of the upper airways after intubation, surgical wounds and the catheterized bladder to bloodstream infections with a mortality rate of 60%.

As many as 80% of CF patients may be colonized in the lung with *P. aeruginosa* and once established it is particularly refractory to antibiotic treatment. Many isolates grow as mucoid colonies but mixtures of different colonial forms are frequently found on primary plates. These variants invariably prove to be genetically identical. *P. aeruginosa* from CF patients is often atypical in growth requirements and may be auxotrophic for specific amino acids, be non motile and a minority may exhibit extreme susceptibility to semi-synthetic penicillins.

For epidemiological studies, isolates of *P. aeruginosa* can be serotyped by slide agglutination of live cultures. There are 21 internationally accepted O serotypes but four types account for approximately 50% of clinical and environmental isolates. DNA fingerprinting using pulsed-field gel electrophoresis of *XbaI* restriction endonuclease digests can achieve further discrimination between serotypes. Other typing systems used include ribotyping and random polymerase chain reaction (PCR) typing (Grundmann *et al.*, 1995).

Compared with the Enterobacteriaceae, *P. aeruginosa* is relatively resistant to antibiotics but the great majority of strains from acute infections are susceptible to ticarcillin, piperacillin, ceftazidime, imipenem, gentamicin, tobramycin, ciprofloxacin and aztreonam. Resistance to antibiotics is conferred by mechanisms including low cell permeability, chromosomal and plasmid-mediated β-lactamases, aminoglycosidases and carbapenemases, altered antibiotic binding proteins, and an efflux system active against a wide range of compounds (Livermore, 2002).

Pseudomonas fluorescens and P. putida

Two other species, *P. fluorescens* and *P. putida*, are fluorescent pseudomonads but these are rare in clinical specimens because they are psychrophiles with optimal growth temperatures below 30°C. *P. fluorescens* has often been implicated as a contaminant of stored blood products, which when transfused into a patient may cause endotoxic shock. The species may occasionally be recovered from the sputum of CF patients. It is also a recognized food spoilage agent, particularly of refrigerated meat, and it may gel UHT milk if this is stored above 5°C. They may be recovered from fresh vegetables or plants, and sinks, taps and drains in the hospital environment. *P. fluorescens* grows at 4°C, produces fluorescein pigment on King's B agar and some isolates will not grow at 37°C. It is resistant to a wide range of antibiotics including imipenem.

Pseudomonas alcaligenes and P. pseudoalcaligenes

Both of these species grow at 42°C but neither grows at 4°C. They are ubiquitous and are opportunist pathogens. They do not liquefy gelatin and may or may not grow on cetrimide agar. Some strains of *P. alcaligenes* do not hydrolyse arginine.

Pseudomonas stutzeri

Colonies are commonly rough, dry and wrinkled and can be removed entire from the medium. Older colonies turn brown. It may hydrolyse arginine and may not grow at 4°C but does grow at 42°C. It is salt tolerant but not halophilic. A ubiquitous organism, it has been found in clinical material but is not regarded as a pathogen.

BURKHOLDERIA SPP.

The genus *Burkholderia* was defined in 1992 by Yabuuchi *et al.* for *Pseudomonas* species formerly belonging to rRNA group II. There are 30 validly named species in the genus but the medically important species are *B. cepacia*, *B. pseudomallei* and *B. mallei*. The organism formerly known as *Pseudomonas pickettii* was reassigned by Yabuuchi *et al.* (1992) to *Burkholderia*, but has subsequently been reclassified as *Ralstonia pickettii*. Occasional clinically significant isolates of *R. pickettii* are recovered from hospital patients but they are most often isolated from the ward environment and as contaminants of antiseptic and disinfectant solutions. It is oxidase and nitrate variable and does not hydrolyse arginine.

Burkholderia cepacia

Burkholderia cepacia is a complex of at least seven different genomovars (phenotypically similar, genetically distinct species). These are *B. cepacia* (genomovars I, III and IV), *B. multivorans* (II), *B. vietnamensis* (V), *B. stabilis* (VI) and *B. ambifaria* (VII). The genus, *Pandoraea*, comprising five species, clusters closely to the complex in phylogenetic trees based on 16S rRNA gene sequences (Coeyne *et al.*, 2001). Genomovar I strains are often isolated from the environment but genomovars II and III, respectively, are principally associated with hospital-acquired infections and lung infections of a minority of CF patients in whom it may prove fatal. The species can survive and multiply in pharmaceuticals and cosmetics, and is able to grow in distilled water.

These organisms grow moderately well on nutrient agar and a variety of non-fluorescent pigments may be produced by some strains. About half of the isolates are oxidase positive; none hydrolyse arginine. They grow slowly at 37°C and extended incubation for 48 h is recommended to optimize their recovery from sputum. Cultures often die rapidly on storage on nutrient agar slopes, but survive remarkably well suspended in sterile distilled water. Selective media, based on their constitutive resistance to colistin and bile salts, have been described but other colistin-resistant, Gram-negative rods may also be recovered on these media. Members of the complex are not differentiated well

by phenotypic tests, although PCR assays specific for individual genomovars have been developed (see Coeyne *et al.*, 2001).

Burkholderia cepacia has high intrinsic resistance to antimicrobial agents and is generally resistant to the antibiotics active against *P. aeruginosa*. It is resistant to aminoglycosides, colistin, ticarcillin, azlocillin and imipenem. Variable susceptibility is shown to temocillin, aztreonam, ciprofloxacin and tetracycline and about two-thirds of strains from CF patients are susceptible to ceftazidime, piperacillin/tazobactam and meropenem.

Burkholderia pseudomallei

Cultures on blood agar and nutrient agar at 37°C give mucoid or corrugated, wrinkled, dry colonies in 1–2 days, and an orange pigment may develop on prolonged incubation. There is no growth on cetrimide agar. Variation between rough and smooth colonies is frequent. Cells may also exhibit bipolar staining in Gram stains. *B. pseudomallei* is a strict aerobe; it is motile, oxidizes glucose and breaks down arginine. Most isolates are reliably identified by API 20NE microgalleries but must be distinguished from non-pigmented strains of *P. aeruginosa*, *P. stutzeri* and *B. mallei*. It is resistant to colistin and gentamicin but isolates are generally susceptible to imipenem, piperacillin, amoxicillin–clavulanic acid, doxycycline, ceftazidime, aztreonam and chloramphenicol.

Caution: B. mallei and B. pseudomallei are in Hazard (Risk) Group 3 and should be handled only under full Biosafety Containment Level 3 conditions. Although odour is said to be a useful characteristic, cultures should never be sniffed. Laboratory-acquired infections are not uncommon.

Burkholderia pseudomallei is an important pathogen of humans (melioidosis) and farm animals in tropical and subtropical areas, where it is endemic in rodents and is found in moist soil, and on vegetables and fruit. A closely related but non-pathogenic species, *B. thailandensis*, has been described from environmental samples in south-east Asia. Cultures should be sent to a reference laboratory for species confirmation. For further information on these organisms and melioidosis, see Dance (1999).

Burkholderia mallei

In exudates this organism may appear granular or beaded and may show bipolar staining. It gives 1-mm, shining, smooth, convex, greenish–yellow, buttery or slimy colonies, which may be tenacious on nutrient agar, no haemolysis on blood agar and it does not grow on MacConkey agar in primary culture. Growth may be poor on primary isolation. It grows at room temperature, does not change the Hugh and Leifson medium, produces no acid in peptone carbohydrates, is nitrate positive, and gives variable urease and oxidase reactions.

Burkholderia mallei is the causative agent of glanders, a rare disease of horses. It has not been isolated from animals or humans in the UK since World War II.

DELFTIA ACIDOVORANS

This organism was reclassified from the genus Comamonas. It is found on occasion in clinical specimens and the hospital environment. Isolates grow as non-pigmented colonies overnight at 37°C but incubation should be extended to 48 h for slow growing strains. Some strains are resistant to colistin and gentamicin and may grow on B. cepacia-selective media. Antimicrobial susceptibility is variable but most isolates are susceptible to ureidopenicillins, tetracycline, the quinolones and trimethoprim–sulphamethoxazole.

BREVUNDIMONAS DIMINUTA AND BREVUNDIMONAS VESICULARIS

These are closely related species previously of rRNA homology group IV of Pseudomonas spp. They grow slowly on nutrient agar and require 48 h incubation at 37°C. B. vesicularis grows as orange pigmented colonies on nutrient agar and gives a weak oxidase reaction. B. diminuta is not pigmented. Both species are negative in Hugh and Leifson's test but B. vesicularis will oxidize glucose in an ammonium salts medium. They are rare in clinical specimens and of doubtful clinical significance.

STENOTROPHOMONAS MALTOPHILIA

Colonies resemble those of P. aeruginosa but a yellow or brown diffusible pigment may be produced; on blood agar they can appear as faint lavender colour. It is usually oxidase negative, does not hydrolyse arginine and does not grow on cetrimide agar. It is the only pseudomonad, along with B. cepacia (63%), that gives a positive lysine decarboxylase reaction. It has been isolated from a variety of hospital environmental sources and is usually of clinical significance in severely immunocompromised patients. The extensive use of imipenem, to which S. maltophilia is resistant, appears to be associated with nosocomial outbreaks. S. maltophilia is increasingly isolated from the sputum of CF patients and is often misidentified as B. cepacia because it grows reasonably well on colistin-containing media. Most strains are susceptible to co-trimoxazole, doxycycline and minocycline, and third-generation cephalosporins but are resistant to aminoglycosides.

SPHINGOMONAS PAUCIMOBILIS

This species produces a non-diffusible yellow pigment and is most likely to be confused with flavobacteria. Motility is poor and best seen in cultures incubated at room temperature. It forms acid from a wide range of sugars in ammonium salts-based medium and is consistently ONPG positive. It has been found in clinical material and recovered from hospital equipment. Most strains are susceptible to erythromycin, tetracycline, chloramphenicol and aminoglycosides.

For further information about the classification and identification of the pseudomonads see Kersters *et al.* (1996) and *Cowan and Steel's Manual* (Barrow and Feltham, 1993) and Pitt (1999).

REFERENCES

Coeyne, T., Vandamme, P., Govan J. R. W. *et al.* (2001) Taxonomy and identification of the *Burkholderia cepacia* complex. *Journal of Clinical Microbiology* **39**: 3427–3436.

Barrow, G. I. and Feltham R. K. A. (eds) (1993) *Cowan and Steel's Manual for the Identification of Medical Bacteria*, 3rd edn. Cambridge: Cambridge University Press.

Dance, D. A. B. (1999) Melioidosis. In: Guerrant, R. L., Walker, D. H. and Weller, P. F. (eds), *Tropical Infectious Diseases: Principles, Pathogens and Practice*, Vol. 1. Philadelphia, PA: Churchill Livingstone, pp. 430–407.

Grundmann, H., Schneider, C., Hartung, D. *et al.* (1995) Discriminatory power of three DNA-based typing techniques for *Pseudomonas aeruginosa*. *Journal of Clinical Microbiology* **33**: 528–534.

Kersters, K., Ludwig, W., Vancanneyt, M. *et al.* (1996) Recent changes in the classification of the pseudomonads; an overview. *Systematic and Applied Microbiology* **19**: 465–477.

Livermore, D. M. (2002) Multiple mechanisms of antimicrobial resistance in *Pseudomonas aeruginosa*: our worst nightmare? *Clinical Infectious Diseases* **34**: 634–640.

Pitt, T. L. (1999) *Pseudomonas, Burkholderia* and related genera. In: Collier, L., Balows, A. and Sussman, M. (eds), *Topley and Wilson's Microbiology and Microbial Infections*, 9th edn, Vol 2, *Systematic Bacteriology*, CD-ROM. London: Arnold, Chapter 47.

Yabuuchi, E., Kosaka, Y., Oyaizu, H. *et al.* (1992) Proposal of *Burkholderia* gen. nov; and transfer of seven species of the *Pseudomonas* homology group II to the new genus, with the type species *Burkholderia cepacia* (Palleroni and Holmes 1981) comb. nov. *Microbiology and Immunology* **36**: 1251–1275.

Yamamoto, S., Kasai, H., Arnold, D. L. *et al.* (2000) Phylogeny of the genus *Pseudomonas*: intrageneric structure reconstructed from the nucleotide sequences of *gyrB* and *rpoD* genes. *Microbiology* **146**: 2385–2394.

23

Acinetobacter, Alcaligenes, Flavobacterium, Chromobacterium and Janthinobacterium, and acetic acid bacteria

ACINETOBACTER

Organisms have been moved into and out of this genus for several years. At present it seems to contain two species – *Acinetobacter calcoaceticus* (formerly *A. anitratum*) and *A. lwoffii* – although some workers use the name *A. baumannii* for both or for various 'subspecies'. It is difficult to distinguish between the two species.

Acinetobacters are Gram-negative, non-motile, obligate aerobes, oxidase and nitrate reduction negative, catalase positive, and are oxidative or give no reaction on Hugh and Leifson medium (see below) and grow on MacConkey agar.

They are ubiquitous in the environment and have been found in clinical material.

Isolation and identification

Plate material on nutrient, blood and MacConkey agar (and *Bordetella* or similar media if indicated). Incubate at 22°C and at 37°C for 24–48 h. Do oxidase and catalase tests. Inoculate nutrient broth for motility, Hugh and Leifson medium, glucose peptone water, nitrate broth and urea medium.

Species of Acinetobacter

Acinetobacter calcoaceticus ssp. anitratus
This gives large non-lactose fermenting colonies on MacConkey and DC agars and grows at room temperature. On Hugh and Leifson medium it is oxidative, producing acid but not gas from glucose. It gives a variable urease reaction.

Infections in a wide variety of sites in humans have been reported. Many strains are resistant to several antibiotics. Hospital cross-infections with resistant strains cause problems. Sometimes in direct films this organism may resemble *Neisseria* spp.

Acinetobacter calcoaceticus ssp. lwoffii
Colonies on blood agar are small and haemolytic, and may be sticky. There is no change in Hugh and Leifson medium; it fails to produce acid from glucose and is urease negative. It can cause conjunctivitis. This and the other two biotypes are distinguished by the API 20NE.

For more information about infections with *Acinetobacter* spp., see Vivian *et al.* (1981) and Bergogne-Bérézin *et al.* (1996).

ALCALIGENES

These organisms resemble acinetobacters in many ways and some workers include them in the genus *Acinetobacter*. They are widely distributed in soil, fresh and salt water, and are economically important in food spoilage (especially of fish and meat). Many strains are psychrophilic. Some species have been isolated from human material and suspected of causing disease.

Isolation and identification

Plate on blood and MacConkey agars. Incubate at 20–22°C (foodstuffs) or 35–37°C (pathological material) for 24–48 h. Subculture white colonies of

Gram-negative rods on agar slopes for oxidase test, in peptone water for motility, in Hugh and Leifson medium, glucose peptone water, bromocresol purple milk, arginine broth and gelatin.

Species of *Alcaligenes*

The genus *Alcaligenes* is restricted here to motile, oxidase-positive, catalase-positive Gram-negative rods which give an alkaline or no reaction in Hugh and Leifson medium, an alkaline reaction in bromocresol purple milk and grow on MacConkey agar. Arginine is not hydrolysed and gelatin is not usually liquefied. Most strains are resistant to penicillin (10-unit disc). There is one species, *A. faecalis*, which now includes *A. odorans* and several strains of doubtful taxonomic status.

Alcaligenes faecalis

This culturally resembles *Bordetella bronchiseptica* but is urease negative. It is a widely distributed saprophyte and commensal. Cultures of some strains, especially those associated with food spoilage, have a fruity odour, e.g. of apples or rotting melons. At least one strain of *Alcaligenes* is associated with ropy milk and ropy bread.

ACHROMOBACTER

This is no longer officially recognized, but food bacteriologists find it a convenient 'dustbin' for non-pigmented Gram-negative rods which are non-motile, oxidative in Hugh and Leifson medium, oxidase positive, produce acid but no gas from glucose, acid or no change in purple milk, do not hydrolyse arginine and usually liquefy gelatin.

FLAVOBACTERIUM

The genus *Flavobacterium* contains many species that are difficult to identify and some have been transferred to other genera (Murray *et al.*, 1999). Food bacteriologists, however, frequently isolate Gram-negative, bacilli-forming, yellow colonies from their samples and, although some undoubtedly belong to other genera, it suits the convenience of many food bacteriologists to call them 'flavobacteria'. Some of these organisms are proteolytic or pectinolytic and are associated with spoilage of fish, fruit and vegetables.

There is one species of medical importance.

Flavobacterium meningosepticum

This has been isolated from cerebrospinal fluid (CSF) in meningitis in neonates, from other human material, and from hospital intravenous and irrigation fluids.

It grows on ordinary media; colonies at 18 h are 1–2 mm, smooth, entire, grey or yellowish and butyrous. There is no haemolysis on blood agar, no growth on DCA and none on primary MacConkey cultures, but growth on this medium may occur after several subcultures on other media. A yellowish–green pigment may diffuse into nutrient agar media.

Flavobacterium meningosepticum is oxidative in Hugh and Leifson medium, non-motile, oxidase and catalase positive, liquefies gelatin in 5 days, does not reduce nitrates and does not grow on Simmons' citrate medium. Rapid hydrolysis of aesculin is a useful characteristic. The API ZYM system is useful. Acid is produced from 10% glucose, mannitol and lactose in ammonium salt medium but not from sucrose or salicin. Indole production seems to be positive by Ehrlich's but negative by Kovac's methods (Snell, 1973). See Table 23.1 for differentiation from pseudomonads and *Acinetobacter*. For more information see Holmes (1987).

CHROMOBACTERIUM AND JANTHINOBACTERIUM

An important characteristic of these organisms is the production of a violet pigment (unpigmented

Table 23.1 *Flavobacterium meningosepticum, Pseudomonas* and *Acinetobacter* spp.

Species	Oxidase	Nitrate reduction	Citrate
F. meningosepticum	+	−	−
Pseudomonas spp.	+	+	+
Acinetobacter spp.	−	−	v

v, varies with subspecies, see text

strains are not unknown), although this may not be apparent until the colonies are several days old. They grow well on ordinary media, giving cream or yellowish colonies that turn purple at the edges. The pigment is soluble in ethanol but not in chloroform or water, and can be enhanced by adding mannitol or meat extract to the medium. Citrate but not malonate is used in basal synthetic medium; the catalase test is positive and the urease test negative. Ammonia is formed. Gelatin is liquefied. The oxidase test is positive but difficult to do if the pigment is well formed.

Chromobacterium violaceum

This species is mesophilic, growing at 37°C but not at 5°C. It is fermentative in Hugh and Leifson medium, hydrolyses arginine and casein but not aesculin (Table 23.2). It is a facultative anaerobe.

The violet pigment is best seen in cultures on mannitol yeast extract agar incubated at 25°C. Colonies of non-pigmented variants may be seen in cultures.

Although usually a saprophyte, cases of human and animal infection have been described in Europe, the USA and the Far East.

Janthinobacterium lividum

This organism is psychrophilic, growing at 5°C but not at 37°C. It is oxidative in Hugh and Leifson medium, does not hydrolyse arginine or casein, and hydrolyses aesculin. It is an obligate aerobe.

The violet pigment is best seen in cultures on mannitol yeast extract agar incubated at 20°C.

More information about the identification of these organisms is given by Logan and Moss (1992).

Table 23.2 *Chromobacterium violaceum* and *Janthinum lividum*

Species	Growth at			Hydrolysis		
	5°C	37°C	HL test[a]	Arginine	Casein	Aesculin
C. violaceum	−	+	F	+	+	−
J. lividum	+	−	O	−	−	+

[a] Hugh and Leifson (fermentation oxidation) test
F, fermentative; O, oxidative

ACETIC ACID BACTERIA

These bacteria are widely distributed in vegetation. They are of economic importance in the fermentation and pickling industries, e.g. in cider manufacture (Carr and Passmore, 1979), as a cause of ropy beer and sour wine, and of off-odours and spoilage of materials preserved in vinegar. Ethanol is oxidized to acetic acid by *Gluconobacter* spp. but *Acetobacter* spp. continue the oxidation to produce carbon dioxide and water. They also differ in that *Acetobacter* produces carbonate from lactate and has peritricate flagella whereas *Gluconobacter* does not produce carbonate and has polar flagella. Seven species of *Acetobacter* and three of *Gluconobacter* are recognized but identification is difficult.

Isolation and identification to genus

Identification to species is rarely necessary. It is usually sufficient to recognize acetic acid bacteria and to determine whether the organisms produce acetic acid and/or destroy it.

Inoculate wort agar or Universal Beer Agar or yeast extract agar containing 10% glucose that has been filter sterilized and added at 50°C and 3% calcium carbonate at pH 4.5. Incubate at 25°C for 24–48 h. Colonies are large and slimy. Subculture into yeast broth containing 2% ethanol and Congo red indicator at pH 4.5. Acid is produced.

Test the ability to produce carbon dioxide from acetic acid in nutrient broth containing 2% acetic acid. Use a Durham tube.

Plant bacterial masses (heavy inoculum, do not spread) on yeast extract agar at pH 4.5 containing 5% glucose and at least 3% finely divided calcium carbonate in suspension. Clear zones occur around masses in 3 weeks but if less chalk is used some pseudomonads can do this.

Culture on yeast extract agar containing 2% calcium lactate. *Acetobacter* spp. grow well and precipitate calcium carbonate. *Gluconobacter* spp. grow poorly and give no precipitate.

Carr and Passmore (1979) describe a medium to differentiate the two genera. It is yeast extract agar containing 2% ethanol and bromocresol green (1 ml 2.2% solution/l) made up in slopes. Both *Gluconobacter* and *Acetobacter* spp. produce acid from ethanol and the indicator changes from blue–green to yellow. *Acetobacter* spp. then use the acid and the colour changes back to green (Table 23.3).

For more information, see Carr and Passmore (1979) and Swings *et al.* (1992).

REFERENCES

Bergogne-Bérézin, E., Joly-Guillou, N. L. and Towner, K.J. (1996) Acinetobacter: *Microbiology, Epidemiology, Infections and Management.* Boca Raton, FL: CRC Press.

Carr, J. G. and Passmore, S. M. (1979) Methods for identifying acetic acid bacteria. In: Skinner, F. A. and Lovelock, D. W. (eds), *Identification Methods for Microbiologists*, 2nd edn. Society for Applied Bacteriology Technical Series No. 14. London: Academic Press, pp. 33–45.

Holmes, B. (1987) Identification and distribution *Flavobacterium meningosepticum* in clinical material. *Journal of Applied Bacteriology* 43: 29–42.

Table 23.3 Acetic acid bacteria

Genus	Acid from ethanol	CO$_2$ from acetic acid	Calcium carbonate precipitation
Gluconobacter	+	−	−
Acetobacter	+	+	+

Logan, N. A. and Moss, M. O. (1992) Identification of *Chromobacterium, Janthinobacterium* and *Iodobacter* species. In: Board, R. G., Jones, D. and Skinner, F. A. (eds), *Identification Methods in Applied and Environmental Microbiology*. Society for Applied Bacteriology Technical Series No. 29. London: Blackwells, pp. 183–192.

Murray, P. R., Baron, J. O., Pfaller, M. A. *et al.* (eds) (1999) *Manual of Clinical Microbiology*, 7th edn. Washington: American Society of Microbiology.

Snell, J. J. S. (1973) *The Distribution and Identification of Non-fermenting Bacteria*. Public Health Laboratory Service Monograph No. 4. London: HMSO.

Swings, J., Gillis, M. and Kersters, K. (1992) Phenotypic identification of acetic acid bacteria. In: Board, R. G., Jones, D. and Skinner, F. A. (eds), *Identification Methods in Applied and Environmental Microbiology*. Society for Applied Bacteriology Technical Series No. 29. London: Blackwells, pp. 103–110.

Vivian, A., Hinchcliffe, E. J. and Fewson, C. A. (1981) *Acinetobacter calcoaceticus:* some approaches to a problem. *Journal of Hospital Infection* **2**: 199–204.

24

Vibrio, Plesiomonas and Aeromonas

The differential properties of the organisms in this group and those of the enterobacteria are summarized in Table 24.1.

Table 24.1 Characteristics that distinguish genera

Genus	O/F test	Oxidase	0/129 150 μg	NaCl enhancement
Vibrio	F	+	S	+
Plesiomonas	F	+	S	−
Aeromonas	F	+	R	−
Enterobacteria	F	−	R	−
Pseudomonas	O/N	+	R	−

F, fermentative; O/N, oxidative/no acid produced; R, resistant; S, sensitive

VIBRIOS

The genus *Vibrio* includes at least 40 species of Gram-negative, non-sporing rods, motile by polar flagella and enclosed within a sheath. Some have lateral flagella and may swarm on solid media. They are catalase positive, use carbohydrates fermentatively and rarely produce gas. All but one species (*Vibrio metschnikovii*) are oxidase and nitrate reductase positive. They are sensitive to 150 μg discs of the vibriostatic agent 2,4-diamino-6,7-di-isopropyl pteridine (0/129). An exception is *V. cholerae* O:139 which is resistant to 0/129. A wide range of extracellular enzymes is produced. Vibrios occur naturally in fresh and salt water. Species include some human and fish pathogens. See Table 24.1 for differentiation of *Vibrio*, *Plesiomonas, Aeromonas* and *Pseudomonas* spp. and enterobacteria.

Isolation

Vibrios grow readily on most ordinary media but enrichment and selective media are necessary for faeces and other material containing mixed flora.

Faeces

Add 2 g of 20-ml amounts of modified alkaline peptone water (MAPW – see p. 75). Inoculate thiosulphate citrate bile salt sucrose agar (TCBS) medium. Incubate MAPW at 37°C for 5–8 h or at 20–25°C for 18 h and subculture to TCBS. Incubate TCBS cultures at 37°C overnight.

Other human material

Vibrios may occur in wounds and other material, particularly if there is a history of sea bathing. Examine primary blood agar plates (see below for colony appearance).

Water and foods

Add 20 ml water to 100 ml MAPW. Add 25 g food to 225 ml MAPW and emulsify in a Stomacher. Incubate at 37°C for 18 h and subculture on TCBS. If *V. parahaemolyticus* is specifically sought, enrich the food sample in salt polymyxin broth (SPB) in addition to MAPW.

For further information on the choice of media combinations for specific vibrios from food and environmental samples, see Donovan and van Netten (1995).

Enumeration of vibrios

Use liquid samples neat and diluted 10^{-1}. Make 10^{-1} homogenates of solid samples. Spread 50 or 100 μl on TCBS medium, incubate 20–24 h and count vibrio-like colonies. Report as colony-forming units (cfu) per gram.

For the three-tube MPN (most probable number) method prepare dilutions of 10^{-1}, 10^{-2} and 10^{-3}. Add 1 ml of each dilution to each of three tubes of APW and/or SPB and incubate at 30°C overnight. Subculture to TCSB and incubate at 37°C for 20–24 h. Examine for vibrio-like colonies, identify them and use the three-tube table for food samples (Roberts *et al.*, 1995).

Colonial morphology

Table 24.2 lists the colony appearances of vibrios and related organisms on TCBS medium. On non-selective media colonial morphology is variable. The colonies may be opaque or translucent, flat or domed, haemolytic or non-haemolytic, smooth or rough. One variant is rugose and adheres closely to the medium.

Identification

The following tests enable the common species to be identified. For more detailed tests, use the API 20E and 20NE systems (Austin and Lee, 1992) but include 0/129 sensitivity and growth on CLED (cystine lactose electrolyte-deficient) medium.

Oxidation/fermentation test

Use Hugh and Leifson method (see p. 96).

Oxidase test

Use the method on p. 99 to test colonies from non-selective medium. Do not use colonies from TCBS or other media containing fermentable carbohydrate because changes in pH may interfere with the reaction.

Arginine dihydrolase

Use arginine broth or Moeller's medium with 1% NaCl (see p. 95).

Decarboxylase tests

Use Moeller's medium containing 1% NaCl, but do not read the results too early because there is an initial acid reaction before the medium becomes alkaline. The blank should give an acid reaction;

Table 24.2 Growth of vibrios on thiosulphate citrate bile salt agar (TCBS), 37°C for 18 h

Species of *Vibrio*	Growth	Acid from Sucrose	Colour	Colony size (mm)
V. cholerae	+	+	Y	2–3
V. parahaemolyticus	+	−	G	2–5
V. vulnificus	+	−	G	2–3
V. alginolyticus	+	+	Y	3–5
V. fluvialis	+	+	Y	2–3
V. furnissi	+	+	Y	2–3
V. metschnikovii	+	+	Y	2–4
V. anguillarum	+	+	Y	1–3
V. hollisae	−	−		−
V. damsela	+	−	G	2–3

G, colonies green; Y, colonies yellow

failure to do so may suggest poor growth and the electrolyte supplement should be added (see p. 75).

Carbohydrate fermentation tests

Use peptone water sugars (see p. 73) with 1% NaCl.

Indole

Use peptone water containing 1% NaCl and Kovac's method (see p. 97).

Voges–Proskauer test

Use a semisolid medium under controlled conditions. If incubation is prolonged and a sensitive method is used almost any vibrio may give positive results (Furniss *et al.*, 1978).

Gelatin liquefaction

Use gelatin agar (see p. 96).

ONPG test

Test for β-galactosidase activity in ONPG broth with 1% NaCl or use discs (see p. 99).

Growth at 43°C

Inoculate nutrient broth with 1% NaCl and incubate in a water bath.

Sensitivity to 2,4-diamino-6,7-di-isopropylpteridine (0/129)

This was originally used as a disc method to distinguish between vibrios (sensitive) and aeromonads (resistant), but the use of two discs (150 μg and 10 μg) enables two groups of vibrios to be distinguished. Make a lawn of the organisms on nutrient agar, and place one disc of each concentration of 0/129 on it and incubate overnight. Do not use any special antibiotic sensitivity-testing medium because the growth of vibrios and the diffusion characteristics of 0/129 differ from those on nutrient agar (Furniss *et al.*, 1978).

Growth on CLED medium

Some vibrios require the addition of sodium chloride to the medium whereas others do not. Culture on an electrolyte-deficient medium, e.g. CLED. This permits two groups, halophilic and non-halophilic vibrios, to be distinguished. Inoculate CLED lightly with the culture and incubate at 37°C overnight.

Salt tolerance

This varies with species. Inoculate peptone water containing 0, 3, 6, 8 and 10% NaCl. This technique must be standardized to obtain consistent results (Furniss *et al.*, 1978).

Toxin detection

There are kit tests (e.g. Oxoid) for detecting cholera toxin in culture filtrates. However, cultures will need to be referred to international reference laboratories in the case of isolates of *V. cholerae* O:1 and O:139.

PROPERTIES OF VIBRIOS, *PLESIOMONAS* AND *AEROMONAS*

These are shown in Tables 24.2 and 24.3.

Species of vibrios

Vibrio cholerae

This is sensitive to 0/129 and oxidase positive, decarboxylates lysine and ornithine, but does not hydrolyse arginine produces acid from sucrose, although no acid from arabinose or lactose. It grows on CLED medium and is therefore nonhalophilic.

All strains possess the same heat-labile *H* antigen but may be separated into serovars by their *O* antigens. There are 198 'O' serotypes recognized at the present time.

Serovar O:1: the cholera vibrio

This is the causative organism of epidemic or Asiatic cholera. It is agglutinated by specific O:1 cholera antiserum. Phage typing (Lee and Furniss, 1991) is of epidemiological value. Non-toxigenic strains of *V. cholerae* O:1 have been isolated and have shown distinctive phage patterns.

Table 24.3 *Vibrio species and allied genera*

Species	Oxidase	O/129		Growth on CLED	NaCl		VP	Lysine decarboxylase	Arginine dihydrolase	Ornithine decarboxylase	ONPG hydrolysis	Acid from sucrose
		10 µg	150 µg		0%	6%						
V. cholerae	+	S	S	+	+	−	+	+	−	+	+	+
V. parahaemolyticus	+	R	S	−	−	+	−	+	−	+	−	−
V. vulnificus	+	S	S	−	−	+	+	+	−	+	+	−
V. alginolyticus	+	R	S	−	−	+	+	+	−	+	−	+
V. fluvialis	+	R	S	−	−	+	−	−	+	−	+	+
V. furnissi	+	R	S	+	>	+	−	−	+	−	+	+
V. metschnikovii	−	S	S	+	+	>	>	>	+	−	>	+
V. anguillarum	+	S	S	>	>	>	+	−	+	−	+	+
V . mimicus	+	S	S	+	+	−	+	+	−	+	+	−
Aeromonas	+	R	R	+	+	−	>	>	+	−	+	+
Plesiomonas	+	S	S	+	+	−	+	+	+	+	+	−

S, sensitive; R, resistant; v, variable

It is possible, by using carefully absorbed sera, to distinguish two subtypes of *V. cholerae* O:1. These are known as Ogawa and Inaba but as they are not completely stable and variation may occur *in vitro* and *in vivo* subtyping is of little epidemiological value.

Although there are two biotypes of *V. cholerae* O:1 – the 'classic' (non-haemolytic) and the 'eltor' (haemolytic) – the former now virtually non-existent. Biotyping is not, therefore, a useful epidemiological tool (Table 24.4).

Table 24.4 *Vibrio cholerae* O:I biovars (biotypes)

	Classic	Eltor
Haemolysis	–	V
VP	–	+
Chick cell agglutination	–	+
Polymyxin B 50 IU	S	R
Classic phage IV	S	R
Eltor phage 5	R	S

S, sensitive; R, resistant

Serovars other than O:I, non-O:I *V. cholerae*

These serovars have the same biochemical characteristics as *V. cholerae* O:1 and the same H antigen but possess different *O* antigens and are not agglutinated by the O:1 serum. They have been called 'non-agglutinating' (NAG) vibrios. This is an obvious misnomer; they are agglutinated both by cholera *H* antiserum and by antisera prepared against the particular *O* antigen they possess. Another term that has been used is 'non-cholera vibrio' (NCV); as both terms have been used in different ways, this has resulted in considerable confusion. It is best if all these vibrios are referred to as non-O:1 *V. cholerae*, or, if a strain has been serotyped, by that designation.

Vibrio cholerae O:139 (Bengal)

In 1992 a new outbreak of cholera occurred in the Indian subcontinent. The causative organism was identified as *V. cholerae* serotype O:139 (Bengal).

This strain produces cholera toxin (CT) and did not type with the existing antisera set (1–138) at that time. The strain is resistant to the vibriostatic agent 0/129 and has not spread from the Indian subcontinent.

Some of the other serotypes are undoubtedly potential pathogens and produce a toxin similar to, if not identical with, that of the cholera vibrio. Some outbreaks have occurred but most isolates have been from sporadic cases. These vibrios are widespread in fresh and brackish water in many parts of the world, including the UK. They do not, however, cause true epidemic cholera.

Vibrio parahaemolyticus

This halophilic marine vibrio will not grow on CLED. It does not ferment sucrose and therefore gives a (large) green colony on TCBS agar. O and *K* antigens may be used to serotype strains. It is associated with outbreaks of gastroenteritis following the consumption of seafood and was recognized as the most common cause of food poisoning in Japan. It is usually present in coastal waters, although only in the warmer months in the UK. The Kanagawa haemolysis test is said to correlate with pathogenicity (see Barrow and Feltham, 1993). Controlled conditions for testing the haemolysis of the human red blood cells are essential. Only laboratories with sufficient experience should do this test. Most environmental strains are negative.

Vibrio vulnificus

These vibrios were first called L+ vibrios after their ability to ferment lactose. They are ONPG positive and sensitive to 10 μg 0/129 disc. Care needs to be taken to distinguish them from *V. parahaemolyticus*. *V. vulnificus* produces sulphatase (opaque halo) on sodium dodecylsulphate polymyxin B sucrose (SPS) agar; *V. parahaemolyticus* does not (Kitaura *et al.*, 1983; Donovan and van Netten, 1995).

Vibrio alginolyticus

A very common marine (halophilic) vibrio of worldwide distribution. It grows well in MAPW and on TCBS and other plating media. It produces large sucrose-fermenting colonies on TCBS and most strains will show swarming growth on marine or seawater agars. It is biochemically similar to *V. parahaemolyticus*, but is distinguished by the ability to ferment sucrose. It does not cause human gastroenteritis. As a result of its presence in most seawater, this vibrio has value as an indicator organism for cooked seafoods.

Vibrio fluvialis

First reported as group F vibrios, this is common in rivers, particularly in the brackish water of estuaries, and may cause gastroenteritis in humans. *V. fluvialis* needs to be distinguished from *V. anguillarum* (see Table 24.3).

Vibrio furnissi

This was previously classified as *V. fluvialis* biovar II. It has been associated with human gastroenteritis in Japan and other parts of the East, and has been isolated from the environment and seafood.

Vibrio metschnikovii

This is both oxidase and nitrate reduction negative, but its other characters are typical of vibrios. It includes the now illegitimate species *V. proteus*. It is frequently isolated from water and shellfish, but there is little evidence of pathogenicity for humans.

Vibrio anguillarum

This is phenotypically far from being a uniform species. Some strains are pathogenic for fish, although not especially for the eel, as its name might suggest. Many strains will not grow at 37°C but do grow at 30°C and would be missed in laboratories that confine their incubation temperatures to about 37°C.

Vibrio hollisae

A halophilic vibrio isolated from human cases of gastroenteritis, this does not grow on TCBS. It is distinguished from other halophilic vibrios from human sources by its negative lysine, arginine and ornithine tests, and the limited range of carbohydrates fermented. It has not been recovered from the environment and a clear relationship with human disease has not been established.

Vibrio damsela

A marine bacterium isolated from human wounds and from water and skin ulcers of damsel fish. It is arginine positive, produces gas from glucose, requires NaCl for growth, is not bioluminescent, and ferments glucose, mannose and maltose. It is distinguished from other named vibrios by DNA–DNA hybridization.

Vibrio mimicus

Resembles non-O:I *V. cholerae* but is VP negative and does not ferment sucrose. Colonies on TCBS media are therefore green. It has been isolated from the environment and is associated with seafoods. It has also been isolated from human faeces. Some strains produce a cholera-like toxin.

Plesiomonas shigelloides

This bacterium has had a varied taxonomic history having been labelled C27 *Paracolon*, *Pseudomonas*, *Aeromonas* and *Vibrio shigelloides*. It is close to the enterobacteria and some strains are agglutinated by *Shigella sonnei* antisera, hence the shigella-like name *shigelloides*. However, this bacterium is oxidase positive and sensitive to 0/129. These characteristics distinguish it from the enterobacteria.

It is now thought to be a cause of gastroenteritis because most isolates are from patients with diarrhoea following travel.

There are no common specific enrichment methods or media for *Plesiomonas* spp. It grows poorly on TCBS, but well on enteric media such as DCA and MacConkey. Colonies are shigella-like and are identified using a range of tests, as shown in Table 24.3, with key characteristics being: sensitive to 0/129 lysine, arginine and ornithine decarboxylase positive, oxidase positive and cross-reactions with shigella antisera.

Aeromonas species

These are Gram-negative rods that are motile by polar flagella. They are oxidase and catalase positive, and resistant to 0/129. They metabolize carbohydrates fermentatively (gas may be produced from glucose), liquefy gelatin and reduce nitrates. The arginine dihydrolase test using arginine broth is positive for most strains. Indole and VP reactions vary with species. Some strains are psychrophilic and most grow at 10°C. See Tables 24.2 and 24.3 for differentiation from other groups.

The genus is divided into two groups, Salmonicida and Hydrophila (Table 24.5).

Table 24.5 Aeromonas groups

Group	Motility	Growth at 37°C	Brown pigment
Hydrophila	+	+	v
Salmonicida	–	–	+

The Salmonicida group

This contains psychrophilic, non-motile aeromonads. They may be divided at the subspecies level and are associated with furunculosis of fish, a disease of economic importance in fish farming (Frerichs and Hendrie, 1985). The aeromonads in the Salmonicida group are not associated with human disease and are not known to be found in food.

The Hydrophila group

Aeromonads in this group are motile, mesophilic and grow at standard laboratory temperatures of 30 and 37°C. This group contains a number of clusters or species. Most workers recognize the species *A. hydrophila*, *A. sobria* and *A. caviae* (Table 24.6). Other species, e.g. *A. media*, *A. veronii* and *A. schuberti*, have been described in recent years. Until satisfactory identification systems are available and validated it is prudent to divide the Hydrophila group only into the species *A. hydrophila*, *A. sobria* and *A. caviae*.

Species within this group may be the cause of human gastroenteritis. Several workers have detected virulence associated factors, e.g. enterotoxin production with clinical isolates. These factors are found mainly in *A. hydrophila* and *A. sobria*, but less frequently with *A. caviae* isolates. In countries with high ambient temperatures, where aeromonads are common in drinking water, they are frequently isolated in both normal and diarrhoeal faecal specimens. Aeromonads in this group have been found in wound infections and as rare isolates from blood cultures.

Aeromonads in the Hydrophila group grow rapidly in food even at refrigeration temperatures (Mattrick and Donovan, 1998a,b) and were found in 54% of salad samples in a European Community Co-ordinated Food Control Programme in England and Wales in 1995 (Little *et al.*, 1997). They are wide spread in nature and are frequently found in water, animals, sewage, dairy products and fish. They may be misidentified as 'coliforms' because lactose-fermenting strains are common.

Isolation

An incubation temperature of 30°C is recommended for all procedures. *Aeromonas* species will grow on non-selective media, such as blood agar, which is suitable for clinical specimens other than faeces.

Table 24.6 Aeromonas Hydrophila group

Species	Aesculin hydrolysis	Gas from glucose	Acid from arabinose	Lysine decarboxylase	VP	Haemolysis	H$_2$S
A. hydrophila	+	v	+	+	+	++	+
A. caviae	+	–	+	–	–	±	–
A. sobria	–	+	–	+	+	++	+
v, variable							

Faeces

Plate on selective agar such as bile salt irgasan brilliant green agar (BSB) or Ryan's modification of XLD (RXLD (Mattrick and Donovan, 1998a). *Aeromonas* spp. produce non-xylose-fermenting colonies on BSB and RXLD. Enrichment is not usually needed for faeces.

Food

Plate food homogenates on BSB or RXLD and starch ampicillin agar (SAA). This will detect large numbers and may be used for quantitative procedures. For enrichment inoculate alkaline peptone water (APW) and tryptone soy broth (TSB) containing 30 mg/l ampicillin and 40 mg/l irgasan (Mattrick and Donovan, 1998a).

Water

Use a membrane filtration method with SAA for the quantitative procedures for both treated and raw water.

Identification

For a presumptive test do oxidase and spot indole tests on suspected colonies. *Aeromonas* spp. give a positive reaction in both.

To distinguish between the psychrophilic Salmonicida group and the mesophilic Hydrophila group, test for oxidase, fermentation on glucose and resistance to 0/129 discs. For separation into the two groups, test for growth at 20 and 37°C, for motility and a brown pigment on media containing 1% tyrosine incubated at 20°C (see Table 24.5).

For identification to species level within the Hydrophila group, test for aesculin hydrolysis, lysine decarboxylase, VP, gas from glucose, H2S production, acid from arabinose and salicin, and arginine dihydrolase test for confirmation of generic identification (see Table 24.6).

REFERENCES

Austin, B. and Lee, J. V. (1992) Aeromonadaceae and Vibrionaceae. In: Board, R. G., Jones, D., Skinner, F. A. (eds), *Identification Methods in Applied and Environmental Microbiology*. Society for Applied Bacteriology Technical Series No. 29. Oxford: Blackwells, pp. 163–183.

Barrow, G. I. and Feltham, R. K. A. (1993) *Cowan and Steel's Manual for the Identification of Medical Bacteria*. Cambridge: Cambridge University Press.

Donovan, T. J. and van Netten, P. (1995) Culture media for the isolation and enumeration of pathogenic *Vibrio* species in foods and environmental samples. *International Journal of Food Microbiology* **26**: 77– 91.

Frerichs, G. N. and Hendrie, M. S. (1985) Bacteria associated with disease of fish. In: Collins C. H. and Grange , J. M. (eds), *Isolation and Identification of Micro-organisms of Medical and Veterinary Importance*. Society for Applied Bacteriology Technical Series No. 21. London: Academic Press, pp. 355–371.

Furniss, A. L., Lee, J. V. and Donovan, T. J. (1978) *The Vibrios*. Public Health Laboratory Service Monograph No. 11. London: HMSO.

Kitaura, T., Doke, S., Azuma, I. *et al.* (1983) Halo production by sulphatase activity in *V. vulnificus* and *V. cholerae* 0:1 on a new selective sodium dodecyl sulphate containing medium: a screening marker in environmental surveillance. *FEMS Microbiology Letters* **17**: 205–209.

Lee, J. V. and Furniss, A. L. (1981) Discussion 1. The phage-typing of *Vibrio cholerae* serovar O1. In: Holme, T., Holmgren, J., Mersom, M. H. *et al.* (eds), *Acute Enteric Infections in Children. New prospects for treatment and prevention*. Amsterdam: Elsevier, pp. 191–222

Little, C. L., Monsey, H. A., Nichols, G. L. *et al.* (1997) The microbiological quality of refrigerated salads and crudités. *PHLS Microbiological Digest* **14**: 142–146.

Mattrick, K. L. and Donovan, T. J. (1998a) Optimisation of the protocol for the detection of *Aeromonas* species in ready-to-eat salads, and its use to speciate isolates and establish their prevalence. *Communicable Disease and Public Health* **1**: 263–266.

Mattrick, K. L. and Donovan, T. J. (1998b) The risk to public health of aeromonas in ready-to-eat salad products. *Communicable Disease and Public Health* **1**: 267–270.

25

Key to the enterobacteria

This informal group of Gram-negative non-sporing rods contains a number of species of Enterobacteriaceae as well as some other species. Only those that are relatively common are considered in the following chapters. They are aerobic, facultatively anaerobic, oxidase negative and fermentative in Hugh and Leifson medium. Details of others may be found in the latest edition of *Topley and Wilson's Microbiology and Microbial Infections* (Collier *et al.*, 1998) and in *Cowan and Steel's Manual for the Identification of Medical Bacteria* (Barrow and Feltham, 1993).

The genera considered here and in the following chapters may be conveniently divided into four groups:

1. *Escherichia, Citrobacter, Klebsiella* and *Enterobacter* spp., which usually produce acid and gas from lactose. For historical reasons they are known collectively as the coliform bacilli (see Chapter 26).
2. *Edwardsiella, Hafnia* and *Serratia* spp., which usually do not ferment lactose, although some workers include them in the coliforms (see Chapter 27).

3. *Salmonella* and *Shigella* spp., which do not ferment lactose and are important intestinal pathogens (see Chapter 28).
4. *Proteus, Providencia* and *Morganella* spp., which do not ferment lactose and differ from the other groups in their ability to deaminate phenylalanine (see Chapter 29).

Table 25.1 shows the general biochemical properties of these genera.

The genus *Yersinia* is included among the enterobacteria by some workers but for convenience is considered in this book in Chapter 34.

REFERENCES

Barrow, G.I. and Feltham, R. K. A. (1993) *Cowan and Steel's Manual for the Identification of Medical Bacteria*, 3rd edn. Cambridge: Cambridge University Press.

Collier, L., Balows, A. and Sussman, M. (eds) (1998) *Topley and Wilson's Microbiology and Microbial Infections*, 9th edn. Vol. 2, *Systematic Bacteriology*. London: Arnold.

Table 25.1 General properties of enterobacteria

Genus	Indole	Motility	Acid from			Citrate utilization	Urease	H₂S	Gelatin liquefaction	PPA*	ONPG hydrolysis	
			Lactose	Mannitol	Inositol							
Escherichia	+	+	+	+	–	–	–	–	–	–	+	⎫
Citrobacter	v	+	v	+	–	+	–	+	–	–	+	⎬ Ch. 26
Klebsiella	v	+	v	+	+	+	(+)	–	–	–	+	
Enterobacter	–	+	v	+	–	+	–	–	–	–	+	⎭
Erwinia	–	+	v	+	–	v*	–	–	v	–	+	⎫
Edwardsiella	+	+	–	–	–	v*	–	+	–	–	–	⎬ Ch. 27
Hafnia	–	+	–	+	–	+	–	–	–	–	+	
Serratia	–	+	–	+	–	+	–	–	+	–	+	⎭
Salmonella	–	+	–	+	+	+	–	+	–	–	–	⎫ Ch. 28
Shigella	v	–	–	v	–	–	–	–	–	–	+	⎭
Proteus	v	+	–	v	v	v	+	v	v	+	–	⎫
Providencia	+	+	–	v	+	+	–	–	–	+	–	⎬ Ch. 29
Morganella	+	+	–	v	–	v	+	–	–	+	–	⎭

v, variable; (+) slow; v*, varies with method; * phenylalanine deaminase

26

Escherichia, Citrobacter, Klebsiella and Enterobacter

These genera are included in the group of lactose-fermenting, Gram-negative, non-sporing bacilli informally known as 'coliform bacilli', and are therefore considered together in this chapter. They are oxidase negative and fermentative (as distinct from *Pseudomonas* spp.) and nutritionally non-exacting, and grow well on ordinary culture media. They also grow in the presence of bile, which inhibits many cocci and Gram-positive bacilli.

Table 25.1 shows their general properties and those that distinguish them from the other Gram-negative bacilli considered in Chapters 27–29.

ISOLATION

Pathological material

Plate stools, intestinal contents of animals, urine deposits, pus, etc., on MacConkey, eosin methylene blue (EMB) or Endo agar and on blood agar. Cystine lactose electrolyte-deficient medium (CLED) is useful in urinary bacteriology. Proteus does not spread on this medium. Incubate at 37°C overnight.

Foodstuffs

The organisms may be damaged and may not grow from direct plating. Make 10% suspensions of the food in tryptone soy broth. Incubate at 25°C for 2 h and then subculture into brilliant green bile broth and lauryl tryptose broth. MacConkey broth should not be used because it supports the growth of *Clostridium perfringens*, which produces acid and gas at 37°C. Minerals modified glutamate broth may be useful for the recovery of 'damaged' coliforms. Incubate at 37°C overnight and plate on MacConkey, Endo, EMB, violet red bile agar or CLED media.

IDENTIFICATION

Coliform bacilli show pink or red colonies, 2–3 mm in diameter, on MacConkey agar. Klebsiella colonies may be large and mucoid. On EMB agar, colonies of *Escherichia coli* are blue–black by transmitted light and have a metallic sheen by incident light. Colonies of klebsiellas are larger, brownish, convex and mucoid, and tend to coalesce. On Endo medium, the colonies of both are deep red and colour the surrounding medium. They may have a golden yellow sheen.

For full investigation test as follows: H_2S production (Triple Sugar Iron agar or Kligler medium), growth on Simmon's citrate agar, lysine and ornithine decarboxylases, arginine dihydrolase, urea hydrolysis, growth in KCN broth, phenylalanine deaminase, gelatin liquefaction and β-glucuronidase (β-GUR – see below and Table 26.1).

The following short cuts are useful.

Strains from clinical material

Examine a Gram-stained film: some cocci and Gram-positive bacilli grow on selective media such as MacConkey and CLED agar.

Table 26.1 *Escherichia, Citrobacter, Klebsiella* and *Enterobacter* spp.

	Indole 37°C	Indole 44°C	Lysine decarboxylase	Ornithine decarboxylase	Arginine dihydrolase	Citrate utilization	H$_2$S	Acid from Adonitol	Acid from Inositol	Urease	ONPG hydrolysis
Escherichia coli	+	+	+	v	v	–	–	–	–	–	v
Citrobacter freundii	v	–	–	–	+	+	+	+	–	v	+
C. koseri	+	–	–	+	+	+	–	+	–	v	+
Klebsiella pneumoniae	v	–	+	–	–	+	–	+	+	+	+
K. oxytoca	+	–	+	–	–	+	–	+	+	+	+
K. ozaenae	–	–	v	–	v	+	–	+	v	–	+
K. rhinoscleromatis	–	–	–	–	–	–	–	+	+	–	–
Enterobacter aerogenes	–	–	+	+	–	+	–	+	+	–	+
E. cloacae	–	–	–	+	+	+	–	v	–	v	–

v, variable

From Donovan (1966)

Inoculate peptone water to test for indole production (at 37 and 44°C), motility and β-glucuronidase (β-GUR) test. Inoculate ONPG broth to test for β-galactosidase.

Non-clinical strains

Examine a Gram-stained film. Inoculate lactose broth containing an indicator and a Durham's tube, and tryptone peptone water for indole test: incubate in a water bath at 44 ± 0.2°C. Include the following controls:

- Stock *E. coli*, which produces gas and is indole positive
- Stock *K. aerogenes* which produces no gas and is indole negative.

Confirm lactose-fermenting colonies on membrane filters and in primary broth cultures by subculturing to tryptone peptone water and lactose broth, testing for indole and gas production at 37 and 44°C. Plate on MacConkey agar for colony appearance. Do identification tests below. Fermentation of adonitol and inositol may be useful (Table 26.1).

β-GUR and MUG tests

These tests, together with the indole test, are very useful in distinguishing between *Escherichia coli* and other coliforms, especially in the examination of foods.

The enzyme β-GUR is produced by over 90% of strains of *E. coli* but not by the O157 strain (see below). It is also produced by some other enterobacteria (Perez *et al.*, 1986).

Inoculate 0.5 ml of a 0.1% solution of *p*-nitrophenyl-β-D-glucopyranoside in 0.067M phosphate buffer with a loopful of an 18 h peptone water culture. Incubate at 37°C for 4 h. A yellow colour indicates the release of *p*-nitrophenyl. A solid medium, allowing several spot tests on one plate, is available (MAST).

MUG is a fluorescent agent – 4-methyl umbelliferyl-β-D-glucuronide – which is commercially available for addition to culture media. Colonies of *E. coli* fluoresce under ultraviolet (UV) light (source not greater then 6 W).

'Short set' method

A modification of the 'short set' system of Donovan (1966) uses tubed media (Table 26.2). There are two API systems: RAPIDEC coli and RAPIDEC GUR, both of which include a β-GUR test.

Multipoint inoculation tests are cost-effective for testing large number of strains. A 'short set' from the MAST range may be used, e.g. β-GUR, inositol, aesculin, hydrogen sulphide or the full MAST 15 system. There are two that are aimed at *E. coli*: the Coli strip (LabM, UK), which uses a fluorogenic substrate and an indole test; and Chromocult Coliform agar (Merck, Germany) which uses β-GUR and indole.

Table 26.2 'Short set' for coliforms

Species	β-GUR	Motility	Inositol	Aesculin	H₂S	Indole
E. coli	+	+	−	−	−	+
Klebsiella spp.	−	−	+	+	−	v
C. freundii	−	+	−	−	+	−
C. koseri	−	+	−	−	−	+
Enterobacter spp.	−	+	−	v	−	−

v, variable; β-GUR, β-glucuronidase

ANTIGENS OF ENTEROBACTERIA

There are four kinds of antigens. The O or somatic antigens of the cell body are polysaccharides and are heat stable, resisting 100°C. The H or flagellar antigens are protein and destroyed at 60°C. The K and Vi antigens are envelope, sheath or capsular antigens and heat labile. There are three kinds of K antigen: L, which is destroyed at 100°C and is an envelope, occasionally capsular; A, which is destroyed at 121°C and is capsular; and B, which is destroyed at 100°C and is an envelope. These Vi and K antigens mask the O, and agglutination with O sera will not occur unless the bacterial suspensions are heated to inactivate them.

SPECIES OF COLIFORM BACILLI

Escherichia coli

This species is motile, produces acid and gas from lactose at 44°C and at lower temperatures, is indole positive at 44°C and 37°C, and fails to grow in citrate. It is MR positive, VP negative and H_2S negative and usually decarboxylates lysine. It does not hydrolyse urea or grow in KCN broth and does not liquefy gelatin. Most strains give positive β-GUR and MUG tests.

This is the so-called 'faecal coli' which occurs normally in the human and animal intestine, and it is natural to assume that its presence in food indicates recent contamination with faeces. It is widespread in nature, however, and, although most strains probably had their origin in faeces, its presence, particularly in small numbers, does not necessarily mean that the food contains faecal matter. It does suggest a low standard of hygiene. It seems advisable to avoid calling them 'faecal coli' and to report the organisms as E. coli.

Although normally present as a commensal in the human and animal intestine, E. coli is also associated with human and animal infections. It is the most common cause of urinary tract infections in humans and is also found in suppurative lesions, neonatal septicaemias and meningitis. In animals it causes mastitis, pyometria in bitches, coli granulomas in fowls and white scours in calves.

Pathotypes of E. coli causing intestinal disease

There are at least four of these (Gorbach, 1986, 1987), described as: enteropathogenic (EPEC), enterotoxigenic (ETEC), enteroinvasive (EIEC) and verotoxigenic (VTEC). A number of different O serotypes in each pathotype have been incriminated. The lists of these change frequently.

The EPEC strains have been associated with outbreaks of infantile diarrhoea and also travellers' diarrhoea, and are identified serologically (see above), but this is necessary only in outbreaks. It is not recommended that single cases of EPEC be investigated in this way.

ETEC strains cause gastroenteritis in both adults (travellers' diarrhoea) and children (especially in developing countries). The vehicles are contaminated water and food. They produce enterotoxins, one of which is heat labile (LT) and the other heat stable (ST).

EIEC strains cause diarrhoea similar to that in shigellosis. The strains associated with invasive dysenteric infections are less reactive than typical E. coli. They may be lysine negative, lactose negative and anaerogenic, and resemble shigellas. The correlation between serotype and pathogenicity is imperfect. Reference laboratories use tissue culture and other specialized tests.

VTEC strains derive their name from their cytotoxic effect on Vero cells in tissue culture. Their alternative name is enterohaemorrhagic E. coli (EHEC). They have been associated with haemolytic–uraemic syndrome and haemorrhagic colitis, and are known to be responsible for foodborne disease (see Chapter 13).

Several kits are available for detecting and identifying these pathogens: Phadebact UK, VET-RPLA (Oxoid), E. COLI ST EIA (Oxoid).

Serological testing of E. coli

Screen first by slide agglutination with the (commercially available) antisera. Heat saline suspensions of positives in a boiling water bath for 1 h to destroy *H* and *K* antigens. Cool, and do tube *O* agglutination tests. A single tube test at 1 : 50 (with a control) is adequate. Read at 3 h and then leave on the bench overnight and read again.

Serotype O157:H7 has been responsible for serious outbreaks in recent years. It does not ferment sorbitol and may be recognized on Sorbitol MacConkey agar and identified with commercial O157 antisera (Oxoid latex).

Kits for the identification of O157 strains include Microscreen (Microgen), Prolex (Prolab) and Captivate (LabM).

Citrobacter freundii

This organism is motile, indole variable, grows in citrate and KCN media, and is ONPG positive. It is H₂S positive but the lysine decarboxylase test is negative. Urea hydrolysis is variable. Non- or late-lactose-fermenting strains occur. This organism occurs naturally in soil. It can cause urinary tract and other infections in humans and animals.

Citrobacter koseri

This differs from *C. freundii* in that it is H₂S negative and does not grown in KCN medium. It is found in urinary tract infections and occasionally in meningitis.

Klebsiella pneumoniae

Also known as Friedlander's pneumobacillus, this is indole negative and gives a positive lysine decarboxylase reaction, but does not produce H₂S. It is urease and β-galactosidase positive. It may produce large, mucoid colonies on media that contain carbohydrates.

It is associated with severe respiratory infections and can cause urinary tract infections.

Klebsiella aerogenes

There has always been some confusion between this organism and *K. pneumoniae* because the latter name was used in the USA. It is now regarded as a subspecies of *K. pneumoniae*. *K. aerogenes* is gluconate positive but *K. pneumoniae* is gluconate negative.

Klebsiella oxytoca

This differs from *K. pneumoniae* in being indole positive.

Klebsiella ozoenae

Differs from *K. pneumoniae* in failing to hydrolyse urea. It is also difficult to identify, but is not uncommonly found in the respiratory tract associated with chronic destruction of the bronchi.

Klebsiella rhinoscleromatis

Differs from *K. pneumoniae* in that it does not grow in citrate media, is β-galactosidase negative and fails to decarboxylate lysine. It is difficult to identify, is rare in clinical material and may be found in granulomatous lesions in the upper respiratory tract.

Enterobacter cloacae

This is motile, indole negative, MR negative and VP positive, grows in citrate and liquefies gelatin (but this property may be latent, delayed and lost). It does not produce H₂S and the lysine decarboxylase test is negative. It is found in sewage and in polluted water.

Enterobacter aerogenes

This resembles *E. cloacae* but the lysine decarboxylase test is positive. Gelatin liquefaction is late. It is often confused with *K. aerogenes*. *E. aerogenes* is motile and urease negative; *K. aerogenes* is non-motile and urease positive (see Table 26.1).

THE NON-LACTOSE-FERMENTING ENTEROBACTERIA

For more detailed information, especially about the many 'new' species not considered above, see Barrow and Feltham (1993), Altwegg and Bockemühl (1999) and Holmes (1999).

REFERENCES

Altwegg, M. and Bockemühl, J. (1999) *Escherichia* and *Shigella*. In: Collier, L., Balows, A. and Sussman, M. (eds) *Topley and Wilson's Microbiology and Microbial Infections*, 9th edn. Vol 2, *Systematic Bacteriology*, CD-ROM. London: Arnold, Chapter 40.

Barrow, G. I. and Feltham, R .K. A. (eds) (1993) *Cowan and Steel's Manual for the Identification of Medical Bacteria*, 3rd edn. Cambridge: Cambridge University Press.

Donovan, T. J. (1966) A Klebsiella screening medium. *Journal of Medical Laboratory Technology* 23: 194–196.

Gorbach, S. L. (1986) *Infectious Diarrhoea*. London: Blackwells.

Gorbach, S. L. (1987) Bacterial diarrhoea and its treatment. *Lancet* ii: 1378–1382.

Holmes, B. (1999) The Enterobacteriaceae: general characteristics. In: Collier, L., Balows, A. and Sussman, M. (eds) *Topley and Wilson's Microbiology and Microbial Infections*, 9th edn. Vol. 2, *Systematic Bacteriology*, CD-ROM. London: Arnold, Chapter 39.

Perez, J. L., Berrocal, C. I. and Berrocal, L. (1986) Evaluation of a commercial beta-glucuronidase test for the rapid and economical identification of *Escherichia coli*. *Journal of Applied Bacteriology* 61: 541–545.

27

Edwardsiella, Hafnia and Serratia

Although most species in these genera do not usually ferment lactose, some workers include them in the coliforms.

They are Gram-negative, motile, oxidase-negative, fermentative (O/F test), non-sporing rods that are catalase and citrate positive, urease negative (with exceptions), and arginine dihydrolase and phenylalanine deaminase negative (see Table 25.1).

IDENTIFICATION

Test for indole, lysine decarboxylase, fermentation of mannitol and inositol, H_2S (with TSI), gelatin liquefaction and ONPG (Table 27.1).

SPECIES

Edwardsiella tarda

This is the only species of medical interest in the genus *Edwardsiella*. It is the only member of the group that produces indole and H_2S (in TSI) and that does not ferment mannitol. It is lysine decarboxylase positive and ONPG negative, and does not liquefy gelatin.

It is a fish pathogen and a rare pathogen of humans, when it may cause wound infection. Although it is occasionally isolated from patients with diarrhoea, its causative role is uncertain.

Hafnia alvei

This gives reliable biochemical reactions only at 20–22°C. It ferments mannitol, decarboxylates lysine, is ONPG positive, but is indole and H_2S negative and does not liquefy gelatin. It is a commensal in the intestines of animals, and present in water, sewage and soil although it is rarely, if ever, pathogenic.

Serratia marcescens

Once known as *Chromobacterium prodigiosus*, this is known for the red pigment it sometimes pro-

Table 27.1 *Edwardsiella, Hafnia* and *Serratia* species

Species	Indole production	Lysine decarboxylase	Acid from		H_2Sᵃ	Gelatin liquefaction	ONPG hydrolysis
			Mannitol	Inositol			
Edwardsiella tarda	+	+	–	–	+	–	–
Hafnia alvei	–	+	+	v	–	–	+
Serratia marcescens	–	v	+	+	–	+	+

ᵃ TSI, Triple Sugar Iron
v, variable

duces on agar media depending on incubation temperature. It ferments mannitol and inositol, liquefies gelatin and is ONPG positive, but does not produce indole or H_2S.

Originally thought to be a saprophyte, it was used in aerosol and filter testing experiments; it is now known to be associated with sepsis, hospital-acquired infections such as meningitis and endocarditis, and urinary tract infections. Several 'new' species of *Serratia* have emerged recently (Holmes and Aucken, 1998).

For more information about the biochemical reactions, etc. of these organisms see Farmer *et al.* (1985), Barrow and Feltham (1993), and Holmes and Aucken (1999).

REFERENCES

Barrow, G. I. and Feltham R. K. A. (1993) *Cowan and Steel's Manual for the Identification of Medical Bacteria*, 3rd edn. Cambridge: Cambridge University Press.

Farmer, J. J., Davis, B. R., Hickman-Brenner, F. W. *et al.* (1985) Biochemical identification of new species and biogroups of Enterobacteriaceae. *Journal of Clinical Microbiology* **21**: 46–76.

Holmes, B. and Aucken, H. M. (1999) *Citrobacter, Enterobacter, Klebsiella, Serratia* and other members of the Enterobacteriaceae. In: Collier, L., Balows, A. and Sussman, M. (eds), *Topley and Wilson's Microbiology and Microbial Infections*, 9th edn. Vol 2, *Systematic Bacteriology*, CD-ROM. London: Arnold, Chapter 42.

28

Salmonella and *Shigella*

These are important intestinal pathogens of mammals and, although they are not related generically, they are considered together in this chapter because the methods for their isolation use the same culture media and methods, and their identification is mainly by agglutination tests using polyvalent and specific antisera. An outline of their antigenic structures is therefore presented first.

SALMONELLA ANTIGENS AND NOMENCLATURE

The Kauffmann–White (see Threlfall *et al.*, 1999) system is used for the identification of salmonellas according to their antigenic components and the tables that are used for identification bear their names.

There are over 60 somatic or O antigens among the salmonellas and these occur characteristically in groups. Antigens 1–50 are distributed among groups A–Z. Subsequent groups are labelled 51–61. This enables more than 1700 *Salmonella* 'species' or serotypes to be divided into about 40 groups with the more common organisms in the first six groups, e.g. in Table 28.1 each of the organisms in group B possesses the antigens 4 and 12 but, although 12 occurs elsewhere, 4 does not. Similarly, in group C all the organisms possess antigen 6; some possess, in addition, antigen 7 and others antigen 8. In group D, 9 is the common antigen. Unwanted antigens may be absorbed from the sera produced against these organisms and single-factor O sera are available that enable almost any salmonella to be placed by slide agglutination into one of the groups. Various polyvalent sera are also available commercially.

To identify the individual organisms in each group, however, it is necessary to determine the H or flagella antigen. Most salmonellas have two kinds of H antigen and an individual cell may possess one or the other. A culture may therefore be composed of organisms, all of which have the same antigens or may be a mixture of both. The alternative sets of antigens are called *phases* and a culture may therefore be in phase 1, phase 2 or both phases simultaneously.

The antigens in phase 1 are given lower-case letters. Unfortunately, after z was reached more antigens were found and so subsequent antigens were named z_1, z_2, z_3, etc. It is important to note that z_1 and z_2 are as different as are a and b; they are not merely subtypes of z.

Phase 2 antigens are given the arabic numerals 1–7 but a few of the lettered antigens also occur in phase 2 cells, mostly in groups, e.g. as e, h; e, n, z_{15}; e, n, x; l, y, etc. Lettered antigens in phase 2, apart from these, are uncommon.

In identification procedures it is usual to find the O group first. Various polyvalent and single-factor O antisera are available commercially for this. The H antigens are then found with a commercial Rapid Salmonella Diagnostic (RSD) sera. These are mixtures of antisera that enable the phase 1 antigens to be determined. The phase 2 antigens are found with a polyvalent serum containing factors 1–7 and then individual sera.

Another antigen is the Vi, found in *S. typhi* and a few other species. This is a surface antigen that masks the O antigen. If it is present the organisms may not agglutinate with O sera unless the suspension is boiled.

As 'new' salmonellas are not infrequently reported, the Kauffmann–White scheme is brought

Table 28.1 Antigenic structure of some of the common salmonellae (Kauffmann–White classification)

Absorbed O antisera available for identification	Group	Name	Somatic (O) antigen	Flagellar (H) antigen Phase I	Phase 2
Factor 2	A	S. paratyphi A	1, 2, 12	a	–
Factor 4	B	S. paratyphi B	1, 4, 5, 12	b	1, 2
		S. stanley	4, 5, 12	d	1, 2
		S. schwarzengrund	4, 12, 27	d	1, 7
		S. saintpaul	1, 4, 5, 12	e, h	1, 2
		S. reading	4, 5, 12	e, h	1, 5
		S. chester	4, 5, 12	e, h	e, n, x
		S. abortus equi	4, 12	–	e, n, x
		S. abortus bovis	1, 4, 12, 27	b	e, n, x
		S. agona	1, 4, 12	gs	–
		S. typhimurium	1, 4, 5, 12	i	1, 2
		S. bredeney	1, 4, 12, 27	l, v	1, 7
		S. heidelberg	1, 4, 5, 12	r	1, 2
		S. brancaster	1, 4, 12, 27	z_{29}	–
Factor 7	C1	S. paratyphi C	6, 7, Vi	c	1, 5
		S. cholerae-suis[a]	6, 7	c	1, 5
		S. typhi-suis[a]	6, 7	c	1, 5
		S. braenderup	6, 7	e, h	e, n, z_{15}
		S. montevideo	6, 7	g, m, s	–
		S. oranienburg	6, 7	m, t	–
		S. thompson	6, 7	k	1, 5
		S. infantis	6, 7, 14	r	1, 5
		S. virchow	6, 7	r	1, 2
		S. bareilly	6, 7	y	1, 5
Factor 8	C2	S. tennessee	6, 8	z	–
		S. muenchen	6, 8	d	1, 2
		S. newport	6, 8	e, h	1, 2
		S. bovis morbificans	6, 8	r	1, 5
		S. hadar	6, 8	z_{10}	e, n, z
Factor 9	D	S. typhi	9, 12, Vi	d	–
		S. enteritidis	1, 9, 12	g, m	–
		S. dublin	1, 9, 12	g, p	–
		S. panama	1, 9, 12	lv	1, 5
Factors 3, 10	E1	S. anatum	3, 10	e, h	1, 6
		S. meleagridis	3, 10	e, h	l, w
		S. london	3, 10	l, v	1, 6
		S. give	3, 10	l, v	1, 7
Factor 19	E4	S. senftenburg	1, 3, 19	g, s, t	–
Factor 11	F	S. aberdeen	11	i	1, 2
Factors 13, 22	G	S. poona	13, 22	z	1, 6

Reproduced by permission of Dr F. Kauffman, formerly Director of the International Salmonella Centre using the original nomenclature. The complete tables, revised regularly, are too large for inclusion here. They are obtainable from Salmonella Centres.
[a] Identical serologically but differ biochemically

up to date every few years by the International Salmonella Centre.

For many years salmonellas were given species status, the specific epithet often being that of the places where they were first isolated. Although this served clinical bacteriologists and public health officers very well, it was not in accord with the rules of the *International Code of Nomenclature of Bacteria* and proposals were made by Popoff and Le Minor (1992) for a single species, with a newly coined name *Salmonella enterica*. Eventually, on the grounds of DNA hybridization studies, the *Salmonella* 'species' were reduced to two: *S. enterica* and *S. bongori*, with six subspecies of *S. enterica*: *enterica*, which includes the mammalian pathogens, and *salamae, arizonae, diarizonae, houtenae* and *indica*, all five of which are found in cold-blooded animals and the environment, and are not considered in this book.

The names of serotypes of subspecies of *S. enterica* are written in roman, not italic, and given upper case initials. Serotypes of the other subspecies no longer have names but are designated by their antigenic formula.

Fortunately it has been generally agreed that correct names such as *S. enterica* ssp. *enterica* serotype Typhimurium may now be written simply as *S.* Typhimurium.

Threlfall *et al.* (1999) give detailed accounts of these changes. Only serotypes in subspecies *enterica* are considered in this book.

SHIGELLA ANTIGENS

The genus *Shigella* is divided into four antigenic and biochemical subgroups, A–D (Table 28.2).

Subgroup A

Shigella dysenteriae

There are 10 serotypes that have distinct antigens and do not ferment mannitol.

Subgroup B

Shigella flexneri

This contains six serotypes (1–6) that can be divided into subserotypes according to their possession of some group factors designated: 3,4; 4; 6; 7; and 7,8.

The X variant is an organism that has lost its type antigen and is left with the group factors 7,8. The Y variant is an organism that has lost its type antigen and is left with the group factors 3,4.

Subgroup C

Shigella boydii

There are 15 serotypes with distinct antigens; all ferment mannitol.

Subgroup D

Shigella sonnei

This subgroup has only one, distinct serotype; this ferments mannitol. It may occur in phase 1 or phase 2, sometimes referred to as 'smooth' and 'rough'. The change from phase 1 to phase 2 is a loss of variation and phase 2 organisms are often

Table 28.2 *Shigella* classification (see also Table 28.4)

Subgroup	Species	Serotypes
A: mannitol not fermented	S. dysenteriae	1–10 all distinct
B: mannitol fermented	S. flexneri	1–6 all related; 1–4 divided into subserotypes
C: mannitol fermented	S. boydii	1–15 all distinct
D: mannitol fermented, lactose fermented late	S. sonnei	1 serotype

reluctant to agglutinate or give a very fine, slow agglutination. Phase 2 organisms are rarely encountered in clinical work but the sera supplied commercially agglutinate both phases.

ISOLATION OF SALMONELLAS AND SHIGELLAS

Choice of media

Liquid media

These are generally inhibitory to coliforms but less so to salmonellas. Selenite F media are commonly used. If they are not overheated selenite F media will permit the growth, but not the enrichment, of *S. sonnei* and *S. flexneri* 6. Mannitol selenite and cystine selenite broths are preferred by some workers for the isolation of salmonellas from foods. Tetrathionate media enrich only salmonellas; shigellas do not grow.

Cultures from these media are plated on solid media.

Solid media

Many media exist for the isolation of salmonellas and shigellas but in our experience the Hynes modification of deoxycholate citrate agar (DCA) and xylose lysine deoxycholate agar (XLD) give the best results for faeces. For the isolation of salmonellas from foods, XLD and brilliant green agar (BGA) are superior to DCA.

Colony appearance

Deoxycholate citrate agar

On the ideal DCA medium, colonies of salmonellas and shigellas are not sticky. Salmonella colonies are creamy brown, 2–3 mm in diameter at 24 h and usually have black or brown centres. Unfortunately, colonies of *Proteus* spp. are similar. Shigella colonies are smaller, usually slightly pink and do not have black centres. *S.* Cholerae-suis grows poorly on this medium. White, opaque colonies are not significant. Some strains of *Klebsiella* grow.

Xylose lysine deoxycholate agar

Salmonella colonies are red with black centres, 3–5 mm in diameter. Shigella colonies are red, 2–4 mm in diameter.

Brilliant Green Agar

Salmonella colonies are pale pink with a pink halo. Other organisms have yellow–green colonies with a yellow halo.

Procedure

Faeces, rectal swabs and urines

Plate faeces and rectal swabs on DCA and XLD agars. Inoculate selenite F broth with a portion of stool about the size of a pea or with 0.5–1.0 ml liquid faeces. Break off rectal swabs in the broth. Add about 10 ml urine to an equal volume of double-strength selenite F broth.

Incubate plates and broths at 37°C for 18–24 h. Plate broth cultures on DCA, XLD or BGA and incubate at 37°C for 18 h.

Drains and sewers

Fold three or four pieces of cotton gauze (approximately 15 × 20 cm) into pads of 10 × 4 cm. Tie with string and enclose in small-mesh chicken wire to prevent sabotage by rats. Autoclave in bulk and transfer to individual plastic bags or glass jars. Suspend in the drain or sewer by a length of wire and leave for several days. Squeeze out the fluid into a jar and add an equal volume of double-strength selenite F broth. Place the pad in another jar containing 200 ml single-strength selenite F broth. Incubate both jars at 43°C for 24 h and proceed as for faeces described by Moore (1948) (see also Vassiliades *et al.*, 1978).

Foods, feedingstuffs and fertilizers

The method of choice depends on the nature of the sample and on the total bacterial count. In general, raw foods such as comminuted meats have very high total counts and salmonellas are likely to be present in large numbers. On the other hand, heat-treated and deep-frozen foods should have lower counts and if any salmonellas are present they may suffer from heat shock or cold shock and require resuscitation and pre-enrichment.

Pre-enrichment

Make a 1 : 10 suspension or solution of the sample in buffered peptone water. Use 25 g food if possible. For dried milk make the suspension in 0.002% brilliant green in sterile water. Use a Stomacher if necessary. Incubate at 37°C for 16–20 h.

Enrichment

Add 10 ml of the pre-enrichment to 100 ml Muller–Kauffman tetrathionate broth (MK–TB), mixing well so that the calcium carbonate stabilizes the pH. Incubate at 43°C for up to 48 h.

Selective plating

At 18–24 h and 42–48 h subculture from the surface of the MK–TB to brilliant green phenol red agar. Incubate at 37°C for 22–24 h.

Salmonella colonies are pink, smooth, low and convex. Subculture at least two colonies for identification.

Additional procedures

Add 0.1 ml of the pre-enrichment to 10 ml Rappaport–Vasiliades (RV) enrichment medium. Subculture at intervals as with MK–TB. This method may increase isolation rates, as will additional cultures on DCA and XLD.

Many rapid and automated methods have been developed for determining the presence of salmonellas in foods. There is a variety of commercially available equipment and instrumentation.

IDENTIFICATION

Kit methods, for screening materials for salmonellas, have proliferated and are useful if all one needs to know is whether these organisms (any serotype) are present, and in epidemic circumstances when large numbers of specimens must be tested. They include: the Tecra (Tecra Diagnostics, USA) which is an immunocapture method; Spectate (Rhône Poulenc, France), coloured latex particles with specific antibodies; Salmonella Rapid (Oxoid, UK), which uses a purpose-designed culture vessel; Rapid Salmonella Latex (Oxoid), latex agglutination; Microscreen (Mercia, UK); and MUCAP, which uses a fluorogenic substrate for testing colonies.

For full identification of salmonellas and shigellas to species/serotype, there are two approaches: agglutination tests supported by screening or biochemical tests; and biochemical tests followed by agglutination tests. The first of these requires considerable expertise in the recognition of colonies and in agglutination procedures, but gives an answer much earlier, often within 24 h from primary plates but certainly by 48 h after enrichment.

Agglutination tests for salmonellas

Polyvalent, specific and RSD sera are available commercially. The last consist of three or four mixtures of specific and non-specific sera; agglutination by combinations indicates the presence of more common antigens or groups of antigens.

Test the suspected organism by slide agglutination with polyvalent O sera. If positive, do further slide agglutinations with the relevant single-factor O sera; only one should be positive. If in doubt, test against 1 : 500 acriflavine; if this is positive, the organisms are unlikely to be salmonellas.

Subculture into glucose broth and incubate in a water bath at 37°C for 3–4 h, when there will be sufficient growth for tube H agglutinations. Tube H agglutinations are more reliable than slide H agglutinations because cultures on solid media may not be motile. The best results are obtained when there are about 6–8×10^8 organisms/ml. Dilute, if necessary, with formol saline. Add 0.5 ml formalin to 5 ml broth culture and allow to stand for at least 10 min to kill the organisms.

Place one drop of polyvalent (phase 1 and 2) salmonella *H* serum in a 75 × 9 mm tube. Add 0.5 ml of the formolized suspension and place in a water bath at 52°C for 30 min. If agglutination occurs, the organism is probably a salmonella.

Identify the *H* antigens as follows. Do tube agglutinations against the RSD sera, *S*. Typhimurium (*i*) if it is not included in the RSD set, phase 2 complex 1–7 EN (*e, n*) complex and L (*l*) complex. The manufacturer's guide will indicate which phase 1 antigens are present according to which sera or groups of sera give agglutination. Check the result, if possible, by doing tube agglutinations with the indicated monospecific serum. If the test indicates EN (*e, n*), G or L (*l*) complexes, send the cultures to a public health or communicable diseases laboratory.

If agglutination occurs in the phase 2 1–7 serum, test each of the single factors, 2, 5, 6 and 7. Note that, with some commercial sera, factor *1* is not absorbed and it may be necessary to dilute this out by using one drop of 1 : 10 or weaker serum.

Changing the *H* phase

If agglutination is obtained with only one phase, the organism must be induced to change to the other phase. Cut a 50 × 20 mm ditch in a well-dried nutrient agar plate. Soak a strip of previously sterilized filter paper (36 × 7 mm) in the *H* serum by which the organism is agglutinated, and place this strip across the ditch at right angles. At one end, place one drop of 0.5% thioglycollic acid to neutralize any preservative in the serum. At the other end, place a filter-paper disc, about 7 mm in diameter, so that half of it is on the serum strip and the other half on the agar. Inoculate the opposite end of the strip with a young broth culture of the organism and incubate overnight. Remove the disc with sterile forceps, place it in glucose broth and incubate it in a water bath for about 4 h, when there should be enough growth to repeat agglutination tests to find the alternative phase. Organisms in the original phase demonstrated will have been agglutinated on the strip. Organisms in the alternate phase will not be agglutinated and will have travelled across the strip.

The phase may also be changed with a Craigie tube. Place 0.1 ml serum and 0.1 ml 0.5% thioglycollic acid in the inner tube of a Craigie tube, and inoculate the inner tube with the culture. Incubate overnight and subculture from the outer tube into glucose broth.

Some organisms, e.g. *S*. Typhi, have only one phase. These should be sent to a reference laboratory.

Antigenic formulae

List the O, phase 1 and phase 2 antigens in that order and consult the Kauffmann–White tables for the identity of the organism. In the table, the antigenic formula is given in full, but as single factor sera for identification it is usually written thus:

S. Typhimurium	4, *i*, 2
S. Typhi	9, *d*, —
S. Newport	6, 8, *e*, *h*, 2

(The — sign indicates that the organism is monophasic.)

The *Vi* antigen

If this antigen is present and prevents the organisms agglutinating with O group sera, make a thick suspension in saline from an agar slope and boil it in a water bath for 1 h. This destroys the *Vi* and permits the O agglutination.

Biochemical screening tests

Pick representative colonies on a urea slope as early as possible in the morning. Incubate in a water bath at 35–37°C. In the afternoon, reject any urea-positive cultures and from the others inoculate the following: Triple Sugar Iron (TSI) medium by spreading on slope and stabbing the butt; broth for indole test; and lysine decarboxylase medium. Incubate overnight at 37°C (Table 28.3) and proceed to serological identification if indicated.

Full biochemical characters

For salmonella serotypes and subgenera it is best to use kits such as the API 20E, Mast ID-15 (Mast,

Table 28.3 Biochemical reactions of Salmonella, Shigella, etc.

Organism	TSI[a] Butt	TSI[a] Slope	H₂S[b] production	Indole production	ONPG hydrolysis	Lysine decarboxylase
S. typhi	A	−	+[b]	−	−	+
S. paratyphi A	AG	−	−	−	−	−
Other salmonellas	AG	−	±	−	−	+
Shigella	A	−	−	±	±	−
Citrobacter	AG	−	+	−	+	−

A, acid; AG, acid and gas

[a] In TSI medium acid in the butt indicates the fermentation of glucose and in the slope that of lactose and/or sucrose

[b] H₂S production by S. typhi may be minimal

UK), Cobas ID and Enterotube (Roche, UK). See Holmes and Costas (1992) for more details of these.

Salmonella species and serotypes

Salmonella Typhi, and in some countries S. Paratyphi A (see national lists) are in Hazard Group 3. If tests suggest that a suspect organism is one of these all further investigations should be conducted under Containment Biosafety Level 3 conditions.

Salmonella Typhi

This produces acid but no gas from glucose and mannitol. It may produce acid from dulcitol but fails to grow in citrate media and does not liquefy gelatin. It decarboxylates lysine but not ornithine and fails to hydrolyse ONPG. Freshly isolated strains may not be motile at first. This organism causes enteric fever.

Other serotypes

These produce acid and usually gas from glucose, mannitol and dulcitol (anaerogenic strains occur), and rare strains ferment lactose. They grow in citrate and do not liquefy gelatin. They decarboxylate lysine and ornithine (except S. Paratyphi A) and fail to hydrolyse ONPG. S. Paratyphi A, B and C may also cause enteric fever. Strains with the antigenic formula 4; b; 1, 2 may be S. Paratyphi B, usually associated with enteric fever, or S. Java which is usually associated with food poisoning. Many other serotypes cause food poisoning.

Reference laboratories

It is not possible to identify all of the hundreds of salmonella serotypes with the commercial sera available, or in the smaller laboratory. The following organisms should be sent to a reference or communicable diseases laboratory:

- Salmonellas placed by RSD sera into group E, G or L.
- Salmonellas agglutinated by polyvalent *H* or monospecific *H* sera but not by phase II sera.
- Salmonellas agglutinated by polyvalent *H* but not by phase I or phase II sera.
- Salmonellas giving the antigenic formula 4; b; 1, 2, which may be either S. Paratyphi B or a subtype associated with food poisoning.
- S. Typhi, S. Paratyphi B, S. Typhimurium, S. Thompson, S. Virchow, S. Hadar and S. Enteritidis should be sent for phage typing.

Note that *Brevibacterium* spp., which are sometimes found in foods, possess salmonella antigens. These organisms are Gram positive.

Agglutination tests for shigellas

Do slide agglutinations with *S. sonnei* (phase 1 and 2), polyvalent Flexner and polyvalent Boyd antisera. If these are negative, test with *S. dysenteriae* antiserum. If there is agglutination with the polyvalent antisera test with the individual antisera: Flexner 1–6 and Boyd 1–5.

Sometimes there is no agglutination although biochemical tests indicate shigellas. This may result from masking by a surface (*K*) antigen. Boil a suspension of the organisms and repeat the tests.

Strains that cannot be identified locally, i.e. give the correct biochemical reactions but are not agglutinated by available sera, should be sent to a reference laboratory.

Note that the motile organism *Plesiomonas shigelloides* (see p. 279) may be agglutinated by *S. sonnei* antiserum.

Biochemical tests

Test for acid production from glucose, lactose, sucrose, dulcitol and mannitol, indole production, lysine and ornithine decarboxylases (Table 28.4). Alternatively use one of the kits for enteric pathogens.

Species of *Shigella*

Shigella dysenteriae (S. shiga)

This forms acid but no gas from glucose, and does not ferment mannitol. Types 1, 3, 4, 5, 6, 9 and 10 are indole negative. Types 2, 7 and 8 are indole positive and were known as *S. schmitzii* (*S. ambigua*). *S. dysenteriae* is responsible for the classic bacillary dysentery of the Far East.

Shigella flexneri

Serotypes 1–5 form acid but no gas from glucose and mannitol and produce indole. Some type 6 strains (Newcastle strains) may not ferment mannitol and does not produce indole. A bubble of gas may be formed in the glucose tube. This serotype is not inhibited by selenite media. Serotypes 1–6 are widely distributed, especially in the Mediterranean area, and often cause outbreaks in mental and geriatric hospitals, nurseries and schools in temperate climes.

Shigella flexneri subserotypes

Some manufacturers do not absorb group 6 factor from their type 3 serum and they also provide X and Y variant sera containing group factors 7, 8 and 3, 4. These sera may therefore be used to determine the Flexner subserotypes for epidemiological purposes (Table 28.4).

Shigella sonnei

This forms acid but no gas from glucose and mannitol, is indole negative and decarboxylates

Table 28.4 Shigellas

Species	Acid from					Indole production	Lysine decarboxylase	Ornithine decarboxylase
	Glucose	Lactose	Sucrose	Dulcitol	Mannitol			
S. dysenteriae	+	–	–	–	–	–/+	–	–
S. flexneri 1–5	+	–	–	–	+	v	–	–
S. flexneri 6 (Boyd 88)	+ (G)	–	–	–	+/–	–	–	–
S. sonnei	+	(+)	(+)	–	+	–	–	–
S. boydii 1–15	+	–	–	–	+	+/–	–	–

+ (G), a small buble of gas may be formed; –/+, most strains negative; +/–, most strains positive

None of these organisms grow in citrate or ferment salicin; none produce H$_2$S, liquefy gelatin or are motile ornithine. It causes the most common and mildest form of dysentery, which mostly affects babies and young children, and spreads rapidly through schools and nurseries. It survives in some selenite media.

Shigella boydii

This forms acid but no gas from glucose and mannitol; indole production is variable. The 15 serotypes of this species are widely distributed but not common and cause mild dysentery.

REFERENCES

Holmes, B. and Costas, M. (1992) Methods and typing of Enterobacteriacea by computerized methods. In: Board, R. G., Jones, D. and Skinner, F. A. (eds), *Identification Methods in Applied and Environmental Microbiology*. Society for Applied Bacteriology Technical Series No. 29. Oxford: Blackwells, pp. 127–150.

Moore, B. (1948) The detection of paratyphoid B carriers by means of sewage examination. *Monthly Bulletin Ministry of Health and PHLS* 6: 241–251.

Popoff, M. Y. and Le Minor, L. (1992) Antigenic formulas of the salmonella serovars, 6th rev. WHO Collaborating Centre for Reference and Research on Salmonella. Paris: Institut Pasteur.

Threlfall, J., Ward, L. and Old, D. (1999) Changing the nomenclature of salmonella. *Communicable Disease and Public Health* 2: 156–157.

Vassiliades, P., Trichopoulos, D., Kalandidi, A. and Xirouchaki, E. (1978) Isolation of salmonellas from sewage with a new enrichment method. *Journal of Applied Bacteriology* 44: 233–239.

29

Proteus, Providencia and *Morganella*

These organisms are widely distributed in nature. Some strains of *Proteus* spread over the surface of ordinary agar media ('swarming'), obscuring colonies of other organisms. This may be prevented by using chloral hydrate agar. Swarming does not take place on cystine lactose electrolyte-deficient (CLED) media and media containing bile salts.

The important characteristic that distinguishes these genera from other enterobacteria is their ability to deaminate phenylalanine (positive PPA test). Not all species are described here. For information about others, see Senior (1999).

ISOLATION

Plate material on nutrient or blood agar to observe swarming, and on CLED and chloral hydrate agar to prevent swarming and reveal any other organisms. Incubate at 37°C overnight. Add heavily contaminated material to tetrathionate broth, incubate overnight and plate as above.

IDENTIFICATION

Test PPA positive strains for motility, indole production, fermentation of mannitol and inositol, citrate utilization, H_2S production, gelatin liquefaction and urease (Table 29.1).

Proteus species

Proteus vulgaris

This is indole positive, produces H_2S (lead acetate paper test) but does not ferment mannitol or inositol. Many strains swarm on ordinary media (see above).

It is found in the intestines of humans and animals and is an opportunist pathogen in the urinary tract, wounds and other lesions. It is a food spoilage organism, widely distributed in soil and sewage-polluted water.

Table 29.1 *Proteus, Providencia* and *Morganella* species

Species	Motility	Indole production	Acid from		Citrate utilization	H_2S production	Gelatin liquefaction	Urease
			Mannitol	Inositol				
Proteus vulgaris	v	+	–	–	v	+	v	+
P. mirabilis	+	–	–	–	+	+	+	+
Providencia alcalifaciens	v	+	–	–	+	v	–	–
P. rettgeri	+	+	+	+	+	+	–	+
P. stuartii	v	+	–	+	+	–	–	–
Morganella morganii	+	+	–	–	v	+	–	+
v, variable								

Proteus mirabilis

Differs mainly from *P. vulgaris* in not producing indole. It is a doubtful agent of gastroenteritis, found in urinary tract and hospital-acquired infections, and septic lesions. It is associated with putrefying animal and vegetable matter.

Providencia species

Providencia alcalifaciens

This differs from *Proteus* spp. in not possessing urease. It is found in clinical material where it may be an opportunist pathogen.

Providencia rettgeri

This ferments mannitol and inositol, produces H_2S (on lead acetate paper but not in Triple Sugar Iron [TSI]) and is urease positive. It is associated with 'fowl typhoid', and urinary tract and opportunist infections in humans.

Providencia stuartii

Ferments inositol but not mannitol, may produce H_2S and is urease negative. It has been reported in hospital infections, especially in catheterized elderly people

Morganella species

Morganella morganii

This is the only species so far described. It does not ferment mannitol or inositol but fails to produce H_2S and is urease positive. It is a commensal in the intestines of humans and other animals, found in hospital infections and may be associated with 'summer diarrhoea' in children.

REFERENCE

Senior, B. W. (1999) *Proteus, Morganella* and *Providencia*. In: Collier L, Balows, A. and Sussman, M. (eds), *Topley and Wilson's Microbiology and Microbial Infections*, 9th edn. Vol. 2, *Systematic Bacteriology*, CD-ROM. London: Arnold, Chapter 43.

30

Key to some miscellaneous aerobic, non-sporing, Gram-negative bacilli of medical importance

The 13 genera considered here may be divided into four groups, on the basis of common characteristics.

1. *Brucella, Bordetella* and *Moraxella* spp. are non-motile, cannot grow anaerobically, are catalase and oxidase positive, do not change Hugh and Leifson medium, and have no special growth requirements (see Chapter 31).
2. *Haemophilus, Gardnerella* and *Streptobacillus* spp. are non-motile, can grow anaerobically, are fermentative in Hugh and Leifson medium, and have special growth requirements (see Chapter 32).
3. *Campylobacter* and *Helicobacter* spp., which are motile, can grow anaerobically, do not change Hugh and Leifson medium, and have no special growth requirements, although they are microaerophilic (see Chapter 33).
4. *Actinobacillus, Pasteurella, Yersinia, Cardiobacterium* and *Francisella* spp. are non-motile, can grow anaerobically, are fermentative in Hugh and Leifson medium, and have no special growth requirements (see Chapter 34).

Table 30.1 shows the general characteristics. Those that assist in identification within the groups are given in the appropriate chapters.

Table 30.1 Gram-negative non-sporing bacilli

Genus	Motility	Facultative anaerobe	Catalase	Oxidase	O/F[a]	Growth factor requirement
Brucella	−	−	+	+[b]	−	−
Bordetella	−	−	+	v	−	−
Moraxella	−	−	+	+	−	−
Haemophilus	−	+	v	+	F	+
Gardnerella	−	+	−	−	F	+
Streptobacillus	−	+	−	−	F	+
Campylobacter	+	v[d]	v	+	−	−
Helicobacter	+	v[d]	+	+	−	−
Actinobacillus	−	+	+	+[c]	F	−
Pasteurella	−	+	+	+	F	−
Yersinia	−	+	+	−	F	−
Cardiobacterium	−	+	−	+	F	−
Francisella	−	+	+	v	F	−

[a] Oxidation/fermentation test, Hugh and Leifson medium
[b] Usually
[c] Except *A. actinomecetemcomitans*
[d] micro-aerophilic
v, variable

31

Brucella, Bordetella and Moraxella

BRUCELLA

The genus Brucella *contains Hazard Group 3 pathogens. All manipulations, especially those that might produce aerosols, should be done in microbiological safety cabinets in Biosafety/Containment Level 3 laboratories.*

There are three important and several other species of small, Gram-negative bacilli in this genus. They are non-motile and reduce nitrates to nitrites, but carbohydrates are not metabolized when normal cultural methods are used. There are several biotypes within some species.

Isolation

Human disease

Do blood cultures. Any system may be used but the Casteñada double-phase method (see p. 157) in which the solid and liquid media are in the same bottle lessens the risk of contamination during repeated subculture, and also minimizes the hazards of infection of laboratory workers. Use Brucella agar plus growth and selective supplements.

Examine Casteñada bottles weekly for growth on the solid phase. If none is seen, tilt the bottle to flood the agar with the liquid, restore it to a vertical position and reincubate. Do not discard until 6 weeks have elapsed.

Treat bone marrow and liver biopsies in the same way. Subculture on brucella-selective media and on serum glucose agar or chocolate blood agar.

Animal material

Plate uterine and cervical swabbings or homogenized fetal tissue on Brucella-selective medium

Milk

Dip a throat swab in the gravity cream and inoculate brucella-selective medium. Incubate all cultures at 37°C under 10% CO_2 and examine daily for up to 5 days.

Identification

Colonies on primary media are small, flat or slightly raised and translucent. Subculture on slopes of glucose tryptone agar with a moistened lead acetate paper in the upper part of the tube to test for hydrogen sulphide production.

Do dye inhibition tests. In reference laboratories several concentrations of thionin and basic fuchsin are used to identify biotypes. For diagnostic purposes inoculate tubes of brucella-selective medium or glucose tryptone serum agar containing (1) 1:50 000 thionin and (2) 1:50 000 basic fuchsin (National Aniline Division, Allied Chemical and Dye Corp.). Inoculate tubes or plates with a small loopful of a 24-h culture.

It is advisable to control Brucella identification with reference strains, obtainable from Type Culture Collections or CDCs: *B. abortus* 554, *B. melitensis* 16M and *B. suis* 1330 (Table 31.1).

Do slide agglutination tests with available commercial sera.

Table 31.1 *Brucella* species

Species	Growth in 1 : 50 000		Needs CO$_2$	H$_2$S
	Thionin	**Basic fuschsin**		
B. melitensis	+	+	−	−
B. abortus	−[a]	+[a]	+[a]	+[a]
B. suis biogroup I	+	−[a]	−	+
B. suis other biogroups	+	−[a]	−	−
B. canis	+	−[a]	−	−

[a] Most strains

Species of *Brucella*

Brucella melitensis

This grows on blood agar and on glucose tryptone serum agar aerobically in 3–4 days. It does not need carbon dioxide to initiate growth. It does not produce hydrogen sulphide and is inhibited by 1 : 50 000 fuchsin or thionin. This organism causes brucellosis in humans, the Mediterranean or Malta fever. The reservoirs are sheep and goats, and infection occurs by drinking goats' milk. There are three biotypes.

Brucella abortus

Grows on blood agar and on glucose tryptone serum in 3–4 days. Most strains require 5–10% carbon dioxide to initiate growth. Strains produce hydrogen sulphide, and vary in their susceptibility to the presence of 1 : 50 000 thionin and fuchsin. There are eight biotypes, some of which resemble *B. melitensis*.

Brucella abortus causes contagious abortion in cattle. Drinking infected milk can result in undulant fever in humans. Veterinarians and stockmen are frequently infected from aerosols released during birth or abortion of infected animals. Laboratory infections are usually acquired from aerosols released by faulty techniques.

Brucella suis

The five biotypes of this species do not require carbon dioxide for primary growth. Biotype 1 pro-duces abundant hydrogen sulphide; the others do not. None is inhibited by 1 : 50 000 thionin but they vary in their inhibition by 1 : 50 000 fuchsin. They cause contagious abortion in pigs and may infect humans, reindeer and hares.

Of the other three species (not considered here), *B. ovis* (sheep), *B. canis* (dogs) and *B. neotomae* (desert wood rats), only *B. canis* appears to have caused disease in humans (Corbel and MacMillan, 1999).

Detection in milk

The milk ring test

The antigen for this test may be obtained from a veterinary or commercial laboratory. It is a suspension of *B. abortus* cells stained with haematoxylin or another dye

The test may be applied to bulk raw milk samples or to those from individual animals. In the latter case take samples from all four quarters of the udder. Store the milk samples overnight at 4°C before testing.

Add one drop (0.03 ml) of the stained antigen to 1 ml well-mixed raw milk in a narrow tube (75 × 9 mm). Mix immediately by inverting several times and allow to stand for 1 h at 37°C.

If the milk contains antibodies, these will agglutinate the antigen and the stained aggregates of bacilli will rise with the cream, giving a blue cream line above a white column of milk. Weak positives give a blue cream line and a blue colour in the milk.

A white cream line above blue milk shows absence of antibodies. It may be necessary to add known negative cream to low-fat milk.

False positives may be obtained with milk collected at the beginning and end of lactation, probably as a result of leakage of serum antibodies into the milk. Recent immunization with the S19 strain may affect results.

The ring test does not give satisfactory results with pasteurized milk or with goats' milk.

Whey agglutination test

This is a useful test when applied to milk from individual cows but is of questionable value for testing bulk milk.

Centrifuge quarter milk and remove the cream. Add a few drops of renin to the skimmed milk and incubate at 37°C for about 6 h. When coagulated, centrifuge and set up doubling dilutions of the whey from 1:10 to 1:2560, using 1-ml amounts in 75 × 9 mm tubes. Add one drop of standard concentrated *B. abortus* suspension and place in a water bath at 37°C for 24 h. Read the agglutination titre. More than 1:40 is evidence of udder infection unless the animal has been immunized recently.

Serological diagnosis

Apart from the standard agglutination test (above) there are others that are outside the scope of this book. They include complement fixation tests and enzyme-linked immunosorbent assay (ELISA). Corbel and Hendry (1985) give useful information about these.

For general reviews on the laboratory diagnosis of brucellosis, see Corbel (1998) and Corbel and MacMillan (1999).

BORDETELLA

The bacilli in this genus are small, 1.5 × 0.3 μm and regular. Nitrates are not reduced and carbohydrates are not attacked.

Isolation and identification

Pernasal swabs (preferably of calcium alginate) are better than cough plates, but the swabs should be sent to the laboratory in one of the commercial transport media. Plate on Bordetella, or Charcoal Agar medium containing cephalexin 40 mg/l (Oxoid Bordetella supplement).

Incubate under conditions of high humidity and at 37°C for 3 days. Examine daily and identify by slide agglutination, using commercially available sera. FA reagents are also available.

Test for growth and pigment formation on nutrient agar, urease, nitrate reductase and motility (Table 31.2).

Table 31.2 *Bordetella* species

Species	Growth on nutrient agar	Brown coloration	Urease	Nitrate reduction	Motility
B. pertussis	−	−	−	−	−
B. parapertussis	+	+	+	−	−
B. bronchiseptica	+	−	+	+	+

Species of *Bordetella*

Bordetella pertussis

Growth on the above media has been described as looking like a 'streak of aluminium paint'. Colonies are small (about 1 mm in diameter) and pearly grey. This organism cannot grow in primary culture without heated blood, but may adapt to growth on nutrient agar on subculture. *B. pertussis* is oxidase positive, nitrate and urease negative. It does not grow on MacConkey agar and shows β-haemolysis on blood agar.

Bordetella pertussis causes whooping cough. For detailed information about the bacteriological diagnosis of this disease, see Wardlaw (1990) and Parton (1998).

Bordetella parapertussis

Growth on blood agar may take 48 h and a brown pigment is formed under the colonies. It grows on nutrient and on MacConkey agar. On Bordetella or similar media, the pearly colonies ('aluminium paint') develop earlier than those of *B. pertussis*. It does not change Hugh and Leifson medium and does not metabolize carbohydrates. It is oxidase and nitrate negative but urease positive.

This organism is one of the causative organisms of whooping cough.

Bordetella bronchiseptica

This organism has been in the genera *Brucella* and *Haemophilus*. It forms small smooth colonies, occasionally haemolytic on blood agar, grows best at 37°C, and is motile, oxidase, nitrate and urease positive.

It causes bronchopneumonia in dogs, often associated with distemper, bronchopneumonia in rodents and snuffles in rabbits. It has been associated with whooping cough.

This genus is discussed in detail by Pittman and Wardlaw (1981), Wardlaw (1990), and Parton (1998).

MORAXELLA

The bacilli are plump (2×1 µm), often in pairs end to end, and non-motile. They are oxidase positive and do not attack sugars. They do not produce indole, do not produce hydrogen sulphide and are sensitive to penicillin. Some species require enriched media.

Plate exudate, conjunctival fluid, etc., on blood agar and incubate at 37°C overnight. Subculture colonies of plump, oxidase-positive, Gram-negative bacilli on blood agar and nutrient agar (not enriched), on gelatin agar (use the plate method) and on Loeffler medium and incubate at 37°C. Do the nitrate reduction and catalase tests (Table 31.3).

Table 31.3 *Moraxella* species

Species	Growth on nutrient agar	Gelatin liquefaction	Nitrate reduction	Catalase	Urease
M. lacunata	v	+	+	+	−
M. nonliquefaciens	v	−	+	+	−
M. bovis	+	+	v	+	−
M. osloensis	+	−	v	+	−
M. catarrhalis[a]	+	−	v		−
K. kingii[b]	+	−	−	−	−

[a] Formerly *Branhamella*
[b] *Kingella*, formerly *Moraxella*
v, variable

Species of *Moraxella*

Moraxella lacunata

Colonies on blood agar are small and may be haemolytic. There is no or very poor growth on non-enriched media. Colonies on Loeffler medium are not visible but are indicated by pits of liquefaction ('lacunae'). Gelatin is liquefied slowly. Nitrates are reduced.

This organism is associated with angular conjunctivitis and is known as the Morax–Axenfeld bacillus.

Moraxella liquefaciens

This is similar to *M. lacunata* and may be a biotype of that species. It liquefies gelatin rapidly.

Moraxella nonliquefaciens

This is also similar to *M. lacunata* but fails to liquefy gelatin.

Moraxella bovis

This requires enriched medium, liquefies gelatin and may reduce nitrates. It causes pink eye in cattle but has not been reported in human disease.

Moraxella osloensis (M. duplex, Mima polymorpha var. oxidans)

An enriched medium is not required. Gelatin is not liquefied and nitrates may be reduced. It is found on the skin, and in the eyes and respiratory tract of humans, but its pathogenicity is uncertain.

Moraxella kingii (now Kingella kingii)

Colonies are haemolytic and may be mistaken for haemolytic streptococci or haemolytic haemophilus. It does not require enriched media. This is the only member of the group that is catalase negative. It has been found in joint lesions and in the respiratory tract.

Moraxella catarrhalis

Formerly known as *Branhamella* (and also as *Neisseria*) *catarrhalis*, this grows well on blood agar, forming non-pigmented, non-haemolytic colonies. It is β-lactamase and DNase positive (Buchanan, 1998). Carbohydrates are not fermented; oxidase and catalase tests are negative. It also degrades tributyrin (a commercial butylase strip test is available).

Moraxella catarrhalis is found in human material. Previously regarded as a commensal of the respiratory tract it is now considered to be an opportunist pathogen of the upper and lower respiratory system and is associated with lung abscesses. Middle-ear infections are not uncommon.

REFERENCES

Buchanan, B. K. (1998) *Moraxella, Branhamella, Kingella* and *Eikenella*. In: Collier L, Balows, A. and Sussman, M. (eds), *Topley and Wilson's Microbiology and Microbial Infections*, 9th edn. Vol. 2, *Systematic Bacteriology*. London: Arnold, Chapter 48.

Corbel, M. J. (1998) *Brucella*. In: Collier L, Balows, A. and Sussman, M. (eds), *Topley and Wilson's Microbiology and Microbial Infections*, 9th edn. Vol. 2, *Systematic Bacteriology*. London: Arnold, Chapter 35.

Corbel, M. J. and Hendry, D. (1985) Brucellas. In: Collins, C. H. and Grange, J. M. (eds), *Isolation and Identification of Micro-organisms of Medical Importance*. Society for Applied Bacteriology Technical Series No. 21. London: Academic Press, pp. 53–82.

Corbel, M. J. and MacMillan, A. P. (1999) Brucellosis. In: Collier L, Balows, A. and Sussman, M. (eds), *Topley and Wilson's Microbiology and Microbial Infections*, 9th edn. Vol. 3, *Bacterial Infections*, CD-ROM. London: Arnold, Chapter 41.

Parton, R. (1998) *Bordetella*. In: Collier L, Balows, A. and Sussman, M. (eds), *Topley and Wilson's Microbiology and Microbial Infections*, 9th edn. Vol. 2, *Systematic Bacteriology*. London: Arnold, Chapter 38.

Pittman, M. and Wardlaw, A. C. (1981) The genus *Bordetella*. In: Starr, M. P., Stolp, H., Truper, G., Balows, A. and Schlegel, A. (eds), *The Prokaryotes: A handbook of habitats, isolation and identification of bacteria*. New York: Springer-Verlag, pp. 1506–1529.

Wardlaw, A. C. (1990) *Bordetella*. In: Parker, M. T. and Duerden, B. I. (eds), *Topley and Wilson's Principles of Bacteriology, Virology and Immunity*, 8th edn, vol. 2. Edward Arnold: London, pp. 321–338.

32

Haemophilus, Gardnerella and Streptobacillus

HAEMOPHILUS

Members of this genus are in general small coccobacilli (1.5 × 0.3 μm), although they may be pleomorphic and sometimes filamentous. They are Gram negative and non-motile, and reduce nitrate to nitrite.

They fail to grow on ordinary media, requiring either or both of two factors that are present in blood: X factor, which is haematin, and V factor, which is nicotinamide adenine dinucleotide (NAD) and can be replaced by coenzymes I or II. The X factor is heat stable and the V factor is heat labile. V factor is synthesized by *Staphylococcus aureus*.

Isolation and identification

Plate sputum, cerebrospinal fluid (CSF), eye swabs, etc., on chocolate agar, GC agar base plus 1% yeast autolysate and 5% sheep blood prepared as chocolate agar (Rennie *et al.*, 1992). Incubate at 37°C for 18–24 h. Colonies are usually 1–2 mm, grey and translucent. Gram staining may show filamentous forms in cultures from spinal fluids.

'Satellitism', i.e. large colonies around colonies of *S. aureus* and other organisms synthesizing V factor, may be observed.

Test for X and V factor requirements as follows: subculture several colonies and spread on nutrient agar to make a lawn. Place X, V and X + V factor discs on the medium and incubate at 37°C overnight. Observe growth around the discs, i.e.

whether the organisms require both X and V or X or V or neither (Table 32.1).

Agglutination and capsular swelling tests

Do slide agglutination tests using commercial sera to confirm identity. These are not satisfactory with capsulated strains because agglutination can occur with more than one serum. To test for capsule swelling with homologous sera, use a very light suspension of the organism in saline, coloured with filtered methylene blue solution. Mix a drop of this with a drop of serum on a slide and cover with a cover slip. Examine with an oil immersion lens with reduced light. Swollen capsules should be obvious, compared with non-capsulated organisms. More information about these tests is given by Turk (1982).

Antigen detection

Haemophilus antigen may be directly detected in CSF by countercurrent immunoelectrophoresis (McIntyre, 1987)

Species of *Haemophilus*

Haemophilus influenzae

This requires both X and V factors, produces acid from glucose but not lactose or maltose, and varies in indole production. It occurs naturally in the nasopharynx and is associated with upper respiratory tract infections, including sinusitis and life-threatening acute epiglottitis. It also causes pneumonia, particularly post-influenzal. In eye

Table 32.1 *Haemophilus* species

Species	Haemolysis on blood agar	Factors required			Growth enhanced by 10% CO$_2$
		X + V	**V only**	**X only**	
H. influenzae	−	+			−
H. parainfluenzae	−		+		−
H. haemolyticus	+	+			−
H. parahaemolyticus	+		+		−
H. haemoglobinophilus	−			+	−
H. ducreyi	v			+	v
H. aphrophilus	−			v	+

v, variable

infections ('pink eye'), it is known as the Koch–Weekes bacillus. It is also found in the normal vagina and is one of the causative organisms of purulent meningitis and purulent otitis media. There are several serological types and growth 'phases'

Haemophilus parainfluenzae

This differs only in not requiring X factor. It produces porphyrins from 6-aminolaevulinic acid. It is normally present in the throat, but may be pathogenic.

Haemophilus haemolyticus and parahaemolyticus

Colonies of these are haemolytic on blood agar and otherwise have the same properties as *H. influenzae* and *H. parainfluenzae*, respectively.

Haemophilus haemoglobinophilus (H. canis)

This requires X but not V factor. It is found in preputial infections in dogs and in the respiratory tract of humans.

Haemophilus ducreyi

Ducrey's bacillus is associated with chancroid or soft sore. In direct films of clinical material, the organisms appear in a 'school of fish' arrangement. They are difficult to grow but may be obtained in pure culture by withdrawing pus from a bubo and inoculating inspissated whole rabbit blood slopes or 30% rabbit blood nutrient agar or Mueller–Hinton medium containing Isovitalex. Culture at 30–34°C but no higher, in 10% carbon dioxide. Colonies at 48 h are green, grey or brown, intact and easily pushed along the surface of the media. It requires X but not V factor.

Haemophilus aphrophilus

Colonies on chocolate agar are small (0.5 mm) at 24 h, smooth and translucent. Better growth is obtained in a 10% carbon dioxide atmosphere. Some strains require X factor. There is no growth on MacConkey agar. Acid is produced from glucose, maltose, lactose and sucrose but not from mannitol. Fermentation media should be enriched with Fildes' extract. It is oxidase and catalase negative. This organism closely resembles *Actinobacillus actinomycetemcomitans* (see p. 318) but it does not produce acid from lactose and sucrose (see Table 34.1).

Human infections, including endocarditis, have been reported. For more information about the genus *Haemophilus*, see Slack and Jordans (1999).

GARDNERELLA

The species *Gardnerella vaginalis* was previously known as *Haemophilus vaginalis* or *Corynebacterium vaginale*. It is Gram variable, pleomorphic, non-motile and facultatively anaerobic.

Direct examination

Giemsa-stained films made from vaginal secretions usually show many squamous epithelial cells covered with large numbers of organisms ('clue cells'). Gram-stained films show large numbers of Gram-indifferent bacilli rather than the usual Gram-positive lactobacilli.

Culture

Place specimens in transport medium (Stuart's or Amies'). Plate on Columbia blood agar plus supplement (Oxoid) and incubate under 5–10% carbon dioxide, e.g. in a candle jar. Colonies at 48 h are very small (1 mm), glistening 'dew drops' and β-haemolytic on human blood agar. Gram-stained films show thin, poorly stained bacilli (unlike the solidly stained diphtheroids). They are Gram positive when young but become Gram negative later.

Make lawns on two plates of Columbia blood agar. On one place a drop of 3% hydrogen peroxide. On the other place discs of trimethoprim (5 μg). Incubate for 48 h. Gardnerellas are inhibited by the peroxide and are sensitive to trimethoprim. Lactobacilli and diphtheroids are resistant. Neither X nor V factor is required. There is no growth on MacConkey agar and, apart from acid from glucose (add Fildes' extract), biochemical tests all seem to be negative.

Subculture on blood agar and apply discs of metronidazole (50 μg) and sulphonamide (100 μg) (Oxoid). *G. vaginalis* is sensitive to metronidazole but resistant to sulphonamide.

Two other tests are useful: SPS sensitivity and hippurate hydrolysis. Place a disc (commercially available) of sodium polyethanol sulphate (SPS) on the blood plate and incubate overnight. Gardnerellas are sensitive and show β haemolysis. For the hippurate test make a suspension of the organism in 1 ml hippurate broth, or saline containing a (commercial) hippurate disc. Incubate for 4 h and then add a few drops of ninhydrin reagent. Hippurate hydrolysis is indicated by a deep blue–purple colour (Forbes *et al.*, 1998).

The AP120 kit is useful. *G. vaginalis* is found in the human vagina and may be associated with non-specific vaginitis (see Easmon and Ison, 1985).

STREPTOBACILLUS

There is only one species, *Streptobacillus moniliformis*, which may be filamentous with or without swelling. It is fastidious, requiring blood or serum for growth. It is a commensal of rodents and a pathogen of mice, and is one of the agents of rat-bite fever in humans. Milk-borne disease (Haverhill fever) has been reported.

Isolation and identification

Blood culture is the usual method of isolation, but media containing Liquoid should be avoided because this inhibits the growth of the organism. Inoculate enriched broth medium with aspirated fluids from joints, etc.

The organism grows slowly in liquid media, forming fluffy balls or colonies that resemble breadcrumbs. Remove these with a Pasteur pipette for Gram staining and subculture on blood agar. Incubate at 37°C for 3 days, when colonies will be raised, granular and 1–4 mm in diameter. L-form colonies may be seen (Barrow and Feltham, 1993).

Streptobacillus moniliformis is non-motile, a facultative anaerobe, catalase and oxidase negative and fermentative in Hugh and Leifson medium (see Table 30.1).

REFERENCES

Barrow, G. I. and Feltham, R. K. A. (1993) *Cowan and Steel's Manual for the Identification of Medical Bacteria*, 3rd edn. Cambridge: Cambridge University Press.

Easmon, C. S. F. and Ison, C. A. (1985) *Gardnerella vaginalis*. In: Collins, C. H. and Grange, J. M. (eds), *Isolation and Identification of Micro-organisms of Medical and Veterinary Importance*. Society for Applied Bacteriology Technical Series No. 21. London: Academic Press, pp. 115–122.

Forbes, B. A., Sahm, D. F. and Weissfeld, A. S. (1998) *Bailey and Scott's Diagnostic Microbiology*, 10th edn. St Louis, MO: Mosby.

McIntyre, M. (1987) Counter-current immunoelectrophoresis for the rapid detection of bacterial polysaccharide antigen in body fluids. In: *Immunological Techniques in Microbiology* Grange, J. M., Fox, A. and Morgan, N. L. (eds), Society for Applied Bacteriology Technical Series No. 24. Oxford: Blackwell, pp. 137–143.

Rennie, R., Gordon, T., Yaschak, Y. *et al.* (1992) Laboratory and clinical evaluations of media for the primary isolation of *Haemophilus* species. *Journal of Clinical Microbiology* 30: 1917–1921.

Slack, M. P. E. and Jordens, J. Z. (1999) *Haemophilus*. In: Collier L, Balows, A. and Sussman, M. (eds), *Topley and Wilson's Microbiology and Microbial Infections*, 9th edn. Vol. 2, *Systematic Bacteriology*, CD-ROM. London: Arnold, Chapter 50.

Turk, D. C. (1982) *Haemophilus influenzae*. Public Health Laboratory Service Monograph. London: HMSO.

33

Campylobacter and *Helicobacter*

These organisms were originally in the genus *Vibrio*, then in *Campylobacter*, from which *Helicobacter* was recently separated because of a number of taxonomic differences.

CAMPYLOBACTER

These small, curved, actively motile rods are microaerophilic, reduce nitrates to nitrites, but do not metabolize carbohydrates. They have become very important in recent years as a cause of food poisoning (see Chapter 13). There are about 20 species but only those that are frequently encountered are considered here.

Isolation

Faeces

Plate emulsions of faeces on blood agar containing (commercially available) campylobacter growth and antibiotic selective supplements. Incubate duplicate cultures at 37 and 42°C for 24–48 h in an atmosphere of approximately 5% oxygen, 10% carbon dioxide and 85% nitrogen, preferably using a gas-generating kit, but otherwise a candle jar.

Foods

These may have been frozen, chilled or heated, all of which may cause sublethal damage to the organisms.

Milk

Filter 200 ml through a cottonwool plug. Place the plug in a sterile jar containing 100 ml campylobac-ter enrichment medium without antibiotics. Incubate at 37°C for 2 h. Then add selective antibi-otics and incubate at 43°C for 36 h. Subculture from the surface layer to selective campylobacter agar and incubate plates at 43°C for 36–48 h under microaerophilic conditions.

Water

Filter 100 ml through a 0.45 μm membrane and place the membrane face down on selective campy-lobacter agar. Incubate at 43°C for 24 h under microaerophilic conditions. Remove the filter and reincubate plates at 43°C for 36–48 h.

Place the filter in 100 ml enrichment broth and proceed as for milk.

Meat, poultry, etc.

Add 10 g to 100 ml campylobacter-selective broth plus antibiotic supplements and incubate at 37°C for 2 h. Then continue incubation at 43°C and pro-ceed as for milk.

Enzyme-linked immunosorbent assay (ELISA) techniques are now available for the detection of campylobacters in food (Fricker and Park, 1987).

Identification

Campylobacter colonies are about 1 mm diameter at 24 h, grey, watery and flat. Examine Gram-stained films and test for growth at 43°C, hippu-rate hydrolysis, H_2S production in triple sugar iron (TSI) medium and growth in the presence of 3.5% NaCl. Use media that are known to support growth of the organisms and include controls with known species (Table 33.1).

Table 33.1 *Campylobacter* species

Species	Growth at		Hippurate hydrolysis	H₂S in TSI	Growth in 3.5% NaCl medium
	25°C	43°C			
C. jejuni	−	+	+	−	−
C. coli	−	+	−	+	−
C. lari	−	+	−	+	−
C. fetus	+	v	−	−	−
C. hyointestinalis	v	+	−	v	−
C. sputorum	−	v	−	+	+

v, variable

The CAMPY kit (BioMerieux) and Microscreen (Microgen) are useful.

Campylobacter species

The species that grow at 43°C are called the 'thermophilic campylobacters' and are those most often isolated from humans.

Campylobacter jejuni

This is an important agent of acute enterocolitis in humans (see 'Campylobacter enteritis in food poisoning', Chapter 13). It is also a commensal in the intestines of many wild and domestic animals.

Campylobacter coli

Also responsible for acute gastroenteritis in humans, this is less common than *C. jejuni*.

Campylobacter lari

This is the least common cause of campylobacter enteritis.

Campylobacter fetus

The subspecies *fetus* is an agent of infectious abortion in cattle and sheep and occasionally infects humans.

Campylobacter hyointestinalis

This is a pathogen of pigs and cattle and is sometimes responsible for diarrhoea in humans.

Campylobacter sputorum

This species is a commensal in the mouth of humans.

There are a number of subspecies and biotypes of campylobacters. For information about biotyping, etc., see Bolton *et al.* (1992).

Campylobacters have received much attention in the last few years (see Griffiths and Park, 1990; Skirrow, 1990; Bolton *et al.*, 1992; Nachamkin and Skirrow, 1999). See also the review in relation to the aquatic environment by Thomas *et al.* (1999).

HELICOBACTER

Helicobacters are motile, curved, Gram-negative, microaerophlic bacilli. There are about 14 species but only one, *H. pylori*, is considered here, because of its association with gastric cancer and peptic ulcers.

Direct microscopy

Stain smears with haematoxylin and eosin or by the Giemsa method and look for characteristic curved or S-shaped rods.

Isolation and identification

Use freshly collected biopsy material (otherwise use a transport medium). Culture on one of the commercial campylobacter media plus supplements. Incubate for 3–5 days under microaerobic conditions (5–20% CO_2) and high humidity as for campylobacters. Colonies are circular, convex and translucent.

Test for oxidase, catalase and urease. *H. pylori* gives positive reactions for all three (campylobacters are urease negative).

Other tests

The Clotest

This a commercial test (CLO = campylobacter-like organisms; Tri-Med Specialities, Utah, USA). It is a plastic tray containing a urea agar media and an indicator. A small piece of biopsy material is embedded in the medium. After 1–24 h, a colour change from yellow to magenta indicates the presence of a helicobacter (Murray *et al.*, 1999).

The urea breath test

The patient is given urea labelled with a carbon isotope. This is hydrolysed to ammonia and carbon dioxide if *H. pylori* is present in the stomach and the latter is exhaled. This is measured with a scintillation counter or mass spectrometer, depending on which carbon isotope was used, indicating the amount of urease activity of the organism in the stomach.

Molecular and ELISA methods are also available. There are at least two useful diagnostic kits – CAMPY, RAPIDEC Pylori (Biomerieux) and Sigma EIA. For more information, see Skirrow (1990), Blaser, (1998) and Glupczynski (1999).

REFERENCES

Blaser, M. J. (1998) *Helicobacter pylori* and gastric disease. *British Medical Journal* **316**: 1507–1510.

Bolton, F. J., Waring, D. R. A., Skirrow, M. B. and Hutchinson, D. N. (1992) Identification and biotyping of campylobacters. In: Board, R. G., Jones, D. and Skinner, F. A. (eds), *Identification Methods in Applied and Environmental Microbiology*. Society for Applied Microbiology Technical Series No. 29. Oxford: Blackwell, pp. 151–162.

Fricker, C. R. and Park, R. W. A. (1987) Competitive ELISA, co-agglutination and haemagglutination for the detection and *serotyping* of campylobacters. In: Grange , J. M., Fox, A. and Morgan, N. L. (eds), *Immunological Techniques in Microbiology*. Society for Applied Bacteriology Technical Series No. 24. Oxford: Blackwells, pp. 195–210.

Glupczynski, Y. (1999) Infections with *Helicobacter*. In: Collier L, Balows, A. and Sussman, M. (eds), *Topley and Wilson's Microbiology and Microbial Infections*, 9th edn. Vol. 2, *Systematic Bacteriology*, CD/ROM. London: Arnold, Chapter 30.

Griffiths, P. L. and Park, R. W. A. (1990) Campylobacters associated with human diarrhoeal disease. *Journal of Applied Bacteriology* **69**: 281–301.

Nachamkin, I. and Skirrow, M. B. (1999) *Campylobacter, Arcobacter* and *Helicobacter*. In: Collier L, Balows, A. and Sussman, M. (eds), *Topley and Wilson's Microbiology and Microbial Infections*, 9th edn. Vol. 2, *Systematic Bacteriology*, CD/ROM. London: Arnold, Chapter 54.

Murray, P. R., Baron, E. J., Pfaller, M.A. *et al.* (1999) *Manual of Clinical Microbiology*, 7th edn. Washington: American Society of Microbiologists.

Skirrow, M. B. (1990) *Campylobacter, Helicobacter* and other motile, curved Gram-negative rods. In: Parker, M. Y. and Duerden, B. I. (eds), *Topley and Wilson's Principles of Bacteriology, Virology and Immunity*, 8th edn. London: Edward Arnold, pp. 531–549.

Thomas, C., Gibson, H., Hill, D. J. and Mabey, M. (1999) *Campylobacter* epidemiology: an aquatic perspective. *Journal of Applied Bacteriology Symposium Supplement* **85**: 168S–177S.

Actinobacillus, Pasteurella, Yersinia, Cardiobacterium and Francisella

The general properties of the organisms described in this chapter are shown in Table 30.1.

ACTINOBACILLUS

This genus contains non-motile, Gram-negative rods that are fermentative but do not produce gas, and are indole negative. Most strains grow on MacConkey agar and are ONPG positive. Some are human and animal pathogens; others are commensals.

Isolation and identification

Plate pus, which may contain small white granules, or macerated tissue on blood agar and incubate at 37°C for 48–72 h under a 10% carbon dioxide atmosphere. Colonies are small, sticky and adherent. Subculture small flat colonies on to MacConkey agar, in Hugh and Leifson medium, and test for motility, catalase, oxidase, acid production from lactose, mannitol and sucrose, and aesculin hydrolysis (Table 34.1, and see Table 30.1).

Species of Actinobacillus

Actinobacillus lignieresii

This species grows on MacConkey agar, is fermentative, and produces acid from lactose (late), mannitol and sucrose.

It is oxidase, catalase and urease positive, but does not hydrolyse aesculin. It is associated with woody tongue in cattle and human infections have been reported.

Actinobacillus equuli

This species resembles A. *lignieresii* but ferments lactose rapidly. Lactose, sucrose and mannitol are

Table 34.1 *Actinobacillus* species

Species	Oxidase	Growth on MacConkey	Acid from			Aesculin hydrolysis	Urease
			Lactose	Mannitol	Sucrose		
A. *lignieresii*	+	+	v	+	+	−	+
A. *equuli*	v	+	+	+	+	−	+
A. *suis*	v	+	+	−	−	+	+
A. *actinomycetecomitans*	+	v	−	v		−	−

v, variable

Table 35.1 Summary of laboratory diagnosis procedures for legionellosis

Test	Sample	Sensitivity (%)	Specificity (%)	Cost	Comment
Culture	Sputum, BAL	++	+++	++	All species
Direct IF	Sputum, BAL	++	++	+	*L. pneumophila* serogroups only
Urinary antigen	Urine	++	+++[a]	++	*L. pneumophila* serogroup 1 only
Serology:					
Fourfold change	Blood	++	++	++	
Single titre > 1 : 256		+	++	++	*L. pneumophila* serogroups only
PCR	Sputum, BAL	+++	+++	+++	All species but not routinely available

BAL, bronchoalveolar lavage fluid; IF, immunofluorescence; PCR, polymerase chain reaction
+ < 30%; ++ 30–90%; +++ > 90%; [a] but see p. 323

MYCOPLASMA

Mycoplasmas are the smallest free-living bacteria and are usually 0.2–0.3 µm in diameter. They lack cell walls and are therefore resistant to penicillin. They can pass through bacteria-excluding filters. Their morphology is variable: cells may be coccoid, filamentous or star shaped. Individual cells may not be stained by Gram's method. Mycoplasmas are usually recognized by their characteristic colonial morphology as observed by low-power microscopy.

Isolation

Specimens

These include sputum, throat swabs, bronchial lavage fluid, lung biopsies and cerebrospinal fluid. Homogenize sputum by shaking it with an equal volume of mycoplasma broth in a bottle containing glass beads. Homogenize tissue in mycoplasma broth. Transport swabs to the laboratory in mycoplasma broth; squeeze the swabs into the broth to express material. Culture other specimens (e.g. cerebrospinal fluid, urine) without pre-treatment or centrifuge them and use the deposit.

Culture media

Mycoplasma culture media are available commercially as agar or broth, or in diphasic form. They include thallous acetate and penicillin to inhibit the growth of other bacteria and amphotericin B to suppress fungi. They may be supplemented with glucose, urea or arginine as indicated in Table 35.2.

Inoculate mycoplasma agar plates, mycoplasma broth and either SP4 broth (mycoplasma broth containing fetal calf serum) or selective media in duplicate, with 0.1 ml of the specimen. Incubate at 37°C in anaerobic-type jars under the atmospheric conditions shown in Table 35.2. Examine agar media for colonies daily for up to 20 days. *M. pneumoniae* or *M. genitalium* colonies develop between 5 and 20 days. Subculture broth media to solid media when the appropriate colour changes have occurred.

Colonies of some species have a dark central zone and a lighter peripheral zone, giving them a 'fried egg' appearance; others, notably *Mycoplasma pneumoniae*, lack this zone.

34

Actinobacillus, Pasteurella, Yersinia, Cardiobacterium and Francisella

The general properties of the organisms described in this chapter are shown in Table 30.1.

ACTINOBACILLUS

This genus contains non-motile, Gram-negative rods that are fermentative but do not produce gas, and are indole negative. Most strains grow on MacConkey agar and are ONPG positive. Some are human and animal pathogens; others are commensals.

Isolation and identification

Plate pus, which may contain small white granules, or macerated tissue on blood agar and incubate at 37°C for 48–72 h under a 10% carbon dioxide atmosphere. Colonies are small, sticky and adherent. Subculture small flat colonies on to

MacConkey agar, in Hugh and Leifson medium, and test for motility, catalase, oxidase, acid production from lactose, mannitol and sucrose, and aesculin hydrolysis (Table 34.1, and see Table 30.1).

Species of Actinobacillus

Actinobacillus lignieresii

This species grows on MacConkey agar, is fermentative, and produces acid from lactose (late), mannitol and sucrose.

It is oxidase, catalase and urease positive, but does not hydrolyse aesculin. It is associated with woody tongue in cattle and human infections have been reported.

Actinobacillus equuli

This species resembles *A. lignieresii* but ferments lactose rapidly. Lactose, sucrose and mannitol are

Table 34.1 Actinobacillus species

Species	Oxidase	Growth on MacConkey	Acid from			Aesculin hydrolysis	Urease
			Lactose	Mannitol	Sucrose		
A. lignieresii	+	+	v	+	+	−	+
A. equuli	v	+	+	+	+	−	+
A. suis	v	+	+	−	−	+	+
A. actinomycetecomitans	+	v	−	v		−	−

v, variable

fermented. It is urease positive but aesculin is not hydrolysed. It is associated with joint ill and sleepy disease of foals.

Actinobacillus actinomycetemcomitans

This organism requires carbon dioxide for primary isolation and some strains require the X factor. Use an enriched medium and place an X factor disc on the heavy part of the inoculum. There is no growth on MacConkey agar and the oxidase test is positive. Lactose and sucrose are not fermented but some strains may produce acid from mannitol. It is aesculin and urease negative.

This organism is difficult to distinguish from *Haemophilus aphrophilus* (see Table 32.1). It is sometimes associated with infections by *Actinomyces israelii* and it has been recovered from blood cultures of patients with endocarditis.

Actinobacillus suis

This hydrolyses aesculin and does not ferment mannitol. It has been isolated from pigs.

For further information about the identification of these organisms, see Barrow and Feltham (1993) and Holmes (1999).

YERSINIA

Yersinias cause plague, pseudotuberculosis and gastroenteritis in humans and animals. They are now classified among the enterobacteria but are retained here for pragmatic reasons. They are catalase positive, fermentative in Hugh and Leifson medium (slow reaction), reduce nitrates to nitrites and fail to liquefy gelatin.

The plague bacillus, Yersinia pestis, *is in Hazard Group 3 and all work should be done in a microbiological safety cabinet in a Biosafety/Containment Level 3 laboratory. After isolation and presumptive identification cultures should be sent to a reference laboratory for any further tests.*

Isolation and identification

Yersinia pestis

In suspected bubonic plague examine pus from buboes, in pneumonic plague, sputum, in septicaemic plague, blood, and in rats the heart blood, enlarged lymph nodes and spleen. Before examining rats immerse them in disinfectant for several hours to kill fleas that might be infected.

Make Gram- and methylene blue-stained films and look for small oval Gram-negative bacilli with capsules. Bipolar staining is often seen in films of fresh isolates stained with methylene blue. Plate on blood agar containing 0.025% sodium sulphite to reduce the oxygen tension, on 3% salt agar and on MacConkey agar. Incubate for 24 h.

Colonies of plague bacilli are flat or convex, greyish white and about 1 mm in diameter at 24 h. Subculture into nutrient broth and cover with liquid paraffin. Test for motility at 22°C, fermentation of salicin and sucrose, aesculin hydrolysis and urease.

Films of colonies on salt agar show pear-shaped and globular forms. *Y. pestis* grows on MacConkey agar and usually produces stalactite growth in broth covered with paraffin. It is non-motile at 37 and 22°C. It produces acid from salicin but not sucrose and hydrolyses aesculin, but is urease negative (Table 34.2).

Reference and communicable diseases laboratories use fluorescence antibody (FA) methods, bacteriophage, specific agglutination tests, precipitin tests and animal inoculations.

Yersinia enterocolitica

Culture faeces on yersinia-selective medium (cefsulodin–irgasan–novobiocin agar, CIN). Incubate at 30°C for 24–48 h and look for 'bulls' eye' colonies. Enrichment is not usually recommended for isolating this organism from faeces. Some strains grow on deoxycholate agar (DCA) as small non-lactose fermenting colonies, even after incubation at 37°C.

Make a 10% suspension of food in peptone water diluent. Incubate some at 32°C for 24 h. Refrigerate the remaining suspension at 4°C for up

Table 34.2 *Yersinia* species

Species	Motility at 22°C	Acid from sucrose	Aesculin hydrolysis	Urease
Y. pestis	−	−	+	−
Y. pseudotuberculosis	+	−	+	+
Y. enterocolitica	+	+	−	+

to 21 days and subculture at intervals on yersinia-selective agar. Media used for other enteric pathogens can be reincubated at 28–30°C for 24 h and examined for *Y. enterocolitica* but some strains of this organism are inhibited by bile salts and deoxycholate.

Gram-stained films show small coccoid bacilli, unlike the longer bacilli of other enterobacteria.

Test for growth on nutrient and MacConkey agar, motility at 22°C, acid from sucrose, aesculin hydrolysis and urease. Kits may be used for all or most of these. Incubate at 30°C rather than at 37°C (see Table 34.2).

There are several serotypes and biotypes. Some are pathogenic for humans (O:3, O:5, O:8, O:9), others for animals. Serological tests for specific antibodies may be of value in diagnosis.

Yersinia enterocolitica causes enteritis in humans and animals. In humans there may also be lymphadenitis, septicaemia, arthritis and erythema nodosum. The reservoirs are pigs, cattle, poultry, rats, cats, dogs and chinchillas. The organism has been found in milk and milk products, water, oysters and mussels. It grows at 4°C and may therefore multiply in cold storage of food. Person-to-person spread occurs, mainly in families.

Yersinia fredericksenii, Y. kristensenii and *Y. intermedia* resemble *Y. enterocolitica* but their roles in disease are uncertain. For further information, see Brewer and Corbel (1985), Mair and Fox (1986), Doyle and Cliver (1990), and Holmes (1999).

Yersinia pseudotuberculosis

Homogenize lymph nodes, etc. in tryptone broth. Inoculate blood agar with some of the suspension.

Retain the remainder at 4°C and subculture to blood agar every 2–3 days for up to 3 weeks.

Emulsify faeces in peptone water diluent and inoculate yersinia-selective medium (CIN). Colonies on blood agar are raised, sometimes umbonate, granular and about 1–2 mm in diameter. Rough variants occur. Growth on MacConkey agar is obvious but poor. *Y. pseudotuberculosis* is motile at 22°C, ferments salicin but not sucrose, hydrolyses aesculin and is urease positive (see Table 34.2).

This organism causes pseudotuberculosis in rodents, especially in guinea-pigs. Human infections occur and symptoms resemble those of infection with *Y. enterocolitica*. *Y. pseudotuberculosis* is rarely isolated from faeces, however, and serological tests for specific antibodies are useful. For more information about yersinias, see Wanger (1999).

PASTEURELLA

Pasteurella multocida

Bipolar staining and pleomorphism are less obvious than in *Yersinia* species. The bacilli are very small (about 1.5 × 0.3 μm) and non-motile. Colonies on nutrient agar (37°C) are translucent, slightly raised, and about 1 mm in diameter at 24 h; on blood agar they are slightly larger, more opaque and non-haemolytic. There is no growth on MacConkey agar. Most strains produce acid from mannitol, indole and urease produce. They do not hydrolyse aesculin (Table 34.3).

Table 34.3 *Pasteurella* species

Species	Growth on MacConkey	Acid from mannitol	Aesculin hydrolysis	Indole production	Urease
P. multocida	−	(+)	−	(+)	−
P. pneumotropica	v	−	−	+	+
P. haemolytica	+	−	v	−	−

v, variable; (+) usually positive

This organism causes haemorrhagic septicaemia in domestic and wild animals and birds, e.g. fowl cholera, swine plague, transit fever. Humans may be infected by animal bites or by inhaling droplets from animal sneezes. It has also been isolated from septic fingers.

It is often given specific names according to its animal host, e.g. *avicida, aviseptica, suilla, suiseptica, bovicida, boviseptica, ovicida, oviseptica, cuniculocida, lepiseptica, muricida, muriseptica*. There is some host interspecificity: cattle strains are also pathogenic for mice but not for fowls; fowl cholera strains are pathogenic for cattle and mice; lamb septicaemia strains are not pathogenic for rodents.

Pasteurella pneumotropica, P. haemolytica and P. gallinarum

These species are not infrequently isolated from veterinary material.

For further information about the genus *Pasteurella*, see Carter (1984), Curtis (1985), Barrow and Feltham (1993), and Holmes (1999).

CARDIOBACTERIUM

Cardiobacterium hominis

This is a facultative anaerobe and requires an enriched medium and high humidity. Growth is best under 7–10% carbon dioxide. Colonies on blood agar at 37°C are minute at 24 h and about 1 mm in diameter after 48 h, and are convex, glossy and butyrous. There is no growth on MacConkey medium, the oxidase test is positive and glucose and sucrose, but not lactose, are fermented without gas production. The catalase, urease and nitrate reduction tests are negative. Hydrogen sulphide is produced but gelatin is not liquefied.

This organism has been isolated from blood cultures of patients with endocarditis but may also be found in the upper respiratory tract.

FRANCISELLA

Francisella tularensis is in Hazard Group 3. It is highly infectious and has caused many laboratory infections. It should be handled with great care in microbiological safety cabinets in Biosafety/Containment Level 3 laboratory.

Culture blood, exudate, pus or homogenized tissue on several slopes of blood agar (enriched), cystine glucose agar and inspissated egg yolk medium. Also inoculate tubes of plain nutrient agar. Incubate for 3–6 days and examine for very small drop-like colonies. If the material is heavily contaminated, add 100 μg/ml cycloheximide or 200 units/ml nystatin, and 2.5–5 μg/ml neomycin.

Subculture colonies of small, swollen or pleomorphic Gram-negative bacilli on blood and nutrient agar, on MacConkey agar and in enriched nutrient broth. Test for acid production from glucose and maltose, motility at 22°C and urease activity.

The bacilli are very small, show bipolar staining and are non-motile. There is no growth on nutrient or MacConkey agar, and poor growth on blood agar with very small, grey colonies; on inspissated egg, at 3–4 days, colonies are minute and drop like. It is a strict aerobe. Acid is formed in glucose and sucrose. It is oxidase positive and produces indole.

F. tularensis is responsible for a plague-like disease (tularaemia) in ground squirrels and other rodents in western USA and Scandinavia.

Cultures and materials in cases of suspected tularaemia should be sent to a reference laboratory. Fluorescent antibody procedures are used for rapid diagnosis.

The genus *Francisella* is described in detail by Eigelsbach and McGann (1984) and Nano (1992).

REFERENCES

Barrow, G. I. and Feltham, R .K. A. (eds) (1993) *Cowan and Steel's Manual for the Identification of Medical Bacteria,* 3rd edn. Cambridge: Cambridge University Press.

Brewer, R. A. and Corbel, M. J. (1985) Yersinia enterocolitica and related species. In: Collins, C. H. and Grange, J. M. (eds), *Isolation and Identification of Micro-organisms of Medical and Veterinary Importance.* Society for Applied Bacteriology Technical Series No. 21. London: Academic Press, pp. 83–104.

Carter, G. R. (1984) Pasteurella. In: Krieg, N. R. and Holt, J. G. (eds), *Bergey's Manual of Systematic Bacteriology*, Vol. 1. Baltimore, MA: Williams & Wilkins, pp. 552–558.

Curtis, P. E. (1985) *Pasteurella multocida.* In: Collins, C. H. and Grange, J. M. (eds), *Isolation and Identification of Micro-organisms of Medical and Veterinary Importance.* Society for Applied Bacteriology Technical Series No. 21. London: Academic Press, pp. 43–52.

Doyle, M. P. and Cliver, D. O. (1990) *Yersinia enterocolitica.* In: Cliver D. O. (ed), *Foodborne Diseases.* New York: Academic Press, pp. 224–229.

Eigelsbach, H. T. and McGann, V. G. (1984) Genus *Francisella.* In: Krieg, N. R. and Holt, J. G. (eds), *Bergey's Manual of Systematic Bacteriology*, Vol. 1. Baltimore, MA: Williams & Wilkins, p. 394.

Holmes, B. (1999) *Actinobacillus, Pasteurella* and *Eikenella.* In: Collier L, Balows, A. and Sussman, M. (eds), *Topley and Wilson's Microbiology and Microbial Infections*, 9th edn. Vol. 2, *Systematic Bacteriology*, CD-ROM. London: Arnold, Chapter 51.

Mair, N. S. and Fox, E. (1986) *Yersiniosis: Laboratory diagnosis, clinical features and epidemiology.* London: Public Health Laboratory Service.

Nano, F. E. (1992) The genus *Francisella.* In: Balows, A., Truper, H. G., Dworkin, M. *et al.* (eds), *The Prokaryotes*, 2nd edn. New York: Springer-Verlag.

Wanger, A. R. (1999) *Yersinia.* In: Collier L, Balows, A. and Sussman, M. (eds), *Topley and Wilson's Microbiology and Microbial Infections*, 9th edn. Vol. 2, *Systematic Bacteriology*, CD-ROM. London: Arnold, Chapter 44.

35

Legionella and *Mycoplasma*

These two genera are considered together here because they are both associated with atypical pneumonia.

LEGIONELLA

This genus of aerobic Gram-negative bacteria includes 42 species, 18 of which are responsible for human diseases such as legionellosis (legionnaires' disease) and Pontiac fever. The organisms are widely distributed in aquatic habitats and soil. For further details on this genus and the diseases that it causes, see Harrison and Taylor (1988), Stout and Yu (1997), and Winn (1999).

Isolation

Culture is the preferred method for laboratory diagnosis. Legionellas have fastidious growth requirements and do not grow on ordinary blood agar.

Homogenize the material (e.g. sputum, bronchial aspirates) in peptone water diluent and plate on buffered charcoal yeast extract agar (BCYE) containing 2-oxoglutarate (α-ketoglutarate) (BCYE-α) to which various antibiotics may be added to suppress the growth of other micro-organisms (commercial legionella-selective medium plus antibiotic supplements).

Incubate under 5% CO_2 and high humidity and examine plates after 3, 5, 7 and 10 days before discarding them. Most strains are recovered after 2–3 days of incubation. Colonies of legionellas are small,

round, up to 3 mm in diameter and grey or greenish brown, and present a 'ground-glass' appearance.

Legionellas may also be isolated by modern blood culture systems (e.g. Oxoid Signal, Difco Sentinel) but additional processing is required, including the inoculation of selective media, and the sensitivity is low.

For the isolation of legionellas from water samples, see Chapter 20.

Identification

Examine Gram-stained films of suspect colonies, test those that are Gram negative with fluorescence antibody (FA) sera, and do agglutination tests with polyvalent antisera. Proceed with specific sera if available.

Subculture on BCYE medium for catalase, oxidase and hippurate hydrolysis tests. Test also for gelatin liquefaction but ensure that the medium contains cysteine and iron salts (growth supplements are available commercially). Legionellas are catalase positive. Other reactions vary with species (see below).

Kits for screening and identifying legionellas include those of Sigma and Microscreen.

It is advisable to send cultures of legionellas to a reference laboratory for full identification and serogrouping.

Species of *Legionella*

Only the relatively common species are listed below. Biochemical properties are of limited use. Identification must rely on agglutination tests.

Legionella pneumophila

This is the most commonly encountered species. It is oxidase positive and the only one that hydrolyses hippurate. It liquefies gelatin.

Legionella bozemanii, *L. dumoffii* and *L. gormanii*

These are oxidase negative, do not hydrolyse hippurate but liquefy gelatin.

Legionella micdadei

This is oxidase positive, does not hydrolyse hippurate and does not liquefy gelatin.

Legionella longbeachii and *L. jordanii*

These are also oxidase positive and do not hydrolyse hippurate, but they do liquefy gelatin.

Direct immunofluorescence

This may be used to detect legionella antigens in clinical material. Although results are available rapidly, the sensitivity of this method is often poor and cross-reactions may occur with other bacteria, giving false positives. This test has a limited role in areas with a low prevalence of legionella infections because of the high frequency of cross-reaction.

Urinary antigen

The detection of legionella antigens in urine is likely to become the preferred diagnostic method as tests continue to improve. Enzyme immunoassay (EIA) kits are now available commercially. Both sensitivity and specificity are high and testing may be performed rapidly. The major limitation of these tests is their unreliability in the diagnosis of legionella infections other than *L. pneumophila* serogroup 1. In many parts of the world other species and serogroups cause a sizeable proportion of cases of legionella pneumonia. Urinary antigen tests may remain positive for weeks to months. A positive result does not, therefore, necessarily reflect acute infection.

Serology

An acute serum sample should be collected and stored. If the results of other tests are negative and a legionella infection is still suspected, a convalescent serum sample should be collected at least 3 weeks later and both samples should be tested in parallel for antibodies. If seroconversion has not occurred at this stage, another serum sample should be collected and tested after a further 3 weeks.

The detection of legionella antibodies in serum is a sensitive method for diagnosing legionella infection, especially when caused by *L. pneumophila* serogroup 1. The main disadvantages of serological tests are that they may provide only a retrospective diagnosis, and cross-reactions with other bacterial species may cause false-positive results. Such cross-reacting bacteria include *Pseudomonas aeruginosa*, *Stenotrophomonas maltophilia*, *Bacteroides* spp., *Campylobacter jejuni* and *Bordetella pertussis*. Many populations have relatively high background seropositivity rates to legionella antigens, presumably as a result of previous (often asymptomatic) infection. Consequently, it is necessary to document a fourfold or greater rise in antibody titre (to > 1 : 128) before legionella infection may be diagnosed. As seroconversion is frequently delayed after legionella infection, it is recommended that a convalescent serum sample be collected at least 3 weeks after an acute sample. There are very few situations where testing of a single serum sample (acute or convalescent) is justified.

Table 35.1 summarizes the laboratory tests for legionellosis.

Table 35.1 Summary of laboratory diagnosis procedures for legionellosis

Test	Sample	Sensitivity (%)	Specificity (%)	Cost	Comment
Culture	Sputum, BAL	++	+++	++	All species
Direct IF	Sputum, BAL	++	++	+	*L. pneumophila* serogroups only
Urinary antigen	Urine	++	+++[a]	++	*L. pneumophila* serogroup I only
Serology:					
Fourfold change	Blood	++	++	++	
Single titre > I : 256		+	++	++	*L. pneumophila* serogroups only
PCR	Sputum, BAL	+++	+++	+++	All species but not routinely available

BAL, bronchoalveolar lavage fluid; IF, immunofluorescence; PCR, polymerase chain reaction
+ < 30%; ++ 30–90%; +++ > 90%; [a] but see p. 323

MYCOPLASMA

Mycoplasmas are the smallest free-living bacteria and are usually 0.2–0.3 μm in diameter. They lack cell walls and are therefore resistant to penicillin. They can pass through bacteria-excluding filters. Their morphology is variable: cells may be coccoid, filamentous or star shaped. Individual cells may not be stained by Gram's method. Mycoplasmas are usually recognized by their characteristic colonial morphology as observed by low-power microscopy.

Isolation

Specimens

These include sputum, throat swabs, bronchial lavage fluid, lung biopsies and cerebrospinal fluid. Homogenize sputum by shaking it with an equal volume of mycoplasma broth in a bottle containing glass beads. Homogenize tissue in mycoplasma broth. Transport swabs to the laboratory in mycoplasma broth; squeeze the swabs into the broth to express material. Culture other specimens (e.g. cerebrospinal fluid, urine) without pre-treatment or centrifuge them and use the deposit.

Culture media

Mycoplasma culture media are available commercially as agar or broth, or in diphasic form. They include thallous acetate and penicillin to inhibit the growth of other bacteria and amphotericin B to suppress fungi. They may be supplemented with glucose, urea or arginine as indicated in Table 35.2.

Inoculate mycoplasma agar plates, mycoplasma broth and either SP4 broth (mycoplasma broth containing fetal calf serum) or selective media in duplicate, with 0.1 ml of the specimen. Incubate at 37°C in anaerobic-type jars under the atmospheric conditions shown in Table 35.2. Examine agar media for colonies daily for up to 20 days. *M. pneumoniae* or *M. genitalium* colonies develop between 5 and 20 days. Subculture broth media to solid media when the appropriate colour changes have occurred.

Colonies of some species have a dark central zone and a lighter peripheral zone, giving them a 'fried egg' appearance; others, notably *Mycoplasma pneumoniae*, lack this zone.

Table 35.2 Isolation of mycoplasmas causing human disease

Species	Clinical specimens[a]	Culture media	Substrate	pH	Culture atmosphere[b]	Colony form	Growth rate
M. pneumoniae	BAL fluid, throat swabs, sputum	Diphasic medium; SP4 medium; horse serum agar and broth; SP4 broth or agar	Glucose	7.8	1 + 2	Mulberry	Slow (over 5 days)
M. genitalium	Urethral swabs, throat swabs, sputum	SP4 broth or agar	Glucose	7.8	1 + 2	Mulberry	Slow (over 5 days)
Ureaplasma urealyticum	Vaginal swabs, cervical swabs, urethral swabs, urine	Urea broth or agar supplemented with $MnCl_2$ which will stain colonies	Urea	6.0	3	Small fried egg (10–15 μm)	Medium (over 2 days)
M. hominis	Vaginal swabs, cervical swabs, urethral swabs	Horse serum broth or agar	Arginine	7.0	1 + 2	Fried egg (50–500 μm)	Rapid (occasionally by 24 h)

[a] Swabs should be transported to the laboratory in a suitable medium (see text)
[b] Atmosphere: 1 = 5% CO_2/95% N_2; 2 = air; 3 = 10–20% CO_2/80–90% N_2
BAL, bronchoalveolar lavage

Identification

Mycoplasmas cannot be removed from agar media for microscopy and subcultured in the usual way with inoculating wires and loops. With a sterile scalpel, cut out a block of agar about 5 mm square and containing one or more colonies. Proceed as indicated below.

Microscopy

Place an agar block, colonies uppermost, on a slide. Place a drop of Diene's stain on a cover glass and invert it over the agar block. Examine under a low-power microscope.

Imprints of mycoplasma colonies are stained dark blue and those of ureaplasma greenish blue. Bacteria are also stained blue, but they lose the colour after 30 min, whereas the mycoplasmas retain it for several hours. This stain differentiates mycoplasmas from artefacts such as crystals which may have a similar appearance.

Subculture

Place an agar block, colonies down, on the surface of mycoplasma agar and move it around. New colonies will develop along the track of the block.

Inhibition by hyperimmune serum

Cut out agar blocks containing colonies and place them in mycoplasma broth. Incubate for 4–10 days and then flood mycoplasma agar plates with the culture. Allow to dry and apply filter-paper discs impregnated with hyperimmune sera. Incubate until colonies are visible. Zones of growth inhibition will be seen around discs that are impregnated with the homologous antibody to the isolate.

Haemadsorption test

Pour a 0.5% suspension of washed guinea-pig erythrocytes in phosphate-buffered saline (PBS) over an agar plate bearing about 100 colonies. Leave at room temperature for 30 min, remove the cell suspension, wash the agar surface with PBS and examine under a low-power microscope for adsorption of erythrocytes, which occurs with colonies of *M. pneumoniae*.

Haemolysis test

Prepare an agar overlay containing 1 ml of a 20% suspension of washed guinea-pig or sheep erythrocytes in PBS and 3 ml mycoplasma agar base (at 45°C). Pour it gently over a plate on which there are colonies of mycoplasmas, allow to solidify and incubate aerobically at 37°C. After 24 h and again at 48 h refrigerate the plates at 4°C for 30 min and examine for haemolysis. Colonies of *M. pneumoniae* are surrounded by a zone of β-haemolysis. Other mycoplasmas are surrounded by zones of α-haemolysis, are smaller and take longer to develop the zone.

Immunofluorescent staining

This is a reliable method for identifying mycoplasma colonies but the specific antibodies raised in rabbits are currently not commercially available.

Species of mycoplasmas

Table 35.3 lists the mycoplasmas responsible for human and animal disease. Those associated with human disease are described briefly below. For details of the clinical aspects of diseases caused by mycoplasmas, see Mårdh (1999) and Taylor-Robinson (2002).

Mycoplasma pneumoniae

This forms 'lumpy' colonies in 4–5 days in media containing glucose, incubated in an atmosphere of 5% CO_2 and 95% N_2 as well as aerobically.

Table 35.3 Mycoplasmas causing disease in humans and animals

Host	*Mycoplasma* sp.	Primary isolation site	Disease association
Human	*M. pneumoniae*	Respiratory tract	Atypical pneumonia
	M. genitalium	Genital and respiratory tract	Non-specific urethritis
	M. hominis	Genital, occasionally respiratory	Septic wound infections
	M. fermentans	Genital tract	Genital (and other?) infections
	Ureaplasma urealyticum	Genital tract and oropharynx	Non-gonococcal urethritis and chronic lung disease
Cattle	*M. mycoides* subsp. *mycoides*	Respiratory tract	Contagious bovine pleuropneumonia
	M. bovis	Joints, milk, genital tract	Mastitis and respiratory disease
	Mycoplasma spp. (bovine group 7)		Contagious agalactia
Sheep/goats	*M. agalactiae*	Joints, milk, blood	Contagious agalactia
	M. mycoides subsp apr I	Respiratory tract	Contagious caprine pleuropneumonia
	M. capricolum	Joints, udder, blood	Septicaemic arthritis
Poultry	*M. gallisepticum*	Respiratory tract	Air sacculitis, sinusitis and arthritis
	M. synoviae	Joints, respiratory tract	Air sacculitis and arthritis
Pigs	*M. hypopneumoniae*	Respiratory tract	Enzootic pneumonia
Rats/mice	*M. artthritidis*	Joints	Arthritis
	M. pulmonis	Respiratory tract	Murine respiratory mycoplasmosis

It is a major cause of primary atypical pneumonia, accounting for up to 15% of hospital-admitted pneumonias. Epidemics occur over a 4-year cycle. During periods of peak infection all available laboratory tests should be used to ensure adequate diagnosis and prompt and appropriate chemotherapy, usually with erythromycin or a tetracycline.

Mycoplasma genitalium

Colonies and growth conditions are the same as for *M. pneumoniae*. Colonies may take 5 days to develop. *M. genitalium* was originally isolated from the urethras of men with non-gonococcal urethritis. Subsequently, it has been isolated from the respiratory tract in association with *M. pneumoniae* with which it has several properties in common, including the ability to metabolize glucose and some antigenic cross-reactions. Its frequency in the urogenital and respiratory tracts has yet to be established.

Mycoplasma hominis

This forms 'fried egg' colonies on media containing arginine under the same atmospheric conditions as *M. pneumoniae*. Growth is rapid, occasionally visible in 24 h.

It is present in the genital tract of about 20% of women and may be regarded as an opportunist pathogen. It is sometimes isolated from blood cultures of women with mild postpartum fever and may also cause pelvic inflammation and infertility.

Ureaplasma urealyticum

Small 'fried egg' colonies appear on media containing urea and incubated in an atmosphere containing 10–20% CO_2 and 80–90% N_2. Colonies may take 2 or more days to develop.

About 60% of healthy women carry this organism in their genital tracts. Like *M. hominis* it may be regarded as an opportunist pathogen because it is also occasionally isolated from blood cultures of women with mild postpartum fever and may cause pelvic inflammation and infertility. It is associated with non-specific urethritis in men and chronic lung disease in infants with low birthweights.

Serological diagnosis

Serological tests have been described for a number of mycoplasma infections, notably those caused by *M. pneumoniae*. Culture, however, remains the most useful diagnostic method at present.

Historical methods, such as the cold agglutination test are relatively insensitive and should be reserved for rapid diagnosis during epidemics. The complement fixation test for *M. pneumoniae* is also of low sensitivity. More sensitive tests include agglutination of antigen-coated gelatin particles and μ-capture enzyme-linked immunosobent assay (ELISA) methods for specific IgM antibodies. During re-infection with *M. pneumoniae*, specific IgM production is poor and consequently, for the highest rates of detection in such patients, IgM assays should be supplemented with those for a specific IgG or IgA.

Polymerase chain reaction

There are several very sensitive commercial molecular assay kits for detecting *M. pneumoniae* nucleic acid by the PCR method. As the organism may reside in the respiratory tract for several weeks after infection interpretation of these assays may be difficult. Methods to assess the quantitative level in secretions are awaited. The diagnostic tests for *M. pneumoniae* are summarized in Table 35.4.

Mycoplasmas as contaminants

Mycoplasmas are not infrequent contaminants of tissue cultures. They may be detected by the cultural methods described above or by a DNA probe (e.g. Gen-Probe, San Diego). If they are present antimicrobial susceptibility tests will indicate the appropriate addition to the tissue culture medium. For further information about mycoplasamas, etc., see Taylor-Robinson and Tully (1999).

Table 35.4 Summary of diagnostic tests for *Mycoplasma pneumoniae*

Test	Sample	Sensitivity (%)	Specificity (%)	Cost	Comment
Culture	Throat swabs, sputum, BAL	++	+++	++	2 weeks for colony development, not widely available
Serology: CFT	Blood	++	++	+	Paired samples required
Serology: specific IgM	Blood	++(+)	++	++	Re-infections remain undetected
Serology: IgM and IgG/A	Blood	+++	++	++(+)	Current and re-infections detected
PCR	Throat swabs, sputum, BAL	+++	+++	+++	Not widely available

BAL, bronchoalveolar lavage fluid; CFT, complement fixation test; Ig, immunoglobulin; PCR, polymerase chain reaction
++ 30–90%; +++ ≥ 90%

REFERENCES

Harrison, T. G. and Taylor, A. G. (1988) *A Laboratory Manual for Legionella*. London: Wiley.

Mårdh, P-A. (1999) Mycoplasma and ureaplasma. In: Armstrong, D. and Cohen, J. (eds), *Infectious Diseases*, Vol. 2, Section 8. London: Mosby, pp. 1–6.

Stout, J. E. and Yu, V. L. (1997) Legionellosis. *New England Journal of Medicine* 337: 682–687.

Taylor-Robinson, D. (2002) Mycoplasmas. Atypical pneumonia: genital tract infection. In: Greenwood, D., Slack, R., Peutherer, J. (eds), *Medical Microbiology*, 16th edn. Edinburgh: Churchill Livingstone, pp. 379–389.

Taylor-Robinson, D. and Tully, J. G. (1999) Mycoplasmas, ureaplasmas, spiroplasmas and related organisms. In: Collier, L., Balows, S. and Sussman, M. (eds), *Topley and Wilson's Microbiology and Microbial Infections*, 9th edn, Vol. 2, *Systematic Bacteriology*, CD-ROM. London: Arnold, Chapter 34.

Winn, W. C. (1999) Legionella. In: Murray, P. R., Baron, E. J., Pfaller, M. A. *et al.* (eds), *Manual of Clinical Microbiology*, 7th edn. Washington DC: ASM, pp. 572–585.

36

Staphylococcus and *Micrococcus*

These two genera are considered together, although they are unrelated. They are frequently isolated from pathological material and foods. Distinguishing between them is important. Some staphylococci are pathogens, some are doubtful or opportunist pathogens, and others, and micrococci, appear to be harmless but are useful indicators of pollution. Staphylococci are fermentative, and capable of producing acid from glucose anaerobically; micrococci are oxidative and produce acid from glucose only in the presence of oxygen. Staphylococci divide in more than one plane, giving rise to irregular clusters, often resembling bunches of grapes.

As with other groups of micro-organisms, recent taxonomic and nomenclatural changes have not exactly facilitated the identification of the Gram-positive cocci that are isolated from clinical material and foods. The term 'micrococci' is still loosely but widely used to include many that cannot be easily identified (see also under *Aerococcus*, p. 342 and *Pediococcus*, p. 344).

ISOLATION

Pathological material

Plate pus, urine, swabs, etc., on blood agar or blood agar containing 10 mg/l each of colistin and nalidixic acid to prevent the spread of *Proteus* spp.; also inoculate salt meat broth. Incubate overnight at 37°C. Plate the salt meat broth on blood agar.

Foodstuffs

Prepare 10% suspensions in 0.1% peptone water in a Stomacher. If heat stress is suspected add 0.1-ml amounts to several tubes containing 10 ml brain–heart infusion broth. Incubate for 3–4 h and then subculture. Use one or more of the following media: milk salt agar (MSA), phenolphthalein phosphate polymyxin agar (PPP), Baird-Parker medium (BP) and tellurite polymyxin egg yolk (TPEY). Incubate for 24–48 h. Test the PPP medium with ammonia for phosphatase-positive colonies (see below). There are also commercial tests for the detection of enterotoxins (Berry *et al.*, 1987).

MSA agar relies on the high salt content to select staphylococci. In PPP medium, the polymyxin suppresses many other organisms. BP medium is very selective but may be overgrown by *Proteus* spp. This may be suppressed by adding sulphamethazine 50 μg/ml to the media.

For enrichment or to assess the load of staphylococci, use salt meat broth or mannitol salt broth. Add 0.1 and 1.0 ml of the emulsion to 10 ml broth and 10 ml to 50 ml. Incubate overnight and plate on one of the solid media. If staphylococci are present in large numbers in the food, there will be a heavy growth from the 0.1-ml inoculum; very small numbers may give growth only from the 10-ml sample.

If counts are required, use the Miles and Misra method (see p. 148) and one of the solid media.

STAPHYLOCOCCAL FOOD POISONING

There is no need to examine the stools of all the patients; a sample of 10–25% is sufficient. Inoculate Robertson's cooked meat medium containing 7–10% sodium chloride (salt meat medium) with 2–5 g faeces or vomit and incubate overnight, and plate on blood agar or one of the special staphylococcal media. Examine all food handlers for lesions on exposed parts of the body and swab these, as well as the noses, hands and fingernails of all kitchen staff. Swab chopping boards and utensils that have crevices and cracks likely to harbour staphylococci. Use ordinary pathological laboratory swabs: cottonwool wound on a thin wire or stick, sterilized in a test tube (these swabs can be obtained from most laboratory suppliers). Plate on the solid medium and break off the cottonwool swab in cooked meat medium.

All strains of *S. aureus* isolated from patients, food, implements and food handlers in incidents of staphylococcal food poisoning should be sent to a reference laboratory. Not all strains can cause the disease and others are usually encountered during the investigation (see Chapter 13).

Identification

Colonies of staphylococci and micrococci on ordinary media are golden, brown, white, yellow or pink, opaque, domed, 1–3 mm in diameter after 24 h on blood agar and are usually easily emulsified. There may be β-haemolysis on blood agar. Aerococci show α-haemolysis.

On Baird-Parker medium after 24 h, *S. aureus* gives black, shiny, convex colonies, 1–1.5 mm in diameter; there is a narrow white margin and the colonies are surrounded by a zone of clearing 2–5 mm in diameter. This clearing may be evident only at 36 h.

Other staphylococci, micrococci, some enterococci, coryneforms and enterobacteria may grow and produce black colonies, but do not produce the clear zone. Some strains of *S. epidermidis* have a wide opaque zone surrounded by a narrow clear zone. Any grey or white colonies can be ignored. Most other organisms are inhibited (but not *Proteus* spp. – see above).

On TPEY medium, colonies of *S. aureus* are black or grey and give a zone of precipitation around and/or beneath the colonies.

Examine Gram-stained films. Do clumping factor and coagulase tests on Gram-positive cocci isolated from clinical material and growing in clusters. This is a short cut: strains that are positive by both tests are probably *S. aureus*.

Clumping factor and coagulase tests

Possession of the enzyme coagulase, which coagulates plasma, is an almost exclusive property of *S. aureus*. The slide test detects 'bound' coagulase ('clumping factor'), which acts on fibrinogen directly. The tube test detects 'free' coagulase, which acts on fibrinogen in conjunction with other factors in the plasma. Either or both coagulases may be present.

Slide, clumping factor test

Emulsify one or two colonies in a drop of water on a slide. If no clumping occurs in 10–20 s dip a straight wire into human or rabbit plasma (EDTA) and stir the bacterial suspension with it. *S. aureus* agglutinates, causing visible clumping in 10 s. Use water instead of saline because some staphylococci are salt sensitive, particularly if they have been cultured in salt media. Avoid excess (e.g. a loopful) plasma because this may give false positives. Check the plasma with a known coagulase-positive staphylococcus.

Tube coagulase test

Do this to confirm the clumping factor test or if it is negative. Add 0.2 ml plasma to 0.8 ml nutrient (not glucose) broth in a small tube. Inoculate with the suspected staphylococcus and incubate at 37°C in a water bath. Examine at 3 h and if negative leave overnight at room temperature and examine again.

Staphylococcus aureus produces a clot, gelling either the whole contents of the tube or forming a loose web of fibrin. Longer incubation may result in disappearance of the clot as a result of digestion (fibrinolysis).

Include known positive and negative controls. In the tube test, citrated plasma may be clotted by any organism that can use citrate, e.g. by faecal streptococci (although these are catalase negative), *Pseudomonas* and *Serratia* spp. It is advisable therefore to use EDTA plasma (available commercially) or oxalate or heparin plasma. Check Gram films of all tube coagulase-positive organisms.

There are several commercial kits for identifying *S. aureus* by latex agglutination, e.g. Prolex (Prolab), Microscreen (Microgen) and Staphytect (Oxoid). Such kits remove the potential hazard of using human plasma. Some combine tests for clumping factor and protein A. Others employ haemagglutination.

Phosphatase test

Inoculate phenolphthalein phosphate agar and incubate overnight. Expose to ammonia vapour. Colonies of phosphatase-positive staphylococci will turn pink. *S. aureus* gives a positive test (but negative strains have been reported).

If identification to species (other than *S. aureus*) is required use the API STAPH-IDENT system or test

for nitrate reduction, susceptibility to novobiocin (5-µg disc), and fermentation of mannitol, trehalose and sucrose (Baird-Parker sugar medium) (Table 36.1 and see below). (For further information, see Jones *et al.*, 1990 and Kloos, 1999.)

STAPHYLOCOCCUS

There are now at least 30 species of *Staphylococcus* but some are of little interest. They may be divided into three groups: (1) coagulase positive; (2) coagulase negative and novobiocin susceptible; and (3) coagulase negative and novobiocin resistant. Only known and opportunist pathogens are mentioned below.

Staphylococcus aureus

This species is coagulase and phosphatase positive, forms acid from mannitol, trehalose and sucrose, and is sensitive to novobiocin. Some strains are haemolytic on horse blood agar, but the zone of haemolysis is relatively small compared with the diameter of the colony (differing from the haemolytic streptococcus).

Production of the golden-yellow pigment is probably the most variable characteristic. Young cultures may show no pigment at all. The colour may develop if cultures are left for 1 or 2 days on the

Table 36.1 Properties of some *Staphylococcus* species

Species	Pigment	Coagulase	Phosphatase	Novobiocin 5-µg disc	Acid from Mannitol	Trehalose	Sucrose
S. aureus	+	+	+	S	+	+	+
S. chromogenes	+	−	+	S	v	+	+
S. hyicus	−	−	+	S	−	+	++
S. intermedius	−	v	+	S	+	+	+
S. epidermidis	−	−	+	S	−	−	+
S. warneri	−	−	−	S	v	+	+
S. saprophyticus	−	−	−	R	+	+	+

v, variable; S, sensitive; R, resistant

bench at room temperature. Pigment production is enhanced by the presence in the medium of lactose or other carbohydrates and their breakdown products.

Staphylococcus aureus is a common cause of pyogenic infections and food poisoning (see Chapter 13). Staphylococci are disseminated by common domestic and ward activities such as bedmaking, dressing or undressing. They are present in the nose, on the skin and in the hair of a large proportion of the population.

Methicillin-resistant staphylococci

Methicillin-resistant *S. aureus* (MRSA) is involved in serious hospital infections and requires early identification. Cultures are usually a mixture of resistant and sensitive organisms.

Inoculate mannitol salt agar containing 4 mg/l methicillin and incubate overnight. Most strains of *S. aureus* that ferment mannitol will grow, but so may some coagulase-negative staphylococci. As direct agglutination tests may not be satisfactory on this medium, subculture several colonies on blood agar and test with the growth as described above or with one the kits mentioned. Other media are available: see Davies (1997) and Davies *et al.* (2000); see also Cookson and Phillips (1990).

Coagulase-negative staphylococci

Staphylococcus epidermidis (S. albus)

This produces no pigment; the colonies are 'china white'. Antibiotic-resistant strains are not uncommon and are often isolated from clinical material, including urine, where they may be opportunist pathogens.

Staphylococcus epidermidis is associated with infections with implanted material, e.g. prosthetic heart valves and joints, ventricular shunts and cannulae. Strains resistant to many antibiotics occur.

Staphylococcus saprophyticus

This causes urinary tract infections in young sexually active women. There is a diagnostic kit (Dermaci, Sweden).

Other species

Staphylococcus chromogenes produces pigment like that of *S. aureus* and ferments sucrose. It is an opportunist pathogen and has been found in pigs and in cows' milk. *S. hyicus* is an opportunist pathogen and has been found in dermatitis of pigs, in poultry and cows' milk. *S. intermedius* is known to cause infections in dogs. *S. warneri* and *S. cohnii* are opportunist pathogens.

Staphylococcal toxins

There are kits for testing for the presence of these in foods and also for investigating toxic shock syndrome which is associated with the use of tampons (see Berry *et al.*, 1987; Arbuthnot *et al.*, 1990).

Staphylococcal typing

Strains isolated from hospital infections and food poisoning incidents should be typed for epidemiological purposes. This is best done at a reference laboratory (see Richardson *et al.*, 1992).

The genus *Staphylococcus* is reviewed in a Society for Applied Bacteriology Symposium (Jones *et al.*, 1990), and by Kloos and Bannerman (1995) and Kloos (1999).

MICROCOCCUS

These are Gram-positive, oxidase- and catalase-positive cocci that differ from the staphylococci in that they use glucose oxidatively or do not produce enough acid to change the colour of the indicator in the medium. They are common saprophytes of air, water and soil, and are often found in foods.

The classification is at present confused: the genus contains the organisms formerly called *Gaffkya* and *Sarcina*, tetrad- and packet-forming cocci. These morphological characteristics vary with cultural conditions and are not considered constant enough for taxonomic purposes. The genus *Sarcina* now includes only anaerobic cocci.

At present, it does not seem advisable to describe newly isolated strains by any of the specific names that abound in earlier textbooks.

REFERENCES

Arbuthnott, J. P., Coleman, D. C. and De Azavedo, J. S. (1990) Staphylococcal toxins in human disease. In: Jones, D., Board, R. G. and Sussman, M. (eds), *Staphylococci*. Society for Applied Bacteriology Symposium Series No. 19. Oxford: Blackwells, pp. 101S–108S.

Berry, P. R., Weinecke, A. A., Rodhouse, J. C. and Gilbert, R. J. (1987) Use of commercial tests for the detection of *Clostridium perfringens* and *Staphylococcus aureus* enterotoxins. In: Grange, J. M., Fox, A. and Morgan, N. L. (eds), *Immunological Techniques in Microbiology*. Society for Applied Bacteriology Technical Series No. 24. London: Blackwells, pp. 245–250.

Cookson, B. and Phillips, I. (1990) Methicillin-resistant staphylococci. In: Jones, D., Board, R. G. and Sussman, M. (eds), *Staphylococci*. Society for Applied Bacteriology Symposium Series No. 19. Oxford: Blackwells, pp. 55S–70S.

Davies, S. (1997) Detection of methicillin-resistant *Staphylococcus aureus*: then evaluation of rapid agglutination methods. *British Journal of Biomedical Science* 54: 13–15.

Davies, S., Zadik, P. M., Mason, C. M. *et al.* (2000) Methicillin-resistant *Staphylococcus aureus*: evaluation of five selective media. *British Journal of Biomedical Science* 57: 269–272.

Jones, D., Board, R. G. and Sussman, M. (eds) (1990) Staphylococci. *Journal of Applied Bacteriology Symposium* No. 19, **69**: 1S–188S.

Kloos, W. E. (1999) *Staphylococcus*. In: Collier, L., Balows, A. and Sussman, M. (eds), *Topley and Wilson's Microbiology and Microbial Infections*, 9th edn. Vol. 2, *Systematic Bacteriology*. London: Arnold, Chapter 27, CD-ROM.

Kloos, W. E. and Bannerman, T. L. (1995) *Staphylococcus* and *Micrococcus*. In: Murrau, P. R., Baron, E. J., Pfaller, M. A. *et al.* (eds), *Manual of Clinical Microbiology*, 6th edn. Washington: ASM Press.

Richardson, J. F., Noble, W. C. and Marples, R. M. (1992) Species identification and epidemiological typing of the staphylococci. In: Board, R. G., Jones, D. and Skinner, F. A. (eds), *Identification Methods in Applied and Environmental Microbiology* Society for Applied Bacteriology Technical Series No. 29. Oxford: Blackwells, pp. 193–220.

37

Streptococcus, Enterococcus, Lactococcus, Aerococcus, Leuconostoc and Pediococcus

This group includes organisms of medical, dental and veterinary importance as well as starters used in the food and dairy industries, spoilage agents and saprophytes. The Gram-positive, catalase-negative cocci have undergone extensive taxonomic revision; several new genera and species have been described since the last edition of this book (Facklam and Elliott, 1995; Hardie and Whiley, 1997; Facklam, 2002).

The important characters that distinguish *Streptococcus*, *Enterococcus*, *Lactococcus*, *Leuconostoc*, *Abiotrophia* and *Aerococcus* spp. are shown in Table 37.1.

STREPTOCOCCUS

Streptococci are Gram-positive cocci that always divide in the same plane, forming pairs or chains; they are catalase and oxidase negative, non-sporing and non-motile, and some are capsulated. They are facultatively anaerobic and some require increased CO_2 concentration.

Isolation

From clinical material

Blood agar is the usual primary medium. If incubated anaerobically, important streptococci will grow and many other organisms will be suppressed. If the material is known to contain many other organisms, place a 30-µg neomycin disc on the heavy part of the inoculum. Alternatively use blood agar containing 10 µg/ml colistin and 5 µg/ml oxolinic acid (COBA medium: Petts, 1984) which is selective for streptococci. (Crystal violet blood agar is not very satisfactory because batch variation in the dye affects colony size and amount of haemolysis.)

Islam's medium (Islam, 1977; available commercially) facilitates the recognition of group B strep-

Table 37.1 General properties

Genus	Haemolysis	Growth at 45°C	Growth in 6.5% NaCl	Aesculin hydrolysis	Gas	Vancomycin (30-µg disc)	PYR
Streptococcus	α β n	−	−	v	−	S	−[a]
Enterococcus	α β n	+	+	+	−	S[b]	+
Lactococcus	α n	−	v	−	−	S	+
Aerococcus	α	−	+	v	−	S	±
Leuconostoc	α n	−	v	v	+	R	−
Abiotrophia	α	−	−	−	−	S	+

PYR, pyrrolidonyl-β-naphthylamide reaction
n, none; v, variable; S, sensitive; R, resistant
[a] *S. pyogenes*, *S. iniae* and *S. porcinus* are positive
[b] Resistant strains occur

tococci, e.g. in antenatal screening. Incubate anaerobically at 37°C.

Dental plaque

Plate on blood agar and trypticase yeast extract cystine agar or mitis salivarius agar.

Bovine mastitis

Use Edwards' medium. The crystal violet and thallous sulphate inhibit most saprophytic organisms; aesculin-fermenting saprophytic streptococci give black colonies and mastitis streptococci pale-grey colonies.

Dairy products

Use yeast glucose agar for mesophiles and yeast lactose agar for thermophiles.

Air

For β-haemolytic streptococci in hospital cross-infection investigations, use crystal violet agar containing 1 : 500 000 crystal violet (satisfactory here) with slit samplers. For evidence of vitiation by oral streptococci use mitis salivarius agar.

Identification of streptococci

Colonies on blood agar are usually small, 1–2 mm in diameter, and convex with an entire edge. They may be 'glossy', 'matt' or 'mucoid'. Growth in broth is often granular, with a deposit at the bottom of the tube.

Primary classification is made on the basis of haemolysis on sheep or horse blood agar. α-Haemolytic streptococci, often referred to as 'viridans' streptococci, produce a small greenish zone around the colonies. This is best seen on chocolate blood agar.

β-Haemolytic streptococci give small colonies surrounded by a much larger, clear haemolysed zone in which all the red cells have been destroyed.

Haemolysis on blood agar is only a rough guide to pathogenicity. The β-haemolytic streptococci include those strains that are pathogenic for humans and animals, but the type of haemolysis may depend on conditions of incubation and the medium used as a base for the blood agar. Some α-haemolytic streptococci show β-haemolysis on Columbia-based blood agar.

Some saprophytic streptococci are β-haemolytic, and so are organisms in other genera having similar colonial morphology. Haemolytic *Haemophilus* spp. are often reported as haemolytic streptococci in throat swabs because Gram-stained films are not made.

Clinical strains

β-Haemolytic streptococci from clinical material are usually identified by determining their Lancefield group (see below) or as *S. pyogenes* (group A) by their sensitivity to bacitracin (see below). *S. pneumoniae* (the pneumococcus) is identified by its sensitivity to optochin (see p. 99) or bile solubility. Other streptococci require further tests.

Streptococcal antigens

Species and strains of streptococci are usually identified by their serological group and type. There are Lancefield groups (A–W, excluding I) characterized by a series of carbohydrate antigens contained in the cell wall. Rabbits immunized with known strains of each group produce serum, which will react specifically *in vitro* by a precipitin reaction with an extract of the homologous organism. The carbohydrate is known as the C substance. The streptococci within *S. pyogenes* (group A) can be divided into serological (Griffith) types by means of two classes of protein antigens: M and T.

M is a type-specific antigen near the surface of the organism that can be removed by trypsinization; it is present in matt and mucoid types of colonies but not in glossy types. It is demonstrated by a precipitin test. T is not type specific, may be present with or independent of the M antigen, and is demonstrated by agglutination tests with appropriate antisera.

Lancefield grouping is usually carried out in the laboratory where the organisms are isolated, using sera or agglutination reagents obtained commercially. Only reagents for groups A, B, C, D and G need be used for routine purposes, although most agglutination kits also contain group F.

Workers should be aware that the groups are not always species specific. Some enterococci possess the G antigen in addition to their D antigen. The species in the anginosus group of streptococci (see below) may react as A, C, F or G and, when haemolytic, may be confused with pyogenic streptococci. Some 'viridans' streptococci may also group as A or C and under certain conditions may appear to be β-haemolytic.

Griffith typing of haemolytic (group A, etc.) strains is necessary for epidemiological purposes. Epidemiological typing schemes exist for most of the major pathogens. These are best done by reference laboratories.

Serological grouping of streptococci

There are two approaches. Commercially available kits for latex agglutination are rapid and give results in less than 1 h of obtaining a satisfactory growth on the primary culture. Precipitin tests take longer but allow a larger number of groups to be identified.

Latex agglutination kit tests

Several kits are available from various suppliers, including Oxoid, Murex, Pharmacia, Prolab and Diagnostic Products. The streptococcal antibody is attached to latex particles. These antibody-coated particles are agglutinated when mixed on a slide with suspensions or extracts of streptococci of the same group. In the latex agglutination methods an enzyme or nitrous acid extract of the organism is mixed with the antibody-coated latex reagent on a slide. Groups, A, B, C, D, F and G may be identified by these methods, but it is recommended that biochemical tests are done on group D streptococci to identify enterococci and *S. bovis*, to identify the

species of strains of group C or G and to identify 'viridans' streptococci.

Precipitin tube methods

These have largely been superseded by the agglutination (kit) methods but the following three methods are still in use.

Centrifuge 50 ml of an overnight culture of the streptococcus in 0.1% glucose broth or use a suspension prepared by scraping the overnight growth from a heavily inoculated blood agar plate.

For Lancefield's method suspend the deposit in 0.4 ml 0.2 mol/l hydrochloric acid and place in a boiling water bath for 10 min. Cool, add 1 drop 0.02% phenol red and loopfuls of 0.5 mol/l sodium hydroxide solution until the colour changes to faint pink (pH 7.0). Centrifuge; the supernatant is the extract.

For Fuller's formamide method suspend the deposit in 0.1 ml formamide and place in an oil bath at 160°C for 15 min. Cool and add 0.25 ml acid–alcohol (95.5 parts ethanol : 0.5 parts 1 mol/l hydrochloric acid). Mix and centrifuge. Remove the supernatant fluid and add to it 0.5 ml acetone.

Acetone may explode during centrifugation.

Mix, centrifuge and discard the supernatant. To the deposit add 0.4 ml saline, 1 drop 0.02% phenol red and loopfuls of 0.2 mol/l sodium hydroxide solution until neutral (pH 7.0). This is the extract. It is not a very good method for group D streptococci.

In Maxted's method the C substance is extracted from streptococcal suspension by incubation with an enzyme prepared from a strain of *Streptomyces*. To prepare this enzyme, obtain *Streptomyces* sp. No. 787 from the National Collection of Type Cultures and grow it for several days at 37°C on buffered yeast extract agar. The medium is best sloped in flat bottles (120-ml 'medical flats'). When there is good growth, place the cultures in a freezer or a bowl containing broken pieces of solid carbon dioxide. After 24 h allow the medium to thaw and remove the fluid; this contains the enzyme. Bottle it in small amounts and store in a refrigerator. To prepare the streptococcal extract, suspend a large loopful of growth from a blood agar culture of the organisms into 0.5 ml of enzyme solution in a small tube and place in a water bath at 37°C for 2 h. Centrifuge and use the supernatant to do precipitin tests as described below. This method does not work well with group D strains

Prepare capillary tubes from Pasteur pipettes. Dip the narrow end in the grouping serum so that a column a few millimetres long enters the tube. Place it in a block of Plasticine and, with a very fine Pasteur pipette, layer extract on the serum so that the two do not mix but a clear interface is preserved. Some practice is necessary in controlling the pipette. If an air bubble develops between the two liquids, introduce a very fine wire, when an interface is usually produced.

A positive result is indicated by a white precipitate that develops at the interface. It is sometimes necessary to dilute the extract 1 : 2 or 1 : 5 with saline to obtain a good precipitate.

The bacitracin disc method

Inoculate a blood agar plate heavily and place a commercial bacitracin disc on the surface and incubate overnight. A zone of inhibition appears around the disc if the streptococci are group A. This test is not wholly reliable and is declining in popularity except as a screening method.

Biochemical tests

There are now so many species that conventional tests become very time-consuming. It is best, therefore, to use commercial identification systems such as the API20 Strep (bioMérieux) or ID 32 STREP (bioMérieux) or RapidID STR (Innovative Diagnostic Systems), and refer to their databases. Tables 37.2 and 37.3 show the characteristics of some of the more common species.

Species and groups of streptococci

Lancefield grouping gives a useful indication of the identity of a streptococcus but it is not always species specific. Taxonomic studies have shown that the streptococci may be divided into six clusters, which reflect their ecological and pathogenic properties (Hardie and Whiley, 1997):

1. *Pyogenic*: mainly β-haemolytic species pathogenic for humans or animals.
2. *Anginosus*: found in the human mouth, gut and genital tract, and may be isolated from wound infections and abscesses.
3. *Mitis*: oral streptococci including *S. pneumoniae*.
4. *Salivarius*: some dairy streptococci and species found in the human mouth.
5. *Bovis*: inhabit the gut in humans and animals.
6. *Mutans*: colonize the surfaces of teeth of humans and some animals.

There are some species that are not included in any group.

Pyogenic group

Streptococcus pyogenes (group A)

These are β-haemolytic and are the so-called haemolytic streptococci of scarlet fever, tonsillitis, puerperal sepsis and other infections of humans. Some strains are capsulated and form large (3-mm) colonies like water drops on the surface of the medium. They are differentiated from other species that are grouped as A by a positive PYR (pyrrolidonyl-β-naphthylamide) reaction for which there are commercial kits.

Streptococcus agalactiae (group B)

This is the causative organism of chronic bovine mastitis and an important cause of neonatal meningitis and septicaemia. It is found in the female genital tract and gut. It has been reported as a cause of urinary tract infections. It is usually β-haemolytic but non-haemolytic strains occur. It grows on media containing 10% bile salt and is CAMP (Christie, Atkins and Munch-Petersen) test positive (see p. 93).

Streptococcus dysgalactiae (group C, G or L)

The group C β-haemolytic streptococcus isolated from humans, '*S. equisimilis*', the α- or β-haemolytic

Table 37.2 Pyogenic streptococci and *S. suis*

Species	Lancefield group	Haemolysis	CAMP	PYR	VP	Hydrolysis of Aesculin	Hippurate	Arginine	Growth on bile agar	Acid from Mannitol	Sorbitol	Lactose	Ribose
S. pyogenes	A	β	−	+	−	v	−	+	−	v	−	+	−
S. agalactiae	B	β	+	−	+	−	+	+	+	−	−	v	+
S. equi ssp. *equi*	C	β	−	−	−	−	−	+	−	−	−	−	−
S. equi ssp. *zooepidemicus*	C	β	−	−	−	v	−	+	−	−	+	+	v
S. dysgalactiae ssp. *dysgalactiae*	C	α, β	−	−	−	−	−	+	−	−	+	v	+
S. dysgalactiae ssp. *equisimilis*	C, G L	β	−	−	−	v	−	+	−	−	−	−	+
S. canis	G	β	−	−	−	v	−	+	−	−	−	−	+
S. inae	a	αβ	±	+	−	+	−		−	−	+	−	−
S. porcinus	E, P, U, V^a	β	+	+	+	+	v	+	−	+	+	v	+
S. uberis	E^a	α	−	+	+	+	+	+	−	+	+	+	+
S. suis	D	α β n	−	−	−	+	−	+	+	−	−	+	−

^a No group or not A, B, C, D, F, G

n, none; v, variable

Table 37.3 Examples of the Anginosus, Salivarius, Mitis, Bovis and Mutans groups and other streptococci***

Species	Lancefield group	Haemolysis	Hydrolysis of Aesculin	Hydrolysis of Arginine	VP	Growth on bile agar	Acid from Mannitol	Acid from Sorbitol	Acid from Lactose
Anginosus group									
S. aginosus	A, C, F, G, *	β, n	+	+	+	–	–	–	+
S. constellatus	C, F, *	β, n	+	+	+	–	–	–	v
S. intermedius	*	β, n	+	+	+	–	–	–	+
Salivarius group									
S. salivarius	*	α	+	–	+	–	–	–	–
S. thermophilus	*	α, n	–	–	v	+	–	–	v
Mitis group									
S. pneumoniae**	*	α	–	–	–	–	–	–	–
S. mitis	*	α, β	–	–	–	+	–	–	v
S. oralis	*	α, β	–	–	–	+	+	–	v
Bovis group									
S. bovis	D	α	+	–	+	+	v	–	+
S. equinus	D	α	+	–	+	+	–	–	–
Mutans group									
S. mutans	*	α	+	–	+	–	+	+	+
Other streptococci									
S. acidominus	*	α	+	–	–	–	+	+	+

* No group or not A, B, C, D, F, G
** Optochin sensitive and bile solubility test positive
*** For further information consult Facklam RR (2002)
v, variable; n, no haemolysis

group C from bovine mastitis, *S. dysgalactiae*, the large colony group G β-haemolytic streptococcus from humans, and the group L streptococcus from animals have been shown to be genetically identical and one species – *S. dysgalactiae*. There are two subspecies, *S. dysgalactiae* ssp. *equisimilis*, β-haemolytic human strains of group C or G, and *S. dysgalactiae* ssp. *dysgalactiae*, α-, β- or non-haemolytic strains of group C or L from animals. These can be differentiated from each other by their effect on human or animal fibrin and plasminogen.

Streptococcus equi and S. zooepidemicus (group C)

These β-haemolytic group C streptococci have also been shown to be genetically identical and are now *S. equi* with two subspecies. These are *S. equi* ssp. *equi* which causes strangles in horses and donkeys, and *S. equi* ssp. *zooepidemicus* which comprises group C strains causing infections and outbreaks in animals and some reported human infections.

Streptococcus canis (group G)

A pathogen of dogs and other animals and also an agent of bovine mastitis.

Streptococcus inae

A pathogen of dolphins and of wild and farmed fish.

Streptococcus porcinus

This is found in the respiratory tract and cervical abscesses of pigs

Streptococcus uberis

This is an animal species and one of the agents of bovine mastitis.

Anginosus group

Previously *S. milleri*, this now consists of three species: *S. anginosus*, *S. constellatus* and *S. intermedius*.

Salivarius group

Streptococcus salivarius is part of the normal flora of the mucous membranes of the human mouth. *S. thermophilus* resembles group D streptococci biochemically but has no D antigen; its optimum growth temperature is 50°C. It is used (along with *Lactobacillus bulgaricus*) in the manufacture of yoghurt.

Mitis group

The most important member of this cluster is *S. pneumoniae*, commonly termed the pneumococcus. It is a causative organism of lobar pneumonia, otitis media and meningitis in humans, and of various other infections in humans and animals, including (rarely) mastitis. In clinical specimens the organism appears as a capsulated diplococcus but it usually grows on laboratory media in chains and then shows no capsule. There are several serological types, but serological identification is rarely attempted nowadays. Strains of serotype III usually produce highly mucoid colonies but most other strains give flat 'draughtsman'-type colonies 1–2 mm in diameter with a greenish haemolysis.

Pneumococci often resemble 'viridans' streptococci on culture. The following tests allow rapid differentiation. Pneumococci are sensitive to optochin (ethylhydrocupreine hydrochloride). Most media manufacturers supply optochin discs. A disc placed on a plate inoculated with pneumococci will give a zone of inhibition of at least 10 mm when incubated aerobically (not in carbon dioxide). Pneumococci are lysed by bile; other streptococci are not. Add 0.2 ml 10% sodium deoxycholate in saline to 5 ml of an overnight broth culture (do not use glucose broth). Incubate at 37°C. Clearing should be complete in 30 min.

Pneumococci are alone among the streptococci in fermenting inulin and are not heat resistant as are many 'viridans' streptococci. These two criteria should not be used alone for differentiation.

Streptococcus sanguis

This is associated with endocarditis in humans as are other members of this cluster.

Streptococcus parasanguis

This has been isolated form humans and is reported as a cause of asymptomatic mastitis in sheep.

Bovis group

Streptococcus bovis (group D)

This is present in the human bowel and animal intestines. In bowel disorders of humans it may enter the bloodstream, leading to endocarditis. A blood culture positive for this species is often associated with intestinal malignancy. It may be differentiated from enterococci by a negative PYR reaction.

Streptococcus equinus (group D)

Inhabits animal intestines (especially horses) and is found in milk. Human infections have been reported.

Mutans group

Streptococcus mutans

This is found in dental plaque and is associated with caries.

Other species

Streptococcus suis

This is associated with meningitis in pigs and may infect humans.

Streptococcus acidominimus

This is a commensal in the genital tract of cows.

ENTEROCOCCUS

This genus contains aerobic Gram-positive cocci that form pairs or short chains. As with streptococci, they are catalase negative, but grow at 45°C, in the presence of 40% bile and 6.5% NaCl and hydrolyse aesculin. Haemolysis on blood agar is variable (α, β or none). They are Lancefield group D and most hydrolyse PYR. There are 17 species, some of which are motile. They are found in the intestines of a wide range of animals and are used as indicators of faecal pollution (Leclerc *et al.*, 1996; Goodfree *et al.*, 1997).

Isolation

Clinical material

Plate on blood agar, MacConkey and Edwards' medium. This contains aesculin which, when hydrolysed by enterococci, gives black colonies.

Identification

Test for growth at 10°C and 45°C, aesculin hydrolysis, arginine dihydrolase, motility, production of a yellow pigment, and acid production from mannitol, sorbitol, sucrose, arabinose and adonitol (Table 37.4).

The API20 STREP system identifies most of the species. Some species may be difficult to identify by traditional methods and polymerase chain reaction (PCR) methods are used (Woodford *et al.*, 1997). For further information on identification of this genus see Facklam *et al.* (1999).

Common *Enterococcus* species

The important biochemical reactions are shown in Table 37.4.

Enterococcus faecalis

This is normally present in the human gut, hence its use as an indicator of faecal pollution. It is also found in the gut of chickens but less often in that of other animals. It can be isolated from wound and urinary tract infections. It is the only enterococcus that reduces 2,3,5-triphenyltetrazolium chloride (TTC; in Mead's medium). It grows in the presence of 0.03% potassium tellurite.

Enterococcus faecium

This is present in the intestines of pigs and other animals and has caused spoilage of 'commercially sterilized' canned ham. TTC is not reduced and it does not grow in media containing 0.03% potassium tellurite. It is more resistant to antibiotics than other enterococci and can be isolated from hospital-associated infections.

Enterococcus durans

This is found in milk and dairy products but is uncommon in clinical material.

Enterococcus avium

This is present in the gut of chickens and some other animals, but is not often found in clinical material.

Table 37.4 *Enterococcus* species

Species	Growth at		Arginine dehydrolase	Motility	Yellow pigment	Acid from				
	10°C	45°C				Mannitol	Sorbitol	Sucrose	Arabinose	Adonitol
E. faecalis	+	+	+	–	–	+	+	+	–	–
E. faecium	+	+	+	–	–	+	v	+	+	–
E. avium	+	+	–	–	–	+	+	+	v	+
E. durans	+	+	+	–	–	–	–	–	–	–
E. casseliflavus	+	+	+	+	+	+	v	+	+	–
E. gallinarum	+	+	+	+	–	+	–	+	+	–
E. malodoratus	+	+	–	–	–	+	+	+	–	+
E. mundtii	+	+	+	–	+	+	–	+	+	–

v, variable

Enterococcus casseliflavus

This is associated with plant material but has been isolated from humans. It is motile and if colonies are touched with a swab, the swab will be stained yellow. It has a natural low-level resistance to vancomycin.

Enterococcus gallinarum

This is found in the gut of chickens, it is motile, but does not produce a yellow pigment and also has natural low-level resistance to vancomycin.

Enterococcus mundtii

This produces yellow pigment but is non-motile.

Enterococcus malodoratus

This is found in cheese.

LACTOCOCCUS

Formerly in *Streptococcus*, these species are associated with milk, dairy products, vegetables and fish.

Isolation and identification

Use blood agar or milk agar and incubate at 10, 25 and 45°C. Test for arginine hydrolysis, fermentation of mannitol and sorbitol, and do PYR and Voges–Proskauer tests (Table 37.5).

Lactococcus species

Lactococcus garveiae (previously Enterococcus seriolicida)

This is a cause of mastitis in buffalos and cows; it has been isolated from human blood and urine.

Lactococcus lactis ssp. lactis

This is present in souring milk, to which it imparts a distinctive flavour; it is used as a cheese starter.

Lactococcus lactis ssp. cremoris

This is used as a starter in the dairy industry.

Lactococcus plantarum

This has been implicated in the spoilage of frozen vegetables.

AEROCOCCUS

These species usually occur in clusters, pairs or short chains.

Isolation

Use blood agar for food homogenates and for air sampling. Incubate at 30–35°C.

Table 37.5 *Lactococcus* species and *S. thermophilus*

Species	Growth at		Arginine hydrolysis	Acid from		PYR	VP
	10°C	45°C		Mannitol	Sorbitol		
L. lactis ssp. lactis	+	−	+	+	−	v	v
L. lactis ssp. cremoris	+	−	−	−	−	−	+
L. plantarum	+	−	−	+	+	−	+
S. thermophilus	−	+	−	−	−	−	+

v, variable

Identification

Aerococci give α-haemolysis on blood agar and resemble other 'viridans' streptococci. They grow on bile agar but, unlike the enterococci, they do not grow at 45°C or possess the D antigen.

Aerococcus species

Aerococcus viridans

This is a common airborne contaminant; it is also found in curing brines. In humans it occurs on the skin, and may infect wounds and be responsible for endocarditis. Lobsters infected with this organism develop a pink discoloration on the ventral surface and the blood loses its characteristic bluish-green colour, becoming pink.

Aerococcus urinae

This has been isolated from urine and cases of endocarditis.

LEUCONOSTOC

This genus contains several species of microaerophilic organisms of economic importance. Some produce a dextran slime on frozen vegetables. Leuconostocs are used as cheese starters. When isolated from human infections, because they are resistant to vancomycin, they may be confused with vancomycin-resistant enterococci (VRE), although they grow at 10°C but not at 45°C and are PYR negative.

Isolation

Culture material on semi-solid yeast glucose agar at pH 6.7–7.0. Incubate at 20–25°C under 5% carbon dioxide.

Identification

Test for acid production from sugars – glucose, lactose, sucrose, mannose, arabinose and xylose – in an inoculate MRS medium (see p. 347) and look for dextran slime on solid medium containing glucose (Table 37.6).

Leuconostoc species

Leuconostoc mesenteroides

This is the most important species. It produces gas and a large amount of dextran slime on media and vegetable matter containing glucose. Colonies on agar media without the sugar are small and grey. Growth is very poor on media without yeast extract. It takes part in sauerkraut fermentation and silage production and is responsible for slime disease of pickles (unless *Lactobacillus plantarum* is encouraged by high salinity). It is also responsible for 'slimy sugar' or 'sugar sickness' in peas and fruit juices.

Table 37.6 *Leuconostoc* species

Species	Acid from						Slime
	Arabinose	Xylose	Glucose	Mannose	Lactose	Sucrose	
L. cremoris	–	–	+	–	+	–	–
L. dextranicum	–	v	+	v	+	+	+
L. lactis	–	–	+	v	+	+	–
L. mesenteroides	+	v	+	+	v	+	++
L. paramesenteroides	v	v	+	+	v	+	–
v, variable							

Leuconostoc paramesenteroides

This is similar to *L. mesenteroides* but it does not produce a dextran slime. Widely distributed on vegetation and in milk and dairy products.

Leuconostoc dextranicum

This produces rather less slime than *L. mesenteroides* and is less active biochemically. It is widely distributed in fermenting vegetables and in milk and dairy products.

Leuconostoc cremoris

This is rare in nature but extensively used as a starter in dairy products.

Leuconostoc lactis

This is not common. It is found in milk and dairy products

PEDIOCOCCUS

Pediococci are non-capsulated microaerophilic cocci that occur singly and in pairs. They are nutritionally exacting and are of economic importance in the brewing, fermentation and food-processing industries.

Isolation

Use enriched media and wort agar for beer spoilage pediococci. Tomato juice media at pH 5.5 is best.

Incubate at temperatures according to species sought under 5–10% carbon dioxide.

Pediococci differ from leuconostocs in hydrolysing arginine and not producing slimy or mucoid colonies. They grow in acetate or acetic acid agar.

Identification

If it is necessary to proceed to species identification, inoculate MRS sugar media: lactose, sucrose, galactose, salicin, maltose, mannitol, arabinose and xylose. Incubate at appropriate temperature. Gas is not produced by any species (Table 37.7).

Pediococcus species

Pediococcus damnosus (P. cerevisiae)

This is found in yeasts and wort. Contamination of beer results in a cloudy, sour product with a peculiar odour – 'sarcina sickness'. *P. damnosus* is resistant to the antibacterial agents in hops.

Pediococcus acidi-lactici

This is found in sauerkraut, wort and fermented cereal mashes. It is sensitive to the antibacterial activity of hops.

Pediococcus pentosaceus

Is also found in sauerkraut as well as in pickles, silage and cereal mashes, but not in hopped beer.

Table 37.7 *Pediococcus species*

Species	Growth at		Acid from							
	37°C	45°C	Lactose	Sucrose	Maltose	Mannitol	Galactose	Salicin	Arabinose	Xylose
P. damnosus	−	−	−	−	+	−	−	−	−	−
P. acidi-lactici	+	+	v	v	−	−	+	−	−	−
P. pentosaceus	+	+	+	v	+	−	+	+	+	+
P. urinae-equi	+	−	+	+	+	−	+	+	v	v

v, variable

Pediococcus urinae-equi

This is a contaminant of brewers' yeasts. (The name indicates that it was originally isolated from urine of a horse.)

ABIOTROPHIA

Formerly called nutritionally variant streptococci, these require blood agar containing 0.001% pyridoxal hydrochloride or 0.01% L-cysteine. They show satellitism when cross-streaked with *Staphylococcus aureus*. They are found in the human mouth and have been isolated from endocarditis, ophthalmic infection and brain abscesses.

REFERENCES

Facklam, R. R. (2002) What happened to the streptococci?: overview of taxonomic and nomenclature changes. *Clinical Microbiology Reviews* 15: 613–630.

Facklam, R. R. and Elliott, J. A. (1995) Identification, classification, and clinical relevance of catalase-negative gram-positive cocci; excluding the streptococci and enterococci. *Clinical Microbiology Reviews* 8: 479–495.

Facklam, R. R., Sahm, D. F, Tiexeira, L. M. (1999) Enterococcus. In: Murray, R.M., Baron, E.J., Pfaller, M. A. *et al.* (eds), *Manual of Clinical Microbiology*, 7th edn. Washington DC: ASM Press, pp. 297–305.

Goodfree, A. F., Kay, D. and Wyer, M. D. (1997) Faecal streptococci as indicators of faecal contamination in water. *Society for Applied Bacteriology Symposium Series no. 26* 83: 110S–119S. London: Blackwells.

Hardie, J. M. and Whiley, R. A. (1997) Classification and overview of the genera Streptococcus and Enterococcus. *Society for Applied Bacteriology Symposium Series no.26* 83: 1S–11S.

Islam, A. K. (1977) Rapid recognition of Group B streptococci. *Lancet* i: 256–257.

Leclerc, H., Devriese, L. A. and Mossel, D. A. A. (1996) Taxonomical changes in intestinal (faecal) enterococci and streptococci: consequences on their use as indicators of faecal contamination in drinking water. *Journal of Applied Bacteriology* 81: 459–466.

Petts, D. N. (1984) Colistin-oxolinic acid blood agar: a new selective medium for streptococci. *Journal of Clinical Microbiology* 19: 4–7.

Woodford, N., Egelton, C. M. and Morrison, D. (1997) Comparison of PCR with phenotypic methods for the speciation of enterococci. *Advances in Experimental Medicine and Biology* 418: 405.

38

Lactobacillus and Erysipelothrix

These two genera contain Gram-positive, non-motile rods that are microaerophilic and facultatively anaerobic. Tables 38.1 and 38.2 show the principal characteristics that distinguish them from one another and from *Listeria*.

LACTOBACILLUS

These Gram-positive bacilli may form chains and may resemble corynebacteria. They are non-motile, catalase negative and fermentative, and have specific growth requirements.

Table 38.1 Some *Lactobacillus* species

Species	Growth at		Acid from				Arginine dihydrolase	Aesculin hydrolysis
	15°C	45°C	Lactose	Salicin	Mannitol	Sorbitol		
L. casei	+	v	v	+	+	+	−	+
L. plantarum	+	−	+	+	+	−	−	+
L. brevis	+	−	v	−	−	−	+	v
L. leichmanhii	−	+	v	+	−	−	v	+
L. acidophilus	−	+	+	+	−	−	−	+
L. delbrueckii	−	+	v	v	−	−	v	−
L. salivarius	−	+	+	v	+	+	−	v
v, variable								

Table 38.2 *Lactobacillus*, *Erysipelothrix* and *Listeria* spp.[a]

Genus	Motility	Catalase	Growth at 5°C	Acid from salicin	H_2S production	Growth requirement[b]
Lactobacillus	−	−	−	+	−	+
Erysipelothrix	−	−	+	−	+	−
Listeria	+	+	+	+	−	−

[a] Added for comparison
[b] E.g. MRS medium

They are important in the food, dairy and brewing industries, commensals of animals, and found in the human gut and vagina. Some species are found in plant material. Although generally considered to be non-pathogenic, some strains are associated with caries and others have been recovered from blood cultures.

Isolation

Plate in duplicate on MRS (deMann, Rogosa and Sharp), Rogosa and similar media at pH 5.0–5.8. LS Differential Medium is particularly useful in the examination of yoghurt. Use Raka-Ray agar for beer. Many strains grow poorly, if at all, at pH 7.0, but plate human pathological material on blood agar (see below). Incubate one set of plates aerobically and the other in a 5 : 95 carbon dioxide–hydrogen atmosphere. Incubate cultures from food at 28–30°C and from human or animal material at 35–37°C for 48–72 h. Examine for very small, white colonies. Few other organisms, apart from moulds, will grow in the acid media.

For provisional identification of *L. acidophilus*, which is the most common species in human material, place the following discs on seeded blood agar plates: (1) a sulphonamide sensitivity disc that has been dipped in saturated sucrose solution, and (2) a penicillin (5 units) disc.

Lactobacillus acidophilus gives a green haemolysis. Growth is stimulated by sucrose; it is resistant to sulphonamides and sensitive to penicillin.

Identification

Identification to species is rarely required. It is not easy; reactions vary according to technique and the taxonomic position of some species is still uncertain.

Subculture if necessary to obtain pure growth and heavily inoculate the modified MRS broth containing the following sugars: lactose, salicin, mannitol (for arginine hydrolysis test) and sorbitol. Incubate 28–30°C for 3–4 days. Inoculate tubes of MRS broth and incubate at 15°C and at 45°C (Table 38.1). Test for arginine dihydrolase and aesculin hydrolysis.

Culture on solid medium low in carbohydrate to test for catalase (high-carbohydrate medium may give false positives). The catalase test is negative (corynebacteria and listeria, which may be confused with lactobacilli, are catalase positive).

Table 38.1 lists the properties of some of the more common lactobacilli, but there are many more species. The API 50CHL system is recommended for full identification of members of this genus.

Species of lactobacilli

Lactobacillus acidophilus and *L. casei*

These are widely distributed and found in milk and dairy products and as commensals in the alimentary tract of mammals. They are used, along with other organisms, to make yoghurt and are also associated with dental caries in humans as a secondary invader. *L. acidophilus* is probably the organism described as the Boas–Oppler bacillus, observed in cases of carcinoma of the stomach in humans, and also Doderlein's bacillus, found in the human vagina.

Colonies of this organism on agar media are described as either 'feathery' or 'crab like'.

Lactobacillus plantarum

This species is widely distributed on plants, alive and dead. It is one of the organisms responsible for fermentation in pickles and is used in the manufacture of sauerkraut and silage.

Lactobacillus delbrueckii

This is common in vegetation and grain. It is used as a 'starter' to initiate acid conditions in yeast fermentation of some grains. This organism grows on agar medium as small, flat colonies with crenated edges. This species now includes the former *L. lactis* and *L. bulgaricus*.

Lactobacillus leichmannii

This is found in most dairy products. It is used in the commercial production of lactic acid and grows on agar as small white colonies.

Other species

For these, see Sharpe (1981) and Kandler and Weiss (1986). Some lactobacilli, isolated from meat, are now placed in the genus *Carnobacterium* (see Collins *et al.*, 1987).

Counting lactobacilli

Enumeration of lactobacilli is a useful guide to hygiene and sterilization procedures in dairy industries. Prepare suitable dilutions of bottle or churn rinsings or of swabbings steeped in 0.1% peptone water. Do pour plates (see 'Plate counts', p. 146) in Rogosa medium or one of the other media recommended for lactobacilli. Incubate plates at 37°C for 3 days or at 30°C for 5 days in an atmosphere of 5% carbon dioxide.

Dental surgeons and oral hygienists sometimes request lactobacillus counts on saliva. The patient chews paraffin wax to encourage salivation, and 10 ml saliva are collected and diluted for plate counts. Usually dilutions of 1 : 10, 1 : 100 and 1 : 1000 are sufficient.

Counts of 10^5 lactobacilli/ml of saliva are thought to indicate the likelihood of caries. There is a simple colorimetric test in which 0.2 ml saliva is added to medium, as a shake tube culture, and a colour change after incubation suggests a predisposition to caries.

For more information about this genus, see Sharpe (1981), and Kandler and Weiss (1986).

ERYSIPELOTHRIX

This genus contains small (2 × 0.3 μm), Gram-positive rods that are found in and on healthy animals and on the scales of fish. Humans may be infected through abrasions when handling such animals, particularly pigs and fish. The resulting skin lesions are called erysipeloid. Epizootic septicaemia of mice has also been reported. Table 38.2 shows the characteristics that distinguish *Erysipelothrix* from *Lactobacillus* and *Listeria* spp.

The two species in this genus are aerobic, facultatively anaerobic, non-motile, and produce acid from glucose, lactose and maltose but not from xylose or mannitol. They produce H_2S, do not hydrolyse aesculin, and are catalase negative and resistant to neomycin. They will grow at 5°C (Table 38.2).

Isolation and identification

Culture tissue or fluid from the edge of a lesion. Emulsified biopsy material is better than swabs of lesions. Plate on blood agar and on crystal violet or azide blood agar and incubate at 37°C in 5–10% carbon dioxide for 24–48 h.

Colonies are either small (0.5–1.0 mm) and 'dew-drop' or larger and granular and show α-haemolysis. Young cultures are Gram variable; older cultures show filaments. Test for motility with a Craigie tube. Test for fermentation of glucose, lactose, maltose, sucrose and mannitol using enriched medium, e.g. broth containing Fildes' extract. Inoculate aesculin broth. Inoculate two tubes of enriched agar and incubate one each at 5 and 37°C. Do a catalase test and test for neomycin sensitivity by the disc method. Inoculate gelatin stab tubes, incubate at 22°C for 2–3 days and look for the 'test-tube brush' growth that is characteristic of the two species.

Species of *Erysipelothrix*

There are two species – *E. rhusiopathiae* and *E. tonsillarum* – which have very similar cultural and biochemical properties, except that the latter produces acid from sucrose and the former does not. *E. rhusiopathiae* is a pathogen of pigs; *E. tonsillarum* may be found in their tonsils but is not pathogenic.

For more information about the genus *Erysipelothrix*, see Jones (1986) and McLauchlin and Jones (1999).

REFERENCES

Collins, M. D., Farrow, J. A., Phillips, B. A., Feresu, S. B. and Jones, D. (1987) Classification of *Lactobacillus divergens, Lactobacillus piscicola* and some catalase-negative, asporogenous, rod-shaped bacteria from poultry in a new genus *Carnobacterium. International Journal of Systematic Bacteriology* 37: 310.

Jones, D. (1986) The genus *Erysipelothrix*. In: Sneath, P. H. A. *et al.* (eds), *Bergey's Manual of Systematic Bacteriology*, Vol. 2. Baltimore, MA: Williams & Wilkins, p. 1245.

Kandler, O. and Weiss, N. (1986) The genus *Lactobacillus*. In: Sneath, P. H. A., Mair, N. H. S., Sharpe, M. E. *et al.* (eds), *Bergey's Manual of Systematic Bacteriology*, Vol. 2. Baltimore, MA: Williams & Wilkins, p. 1245.

McLauchlin, J. and Jones, D. (1999) *Erysipelothrix* and *Listeria*. In: Collier, L., Balows, A. and Sussman, M. (eds), *Topley and Wilson's Microbiology and Microbial Infections*, 9th edn. Vol. 2, *Systematic Bacteriology*, CD-ROM. London: Arnold, Chapter 30.

Sharpe, M. E. (1981) The genus *Lactobacillus*. In: Starr, N. P., Stolph, H., Trüper, H. G. *et al.* (eds), *The Prokaryotes: a Handbook on Habitats, Isolation and Identification of Bacteria*. New York: Springer, p. 1653.

39

Corynebacteria

These organisms are frequently club shaped, thin in the middle with swollen ends that contain metachromatic granules. These are best seen when Albert's or Neisser's stain is used. The stained bacilli may also be barred, with a 'palisade' or 'Chinese letter' arrangement, as a result of a snapping action in cell division.

THE DIPHTHERIA GROUP AND RELATED ORGANISMS

Diphtheria is now a very rare disease in developed countries and may be clinically atypical.

If an organism is suspected of being a diphtheria bacillus because of its colonial and microscopic appearance, inform the physician at once and before proceeding to identify the bacillus. If possible send a culture to an appropriate reference laboratory.

It is in the best possible interests of the patient that this is done. No harm will result if the final result is negative, but a delay of 24 h or more to confirm a positive case may have serious consequences, especially in an unimmunized child, apart from the possibility of an outbreak of the disease. A case of diphtheria will usually generate large numbers of swabs from contacts. For the logistics of mass swabbing and the examination of large numbers of swabs in the shortest possible time, see Collins and Dulake (1983).

'Diphtheroids' is an informal name traditionally given by clinical microbiologists to microorganisms which, on primary culture, superficially resemble the diphtheria bacillus.

Isolation and identification of the diphtheria organism and related species

Plate throat, nasal swabs, etc., on Hoyle's or other tellurite media and on blood agar. Incubate at 37°C and examine at 24 and 48 h. For colonial characteristics, see species descriptions.

Make Gram-stained films of suspicious colonies. This is better than Albert's or other stains for the examination of colonies from tellurite media. *C. diphtheriae* tends to decolourize easily and appears Gram negative compared with diphtheroids that stain solidly Gram positive. Stain a control film of staphylococci on the same slide.

Subculture a colony from the tellurite to a Loeffler medium/slope as soon as possible. Incubate at 37°C for 4–6 h. The best results are obtained with Loeffler medium containing an appreciable amount of water of condensation and when the tube has a cottonwool plug or the cap is left loose. Examine Gram- and Albert-stained films for typical beaded bacilli, which may show metachromic granules at their extremities. Inoculate Robinson's serum water sugars (glucose, sucrose and starch) and a urea slope. But do not wait for the result before doing the plate toxigenicity test.

Note that Robinson's sugar medium is buffered and contains horse serum. Unbuffered horse serum medium may give false-positive results because of fermentation of the small amount of glycogen it contains, which is metabolized by many corynebacteria. Similarly, media that contain unheated rabbit serum may give false-positive starch fermentation reactions. Natural amylases hydrolyse starch, forming glucose, which is fermented by the

Table 39.1 Corynebacteria from human sources

Species	Acid from			β-Haemolysis	Urease
	Glucose	Sucrose	Starch		
C. diphtheriae					
gravis	+	–	+	–	–
mitis	+	–	–	+	–
intermedius	+	–	–	–	–
C. ulcerans	+	–	+	–	+
C. pseudodiphtheriticum	–	–		–	–
C. xerosis	+	+	+	–	–
C. jeikanum	+	–		–	–

organism. The biochemical reactions are given in Table 39.1. The API CORYNE kit is useful.

Plate toxigenicity test

Elek's method, as modified by Davies (1974) gives excellent results. This requires control strains, one toxigenic (e.g. NCTC 10648) and one non-toxigenic (e.g. NCTC 10356) and commercial diphtheria antitoxin. These may be maintained on Dorset egg medium slopes. The titres of available antisera vary, so laboratories should titrate each new batch against stock cultures, using concentrations between 500 and 1000 units/ml. This should be done in anticipation, not when a diphtheria investigation is imminent.

For reasons not yet understood, certain batches of medium and serum give poor results. To test the medium and serum, use the control strains noted above and also NCTC 3894, which is weakly toxigenic. The strength of the antitoxin used is critical.

To be of value to the clinician, the plate toxigenicity test must give clear results in 18–24 h. If the toxigenic control does not show lines at 24 h, the medium or serum is unsatisfactory. The test requires a certain amount of experience, and the medium and serum used must be tested frequently.

Inoculate a moist slope of Loeffler's medium with a colony from the primary plate as early as possible in the morning. Also inoculate Loeffler slopes with the stock control cultures. Incubate at 37°C for 4–6 h.

Melt two 15-ml tubes of Elek's medium and cool to 50°C. Add 3 ml sterile horse serum to each and pour into plastic or very clear glass Petri dishes. Place immediately, in the still liquid medium in each plate, a strip of filter paper (60 × 15 mm) that has been soaked in diphtheria antitoxin. Place the strip along a diameter of the plate. Dry the plates. These may be stored in a refrigerator for several days.

On each plate, streak the 4–6 h Loeffler cultures at right-angles to the paper strip so that the unknown strain is between the toxigenic and non-toxigenic controls as in Figure 39.1, approximately 10 mm apart. Incubate at 37°C for 18–24 h. If no lines are visible re-incubate for a further 12 h. Examine by transmitted light against a black background. If the unknown strain is toxigenic, its precipitation lines should join those of the toxigenic control, i.e. a reaction of identity. Strains that produce little toxin may not show lines but turn the lines of adjacent toxigenic strains.

An alternative method is that of Jamieson (1965). Prepare the Loeffler cultures as above. Pour the plates but do not add the antitoxin strip. Dry the plates and place along the diameter of each a filter paper strip (75 × 5 mm) that has been soaked in diphtheria antitoxin (for concentration see above). Place the plates over a cardboard template marked out as in Figure 39.2 and heavily inoculate them from the Loeffler cultures in the shape of the arrowheads. Place the unknown strain between the two controls. Incubate at 37°C for 18–24 h and examine for reactions of identify.

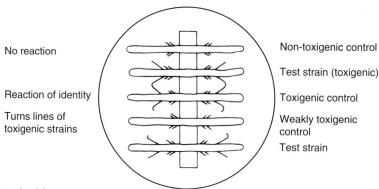

Figure 39.1 Elek plate toxigenicity test

Labels (left, top to bottom): No reaction; Reaction of identity; Turns lines of toxigenic strains

Labels (right, top to bottom): Non-toxigenic control; Test strain (toxigenic); Toxigenic control; Weakly toxigenic control; Test strain

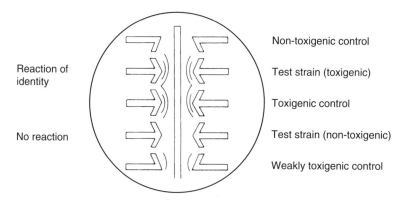

Figure 39.2 Jamieson's method for plate toxigenicity test

Labels (left): Reaction of identity; No reaction

Labels (right, top to bottom): Non-toxigenic control; Test strain (toxigenic); Toxigenic control; Test strain (non-toxigenic); Weakly toxigenic control

Report the organism as toxigenic or non-toxigenic *C. diphtheriae* cultural type *gravis*, *mitis* or *intermedius*.

Note that, in any outbreak of diphtheria caused by a toxigenic strain, a proportion of contacts may be carrying non-toxigenic *C. diphtheriae*.

Species of corynebacteria from human sources

The following general descriptions of colony appearance and microscopic morphology of diphtheria bacilli and diphtheroids is a guide only. The appearance of the colony varies considerably with the medium and it is advisable to check this period-ically with stock (NCTC and ATCC) organisms and, if possible, strains from recent cases of the disease. Even then, strains may be encountered that present new or unusual colony appearances.

Corynebacterium diphtheriae

Three biotypes, *gravis*, *mitis* and *intermedius*, were originally given these names because of their respective association with the severe, mild and intermediate clinical manifestations of the disease. These names have become attached to the colonial forms and related to starch fermentation. These properties do not always agree and it is more important to carry out the toxigenicity test than to attempt to interpret the colonial and biochemical properties, which should be used only to establish that the organism belongs to the species. The following colony appearances relate to tellurite medium of the Hoyle type.

Corynebacterium diphtheriae, gravis type

At 18–24 h, the colony is 1–2 mm in diameter, pearl grey with a darker centre; the edge is slightly crenated. It fractures easily when touched with a wire but is not buttery. At 48 h, it has enlarged to 3–4 mm and is much darker grey, with the edge markedly crenated and with radial striations. This is the 'daisy head' type of colony. The organism does not emulsify in saline. Films stained with methylene blue show short, often barred or beaded, bacilli arranged in irregular 'palisade' form or 'Chinese letters'. Well-decolourized Gram films compared on the same slide with staphylococci or diphtheroids appear relatively Gram negative. Glucose and starch are fermented, but not sucrose. There is no haemolysis on blood agar.

Corynebacterium diphtheriae, mitis type

At 18–24 h, colonies are 1–2 mm across and are dark-grey, smooth and shining. The colonies fracture when touched but are much more buttery than the gravis type. At 48 h, they are larger (2–3 mm diameter), less shiny and darker in the centre, which may be raised, giving the 'poached egg' appearance. The organisms emulsify easily in saline. Films show long, thin bacilli, usually beaded and with granules, which may not be very Gram positive (see above). Glucose is fermented but not starch or sucrose. Colonies on blood agar are haemolytic.

C. diphtheriae, intermedius type

Colonies at 18–24 h are small (1 mm diameter), flat with sharp edges, black and not shining or glossy. At 48 h, there is little change. They are easily emulsified and films show beaded bacilli, intermediate in size between gravis and mitis types. Glucose is fermented but not starch or sucrose. Colonies on blood agar are non-haemolytic.

Corynebacterium ulcerans

This organism resembles the gravis type of diphtheria bacillus and is often mistaken for it. It stains more evenly, however, and the bacilli are usually shorter than those of C. diphtheriae. Colonies tends to be more granular in the centre. Glucose and starch are fermented but not sucrose. C. ulcerans splits urea; C. diphtheriae does not. C. ulcerans also grows quite well at 25–30°C. It gives a reaction of incomplete identity with the plate toxigenicity test and C. diphtheriae antitoxin. This organism does not cause diphtheria and is not readily communicable, but can cause ulcerated tonsils.

Corynebacterium pseudodiphtheriticum

Formerly known as Hoffman's bacillus, colonies of this species at 18–24 h are 2–3 mm in diameter, domed, shiny, and dark-grey or black. They are buttery (occasionally sticky) and easily emulsified, and films show intensely Gram-positive bacilli that are smaller and much more regular in shape, size and arrangement than diphtheria bacilli. Carbohydrates are not fermented. These organisms are common and harmless commensals.

Corynebacterium xerosis

At 18–24 h, colonies are 1–2 mm in diameter, flat, grey and rough, with serrated edges. At 48 h, they are 2–3 mm in diameter. Films show beaded bacilli, very Gram positive. This is a commensal of the conjunctivae and often found in the female genitalia. It is not a pathogen but resembles very closely the diphtheria bacillus in microscopic (but not colonial) morphology. It is, however, much more strongly Gram positive than C. diphtheriae and produces acid from glucose, usually sucrose but not starch.

Corynebacterium jeikeium

Formerly the 'JK diphtheroids', these have been isolated from blood and 'sterile' body fluids. They may colonize prostheses and are resistant to many antibiotics (see Jackman et al., 1987).

CORYNEBACTERIA FROM ANIMAL SOURCES

Isolation and identification

Culture, pus, etc., on blood agar and tellurite medium, incubate at 37°C and examine at 24 and

Table 39.2 Corynebacteria from animal sources

Species	Gelatin liquefaction	Acid from					β-Haemolysis	Urease
		Glucose	Sucrose	Mannitol	Maltose	Lactose		
C. pseudotuberculosis	v	+	v	–	+	v	v	+
C. kutscheri	–	+	+	–	+	–	–	v
C. renale	–	+	–	–	v	–	–	+
C. bovis	–	+	–	v	+	v	–	–

v, variable

48 h for colonies of beaded Gram-positive bacilli. Some corynebacteria are haemolytic on blood agar. Small grey–black colonies are usually found on tellurite medium.

Subculture on Loeffler medium and examine Albert-stained films at 6–18 h for clubbed or beaded bacilli which may have metachromatic granules. From the Loeffler medium inoculate gelatin medium, nutrient agar, Robinson's serum water with glucose, sucrose, mannitol, maltose, lactose and urea medium. Incubate at 37°C for 24–48 h (Table 39.2). The API CORYNE kit is useful.

Corynebacterium pseudotuberculosis (C. ovis)

Colonies are small, usually grey but may be yellowish and are haemolytic, especially in anaerobic cultures. Loeffler serum may be liquefied. The bacilli are slender and clubbed, very like C. diphtheriae. It is responsible for epizootics in rodents and is a nuisance in laboratory animal houses.

This is the Preisz–Nocard bacillus that causes pseudotuberculosis in sheep, cattle and swine and lymphangitis in horses.

Corynebacterium kutscheri (C. murium)

Colonies on blood agar are flat, grey and non-haemolytic. The bacilli are slender, often filamentous. Loeffler medium is not liquefied. It is responsible for fatal septicaemia in mice and rats.

Corynebacterium renale

This is now believed to be three separate species – C. renale, C. pilosum and C. cystitidis – all associated with urinary tract infections of cattle.

Corynebacterium bovis

This is found in cows' udders and may cause mastitis.

Two organisms formerly regarded as corynebacteria have been transferred to other genera. 'C. equi' is now Rhodococcus equi (see p. 402) and 'C. pyogenes' is now Actinomyces pyogenes (see p. 406).

'CORYNEFORM BACILLI'

This informal name is given to various bacilli that resemble corynebacteria in morphology and are found in association with animals and dairy products. They are relatively heat resistant (70°C for 15 min). One such is Microbacterium lacticum, which is thermoresistant and a strict aerobe. It is non-motile, grows slowly at 30°C on milk agar, forms acid from glucose and lactose, is catalase positive, hydrolyses starch, is lipolytic and does not liquefy gelatin in 7 days. A pale-yellow pigment may be formed. These organisms may cause spoilage in milk and indicate unsatisfactory dairy hygiene. They are also found in deep litter in hen houses.

The genus Corynebacterium is reviewed by von Graevenitz et al. (1998). See also Clarridge et al. (1999).

REFERENCES

Clarridge, J. E., Popovic, T. and Inzana, T. J. (1999) Diphtheria and other corynebacterial and coryneform infections. In: Collier, L., Balows, A. and Sussman, M. (eds), *Topley and Wilson's Microbiology and Microbial Infections*, 9th edn. Vol. 3, *Bacterial Infections*, CD-ROM. London: Arnold, Chapter 19.

Collins, C. H. and Dulake, C. (1983) Diphtheria: the logistics of mass swabbing. *Journal of Infection* **6**: 227–230.

Davies, J. R. (1974) Identification of diphtheria bacilli. In: Collins, C. H. and Willis, A. T. (eds), *Laboratory Methods 1*. Public Health Laboratory Service Monograph No. 4. London: HMSO.

Jackman, P. J. H., Pitcher, D. G., Pelcznska, S. and Borman, P. (1987) Classification of corynebacteria associated with endocarditis (Group JK) as *Corynebacterium jeikeium* sp. nov. *Systematic and Applied Microbiology* **9**: 83–90.

Jamieson, J. E. (1965) A modified Elek test for toxigenic *Corynebacterium diphtheriae*. *Monthly Bulletin of the Ministry of Health* **24**: 55–58.

von Graevenitz, A. W. C., Coyle, M. B. and Funke, G. (1998) Corynebacteria and rare coryneforms. In: Collier, L., Balows, A. and Sussman, M. (eds), *Topley and Wilson's Microbiology and Microbial Infections*, 9th edn. Vol. 2, *Systematic Bacteriology*. London: Arnold, Chapter 25.

40

Listeria and *Brochothrix*

These are aerobic, Gram-positive, catalase-positive, non-sporing rods. Both are of interest to the food industry. *Listeria* is associated with food-borne disease (see Chapter 13) and *Brochothrix* is a food-spoilage organism. Table 40.1 shows some distinguishing characteristics.

Table 40.1 *Listeria* and *Brochothrix* species

Genus	Growth	
	At 37°C	On Gardner's STAA
Listeria	+	−
Brochothrix	−	+

LISTERIA

These are aerobic, facultative anaerobic, Gram-positive rods that are catalase positive and ferment carbohydrates. They do not grow on Gardner's STAA medium.

They are found in grass, silage, soil, sewage and water. There are seven species, two of which, *L. monocytogenes* and *L. ivanovii*, are pathogenic for humans and animals. Listerias have become important food-borne pathogens (see Chapter 13). They survive in cold storage and may multiply at low temperatures.

Isolation

Clinical material

Plate cerebrospinal fluid (CSF), centrifuged deposits and blood cultures on blood agar and (Oxford) listeria-selective agar. There are other selective commercially available media, which contain various antibiotics (e.g. cycloheximide colistin sulphate, cefotetan, fosfomycin) that suppress the growth of other organisms.

Homogenize tissues in peptone water diluent and plate on blood agar and listeria-selective agar. Treat the remainder of the homogenate as for faeces (see below). Incubate primary plates at 37°C for 24–48 h.

Homogenize faeces in peptone water. Add 1 volume of homogenate to 9 volumes of listeria selective broth. Subculture at 4 and 24 h to selective media. At 24 h add 0.1 ml to secondary listeria-selective broth. Incubate at 37°C for 24–48 h and plate on selective media.

Food

Homogenize 25 g of the material in 225 ml commercial listeria-enrichment broth. Incubate at 30°C for 48 h. Subculture to listeria-selective (Modified Oxford – MOX – or PALCAM) agar and incubate at 37°C for 48 h under microaerophilic conditions (e.g. with campylobacter gas-generating kit, Oxoid). Colonies on MOX are black, and those on PALCAM are grey–green or black, with shrunken centres and a black halo. Neither medium is wholly selective for listerias.

Identification

Colonies on blood agar are small (1 mm diameter), smooth and surrounded by a narrow zone of β-haemolysis. On (Oxford) selective agar they are the same size but are black, with a black halo and a sunken centre. Colonies on translucent media, viewed by light reflected at 45° from below (Henry method, see Prentice and Neaves, 1992), are grey–blue in colour and finely textured. Colonies of most other organisms are opaque and/or otherwise coloured.

Examine for typical 'tumbling' motility at 20–22°C (no higher, because listerias are non-motile above 25°C). Test for acid production from glucose, rhamnose, xylose and mannitol (10% carbohydrate in purple broth) and do the CAMP (Christie, Atkins and Munch-Petersen) test.

CAMP test

Use blood agar prepared from fresh, washed, defibrinated sheep blood. Make dense suspensions from colonies of the suspected listerias and of *Staphylococcal aureus* (NCTC 1803). Make streaks of the *S. aureus* across the blood agar, about 2–3 cm apart. Streak the suspect listerias at right angles but do not touch the other streaks. Incubate at 37°C for 12–18 h (not longer).

Listeria monocytogenes and *L. seeligeri* give narrow zones of β-haemolysis near the staphylococcal cross steaks. Other species give no reactions (Table 40.2).

Kit and commercial tests

There are several commercial kits that facilitate the testing of food samples. These include the Rosco

Listeria ZYM (Lab M), Pathalert (Merck), Microscreen (Microbact), Mastalist (Mast), Micro-ID and the API listeria strip. There is an enzyme-linked immunosorbent assay (ELISA) kit that uses monoclonal antibodies (Listeria-TEC, Organon), and gene probe systems (Gene-Trak and Gen-Probe) For a review of some of these, see Beumer *et al.* (1996).

Listeria species

Listeria monocytogenes

This is responsible for human listeriosis, usually food borne. Symptoms include meningitis (especially in neonates) and septicaemia.

Listeria ivanovii

This causes septicaemia, encephalitis and abortion in farm animals and also disease in humans.

The other species, *L. innocua*, *L. seeligeri*, *L. welshimeri*, *L. grayi* and *L. murrayi*, are not clinically significant but are included in Table 40.2 because it is important to distinguish them from the pathogens.

For a review of listeriosis, see Bahk and Marth (1990), and for more detailed information about isolation and identification, see Prentice and Neaves (1992).

Table 40.2 *Listeria* species

Species	β-Haemolysis	CAMP test (*S. aureus*)	Acid from		
			Rhamnose	Xylose	Mannitol
L. monocytogenes	+	+	+	−	−
L. ivanovii	+	−	−	+	−
L. innocua	+	−	v	−	−
L. seeligeri	w	w	−	v	−
L. welshimeri	−	−	v	+	−
L. grayi	−	−	−	−	+
L. murrayi	−	−	−	−	+

w, weak; v, variable

BROCHOTHRIX

These are aerobic, facultative anaerobic, Gram-positive, non-motile rods that occur singly or in short chains or filaments, often with pleomorphic forms. They are catalase positive, ferment carbohydrates, and grow at 4°C and 25°C, but not at 37°C. They grow poorly on MRS but well on STAA media.

There are two species: *B. thermosphacta* is a meat spoilage organism, found in souring ('British fresh') sausages at low temperatures, in film-packaged products, milk and cheese. *B. campestris* is found in grass and soil.

Isolation

Culture on Gardner's STAA medium and incubate at 20–25°C.

Identification

The most important characteristic is growth on STAA medium. Test for growth in the presence of 8% NaCl and for hippurate hydrolysis. *B. thermosphacta* grows in the NaCl medium but is hippurate negative. *B. campestris* fails to grow in NaCl medium but hydrolyses hippurate.

For more information, see Gardner (1966) and Dodd and Dainty (1992).

REFERENCES

Bahk, J. and Marth, E. H. (1990) Listeriosis and *Listeria monocytogenes*. In: Cliver, D. O. (ed.), *Foodborne Diseases*. New York: Academic Press, pp. 248–259.

Beumer, R. R., te Giffel, M. C., Kok, M. T. C. *et al.* (1996) Confirmation and identification of *Listeria* species. *Letters in Applied Microbiology* **22**: 448–452.

Dodd, C. E. R. and Dainty, R. H. (1992) Identification of *Brochothrix* by intracellular and surface biochemical composition. In: Board, R. G., Jones, D. and Skinner, F. A. (eds), *Identification Methods in Applied and Environmental Microbiology*. Society for Applied Bacteriology Technical Series No. 29. Oxford: Blackwells, pp. 297–310.

Gardner, G. A. (1966) A selective medium for the enumeration of *Microbacterium thermosphactum* in meat and meat products. *Journal of Applied Bacteriology* **29**: 455–457.

Prentice, G. A. and Neaves, P. (1992) The identification of *Listeria* species. In: Board, R.G., Jones, D. and Skinner, F.A. (eds), *Identification Methods in Applied and Environmental Microbiology*. Society for Applied Bacteriology Technical Series No. 29. Oxford: Blackwells, pp. 283–296.

41

Neisseria

Members of *Neisseria* species are aerobic, oxidase and catalase positive, Gram-negative cocci. They occur mostly as oval or kidney-shaped cocci arranged in pairs, with their long axes parallel. They are non-motile, non-sporing and oxidase positive, and reduce nitrates to nitrites. The two important pathogens are *N. gonorrhoeae* (the gonococcus), causative organism of gonorrhoea, and *N. meningitidis* (the meningococcus), one of the organisms causing meningitis. Both require media containing blood or serum. Colonies are non-pigmented and translucent.

There are about 17 other species that are commensals or opportunist pathogens. Most of them grow at 22°C on nutrient agar with opaque and often yellow-pigmented colonies.

THE TWO IMPORTANT PATHOGENS

Neisseria gonorrhoeae

Examine Gram-stained films of exudates (urethra, cervix, conjunctivae, etc.) for intracellular Gram-negative diplococci. Exercise caution in reporting the presence of gonococci in vaginal and conjunctival material and in specimens from children.

Culture at once or use one of the transport media. Culture on deep plates (20–25 ml of medium) of New York City (NYC) or Thayer Martin (TM) media with commercial selective supplements. Incubate all cultures 35–36°C (better than 37°C) in 5–10% CO_2 and 70% humidity. Examine at 24 and 48 h. Colonies of *N. gonorrhoeae* are transparent discs about 1 mm in diameter, later increasing in size

and opacity, when the edge becomes irregular. Test suspicious colonies by the Gram film and oxidase test. Typical morphology and a positive oxidase test are presumptive evidence of the gonococcus in material from the male urethra, but not from other sites.

For rapid results test colonies by FA method, co-agglutination (Phadebact) or rapid biochemical tests (Gonocheck II, Rapid IDNH).

Subculture oxidase positive colonies on chocolate agar and incubate overnight in a 10% carbon dioxide atmosphere. Immediate subculture of colonies on primary media into carbohydrate test media may not be satisfactory because such media contain antibiotics; other organisms may grow. Repeat the oxidase test on the subculture and emulsify positive colonies in about 1 ml serum broth. Use this to test for acid production from glucose, maltose, sucrose and lactose in a serum-free medium. Some serum media may contain a maltase that may give a false-positive reaction with maltose. The best results are obtained with a semisolid medium, as gonococci do not like liquid media (Table 41.1). The API QUAD-FERM and API WH are useful.

Confirm fermentation results by agglutination or co-agglutination tests (Phadebact).

Exercise caution in reporting these organisms without adequate experience, especially if they are from children, eye swabs, anal swabs or other sites. They may be meningococci or other *Neisseria* that are genital commensals, which do not cause sexually transmitted disease. Sugar reactions and colonial morphology are not entirely reliable. Certain other organisms, *Moraxella* (see p. 307) and *Acinetobacter calcoaceticus*, may resemble gonococci in direct films and on primary isolation. *Moraxella*, like *Neisseria*, is oxidase positive but *Acinetobacter* is oxidase negative.

Table 41.1 *Neisseria* species

Species	Growth on nutrient agar	Growth at 22°C	Acid from			
			Glucose	Maltose	Sucrose	Lactose
N. gonorrhoeae	–	–	+	–	–	–
N. meningitidis	–	–	+	+	–	–
N. lactamica	+	–	+	+	–	+

For further information on gonococci, see Jephcott and Egglestone (1985), Jephcott (1987) and Morse and Genco (1999).

Neisseria meningitidis

Examine films of spinal fluid for intracellular Gram-negative diplococci. Inoculate blood agar and chocolate agar with cerebrospinal fluid (CSF), or its centrifuged deposit. Incubate, preferably in a 5% carbon dioxide atmosphere, at 37°C overnight. Incubate the remaining CSF overnight and repeat culture on chocolate agar.

Colonies of meningococci colonies are transparent, raised discs about 2 mm in diameter at 18–24 h. In CSF cultures, the growth will be pure whereas, in eye discharges and vaginal swabbings of young children, etc., other organisms will be present. Subculture oxidase-positive colonies of Gram-negative diplococci in semisolid serum sugar media (glucose, maltose, sucrose). Meningococci give acid in glucose and maltose (see Table 41.1).

Agglutinating sera are available but should be used with caution. There are at least four serological groups. Slide agglutinations may be unreliable. It is best to prepare suspensions in formol saline for tube agglutination tests.

OTHER SPECIES AND SIMILAR ORGANISMS

These grow on nutrient agar. Colonies are variable: 1–3 mm, opaque, glossy or sticky, fragile or coherent; others may be rough and granular.

Neisseria lactamica is important in that it might be confused with pathogenic species because it is oxidase positive and does not grow at 22°C. It is fermentative and produces acid from lactose, but this may be delayed. The pathogens do not change lactose and are oxidative.

Other species include *N. flavescens*, which produces a yellow pigment, *N. mucosa*, *N. elongata* and *N. subflava*. For the properties of these see *Cowan and Steel's Manual* (Barrow and Feltham, 1993) and Johnson (1983) for potential pathogenicity.

Gemella haemolysans is sometimes confused with neisserias because, although it is Gram positive, it is easily decolourised. It differs from the neisserias in being β-haemolytic, oxidase and catalase negative.

REFERENCES

Barrow, G. I. and Feltham, R. K. A. (eds) (1993) *Cowan and Steel's Manual for the Identification of Medical Bacteria*, 3rd edn. Cambridge: Cambridge University Press.

Jephcott, A. E. (1987) Gonorrhoea. In: *Sexually Transmitted Diseases*. London: Public Health Laboratory Service, pp. 24–40.

Jephcott, A. E. and Egglestone, S. I. (1985) *Neisseria gonorrhoeae*. In: Collins, C. H. and Grange, J. M. (eds), *Isolation and Identification of Microorganisms of Medical and Veterinary Importance*. Society for Applied Bacteriology Technical Series No. 21. London: Academic Press, pp. 143–160.

Johnson, A. P. (1983) The pathogenic potential of

commensal species of *Neisseria*. *Journal of Clinical Pathology* **36**: 213–223.

Morse, S. A. and Genco, C. A. (1999) *Neisseria*. In: Collier, L., Balows, S. and Sussman, M. (eds), *Topley and Wilson's Microbiology and Microbial Infections*, 9th edn, Vol. 2, *Systematic Bacteriology*, CD-ROM. London: Arnold, Chapter 37.

42

Bacillus

This is a large genus of Gram-positive, spore-bearing bacilli that are aerobic, facultatively anaerobic and catalase positive. Many species are normally present in soil and in decaying animal and vegetable matter. One species, *Bacillus anthracis*, is responsible for anthrax in humans and animals. Several other species are now known to cause disease in humans (e.g. food poisoning) and animals. Some are responsible for food spoilage.

The bacilli are large (up to 10×1 µm) and commonly adhere in chains. Most species are motile, some are capsulated, and some produce a sticky levan on media that contain sucrose. Spores may be round or oval, central or subterminal.

Physiological characters vary. There are strict aerobes and species that are facultative anaerobes. A few are thermophiles. In old cultures, Gram-negative forms may be seen and some species are best described as 'Gram indifferent'.

ANTHRAX

Bacillus anthracis causes a fatal septicaemia in animals. It is transmissible to humans, producing a localized cutaneous necrosis ('malignant pustule') or pneumonia (once known as woolsorters' disease) as a result of the inhalation of spores. Such inhalation may also result in a 'flu-like' illness, as has happened when the organisms were used as a terrorist weapon.

Imported bone meal, meat meal and other fertilizers, and sometimes feeding-stuffs may be infected. These are often prepared from the remains of animals that died of anthrax. In most developed countries, an animal dying of anthrax must be buried in quicklime under the supervision of a veterinarian and below the depth at which earthworms are active.

There are regulations governing the importation of hairs and hides from foreign countries where anthrax is endemic.

Isolation of *B. anthracis* from pathological material

Veterinarians often diagnose anthrax in animals on clinical grounds and by the examination of blood films stained with polychrome methylene blue. These show chains of large, square-ended bacilli with the remains of their capsules forming pink-coloured debris between the ends of adjacent organisms. This is M'Fadyean's reaction.

Culture animal blood, spleen or other tissue ground in a Griffith's tube or macerator. In humans, culture material from the cutaneous lesions, faeces, urine and sputum.

Plate on blood agar and on PLET medium: i.e. brain–heart infusion agar containing (per millilitre): polymyxin, 30 units; lysozyme, 40 µg; EDTA, 300 µg; thallous acetate, 40 µg. Incubate overnight. Subculture woolly or waxy 'medusa head' colonies of Gram-positive rods into Craigie tubes to test for motility and on chloral hydrate blood agar. It may be difficult to obtain pure growths without several subcultures. For identification, see below.

Isolation of B. anthracis from hairs, hides, feeding-stuffs and fertilizers

The sample should be in a 200- to 300-ml screw-capped jar, and should occupy about 110 ml.

Add sufficient warm 0.1% peptone solution to cover it. Shake, stand at 37°C for 2 h, decant and heat the fluid at 60°C for 15–30 min. If the suspension is very turbid, centrifuge it, retain the deposit and make 10^{-1}, 10^{-2} and 10^{-3} dilutions in sterile water. Plate these and the deposit on PLET medium (see above). Incubate at 37°C for 36–48 h. Subculture characteristic colonies as described above.

In addition, place 0.1- to 0.5-ml amounts of the dilutions and a few loopfuls of the deposit in a Petri dish, pour on 15–20 ml melted PLET medium at 40–50°C and incubate as above. Examine under a low-power binocular (plate) microscope. Ignore surface colonies and look for deep colonies that resemble dahlia tubers or Chinese artichokes. Subculture these as above.

Identification of B. anthracis

The growth of B. anthracis on agar media has a characteristic woolly or waxy nature when touched with a wire (the 'tenacity test').

Bacillus anthracis is non-motile. The Craigie tube is the best way of demonstrating this. Most other Bacillus species are motile. There is no haemolysis on standard blood agar media. It may be readily identified by immunofluorescence. Sera are available commercially. If the anthrax w (or γ) bacteriophage is available (e.g. from a research or commercial institution), make a lawn of the suspect culture on blood agar, place one drop (0.02 ml) of the phage (titre about 10^6 plaque-forming units [pfu]/ml) on the surface and incubate at 37°C overnight. B. anthracis gives a clear zone of lysis (so do some other Bacillus spp.). Some research and commercial institutions keep stocks of the phage but it is not available commercially.

Bacillus anthracis is sensitive to penicillin (10-unit disc on nutrient agar) and gives a positive 'string-of-pearls' appearance in a wet preparation after growing in nutrient broth containing 0.5 unit penicillin per ml for 3–6 h at 37°C.

The biochemical properties given in Table 42.1 may be helpful but not altogether reliable because of much variation within and between species. The API 50 CHB kit is probably better.

For further information on B. anthracis, see Turnbull (1990), Turnbull and Kramer (1995), and Quinn and Turnbull (1999).

FOOD POISONING

Several species of Bacillus are associated with food poisoning (see Chapter 13). They are B. cereus, and, less frequent, B. subtilis, B. licheniformis, B. pumilus and B. sphaericus (Turnbull, 1997; see also Chapter 13).

Isolation from food, etc.

Make 10% suspensions of the food, faeces and vomit in 0.1% peptone water, using a Stomacher. Inoculate glucose tryptone agar and Bacillus cereus-Selective Agar. Incubate at 30°C for 24–48 h. This medium shows the egg yolk reaction and mannitol fermentation, and encourages sporulation (see below). Colonies of B. cereus are yellow, whereas other species, which ferment mannitol, are pink.

Prepare 10^{-1}, 10^{-2} and 10^{-3} dilutions of these and plate 0.1-ml amounts on B. cereus-selective agar. Incubate at 37°C for 24 h and then at 25–30°C for a further 24 h. Count colonies of presumptive B. cereus; more than 10^5 cfu/g is suggestive.

Rapid staining method for B. cereus

Make films from colonies on B. cereus-selective medium, which encourages sporulation, and dry in air. Place the slide over boiling water and stain with 5% malachite green for 2 min. Wash and blot dry.

Table 42.1 Some *Bacillus* species

Species	Pathogenicity[a]	Food spoilage	Motility	VP[b]	Anaerobic growth	Growth at 5°C	Growth at 60°C	Acid from glucose[c]	Starch hydrolysis	Spores
B. anthracis	+	–	–	+	+	–	–	+	+	CO
B. cereus	+ FP	+	+	+	+	–	–	–	+	CO
B. subtilis	(+) FP	+	+	+	–	+	–	–	+	CO
B. pumilus	(+) FP	–	+	+	–	+	–	–	–	CO
B. coagulans	+	–	+	+	+	+	–	–	+	C(T)O
B. macerans	+	–	+	–	+	+	–	+	+	CO
B. circulans	+	–	+	–	v	+	–	–	+	C(T)O
B. brevis	+	–	+	–	–	+	–	–	–	CO
B. licheniformis	+ FP	–	+	+	+	+	–	–	+	CO
B. sphaericus	+ FP	–	+	–	–	–	–	–	–	TR
B. stearothermophilus	–	+	+	–	–	+	+	–	+	C(T)O
B. megaterium	–	–	+	–	–	–	–	–	+	CO
B. polymyxa	–	+	+	+	+	–	–	+	+	CO

[a] From Turnbull (1997)
[b] In 1% NaCl glucose broth
[c] In ammonium salt sugars

FP, food poisoning; CO, central, oval; TR, terminal, round; C(T)O, central, occasionally terminal, oval

Stain with 0.3% Sudan black in 70% alcohol for 15 min. Wash with xylol and blot dry. Counterstain with 0.5% safranin for 20 s. Wash and dry. *B. cereus* spores are oval, central or sub-terminal, do not swell the cells and are stained green. The vegetative cells stain red and contain black granules.

Food spoilage

Examine Gram-stained films for spore-bearing, Gram-positive bacilli. Make 10% suspensions of food in 0.1% peptone water, and pasteurize some of the suspension at 75–80°C for 10 min to kill vegetative forms.

Inoculate glucose tryptone agar and *B. cereus*-selective media with both unheated and pasteurized material.

Incubate at 25–30°C overnight. If the food was canned, incubate replicate cultures at 60°C. Identify colonies of Gram-positive bacilli.

Cold-tolerant spore-formers are important spoilage organisms. Incubate at 5°C for 7 days.

Identification

Test colonies for VP reaction (use 1% NaCl glucose broth), anaerobic growth, growth at 5 and 60°C, production of acid from glucose (ammonium salt sugar media) and starch hydrolysis (see Table 42.1). The API 20E and API50 CHB will identify more species.

BACILLUS SPECIES

Bacillus cereus

General properties are described above. This organism is responsible for 'bitty cream', particularly in warm weather. It may cause food poisoning (see above). Bacteraemia and pneumonia associated with *B. cereus* have been reported. It is a common contaminant on shell eggs. It does not ferment mannitol, unlike *B. megaterium*, which it may otherwise resemble.

Bacillus subtilis ('B. mesentericus')

Colonies on glucose tryptone agar are wrinkled or smooth and folded colonies, 4–5 mm across, with yellow halo (acid production). It is responsible for spoilage in dried milk and in some fruit and vegetable products. It causes 'ropey' bread. *B. subtilis* is used in industry to manufacture enzymes for biological washing products.

Bacillus stearothermophilus

This and associated thermophiles form large colonies (4 mm in diameter) on glucose tryptone agar with a yellow halo caused by acid production at 60°C. They grow poorly at 37°C and not at 20°C. This organism is responsible for 'flat-sour' (i.e. acid but no gas) spoilage in canned foods. For more detailed investigation, see Chapter 16.

Bacillus licheniformis

This may cause food poisoning.

Bacillus megaterium

This is a very large bacillus, widely distributed. It may cause *B. cereus*-like food poisoning.

Bacillus pumilus

This is found on plants and is a common contaminant of culture media.

Bacillus coagulans

This is widely distributed, and found in canned foods.

Bacillus anthracis

See above (p. 362).

Bacillus polymyxa

This is widely distributed in soil and decaying vegetables (possesses a pectinase). It is the source of the antibiotic polymyxin.

Bacillus macerans

This is common in soil and found in rotting flax.

Bacillus circulans

This forms motile colonies with non-motile variants. It is found in soil.

Bacillus sphaericus

This has large spores, and resembles *C. tetani* in appearance. It forms motile colonies and is a soil organism.

There are many other species (see Claus and Berkeley, 1986; Quinn and Turnbull, 1999).

REFERENCES

Claus, D. and Berkeley, R. C. W. (1986) The genus *Bacillus*. In: Sneath, P. H. A., Mair, N. S. and Sharpe, M. E. (eds), *Bergey's Manual of Determinative Bacteriology*. Baltimore, MA: Williams & Wilkins, pp. 1105–1139.

Quinn, C. P. and Turnbull, P. C. B. (1999) Anthrax. In: Collier, L., Balows, A. and Sussman, M. (eds), *Topley and Wilson's Microbiology and Microbial Infections*, 9th edn. Vol. 3, *Bacterial Infections*, CD-ROM. London: Arnold, Chapter 40.

Turnbull, P. C. B. (ed.) (1990) *Proceedings of an International Workshop on Anthrax, Winchester, 1989. Salisbury Medical Bulletin* No. 68. Special Supplement. Porton Down, Salisbury: PHLS.

Turnbull, P. C. B. (1997) The role of the *Bacillus* genus in infection. *Culture* **18**: 5–8.

Turnbull, P. C. B. and Kramer, J. M. (1995) Intestinal carriage of *Bacillus cereus*: faecal isolation studies in three population groups. *Journal of Hygiene, Cambridge* **95**: 629–638.

43

Gram-negative anaerobic bacilli and cocci

GRAM-NEGATIVE ANAEROBIC BACILLI

Gram-negative anaerobic bacilli are found in the alimentary and genitourinary tracts of humans and other animals and in necrotic lesions, often in association with other organisms. They are pleomorphic, often difficult to grow and die easily in cultures. Identification is often difficult. Reference experts use gas–liquid chromatography.

There have been changes in the classification of these organisms. The genus *Bacteroides* now contains only the *B. fragilis* group. The original *B. melaninogenicus* group now contains two new genera: *Prevotella* (saccharolytic) and *Porphyromonas* (asaccharolytic). Other new genera include *Tissierella*, *Sebaldella*, *Rikenella*, *Fibrobacter*, *Megamonas* and *Mitsuokella*. Another new genus, *Bilophila*, has been found in gangrenous appendices. *Eikenella* species are not strict anaerobes and will grow in 10% CO_2 in air. The organism known as *E. corrodens* may be confused with another corroder – *Bacteroides urealyticus*.

In pathological material they are not uncommonly mixed with other anaerobes and capnophilic organisms. In general, primary infections above the diaphragm are caused by *Fusobacterium* species, but *Bacteroides* species may occur anywhere and are frequently recovered from blood cultures.

Motile, anaerobic, Gram-negative bacilli probably belong to newly described genera such as *Selenomonas* or *Wolinella*, or to *Mobiluncus*. *Capnocytophaga* spp. are CO_2 dependent but grow anaerobically; they are found in the human and animal mouth.

Isolation

Examine Gram films of pathological material. Diagnosis of Vincent's angina can be made from films counterstained with dilute fuchsin, when fusiform bacilli can be seen in large numbers, associated with spirochaetes (*B. vincenti* – see Chapter 49). In pus from abdominal, brain, lung or genital tract lesions, large numbers of Gram-negative bacilli of various shapes and sizes, with pointed or rounded ends, suggest infection with *Bacteroides* spp. Such material often has a foul odour.

Infections involving *Prevotella melaniniogenica* and other members of the genus *Prevotella* are rapidly diagnosed by viewing the pus under long-wave ultraviolet (UV, 365 nm) light. These and *Porphyromonas* spp. produce porphyrins which give a bright red fluorescence.

Culture pus and other material anaerobically, aerobically, and aerobically under 5–10% carbon dioxide. For anaerobic cultures use both non-selective and enriched selective media, e.g. haemin, menadione, sodium pyruvate, cysteine-HCl, or use commercial fastidious anaerobe (FA) agar. The usual selective agents are neomycin 75 µg/ml, nalidixic acid 10 µg/ml and vancomycin 2.5 µg/ml. Various combinations may be used to restrict contaminating facultative flora from certain specimens, e.g. neomycin and vancomycin for pus from the upper respiratory tract.

A selective medium for fusobacteria containing josamycin, vancomycin and norfloxacin at 3, 4 and 1 µg/ml, respectively, is promising (Brazier *et al.*, 1991).

Place a 5-µg metronidazole disc on the streaked-out inoculum for the rapid recognition of obligate anaerobes, all of which are sensitive. Incubate at 37°C for 48 h.

Compare growths on selective and non-selective media and compare the anaerobic with aerobic cultures.

Identification

The Gram stain is important but may give equivocal results. Full identification of these organisms is difficult and rarely necessary. All that is required in a clinical laboratory is confirmation that anaerobes are involved. A simple report that 'mixed anaerobes are present – sensitive to metronidazole' usually suffices.

Identification is more important when anaerobes are present in blood cultures and other normally sterile sites. These simple presumptive tests are useful.

Potassium hydroxide test

Emulsify a few colonies in a drop of 3% KOH on a slide. *Bacteroides* spp. tend to become 'stringy' and form long strands when the loop is moved away from the drop.

UV fluorescence

Examine under long-wave UV light. Red fluorescence suggests *Prevotella* or *Porphyromonas* spp. and develops much earlier than the characteristic black pigment. Yellow fluorescent colonies of Gram-negative rods are probably fusobacteria (Brazier, 1986).

Phosphomycin test

Place a disc, soaked in phosphomycin 300 µg/ml, on a seeded plate and incubate. Fusobacteria are usually sensitive and *Bacteroides* resistant (Bennett and Duerden, 1985).

Carbohydrate fermentation

For information on this, see Phillips (1976) and Holdeman *et al.* (1977).

Antibiogram and commercial identification kits

Antibiogram patterns give a low level of accuracy and offer only a limited range of identifications. Commercial kits are an improvement but results based on a generated number may be misleading. Much depends on the database on which the scheme is based. The API 32 ATB kit is one of the most useful.

ANAEROBIC COCCI

These may occur in cultures from a wide variety of human and animal material. They have undergone recent taxonomic changes and are now described as *Peptostreptococcus*, *Peptoniphilus*, *Anaerococcus*, *Finegoldia* and *Peptococcus*.

It is difficult to separate these genera. Cocci occur singly, in pairs, clumps or short chains. They grow on blood agar at 37°C and are non-haemolytic, but grow better on Wilkins–Chalgren agar containing a supplement for non-sporing anaerobes.

These organisms are sensitive to metronidazole (5-µg disc) and grow slowly. There is no need to distinguish between them.

Peptostreptococci may be pathogenic for humans and animals, and associated with gangrenous lesions. They are commensals in the intestine and have been isolated from the vagina in health and in puerperal fever. They produce large amounts of hydrogen sulphide from high-protein media, e.g. blood broth, and a putrid odour in ordinary media. The API system is useful for identifying these organisms.

The organism formerly known as *Gaffkya tetragenus* is now called *Peptostreptococcus tetradius*. The genus *Peptococcus* now contains only one species, *P. niger*.

Sarcina

These cocci are in packets of eight. They are obligate anaerobes, requiring incubation at 20–30°C and carbohydrate media. They are found in air, water and soil, and occasionally in clinical material.

Sarcina ventriculi is a large coccus arranged in clumps of four to eight. It is found in the human gut, particularly in vegetarians.

Veillonella

The members of this genus are obligate anaerobes and occur as irregular masses of very small Gram-negative cocci, cultured from various parts of the respiratory and alimentary tracts and genitalia. They do not appear to cause disease.

Growth may be improved by adding 1% sodium pyruvate and 0.1% potassium nitrate to the basal medium.

Taxonomists now recognize seven species: *Veillonella atypica*, *V. dispar* and *V. parvula* are found in humans; and *V. caviae*, *V. criceti*, *V. ratti* and *V. rodentium* in rodents. In clinical laboratories it is not necessary to identify species.

Fluorescence under UV light is generally weak, fades rapidly and is medium dependent. It works only on brain infusion agars, except for *V. criceti* which fluoresces on a range of media (Brazier and Riley, 1988). Phenotypic differentiation is based on catalase, fermentation of fructose and the origin of the strain. For further information, see Mays *et al.* (1982).

REFERENCES

Bennett, K. W. and Duerden, B. I. (1985) Identification of fusobacteria in a routine diagnostic laboratory. *Journal of Applied Bacteriology* 59: 171–181.

Brazier, J. S. (1986) Yellow fluorescence of fusobacteria. *Letters in Applied Microbiology* 2: 125–126.

Brazier, J. S. and Riley, T. V. (1988) UV red fluorescence of *Veillonella* spp. *Journal of Clinical Microbiology* 26: 383–384.

Brazier, J., Citros, D. M. and Goldstein, E. J. C. (1991) A selective medium for *Fusobacterium* spp. *Journal of Applied Bacteriology* 71: 343–346.

Holdeman, L. V., Cato, E. P. and Moore, W. E. C. (eds) (1977) *Anaerobic Laboratory Manual*, 4th edn. Blacksburg, VA: Virginia State University.

Mays, T. D., Holdeman, L. V., Moore, W. E. C. and Johnson, J. L. (1982) Taxonomy of the genus *Veillonella* Prevot. *International Journal of Systematic Bacteriology* 32: 28–36.

Phillips, K. D. (1976) A simple and sensitive technique for determining the fermentation reactions of nonsporing anaerobes. *Journal of Applied Bacteriology* 41: 325–328.

44

Clostridium

This is a large genus of Gram-positive (but see below), spore-bearing anaerobes (a few are aerotolerant) that are catalase negative. They are normally present in soil; some are responsible for human and animal disease, whereas others are associated with food spoilage.

Clostridia are classified according to the shape and position of the spores (Table 44.1) and by their physiological characteristics. Most species are mesophiles; there are a few important thermophiles and some psychrophiles and psychrotrophs. They may be either predominantly saccharolytic or proteolytic in their energy-yielding activities. Saccharolytic species decompose sugars to form butyric and acetic acids, and alcohols. The meat in Robertson's medium is reddened and gas is produced. Proteolytic species attack amino acids. Meat in Robertson's medium is blackened and decomposed, giving the culture a foul odour.

The clostridia are conveniently considered under four headings: food poisoning, tetanus and gas gangrene, pseudomembranous colitis and antibiotic related diarrhoea, and food spoilage.

Table 44.1 Morphology and colonial appearance of some species of *Clostridium*

Species	Spores	Bacilli	Haemolysis	Colony appearance on blood agar
C. botulinum	OC or S	normal	+	large, fimbriate, transparent
C. perfringens	[a]	large thick	+	flat, circular, regular
C. tetani	RT	normal	+	small, grey, fimbriate, translucent
C. novyi	OS	large	+	flat, spreading, transparent
C. septicum	OS	normal	+	irregular, transparent
C. fallax	OS	thick	−	large, irregular, opaque
C. sordellii	OC or S	large thick	+	small, crenated
C. bifermentans	OC or S	large thick	+	small, circular, transparent
C. histolyticum	OS	normal	−	small, regular, transparent
C. sporogenes	OS	thin	+	medusa head, fimbriate, opaque
C. tertium	OT	long thin	−	small, regular, transparent
C. cochlearium		thin	−	circular, transparent
C. butyricum	OC	normal	−	white, circular, irregular
C. nigrificans		normal		black
C. thermosaccharolyticum		normal		granular, feather edges
C. difficile	OS	normal	−	grey, translucent, irregular

Spores: O, oval; R, round; S, subterminal; C, central; T, terminal
[a] not usually seen

FOOD POISONING

Botulism

This is the least common but most often a fatal kind of food poisoning (see Chapter 13); it is caused by *Clostridium botulinum*. This is a strict anaerobe which requires a neutral pH. It is difficult organism to isolate in pure culture because it competes poorly with other micro-organisms that may be present in food (e.g. canned, underprocessed).

Emulsify the suspected material in 0.1% peptone water and inoculate several tubes of Robertson's cooked meat medium, which have been heated to drive off air and then cooled to room temperature. Heat some of these tubes at 75–80°C for 30 min and cool. Incubate both heated and unheated tubes at 35°C for 3–5 days. Plate them on blood agar that has been pre-reduced in an anaerobic jar or cabinet and on egg yolk agar, and incubate under strict anaerobic conditions at 35°C for 3–5 days.

If the material is heavily contaminated, heat some of the emulsion as described above, make several serial dilutions of heated and unheated material and add 1-ml amounts to 15–20 ml glucose nutrient agar melted and at 50°C in 152 × 16 mm tubes (Burri tubes, see p. 377). Mix, cool and incubate. Cut the tube near the sites of suspected colonies, aspirate these with a Pasteur pipette and plate on blood agar or on egg yolk agar containing (per millilitre): cycloserine, 250 µg; sulphamethoxazole, 76 µg; trimethoprim, 4 µg (Dezfullian *et al.*, 1981).

Clostridium botulinum gives a positive egg yolk and Nagler (half-antitoxin plate) reaction (see below). Types A, B and F liquefy gelatin, blacken and digest cooked meat medium, and produce H_2S. Types C, D and E do none of these. All strains produce acid from glucose, fructose and maltose and are indole negative (Table 44.2, and see p. 374 and Table 44.1).

It is best to use fluorescent antibody methods and to send suspected cultures to a reference laboratory.

Type A *C. botulinum* is usually associated with meat. Type E is found in fish and fish products and estuarine mud, and is psychrotrophic.

Clostridium perfringens food poisoning

Clostridium perfringens is a common cause of food poisoning (see Chapter 13). However, not all strains are capable of causing symptoms. About 5% of strains carry the gene responsible for the production of *C. perfringens* enterotoxin (CPE) that causes abdominal pain and diarrhoea.

Faeces

It is pointless to examine stools of food handlers and 'contacts' in outbreaks of *C. perfringens* food poisoning because the disease is not spread by human agency.

C. perfringens food poisoning may be diagnosed by demonstrating CPE in the faeces using an enzyme immuno-assay kit (C. perfringens Enterotoxin WIA, Bioconnections Ltd., Leeds) or a latex kit (Berry *et al.* 1987).

The ethanol shock method may be useful. Emulsify the faeces in ethanol (industrial grade) to give a 1:1 suspension. Mix well and stand for 1h. Inoculate culture media, incubate anaerobically and proceed as under Identification (see p. 373).

Food

Outbreaks involving both heat-resistant and heat-sensitive strains have been reported (see Chapter 13).

Use a Stomacher to prepare a 10% suspension of the food in peptone water diluent. Make dilutions of 10^{-1} to 10^{-3} and plate both suspensions and dilutions on Willis and Hobbs medium, tryptose sulphite cycloserine agar and neomycin blood agar. Place a metronidazole disc on the inoculum. Incubate anaerobically at 37°C overnight.

To count the clostridia do pour plates with the dilutions in oleandomycin polymyxin sulphadiazine perfringens (OPSP) agar. Alternatively, use the Miles and Misra method (see p. 148) with neomycin blood agar. Incubate duplicate plates aerobically and anaerobically to distinguish between clostridia and other organisms.

Add some of the suspension to two tubes of Robertson's cooked meat medium. Heat one tube

Table 44.2 Cultural and biochemical properties of some species of clostridia

Species	RCM				Litmus milk	Acid from				Indole production	Gelatin liquefaction	Lecithinase
	Colour	Digestion	Odour	Gas		Glucose	Sucrose	Lactose	Salicin			
C. botulinum	black	+	–	+	D	+	>	–	–	–	+	–
C. perfringens	black	+	+	+	CD	+	+	+	>	–	+	+
C. tetani	black	+	+	–	C	–	+	–	–	+	+	–
C. novyi	red	–	–	+	GC	+	–	–	–	>	+	>
C. septicum	red	–	–	+	AC	+	–	+	+	–	+	–
C. fallax	red	–	–	+	AC	+	+	+	+	–	–	–
C. sordellii[a]	black	+	+	+	CD	+	–	–	–	+	+	+
C. bifermentans[a]	black	+	+	+	CD	+	–	–	–	+	+	+
C. histolyticum	black	+	+	–	D	–	–	–	>	–	+	–
C. sporogenes	black	+	+	+	D	+	–	–	–	–	+	–
C. tertium		–	–	+	AC	+	+	+	>	–	–	–
C. cochlearium	red	–	–	+		–	–	–	+	–	–	–
C. butyricum		–	–	+	ACG	+	+	+	+	–	–	–
C. nigrificans		–	–	–		–	–	–	–	–	–	–
C. thermosaccharolyticum		–	–	+		+	+	+	+	–	–	–

RCM: Robertson's cooked meat medium

Litmus milk: A, acid; C, clot; D, digestion; G, gas

[a] C. sordellii also splits urea; C. bifermentans does not

for 10 min at 80°C. This helps to select the heat-resistant strains that are most likely to be associated with food poisoning. Incubate overnight and plate out as indicated above.

Commercial kits are available for the detection of enterotoxins (see Berry *et al.*, 1987).

Identification

It is usually sufficient to identify *C. perfringens* by the Nagler half-antitoxin test (see below) but other properties are given in Tables 44.1 and 44.2. No great reliance should be placed on the 'stormy fermentation' of purple milk. General cultural methods are given on p. 371.

Nagler half-antitoxin plate test

Spread good quality antitoxin to *C. perfringens* alpha toxin on one half of a plate of nutrient agar containing 10% egg yolk. Dry the plates and inoculate the organism across both halves of the plate. Incubate anaerobically overnight at 37°C.

The lecithinases of *C. perfringens* and *C. novyi* produce white haloes round colonies on the untreated half of the plate. This activity is inhibited on the half treated with antitoxin. The lecithinases of *C. bifermentans* and *C. sordellii* are partially inhibited by *C. perfringens* antitoxin, but they may be distinguished by a positive indole reaction and the presence of spores, rarely seen *in vitro* with *C. perfringens*.

On egg yolk medium, some clostridia show a 'pearly layer' resulting from lipolysis. This is a useful differential criterion for species of *Clostridium*.

TETANUS AND GAS GANGRENE

Tetanus is caused by toxigenic strains of *C. tetani*. Clostridia associated with gas gangrene include *C. novyi*, *C. perfringens*, *C. septicum* and *C. sordellii*. *C. histolyticum* is a possible pathogen and some other species, such as *C. sporogenes*, *C. tertium*

and *C. bifermentans*, may also be found in wounds but are not known to be pathogenic.

Examine Gram- and fluorescent antibody-stained films of wound exudates. Bacilli with swollen terminal spores ('drumstick') suggest *C. tetani* but are not diagnostic. Thick, rectangular, box-like bacilli suggest *C. perfringens*. Very large bacilli might be *C. novyi*.

Inoculate two plates of blood agar, one for anaerobic and the other for aerobic culture. Inoculate two tubes of Robertson's cooked meat medium. Heat one at 80°C for 30 min and cool. Incubate blood agar and all the other media anaerobically at 37°C for 24 h. Plate out the Robertson's medium on blood agar and other clostridial media, and incubate anaerobically with 10% carbon dioxide.

Identification

Commercially available kits for the identification of anaerobes, including clostridia, are commonly used but strains giving equivocal results should be sent to a reference laboratory.

PSEUDOMEMBRANOUS COLITIS AND ANTIBIOTIC-ASSOCIATED DIARRHOEA

Pseudomembranous colitis is almost exclusively caused by *C. difficile* but antibiotic-associated diarrhoea may also be caused by CPE-producing strains of *C. perfringens*.

Treat faeces by the ethanol shock method (see p. 371). Plate on selective agar containing cefoxitin 8 μg/ml and cycloserine 250 μg/ml (CCY agar, Bioconnections Ltd., UK) and inoculate cooked meat medium. Alternatively plate on Columbia agar containing those antibiotics and 5% egg yolk so that lecithinase-producing colonies may be ignored (*C. difficile* is lecithinase negative). Incubate anaerobically for 48 h. Enterotoxigenic *C. perfringens* may be identified by ELISA as described above.

Colonies of *C. difficile* are 2–5 mm in diameter,

irregular and opaque. Confirm by latex agglutination (Mercia Diagnostics). There may be cross-reaction with *C. glycolicum*.

Examine blood agar cultures under long-wave ultraviolet (UV) light. *C. difficile* colonies (and those of some other organisms) fluoresce yellow–green.

In most diagnostic laboratories, faecal toxin detection is used to identify *C. difficile* disease. It is recommended that either tissue culture methods or diagnostic kits that detect both A and B toxins (Bioconnections Ltd., UK and Premier Meridian Diagnostics, USA) be used, as toxin A-negative, B-positive strains have been reported.

There are diagnostic kits for the toxin (Porton, Cambridge, UK) and Premier *C. difficile* Toxin A (Meridian Diagnostics, USA).

SPECIES OF *CLOSTRIDIUM* OF MEDICAL IMPORTANCE

Clostridium botulinum

This is a strict anaerobe that requires a neutral pH and absence of competition to grow in food (e.g. canned, underprocessed). Free spores may be present. Cultural characteristics are variable within and between strains. There are six antigenic types (A–F). Types A, B and F are proteolytic, causing clearing on Willis and Hobbs medium; types C, D and E are not. It is the causative organism of botulism, one of the least common but the most fatal kind of food poisoning. Type A is usually associated with meat; type E (a psychrotroph) is associated with fish products and is found in soil and marine mud.

Clostridium perfringens

This highly aerotolerant anaerobe may grow in broth, e.g. MacConkey broth inoculated with water. Spores are rarely seen in culture (a diagnostic feature) but can be obtained on Ellner's medium and medium with added bile, bicarbonate and quinoline (Phillips, 1986). There are six antigenic types (A–F). Type A is associated with gas gangrene and with food poisoning and antibiotic-associated diarrhoea. It causes lamb dysentery, sheep 'struck' and pulpy kidney in lambs. It is found in soil, water and animal intestines.

Clostridium tetani

A strict anaerobe, this dies readily on exposure to air. It swarms over the medium, but this may be difficult to see. It is, however, an important diagnostic characteristic. It is found in soil, especially animal manured, and in the animal intestine. It is the causative organism of tetanus, rarely seen in western countries where the disease is largely prevented by immunization and prophylaxis. It may, however, present as jaw stiffness in the elderly with a history of recent, often minor, trauma.

Clostridium novyi (C. oedematiens)

This strict anaerobe dies rapidly in air. There are four antigenic types (A–D). Types A and B cause gas gangrene, and type D bacillary haemoglobinuria in cattle and 'black disease' of sheep.

In 2000, *C. novyi* type A was implicated in an outbreak of serious illness, with deaths, among injecting drug users in the UK. A contaminated batch of heroin was believed to be the source. The identification of clostridia from such sources is discussed by Brazier *et al.* (2002).

Clostridium septicum and C. chauvoei

Both are strict anaerobes; once considered to be a single species, they are now known to be antigenically distinct. *C. septicum* causes gas gangrene in humans, and braxy and blackleg in sheep. *C. septicum* is often isolated from blood cultures in patients with malignant disease of the colon. *C. chauvoei* is not pathogenic for humans but causes quarter evil, blackleg, and 'symptomatic anthrax' of cattle and sheep. Both can be identified by fluorescent antibody methods.

Clostridium bifermentans and C. sordellii

These are separate species, distinguished biochemically, but they are conveniently considered together because both have a toxin (lecithinase) similar to that of C. *perfringens* and give the same reaction as that organism on half-antitoxin plates with C. *perfringens* type A antiserum. C. *sordellii* is urease positive, C. *bifermentans* urease negative, C. *bifermentans* not pathogenic, but C. *sordellii* is associated with wound infections and gas gangrene. Both are found in soil.

Clostridium fallax

This is a strict anaerobe associated with gas gangrene. Spores are not very resistant to heat and the organism may be killed in differential heating methods. It is found in soil.

Clostridium difficile

This may cause a wide spectrum of predominantly nosocomial diarrhoeal disease ranging from mild antibiotic-associated diarrhoea to full-blown pseudomembranous colitis that may be fatal.

SPECIES OF *CLOSTRIDIUM* OF DOUBTFUL OR NO SIGNIFICANCE IN HUMAN PATHOLOGICAL MATERIAL

Clostridium histolyticum

This is not a strict anaerobe. Filamentous forms may grow as surface colonies on blood agar

aerobically. Associated with gas gangrene, but usually in mixed infection with other clostridia. It is found in soil, in bone products such as gelatin and in the intestines of humans and animals.

Clostridium sporogenes

This is not pathogenic but is a common contaminant of wounds. It is widely distributed, and found in soil and animal intestines. It is phenotypically identical with proteolytic strains of C. *botulinum*. Differentiation is by the demonstration (in a reference laboratory) of toxin production.

Clostridium tertium

This is aerotolerant and non-pathogenic.

Clostridium cochlearium

The only common clostridium that is biochemically inert. It is not pathogenic.

FOOD-SPOILAGE CLOSTRIDIA

Clostridium species are known to be responsible for spoilage of a wide variety of foods, including milk, meat products, fresh water fish and vegetables. It also causes spoilage in silage.

Prepare 10% emulsions of the product in 0.1% peptone water using a Stomacher or blender. Heat some of this emulsion at 75–80°C for 30 min (pasteurized sample). Inoculate several tubes of Robertson's cooked meat medium and/or liquid reinforced clostridium medium with pasteurized and unpasteurized emulsion and incubate pairs at various temperatures, e.g. 5–7°C for psychrophiles, 22 and 37°C for mesophiles, and 55°C for thermophiles. Inoculate melted reinforced clostridial agar (RCM) in deep tubes with dilutions of the emulsions. Allow to set. Incubate at the desired temperature and look for black colonies, or blackening of the medium, which suggests clostridia. Inoculate several tubes of litmus/purple milk medium each with 2 ml suspension and incubate at the required temperatures. This gives a useful guide to the identity of the clostridias.

Appearance in litmus milk medium

- *C. putrifaciens, C. sporogenes, C. oedematiens* and *C. histolyticum* are indicated by slightly alkaline reaction, gas, soft curd subsequently digested leaving clear brown liquid, a black sediment and a foul smell.
- *C. sphenoides* gives a slightly acid reaction, soft curd, whey and some gas.
- *C. butyricum* produces acid, firm clot and gas.
- *C. perfringens* usually gives a stormy clot.
- *C. tertium* usually gives a stormy clot.

Note that aerobic spore bearers may give similar reactions, so plate out and incubate both aerobically and anaerobically.

Thermophiles associated with canned food spoilage

Two kinds of clostridial spoilage occur as a result of underprocessing: 'hard swell' and 'sulphur stinkers'.

Examine Gram-stained films for Gram-positive spore bearers. Inoculate reinforced clostridial medium or glucose tryptone agar and iron sulphite medium in deep tube cultures by adding about 1-ml amounts of dilutions of a 10% emulsion of the food material in peptone water diluent to 152 × 16 mm tubes containing 15–20 ml medium melted and at 50°C. Either solid or semisolid media may be used. Incubate duplicate sets of tubes for up to 3 days at 60 and 25°C.

Clostridium thermosaccharolyticum

This produces white lenticular colonies in all three media and changes the colour of glucose tryptone agar from purple to yellow. This organism is responsible for 'hard swell'.

Clostridium nigrificans

This produces black colonies, particularly in media that contain iron. It causes 'sulphur stinkers'.

Neither organism grows appreciably at 25°C (Table 44.3).

Table 44.3 Food-spoilage species

Species	Comments
Thermophiles	
C. nigrificans	Sulphur stinkers
C. thermosaccharolyticum	Hard swell
Mesophiles	
C. butyricum	Cheese disorders; butter and milk products
C. sporogenes	
C. sphenoides	
C. novyi	Meat and dairy products
C. perfringens	
Psychophiles	
C. putrefaciens	Bone taint, off-odours

COUNTING CLOSTRIDIA IN FOODS

Emulsify 10 g of the food in 90 ml 0.1% peptone water in a homogenizer. Divide into two portions and heat one at 75°C for 30 min.

Do MPN tests with 10-, 1- and 0.1-ml amounts and using the five- or three-tube method (see Chapter 10), on each portion, in liquid differential reinforced clostridial medium (DRCM). Clostridia turn this medium black ('black tube method'). The unheated portion gives the total count, the heated portion the spore count.

For the pour-plate method add 0.1-ml amounts of serial dilutions of the emulsion to melted OPSP agar. Incubate anaerobically for 24 h and count the large black colonies.

It may be necessary to dilute the inoculum 1 : 10 or 1 : 100 if many clostridia are present. If the load of other organisms is heavy, add 75 units/ml of polymyxin to the medium used for the unheated count.

To recover individual colonies, Burri tubes may be used. These are open at both ends and are closed with rubber bungs. Use solid DRCM and, after inoculation, cover the medium at both ends with a layer of melted paraffin wax about 2 cm deep. These tubes need not be incubated anaerobically.

After incubation and counting, remove both stoppers and extrude the agar cylinder. Cut it with a sterile knife and aspirate colonies with a Pasteur pipette for subculture. Membrane filters may be used for liquid samples. Roll up the filters and place in test tubes; cover with melted medium.

DRCM is suitable for most counts but for *C. nigrificans* iron sulphite medium is good. Select incubation temperatures according to the species to be counted.

General identification procedure

Subculture each kind of colony of anaerobic Gram-positive bacillus in the following media: Robertson's cooked medium; purple milk; peptone water for indole production; gelatin medium; and glucose, lactose, sucrose and salicin peptone waters. These media should be in cottonwool-plugged test tubes, not in screw-capped bottles. Omit the indicator from these because it may be decolourized during anaerobiosis. Test for acid production after 24–48 h with bromocresol purple. It may be necessary to enrich some media with Fildes' extract. An iron nail in each tube assists anaerobiosis. Gas production in sugar media is not very helpful and Durham's tubes can be omitted. The API 20A system for anaerobes is very helpful.

Do half-antitoxin (Nagler) plates as above using *C. perfringens* type A and *C. novyi* antitoxin (see Tables 44.1 and 44.2).

'Gram-negative clostridia'

Some *Clostridium* spp. appear to be Gram negative. Clostridias, however, are sensitive to a 5-µg vancomycin disc whereas true Gram-negative bacteria are resistant (Engelkirk *et al.*, 1992). For further information on clostridia see Willis (1977), Holdeman *et al.* (1977) and Phillips *et al.* (1985).

REFERENCES

Berry, P. R., Weinecke, A. A., Rodhouse, J. C. *et al.* (1987) Use of commercial tests for the detection of *Clostridium perfringens* and *Staphylococcus aureus* enterotoxins. In: Grange, J. M., Fox, A. and Morgan, N. L. (eds), *Immunological Techniques in Microbiology*. Society for Applied Bacteriology Technical Series No. 24. London: Blackwells, pp. 245–250.

Brazier, J. S., Duerden, B. I., Hall, V. *et al.* (2002) Isolation and identification of *Clostridium* spp. from infections associated with the injections of drugs; experiences of a microbiological investigation team. *Journal of Medical Microbiology* 51, 985–989.

Dezfullian, M., McCroskey, C. L., Hatheway, C. L. and Dowell, V. R. (1981) Isolation of *Clostridium botulinum* from human faeces. *Journal of Clinical Microbiology* 13: 526–531.

Engelkirk, P. G ., Duen-Engelkirk, J. and Dowell, V. R. (1992) *Principles and Practice of Anaerobic Bacteriology*. Belmont, NY: Star.

Holdeman, L. V., Cato, E. P. and Moore, W. E. C. (eds) (1977) *Anaerobic Laboratory Manual*, 4th edn. Blacksburg, VA: Virginia State University.

Phillips, K. D. (1986) A sporulation medium for *Clostridium perfringens. Letters in Applied Microbiology* **3**: 77–79.

Phillips, K. D., Brazier, J. S., Levett, P. N. *et al.* (1985) Clostridia. In: Collins, C.H and Grange, J.M. (eds), *Isolation and Identification of Micro-organisms of Medical and Veterinary Importance*. Society for Applied Microbiology Technical Series No. 21. London: Academic Press, pp. 215–236.

Willis, A. T. (1977) *Anaerobic Bacteriology*, 3rd edn. London: Butterworths.

45

Mycobacterium

The mycobacteria are acid fast: if they are stained with a strong phenolic solution of an arylmethane dye, e.g. carbol–fuchsin, they retain the stain when washed with dilute acid. Other organisms are decolourized. There are no grounds for the commonly held belief that some mycobacteria are acid and alcohol fast, but others are only acid fast. These properties vary with the technique and the organism's physiological state.

The most important members of this genus are obligate parasites. They include the *Mycobacterium tuberculosis* complex, containing various species as described below; and the leprosy (Hansen's) bacillus *M. leprae*. In addition, around 100 named species of mycobacteria, and an unknown number of unnamed ones, live freely in the environment and some of these have been identified as opportunist pathogens of humans and animals. These species are often therefore termed 'environmental mycobacteria' (EM) and are also termed 'non-tuberculous', 'opportunistic' and 'MOTT' ('mycobacteria other than typical tubercle') bacilli.

The M. tuberculosis *complex, and in some countries certain other species of mycobacteria (see national lists), are Risk Hazard Group 3 pathogens and infectious by the airborne route. All manipulations that might produce aerosols should be done in microbiological safety cabinets in Biosafety Containment Level 3 laboratories.*

THE *M. TUBERCULOSIS* COMPLEX (TUBERCLE BACILLI)

Direct microscopic examination

Sputum

This is the most common material suspected of containing tubercle bacilli. It is often viscous and difficult to manipulate. To make direct films use disposable 10-µl plastic loops. These avoid the use of Bunsen burners in safety cabinets. Discard them into disinfectant. Ordinary bacteriological loops are unsatisfactory and unsafe. Spread a small portion of the most purulent part of the material carefully on a slide, avoiding hard rubbing, which releases infected airborne particles. Allow the slides to dry in the safety cabinet then remove them for fixing and staining. This is quite safe; aerosols are released during the spreading, not from the dried film, although these should not be left unstained for any length of time. Fixing may not kill the organisms and dried material is easily detached.

For direct microscopy of centrifuged deposits after homogenizing sputum for culture spread a 10-µl loopful over an area of about 1 cm. Dry slowly and handle with care. The material tends to float off the slide while it is being stained.

Cerebrospinal fluid

Make two parallel marks 2–3 mm apart and 10 mm long in the middle of a microscope slide.

Place a loopful of the centrifuged deposit between the marks. Allow to dry and superimpose another loopful. Do this several times before drying, fixing and staining. Examine the whole area.

Urine

Direct microscopy of the centrifuged deposit is unreliable because urine frequently contains environmental acid-fast bacilli, which have entered the specimen from the skin or the inanimate environment.

Aspirated fluids, pus

Centrifuge if possible and make films from the deposit. Otherwise prepare thin films. Thick films float off during straining. Material containing much blood is difficult to examine and may give false positives.

Gastric lavage

Direct microscopy is unreliable because environmental mycobacteria are frequently present in food and hence in the stomach contents.

Laryngeal swabs

Direct microscopy is unrewarding.

Blood, faeces and milk

Direct microscopy is unreliable because acid-fast artefacts and environmental mycobacteria may be present. Microscopic examination of faeces and centrifugates of lysed blood may be useful in the provisional diagnosis of disseminated mycobacterial disease, usually caused members of the *M. avium* complex (MAC) in patients with acquired immune deficiency syndrome (AIDS; Kiehn *et al.*, 1985).

Tissue

Cut the tissue into small pieces. Work in a safety cabinet. Remove caseous material with a scalpel and scrape the area between caseation and soft tissue. Spread this on a slide. It is more likely to contain the bacilli than other material.

Staining

Stain direct films of sputum or other material by the Ziehl–Neelsen (ZN) method. The bacilli may be difficult to find and it is often necessary to spend several minutes examining each film. Fluorescence microscopy, using the auramine phenol (AP) method, is popular in some clinical laboratories. A lower-power objective may be used and more material examined in less time. False positives are not uncommon with this method and the presence of acid-fast bacilli should be confirmed by overstaining the film with ZN stain. These staining methods are described in Chapter 7.

Reporting

Report 'acid-fast bacilli seen/not seen'. If they are present report the number per high power field. Table 45.1 gives a reporting system in common use in Europe for sputum films.

Table 45.1 Scale for reporting acid-fast bacilli (AFB) in sputum smears examined microscopically

No. of bacilli observed	Report
0 per 300 fields	Negative for AFB
1–2 per 300 fields	(±) Repeat test
1–10 per 100 fields	+
1–10 per 10 fields	+ +
1–10 per field	+ + +
≥ 10 per field	+ + + +

Note that this is a logarithmic scale; this facilitates plotting.

'False' positives

Single acid-fast bacilli should be regarded with caution. They may be environmental contaminants, e.g. from water. They may be transferred from one slide to another during staining or if the same piece of blotting paper is used for more than one slide.

Isolation from, or detection in, pathological material

Traditionally, tubercle bacilli are isolated by couture on solid egg-based media but several weeks elapse between setting up the culture and the appearance of colonies visible to the naked eye. Other techniques have therefore been developed and the most widely used are automated radiometric and non-radiometric systems, and those based on nucleic acid-based technology such as the polymerase chain reaction (PCR). Other techniques used in certain specialized reference centres include the detection of specific mycobacterial lipids by mass spectrometry (French *et al.*, 1987) and mycobacterial antigen in clean specimens such as cerebrospinal fluid (CSF) by immunological methods (Drobnieski *et al.*, 2001).

In most tuberculosis laboratories world wide, the traditional culture methods are still in use so emphasis will be given to these in this chapter.

Preparation of pathological material for culture

Mycobacteria are often scanty in pathological material and not uniformly distributed. Many other organisms may be present and the plating methods used in other bacteriological examinations are useless. Acid-fast bacilli usually grow at a very much slower rate than other bacteria and are overgrown on plate cultures. Some specimens are from normally sterile sites (blood, bone marrow, CSF, certain tissues) and may be inoculated directly on culture media. Others are likely to contain many bacteria and fungi and must first be treated with a reagent that is less lethal for mycobacteria than for other organisms, and that reduces the viscosity of the preparation so that it may be centrifuged. No ideal reagent is known; there is a choice between 'hard' and 'soft' reagents.

Hard reagents are recommended for specimens that are heavily contaminated with other organisms. Exposure times are critical or the mycobacteria are also killed. Soft reagents are to be preferred when there are fewer other organisms, e.g. in freshly collected sputum. The reagent can be left in contact with the specimen for several hours without significantly reducing the numbers of mycobacteria. Soft

reagents may permit the recovery of mycobacteria other than tubercle bacilli, which are often killed by hard reagents. Alternatively, specimens may be inoculated on media containing a 'cocktail' of antimicrobial agents that will kill virtually all micro-organisms other than mycobacteria.

The methods described below are intended to reduce the number of manipulations, particularly in neutralization and centrifugation, so that the hazards to the operator are minimized.

Culture of the centrifuged deposit after treatment gives optimum results, but is avoided by some workers because of the dangers of centrifuging tuberculous liquids. Scaled centrifuge buckets (safety cups) overcome this problem (see Chapter 1). These considerably reduce the hazards of aerosol formation should a bottle leak or break in the centrifuge. They should be opened in a safety cabinet.

If material is not centrifuged, more tubes of culture media should be inoculated and a liquid medium should be included. This should contain an antibiotic mixture (see p. 383).

Sputum culture

Encourage the sputum from the specimen container into the preparation bottle with a pipette made from a piece of sterile glass tubing (200 × 5 mm) with one end left rough to cut through sputum strands and the other smoothed in a flame to accept a rubber teat.

Hard method

Add about 1 ml sputum to 2 ml 4% NaOH solution or a mixture of 2% NaOH and 1% *N*-acetyl-L-cysteine (a mucolytic agent) in a 25-ml screw-capped bottle. Stopper securely, place in a self-sealing plastic bag to prevent accidental dispersal of aerosols and shake mechanically, but not vigorously, for not less than 15 min but not more than 30 min. Thin specimens require the shorter time. Incubation does not help. Ten per cent of mycobacteria survive.

Remove the bottle from the plastic bag and add 3 ml 14% (approximately 1 mol/l) monopotassium hydrogen orthophosphate (KH_2PO_4) solution containing enough phenol red to give a yellow colour. This neutralizing fluid should be dispensed ready for use and sterilized in small screw-capped bottles. It should not be pipetted from a stock bottle.

The colour change of the phenol red from yellow to orange pink indicates correct neutralization. Re-stopper, mix gently and centrifuge with the bottles in sealed centrifuge buckets ('safety cups') for 15 min at 3000 rev./min. Pour off the supernatant fluid carefully into disinfectant. Wipe the neck of the bottle with a piece of filter paper, which is then discarded into disinfectant. Culture the deposit.

Soft method

To 2–4 ml sputum in a screw-capped bottle, add an equal volume of 23% trisodium orrthophosphate solution. Mix gently and stand at room temperature or in a refrigerator for 18 h. Add the contents of a 10-ml bottle of sterile distilled water to reduce the viscosity, centrifuge, decant and culture the deposit using the methods and precautions described above.

Laryngeal and other swabs

Place the swab in a tube containing 1 mol/l sodium hydroxide solution for 5 min. Remove, drain and place in another tube containing 14% potassium dihydrogen orthophosphate (KH_2PO_4) solution for 5 min. Drain and inoculate culture media.

Urine culture

Use fresh, early morning mid-stream specimens. Do not bulk several specimens because this often leads to contamination of cultures. Centrifuge 25–50 ml urine at 3000 rev./min and assess the number of organisms in the deposit by examining a Gram-stained film. Suspend the deposit in 2 ml 4% sulphuric acid for 15–40 min depending on the load of other organisms. Neutralize with 15 ml distilled water, centrifuge and culture the deposit.

Alternatively plate some of the urine on blood agar, incubate overnight and, if this is sterile, filter 50–100 ml of the urine through a membrane filter. Cut the filter into strips and place each strip on the surface of the culture media in screw-capped bottles.

Blood and bone marrow

Since the advent of AIDS it is often necessary to detect disseminated mycobacterial disease by examination of blood and bone marrow. Collect 8.5 ml blood aseptically and add to a tube containing 1.5 ml 0.35% sodium polyethanolium sulphate (SPS; this is the least toxic of the anticoagulants for mycobacteria and lyses white cells). Mix to avoid coagulation and inoculate 10–20 volumes of liquid media, e.g. Middlebrook 7H9 or 13A liquid medium containing 0.025% SPS and, if necessary, an antibiotic cocktail (see p. 383). The Isolator-10 lysis centrifugation system (Du Pont Co., Wilmington, DE, USA) is also suitable (Kiehn *et al.*, 1985). Or collect 10 ml blood into a tube that contains an anticoagulant and saponin which lyses all blood cells. Centrifuge at 3000 × g for 30 min and inoculate deposit on Middlebrook 7H11 agar.

Alternatively, add blood or bone marrow directly to the medium. Check broth weekly for bacterial growth. Confirm acid fastness by ZN smear and inoculate solid media.

Automated techniques are ideal for blood culture and are used according to the manufacturer's instructions (see Chapter 8).

Faeces

Culture of faeces of AIDS patients with suspected intestinal disease caused by the *M. avium* complex may be required (Kiehn *et al.*, 1985). Suspend 1 g faeces in 5 ml Middlebrook 7H9 broth, decontaminate as for sputum by the 'hard' method (see p. 000) and inoculate on Mitchison's antibiotic medium (see p. 383).

Culture of CSF, pleural fluids, pus, etc.

Plate original specimen or centrifuged deposit on blood agar and incubate overnight. Keep the remainder of specimen in a deep-freeze. If the blood agar culture is sterile, inoculate media for mycobacteria without further treatment. Use as much material and inoculate as many tubes as possible. If the blood agar culture shows the presence of other organisms, proceed as for sputum and/or culture directly into media containing antibiotics (see below). Half-fill the original container with

Kirchner medium and incubate. Some tubercle bacilli may adhere to the glass or plastic.

Culture of tissue

Grind very small pieces of tissue, e.g. endometrial curettings, or emulsify them in sterile water with glass beads on a Vortex mixer. Homogenize larger specimens with a Stomacher or blender. Check the sterility and proceed as for CSF, etc. Keep some of the material in a deep-freeze in case the cultures are contaminated and the tests need repeating.

Isolation from the environment

This may be done to trace the sources of environmental mycobacteria that contaminate clinical material and laboratory reagents, and also to study the distribution of these mycobacteria in nature.

Water

Pass up to 2 litres water from cold and hot taps through membrane filters (as for water bacteriology, see p. 254). Drain the membranes and place them in 3% sulphuric or oxalic acid for 3 min and then in sterile water for 5 min. Cut them into strips and place each strip on the surface of the culture medium in screw-capped bottles. Use Lowenstein–Jensen medium and Middlebrook 7H11 medium containing antibiotics (see below).

Swab the insides of cold and hot water taps and treat them as laryngeal swabs.

Milk

Centrifuge at least 100 ml from each animal. Examination of bulk milk is useless. Treat cream and deposit separately by the NaOH method and culture deposit on several tubes of medium.

Dust and soil

Place 2–5 g samples in 1 litre sterile distilled water containing 0.5% Tween 80. Mix gently to avoid too much froth. Pass through a coarse filter to remove large particles and allow to settle in a refrigerator for 24 h. Decant the supernatant and pass it through membrane filters as described above for water. Centrifuge the sediment and treat it with NaOH as for sputum. Neutralize, centrifuge and inoculate several tubes of medium.

Culture media

Egg media are most commonly used and there seems little to choose between any of them. In the UK, Lowenstein–Jensen medium is most popular but, in the USA, ATS or Piezer medium is often used. Both glycerol medium (which encourages the growth of the human tubercle bacillus) and pyruvate medium (which encourages the bovine tubercle bacillus and *M. malmoense*) should be used. Pyruvate medium should not be used alone because some opportunist mycobacteria grow very poorly on it.

Some workers prefer agar-based media, e.g. Middlebrook 7H10 or 7H11. Kirchner liquid medium is also useful for fluid specimens that cannot be centrifuged or when it is desirable to culture a large amount of material. Aspirated pleural, pericardial and peritoneal fluids may be added directly to an equal volume of double-strength Kirchner medium at the bedside.

The antibiotic media of Mitchison *et al.* (1987) are useful for contaminated specimens, even after treatment with NaOH or other decontaminating agents, etc. They offer a safety net for non-repeatable specimens such as biopsies. These media should be used as well as, not in place of, egg media for tissues, fluids and urines. They are prepared as follows. Add the following to complete Kirchner or Middlebrook 7H11 media: polymyxin 200 units/ml, carbenicillin 100 mg/1, trimethoprim 10 mg/1, and amphotericin 10 mg/1. Dispense fluid medium in 10-ml amounts and make slopes of solid medium.

An alternative antibiotic mixture (PANTA) is used, particularly in automated systems. This contains (final concentrations) polymyxin 50 units/ml, amphotericin 5 mg/l, nalidixic acid 20 mg/1, trimethoprim 5 mg/1 and azlocillin 10 mg/l.

Inoculating culture media

Use an inoculum of at least 0.2 ml (not a loopful) for each tube of medium. Inoculate two tubes each of egg medium, one containing glycerol and the other pyruvic acid and other media as indicated above.

Incubation

Incubate all cultures at 35–37°C and cultures from superficial lesions also at 30–33°C for at least 8 weeks. Prolonging the incubation for a further 2–4 weeks may result in a small increment of positives, especially from tissues and for some species. Examine weekly.

Automated culture

Commercial instruments (e.g. BACTEC, Becton Dickinson) have been developed for detecting the early growth of mycobacteria by a radiometric or infrared method (see Chapter 8).

Sputum or other homogenates, decontaminated if necessary, are added to Middlebrook broth medium containing antibiotics, to discourage the growth of other organisms, and ^{14}C-labelled palmitic acid. The medium is prepared commercially (Becton Dickinson) in rubber-sealed bottles and is inoculated with a syringe and hypodermic needle. If growth occurs, $^{14}CO_2$ is evolved. The air space above the medium in each bottle is sampled automatically at fixed intervals and the amount of radioactive gas is estimated and recorded. Growth of mycobacteria may be detected in 2–12 days, but positive results require further tests to distinguish between tubercle bacilli and other mycobacteria. In the BACTEC system, p-nitro-α-acetylamino-β-propiophenone (NAP) is used and this takes another 2–5 days. Mycobacteria that grow in media containing this substance are subcultured and identified by traditional methods (see Heifets, 1986, and below) or by the use of DNA probes (see below).

Disposal of radioactive waste is a major problem with the radiometric system and in many centres it has been replaced by non-radiometric systems including the Mycobacterial Growth Indicator Tube (MGIT Bactec 960, Becton Dickinson, Franklin Lakes, USA) and the MB BacT Alert (Organon Technika, Boxtel, The Netherlands) systems. These permit detection of bacterial growth within 12–18 days and are less demanding on staff time than the radiometric system. They are based on a change in the colour of a dye on release of CO_2 (MB BacT Alert) or unquenching of a fluorescent substance on consumption of oxygen during bacillary growth (MGIT). The systems are used according to the manufacturers' instructions.

Ideally, these automated systems should be used in conjunction with solid media-based culture (Drobniewski *et al.*, 2001). Automated systems based on nucleic acid amplification are considered in Chapter 8.

Identification of tubercle bacilli

In most parts of the world, the majority of mycobacteria cultured from pathological material will be tubercle bacilli. No growth will be evident on egg media until 10–14 days. Colonies of the human tubercle bacillus, *M. tuberculosis*, are cream coloured, dry and look like bread crumbs or cauliflowers. Those of the bovine tubercle bacillus, *M. bovis*, are smaller, whiter and flat. Growth of this organism may be very poor; it grows much better on pyruvate medium than on glycerol medium. (A smooth tubercle bacillus, *M. canetti*, has been described but is exceedingly rare.)

On Middlebrook agar media, colonies of both *M. tuberculosis* and *M. bovis* are flat and grey. In Kirchner fluid media, colonies are round and granular, and may adhere to one another in strings or masses. They grow at the bottom of the tube and settle rapidly after the fluid has been shaken.

Make ZN films to check for acid fastness. Some yeasts and coryneform organisms grow on egg media and their colonies may resemble those of tubercle bacilli. Make the films in a drop of saturated mercuric chloride as a safety measure against dispersing live bacilli in aerosols, and spread the films gently. Note whether the organisms are difficult or easy to emulsify. Check the morphology microscopically. Tubercle bacilli may be arranged in serpentine cords, usually uniformly stained and usually 3–4 μm in length.

Make suspensions of the organisms as follows. Prepare small, glass, screw-capped bottles containing two wire nails shorter than the diameter of the base of the bottles, a few glass beads 2–3 mm in diameter and 2 ml phosphate buffer pH 7.2–7.4. (The nails will rust if water is used; wash both nails and beads in dilute HCl and then in water before use.) Sterilize by autoclaving. Place several colonies of the organisms in a bottle and mix on a magnetic stirrer. Allow large particles to settle and use the supernatant.

Inoculate two tubes of egg medium and one tube of the same medium containing 500 μg/ml *p*-nitrobenzoic acid (PNBA medium). To make the *p*-nitrobenzoic acid stock solution, dissolve 0.5 g of the compound in 50 ml water to which a small volume of 1 mol/l NaOH solution has been added. Carefully neutralize with HCl and make up to 100 ml with water. This solution keeps for several months in a refrigerator. Incubate one egg slope at 25 ± 0.5°C, the other in an incubator with an internal light at 35–37°C. Incubate the *p*-nitrobenzoic acid egg slope at 35–37°C. If an internally illuminated incubator is not available, grow the organisms for 14 days and then expose to the light of a 25-W lamp at a distance of 1–1.5 m for several hours and re-incubate for 1 week.

Tubercle bacilli have the following characteristics:

- Not easily emulsified in water (or concentrated HgCl$_2$ which is safer).
- Regular morphology, 3–4 μm in length and usually showing serpentine cords.
- Relatively slow growing, taking 10 days or more to show visible growth.
- Produce no yellow, orange or red pigment.
- Fail to grow in the presence of 500 μg/ml *p*-nitrobenzoic acid (the bovine organism may show a trace of growth).
- Fail to grow at 25°C (the bovine organism may show a trace of growth).

Acid-fast organisms which grow in 1 week or less, or yield yellow or red colonies, or emulsify easily, or are morphologically small, coccoid, or long, thin and beaded or irregularly stained, are unlikely to be tubercle bacilli.

Variants of tubercle bacilli and BCG

Although the identification of 'species' within the *M. tuberculosis* complex is not essential for the management of the patient, it may be of value for epidemiological work, such as determining the role of cattle as a source of human tuberculosis in some regions. It is also useful to identify BCG, particularly if it may have been responsible for disseminated disease as a result of immunization of neonates, HIV infection or the treatment of malignancy. Although nucleic acid-based methods are available for such typing, a simple system based on cultural characteristics and advocated by the World Health Organization (Grange *et al.*, 1996) differentiates between the major variants within the *M. tuberculosis* complex.

Use suspensions prepared as described above and inoculate media for the following tests and incubate at 37°C.

TCH susceptibility

Use Lowenstein–Jensen medium containing 5 μg/ml thiophen-2-carboxylic acid hydrazide. Read at 18 days.

Pyruvate preference

Compare the growth on Lowenstein–Jensen medium containing glycerol with that containing pyruvate after 18 days' incubation.

Nitrate reduction test

Use Middlebrook 7H10 broth and test after 18 days by the second method described under 'Nitrate reductase test' in Chapter 7 (see p. 98).

Oxygen preference

Use Kirchner medium made semisolid with 0.2% agar. Inoculate with about 0.02 ml suspension.

Mix gently to avoid air bubbles and incubate undisturbed for 18 days. Aerobic growth occurs at or near the surface, and microaerophilic growth as a band 1–3 cm below the surface, sometimes extending upwards.

Pyrazinamide susceptibility and pyrazinamidase activity

Use the method described on p. 397 for pyrazinamide susceptibility testing. As an alternative to the rather complex susceptibility test, pyrazinamidase activity (which correlates very closely with susceptibility to pyrazinamide) may be detected. Inoculate butts of Wayne's medium (Middlebrook 7H9 broth, 1000 ml; pyrazinamide, 100 mg; sodium pyruvate, 2 g; agar, 15 g. Dissolve by steaming, dispense on 5-ml amounts in screw-cap bottles and autoclave at 115°C for 15 min. Allow to cool in the upright position to form butts rather than slopes). Incubate at 37°C for 7 days, add 1 ml freshly prepared 1% w/v ferrous ammonium sulphate in distilled water and refrigerate at 4°C for 4 h. A distinct pink band in the upper part of the butt indicates pyrazinamidase activity. Faint pink bands, seen with some strains of *M. bovis*, should be ignored. Include controls of *M. tuberculosis* (positive) and *M. bovis* (negative) (Table 45.2). Identification of a strain as 'Asian' or 'African' does not indicate origin of strain or ethnic group of the patient. The variants are widely distributed.

The niacin test

It was originally claimed that this test distinguished between human tubercle bacilli, which synthesize niacin, and bovine tubercle bacilli, which do not. Niacin test strips are commercially available. Unfortunately some human strains give negative niacin results and some environmental mycobacteria do synthesize niacin. It is an unreliable test and has little to commend it.

Cycloserine resistance

This test is useful for identifying BCG because it is more resistant to cycloserine than other members of the *M. tuberculosis* complex. Use Lowenstein–Jensen medium containing 20 µg/ml cycloserine. Read after 18 days.

Species within the *M. tuberculosis* complex

Mycobacterium tuberculosis

This, the human tubercle bacillus, species grows well (eugonic) on egg media containing glycerol or pyruvate. Colonies resemble breadcrumbs and are cream coloured. Films show clumping and cord formation – the bacilli are orientated in ropes or cords, especially on moist medium. There is no growth at 25 or 42°C or on PNBA medium. It is

Table 45.2 Variants within the *M. tuberculosis* complex

Species	TCH	Nitrate reductase	Oxygen preference	Pyrazinamide resistance[a]
M. bovis	S	–	M	R
M. africanum type I	S	–	M	S
M. africanum type II	S	+	M	S
M. tuberculosis Asian	S	+	A	S
M. tuberculosis Classical	R	+	A	S
BCG[b]	S	–	A	R

S, sensitive; R, resistant; M, microaerophilic; A, aerobic; TCH, thiophen-2-carboxylic acid
[a]Pyrazinamidase activity correlates with pyrazinamide susceptibility
[b]Differs from the others in being resistant to cycloserine

usually resistant to TCH (the Asian variants are sensitive), is nitrate reduction positive and aerobic, and usually susceptible to pyrazinamide. It causes tuberculosis in humans and may infect domestic and wild animals (usually directly or indirectly from humans). Simians are particularly likely to become infected.

Mycobacterium bovis

This, the bovine tubercle bacillus, grows poorly (dysgonic) on egg medium containing glycerol but growth is usually enhanced on pyruvate medium. Colonies are flat and grey or white; growth may be effuse. Cords are present in films. There is no or very poor growth at 25°C and on PNBA medium compared with a control slope at 37°C. It is sensitive to TCH, nitrate reduction negative, microaerophilic and resistant to pyrazinamide. It was common in dairy cows before eradication schemes were introduced in the industrialized countries, but it is still present in many developing countries. It causes human tuberculosis, both pulmonary and extrapulmonary, the latter usually resulting from the consumption of infected milk. In the industrialized countries, the few cases encountered usually represent reactivation of infections acquired much earlier.

Mycobacterium africanum

This name is given to the African group of strains with properties intermediate between those of the human and bovine tubercle bacillus (see Collins *et al.*, 1982). Growth is usually dysgonic, enhanced by pyruvate. It is susceptible to TCH and microaerophilic, and susceptible to pyrazinamide. The nitrate reduction test may be negative (most West African strains) or positive (most East African strains). It causes tuberculosis, clinically indistinguishable from that caused by the other tubercle bacilli.

BCG

This is the bacillus of Calmette and Guérin, an organism of attenuated virulence used in immu-

nization against tuberculosis. It is eugonic; growth is not enhanced by pyruvate. It is susceptible to TCH, nitrate reduction negative, aerobic, resistant to pyrazinamide and (unlike other strains in this group) resistant to cycloserine.

Other variants within the *M. tuberculosis* complex include: *M. microti*, a cause of disease in the vole and other small mammals; '*M. canetti*', a very rarely encountered variant *of M. tuberculosis* producing smooth colonies on solid media; the caprine (goat) variant of *M. bovis* which has caused disease in a few veterinary surgeons; and some unnamed strains with unusual properties isolated from seals and the rock hyrax.

ENVIRONMENTAL MYCOBACTERIA

Many species of mycobacteria inhabit the natural environment, and are especially associated with watery environments: marshes, streams, rivers and industrial and domestic water supplies. Around 100 species have been named and new ones are regularly being described. Some species have often been incriminated as opportunist pathogens of humans and animals, others rarely or never. There is, however, no clear distinction between potential pathogens and non-pathogens, and it is probable that any mycobacterial species, given a suitable opportunity, could cause disease. Environmental mycobacteria may contaminate specimens or be introduced during collection and/or laboratory processing. It is also difficult to assess the significance of opportunists. Determining whether an isolate is the cause of disease or a contaminant is not easy, especially in the case of isolates from sputum.

Identification of some of the more unusual environmental mycobacteria requires the use of nucleic acid-based technology, such as sequencing of the 16-S ribosomal RNA. Most of the species isolated in clinical laboratories may be identified with the aid of the following tests and Table 45.3.

Table 45.3 Usual properties of some opportunist mycobacteria and others that may be encountered in clinical material

Species	Pigment	TZ	Nitrate reductase	Tween hydrolysis	Growth at (°C)					Arylsulphatase[a]		Catalase[b]	Tellurite[c] reduced	Growth on N medium	Rapid growth
					20	25	33	42	44	3 days	21 days				
M. kansasii	P	S	+	+	–	+	+	v	–	–	+	+++	–	–	–
M. marinum	P	R	–	+ (late)	+	+	+	–	–	–	++	++	–	v	+
M. xenopi	–/Sc	R	–	–	–	–	–	+	+	–	+++	–	–	–	–
M. avium-intracellulare	–/Sc	R	–	–	–	+	+	+	v	–	v	+	+	–	–
M. scrofulaceum	Sc	v	–	–	v	+	+	–	–	–	v	+++	–	–	–
M. malmoense	–	R	+	+ (late)	–	+	+	–	–	–	–	–	–	–	–
M. simiae	P	R	–	–	–	+	+	–	–	–	+	++	–	–	–
M. szulgai	Sc/P[d]	R	+	+	–	+	+	v	–	–	+++	+	–	–	–
M. fortuitum	–	R	+	–	+	+	+	–	–	+++	+++	v	+	+	+
M. chelonei	–	R	–	–	+	+	+	–	–	+++	+++	v	+	+	+
M. gordonae	Sc	R	–	+	+	+	+	–	–	–	+	++	–	–	–
M. flavescens	Sc	R	+	+	+	+	+	–	–	–	+	+	–	+	+
M. gastri	–	R	–	+	+	+	+	–	–	–	+	+	–	–	–
M. terrae	–	R	+	+	v	+	+	–	–	–	+	+++	–	–	–
M. triviale	–	R	–	+	–	+	+	–	–	–	+	+	–	–	–
M. nonchromogenicum	–	R	–	+	v	+	+	–	–	–	+	+++	–	–	–
M. smegmatis	–	R	+	+	+	+	+	+	–	–	+	++	+	+	+
M. phlei	Sc	R	+	+	+	+	+	+	+	–	+	++	+	+	+
M. ulcerans	–/Sc	R	–	–	–	–	+	–	–	–	–	–	–	–	–

TZ, thiacetazone; P, photochromogen; Sc, scotochromogen; S, sensitive; R, resistant; +, usually positive; –, usually negative or none; v, variable

[a] Sulphatase: +++, deep pink; ++, pink; +, pale pink

[b] Catalase: Amount of foam, +++, > 20 mm; ++, 10–20 mm; +, 5–10 mm; –, < 5 mm

[c] Test not done on pigmented strains

[d] M. szulgai: At 25°C = photochromogen; at 37°C = scotochromogen

Inoculation

Unless otherwise stated, inoculate the media with one loopful or a 10-μl drop of a suspension prepared as described on p. 385.

Pigment production

Inoculate two egg medium slopes. Incubate both at 35–37°C, one exposed to light and the other in a light-proof box. Examine at 14 days. Photochromogens show a yellow pigment only when exposed to light. Scotochromogens are pigmented in both light and dark, but the culture exposed to light is usually deeper in colour. False photochromogenicity may arise if the inoculum is too heavy and pigment precursor is carried over. Continue to incubate the slope exposed to light and look for orange-coloured crystals of carotene pigment that may form in the growth.

Temperature tests

Inoculate egg media slopes with a single streak down the centre and incubate at the following (exact) temperatures: 20, 25, 42 and 44°C. Examine for growth at 3 and 7 days and thereafter weekly for 3 weeks.

Thiacetazone susceptibility

Inoculate egg medium slopes containing 20 μg/ml thiacetazone (TZ). Make a stock 0.1% solution of this agent in formdimethylamide (*caution*). This solution keeps well in a refrigerator. Incubate slopes for 18–21 days and observe the growth in comparison with that on an egg slope without the drug.

Nitrate reduction test

Use the second method on p. 98 of Chapter 7 (under 'Nitrate reductase test'). Strains that reduce nitrite may not accumulate enough from nitrate

(nitrate reductase present) to score as nitrate reductase positive.

Sulphatase test

Use the method described on p. 100 of Chapter 7. If the organism is a rapid grower, add the ammonia after 3 days.

Catalase test

Prepare the egg medium in butts in screw-capped tubes (20 × 150 mm). Inoculate and incubate at 35–37°C for 14 days with the cap loosened and then add 1 ml of a mixture of equal parts 30% hydrogen peroxide (*caution*) and 10% Tween 80. Allow to stand for a few minutes and measure the height in millimetres of the column of bubbles. More than 40 mm of froth is a strong positive result (+++).

Tween hydrolysis

Add 0.5 ml Tween 80 and 2.0 ml of a 0.1% aqueous solution of neutral red to 100 ml M/15 phosphate buffer. Dispense this in 2-ml lots and autoclave at 115°C for 10 min. This solution keeps for about 2 weeks in the dark and should be a pale-straw or amber colour. Add a large loopful of culture from solid medium and incubate at 35–37°C. Examine at 5 and 10 days for a colour change from amber to deep pink.

Tellurite reduction

Grow the organisms in Middlebrook 7H10 medium for 14 days or until there is a heavy growth. Add four drops of sterile 0.02% aqueous solution of potassium tellurite (*caution*) and incubate for a further 7 days. If tellurite is reduced there will be a black deposit of metallic tellurium. Ignore grey precipitates. This test is of no value for highly pigmented mycobacteria.

Growth in N medium

Inoculate the medium with a straight wire and incubate. Rapid growers give a turbidity with or without a pellicle in 3 days. 'Frosting', i.e. adherence of a film of growth to the walls of the tube above the surface of the medium, is often seen. Ignore a very faint turbidity.

Resistance to antibacterial agents

Methods of doing these tests are described below. Some mycobacteria exhibit consistent patterns, which may be useful for identification.

Aerial hyphae

Some *Nocardia* spp. are partially acid fast and resemble rapidly growing mycobacteria. Do a slide culture (see p. 133) and look for aerial hyphae. In very young cultures of mycobacteria, a mycelium may be seen, but it fragments early into bacilli. Aerial hyphae are not formed except by *M. xenopi*.

Emulsifiability and morphology

When making films for microscopic examination, note whether the organisms emulsify easily. Note the morphology, whether very short, coccobacilli, long, poorly stained filaments or beaded forms.

Other more specialized techniques for identifying environmental mycobacteria include lipid chromatography (Jenkins, 1981) and pyrolysis mass spectrometry (Sisson *et al.*, 1991). DNA probes for some of the more commonly encountered species are available commercially (see p. 398).

Species of opportunist and other mycobacteria

Mycobacterium kansasii

This is a photochromogen (but see below). Optimum pigment production is observed if the inoculum is not too heavy, if there is a plentiful supply of air (loosened cap) and if exposure to light is continuous. Young cultures grown in the dark and then exposed to light for 1 h will develop pigment, but old cultures that have reached the stationary phase may not show any pigment after exposure. Continuous incubation in the light results in the formation of orange-coloured crystals of carotene. The morphology is distinctive; the bacilli are long (5–6 μm) and beaded. *M. kansasii* is nitrate reduction positive, sensitive to 20 μg/ml thiacetazone, hydrolyses Tween 80 rapidly, and grows at 25°C but not at 20°C; some strains grow poorly at 42°C but not at 44°C. The arylsulphatase test is weakly positive and the catalase strongly positive (more than 40 mm by the method described above). It is resistant to streptomycin, isoniazid and *p*-aminosalicylic acid (PAS), but susceptible to ethionamide, ethambutol and rifampicin. Some strains are susceptible to amikacin and erythromycin by the disc technique.

Occasional nonchromogenic or scotochromogenic strains have been reported but these may be recognized by their biochemical reactions and morphology.

This organism is an opportunist pathogen, associated with pulmonary infection. It is rarely significant when isolated from other sites. It has been found in water supplies.

The *Mycobacterium avium* complex

The species, *M. avium* and *M. intracellulare*, are usually grouped together as the *M. avium* complex (MAC). The two closely related species may be distinguished by DNA probes or agglutination serology in some reference laboratories. The complex also contains *M. avium* ssp. *paratuberculosis* (previously *M. paratuberculosis*), the cause of hypertrophic enteritis or Johne's disease in cattle and other ruminants, and *M. lepraemurium*, the cause of skin lesions in rats and other small rodents and cats.

MAC bacilli are small, almost coccoid. Although usually non-chromogenic, a feeble yellow pigment is produced by some isolates. Some strains are susceptible to thiacetazone. All are nitrate reduction negative, catalase negative or weakly positive, reduce tellurite and do not hydrolyse Tween 80. The sulphatase reaction varies from strongly positive to negative. There is growth on egg medium at 25°C; growth at 20°C and 42°C is variable. Resistance to antituberculosis drugs is usual, but some strains are susceptible to ethionamide.

MAC bacilli are opportunist pathogens of humans, associated with cervical adenitis (scrofula), especially in young children. Pulmonary disease, usually but not always with predisposing lung conditions, occurs in adults. They frequently cause opportunist disease in patients with the AIDS. Such disease is often disseminated and the organisms may be isolated from many sites including blood, bone marrow and faeces (Kiehn *et al.*, 1985). They are also opportunist pathogens of pigs and birds and have been found in soil and water. There have been claims that *M. avium* ssp. *paratuberculosis* is a cause of Crohn's disease in humans but these claims require substantiation.

Strains of *M. avium* ssp. *paratuberculosis* and some strains of *M. avium*, notably those isolated from wood pigeons, require the addition of mycobactin, an iron-binding lipid extracted from mycobacterial cell walls, to media for their growth.

Mycobacterium xenopi

Pigment, if any, is more obvious on cultures incubated at 42–44°C, which become pale yellow. Morphologically it is easily recognized; it stains poorly and the bacilli are long (5–6 µm) and filamentous. It is nitrate reduction negative, resistant to thiacetazone and does not hydrolyse Tween 80. It is a thermophile, growing well at 44°C, slowly at 35°C and not at 25°C. (Almost all other environmental mycobacteria grow at 25°C.)

The arylsulphatase test is strongly positive, and the catalase test negative. It does not reduce tellurite. It is more susceptible to isoniazid than other opportunist mycobacteria, usually giving a resistance ratio of 4. It is susceptible to ethionamide, but its susceptibility to other antituberculosis drugs varies between strains. Some strains are susceptible to amikacin and erythromycin by the disc method.

It is an opportunist pathogen in human lung disease and is rarely significant in other sites. It is a frequent contaminant of pathological material, especially urine, and has been found in hospital hot water supplies. 'Outbreaks' of laboratory contamination are not uncommon. This organism is very common in south-east England and north-west France.

Mycobacterium celatum

Strains of this species were originally isolated in Finland and Somalia, principally from respiratory tract specimens. Strains grow at 25°C and 45°C, they do not hydrolyse Tween 80 and some develop a light-yellow pigment in the dark. They resemble *M. xenopi* in being strongly arylsulphatase positive, but differ from this species in being susceptible to ethambutol and ethionamide. Another species, *M. branderi*, is phenetically similar but differs in its drug susceptibility pattern.

Mycobacterium genavense

This species is an occasional cause of disseminated disease in AIDS patients and has been isolated from pet and zoo birds. It is cultivable only with great difficulty: slow and feeble growth occurs in Bactec 13A medium used for the radiometric detection of mycobacterial growth. It is identifiable by unique base sequences in its ribosomal 16-S RNA.

Mycobacterium scrofulaceum

This, the 'scrofula scotochromogen', is phenetically similar to the MAC bacilli but is scotochromogenic and fails to reduce tellurite to tellurium. It may also be differentiated from MAC by agglutination serology.

Mycobacterium marinum

This is an opportunist pathogen responsible for superficial skin infections, known as swimming pool granuloma, fish tank granuloma or fish-fanciers' finger. It is found in sea-bathing pools and in tanks where tropical fish are kept, and is a pathogen of some fish (see the review by Collins *et al.*, 1985). It is missed in clinical laboratories that do not culture material from superficial lesions at 30–33°C. Primary cultures do not grow at

35–37°C. It is a photochromogen. Beading or banding similar to that of *M. kansasii* may be evident. It is nitrate reduction negative and resistant to thiacetazone, and hydrolyses Tween 80. After laboratory subculture its temperature range is modified so that it grows at 25°C and 37°C but not at 44°C. The catalase and arylsulphatase tests are weakly positive and growth may occur in N medium. Resistance to streptomycin and isoniazid is usual; resistance to other drugs is variable. Some strains are susceptible to co-trimoxazole, erythromycin and amikacin by the disc method.

Mycobacterium gordonae

This is also known as the tap-water scotochromogen, although other scotochromogens are found in water supplies. Organisms in this group grow slowly, produce a deep-orange pigment in light and usually a yellow pigment in the dark. No crystals of carotene are formed. If such crystals are seen, the organism may be a scotochromogenic strain of *M. kansasii*. Morphology is not distinctive. It is nitrate reduction negative, resistant to thiosemicarbazone and hydrolyses Tween 80. Growth occurs at 20°C but not at 44°C. Some strains are psychrophilic. The arylsulphatase reaction is weak and the catalase test is usually strongly positive. Growth does not occur in N medium. They are usually resistant to isoniazid and PAS but susceptible to streptomycin and other antituberculosis drugs.

The organisms in this group are rarely associated with human disease. They are not infrequent contaminants of pathological material but usually appear as single colonies on egg medium. They may be found in tap water, dust and soil.

Mycobacterium szulgai

This scotochromogen differs from *M. gordonae* in being nitrate reductase positive and strongly arylsulphatase positive, and giving a weak catalase reaction. It hydrolyses Tween 80. It is a rare human opportunist pathogen.

Rapidly growing scotochromogens

There are many species in this group, e.g. *M. flavescens*, *M. aurum*, *M. gilvum*, *M. duvalii*

and *M. vaccae*. With very rare exceptions they are not known to cause human disease but occur in the environment and sometimes contaminate pathological material.

Mycobacterium fortuitum and M. chelonae

These two non-pigmented rapid growers are conveniently considered together. There have been taxonomic problems: *M. fortuitum* has subdivided into *M. fortuitum* and *M. peregrinum* whereas *M. chelonae* has been subdivided into *M. chelonae* and *M. abscessus*. On subculture, growth occurs within 3 days on most media, but on primary isolation from clinical or environmental specimens growth may, paradoxically, not be apparent for several weeks or months. The bacilli tend to be rather fat and solidly stained. The nitrate reduction test is positive (*M. fortuitum* and *M. peregrinum*) or negative (*M. chelonae* and *M. abscessus*). Tween 80 is not hydrolysed. There is growth at 20°C and in some strains at 42°C. Psychrophilic strains, which fail to grow at 37°C, are not uncommon, particularly on primary isolation. The arylsulphatase test is positive at 3 days and the catalase reaction is variable. Tellurite is reduced rapidly. There is growth in N medium in 3 days. There is a general resistance to all antituberculosis drugs, except that *M. fortuitum* and *M. peregrinum* are usually susceptible to ethionamide and the quinolones, but *M. chelonei* and *M. abscessus* are resistant.

These are opportunist pathogens usually occurring in superficial infections (e.g. injection abscesses) and occasionally as secondary agents in pulmonary disease. They are common in the environment and frequently appear as laboratory contaminants.

Mycobacterium smegmatis and M. phlei

These two rapidly growing saprophytes are more common in textbooks than in clinical laboratories. Growth occurs in 3 days. They are nitrate reduction positive and hydrolyse Tween 80 in 10 days. They grow at 20°C and 44°C; *M. phlei* grows at 52°C. The 3-day sulphatase test is negative, but longer incubation gives positive reactions. The catalase test is strongly positive. Tellurite is reduced. There is growth in N medium.

They are not associated with human or animal disease. They have contributed to the myth of acid-fast versus acid- and alcohol-fast bacilli (see p. 379).

Slowly growing nonchromogens

There are at least four species: *M. terrae*, *M. nonchromogenicum*, *M. triviale* and *M. gastri*. All grow at 25°C but poorly or not at all at 42°C. All hydrolyse Tween 80 and none reduces tellurite, which permits differentiation from MAC bacilli. *M. terrae* and *M. triviale* give a positive nitrate reduction test but *M. nonchromogenicum* and *M. gastri* are negative. *M. gastri* is occasionally photochromogenic and is usually susceptible to thiacetazone and may therefore be confused with *M. kansasii* to which, although non-pathogenic, it is genetically closely related. Susceptibility to anti-tuberculosis drugs is variable.

Mycobacterium ulcerans

This causes Buruli ulcer, a serious skin infection in certain regions within the tropics and in Australia. It is difficult to cultivate because it has a very narrow temperature range of growth, 31–34°C, and growth is very slow, taking 10–12 weeks to give small colonies resembling those of *M. bovis*. It is biochemically inert and undistinguished.

Mycobacterium simiae

Originally isolated from monkeys, this is a photochromogen and is nitrate reduction negative, and does not hydrolyse Tween 80. Pulmonary and disseminated human disease has been reported.

Mycobacterium malmoense

This resembles the MAC organisms, grows very slowly and is sometimes confused with *M. bovis* by inexperienced workers. This suggests that it is more common than might be expected. It is difficult to identify except by lipid chromatography. It causes pulmonary disease in adults and cervical adenopathy in children.

Mycobacterium haemophilum

This is another rare species, possibly missed because it grows poorly or not at all on media that do not contain iron or haemin. It grows at 30°C on LJ medium containing 2% ferric ammonium citrate and on Middlebrook 7H11 medium containing 60 µg/ml haemin. All the usual biochemical tests are negative. It has been isolated from the skin and subcutaneous tissues of immunologically compromised patients and from cases of cervical lymphadenopathy in otherwise healthy children.

DRUG SUSCEPTIBILITY TESTS

Although most cases of tuberculosis yield tubercle bacilli that are sensitive to the commonly used anti-tuberculosis drugs, drug and multidrug resistance is an increasing problem, particularly in certain 'hotspot' regions. Where facilities are available, drug susceptibility tests are of value, particularly if the patient's condition does not improve as a result of therapy. Susceptibility tests for environmental mycobacteria causing disease are of questionable value because there is a poor correlation between *in vitro* results and response to therapy.

Treatment of tuberculosis with several drugs is essential to prevent the emergence of drug resistance. The WHO has made definitive recommendations on the therapy of tuberculosis. Most patients receive four drugs – rifampicin, isoniazid, ethambutol and pyrazinamide – for 2 months (the intensive phase) and two of these – rifampicin and isoniazid – for a further 4 months (the continuation phase). Streptomycin, being an injectable drug, is now usually only used in regimens for the re-treatment of cases showing bacteriological relapse. Patients with multidrug resistance (by definition, resistance to rifampicin and isoniazid with or without resistance to other drugs) require treatment with second-line drugs including fluoroquinolones and ethionamide.

Susceptibility tests with mycobacteria are complex and are usually done in specialist and reference laboratories. There are four principal methods:

1. The absolute concentration method which is popular in Europe

2. The proportion method, also used in Europe and popular in America
3. The resistance ratio method used in the UK
4. Radiometric and non-radiometric automated methods.

Special techniques are used for pyrazinamide susceptibility tests and nucleic acid-based methods are available for the rapid determination of resistance to certain agents.

The absolute concentration method

Carefully measured amounts of standardized inocula are placed on control media and media containing varying amounts of drugs, and the lowest concentration that will inhibit all, or nearly all, of the growth is reported. It is very difficult, however, to standardize the active concentration of drugs in the media and the method gives different results in different laboratories. It is not possible to use a medium that must be heated after the addition of the drug (e.g. egg medium), because some drugs are partially heat labile and heating times are rarely constant even in the same laboratory. Middlebrook 7H10 or 7H11 media are used but this method has not found much favour in the UK.

The proportion method

Several dilutions of the inoculum are made and media containing no drug and standard concentration of drugs are inoculated. The number of colonies growing on the control from suitable dilutions of the inoculum are counted, and also the number growing on drug-containing medium receiving the same inoculum. Comparison of the two shows the proportion of organisms that are resistant. This is usually expressed as a percentage.

This method is popular in the USA and Europe, but it is technically very difficult and, as it is usually done in Petri dishes, we regard it as highly hazardous. There are also more risks attached to standardizing the inocula than with the resistance ratio method.

The resistance ratio method

The minimal inhibitory concentrations of the drug that inhibit test strains are divided by those that inhibit control strains to give the resistance ratios. Ratios of 1 and 2 are considered susceptible, and those of ≥ 4 resistant (see below). These results usually correlate with the clinical findings. This method is used extensively in the UK and is described below.

Each of the above methods may be used for 'direct' tests on sputum homogenates that contain enough tubercle bacilli to give positive direct films as well as for 'indirect' tests on cultures. The latter are more reliable and reproducible.

For background information about these tests, see Canetti *et al.* (1969) and Vestal (1975). For a discussion on the design of resistance ratio tests, see Collins *et al.* (1997).

Technique of resistance ratio method

Dilutions of drugs are incorporated in Lowenstein–Jensen medium, which is then inspissated (i.e. heat coagulated). As some of these drugs are affected by heat, it is essential that the inspissation procedure be standardized. The inspissator must have a large circulating fan so that all tubes are raised to the same temperature in the same time. The load (i.e. number of tubes) and the time of exposure must be constant. To ensure that the drug-containing medium is not overheated, the machine should be raised to its correct temperature before it is loaded and racks can be devised that allow loading in a few seconds. A period of 45 min at 80°C is usually sufficient to coagulate this medium and no further heat is necessary if it is prepared with a reasonably aseptic technique.

Control strains

The resistance ratio method compares the minimum inhibitory concentration (MIC) of the unknown strain with that of control strains on the same batch of medium.

Some workers use the H37Rv strain of *M. tuberculosis* as the control strain, but the susceptibility of this to some drugs does not parallel that of 'wild' tubercle bacilli and may give misleading ratios. It is better to use the modal resistance method of Leat and Marks (1970), in which the unknown strains are compared with the modal resistance (i.e. that which occurs most often) of a number of known susceptible strains of recent origin.

Drug concentrations

Each laboratory must determine its own ranges, because these will vary slightly according to local conditions. Initially use those suggested in Table 45.4 and inoculate at least 12 sets with known susceptible organisms to arrive at a baseline for future work. A drug-free control slope must be included.

Stock solutions of the drugs are conveniently prepared as 1% solutions in water, except for rifampicin, which should be dissolved in formdimethylamide (*caution*). All stock solutions keep well at –4°C.

Bacterial suspensions

Smooth suspensions must be used. Large clumps or rafts of bacilli give irregular results and make readings difficult.

Sterilize 7-ml (bijou) screw-capped bottles containing a wire nail (shorter than the diameter of the bottle), a few glass beads and 2 ml phosphate buffer (13.3 g anhydrous Na_2HPO_4 and 3.5 g KH_2PO_4 in 2 litres water, pH 7.4). Into each bottle place a scrape of growth equal to about three or four large colonies and place the bottle on a magnetic stirrer for 3–4 min. Allow to stand for a further 5 min for any lumps to settle and use the supernatant to inoculate culture media.

Methods of inoculation

Place the bottle containing the suspension on a block of Plasticine at a suitable angle. Inoculate each tube with a 3-mm loopful of the suspension, withdrawn edgewise. Plastic disposable 10-µl loops are best.

A better and quicker method uses an automatic micropipette (the Jencon Micro-Repette is ideal) fitted with a plastic pipette tip. Deposit 10 µl of the suspension near to the top of each slope of medium. As it runs down it spreads out to give a suitable inoculum. The plastic tips are sterilized individually in small, capped test tubes and after use are deposited in a jar containing glutaraldehyde, which is subsequently autoclaved. The tips may be re-used.

Incubation

Incubate at 37°C for 18–21 days.

Reading results

Examine tubes with a hand lens and record as follows: confluent growth, CG; innumerable discrete colonies, IC; between 20 and 100 colonies, +; < 20 colonies, 0.

The drug-free control slope must give growth equal to CG or IC. Less growth invalidates the test. The modal resistance, i.e. the MIC occurring most frequently in the susceptible control strains, should give readings of CG or IC on the first two tubes

Table 45.4 Susceptibility testing by the resistance ratio method: suggested concentrations of drugs in Lowenstein–Jensen medium

Drug	Final concentration (µg/ml)						
Isoniazid	0.007	0.015	0.03	0.06	0.125	0.25	0.5
Ethambutol	0.07	0.15	0.31	0.62	1.25	2.5	5
Rifampicin	0.53	1.06	3.12	6.25	12.5	25	50
Streptomycin	0.53	1.06	3.12	6.25	12.5	25	50

These are for the preliminary titration. Choose six that give confluent or near confluent growth in the two lowest and no growth in the next four

Table 45.5 'Modal resistance'

Strain	Drug concentrations in tube no.					
	1	**2**	**3**	**4**	**5**	**6**
Strain A	CG	CG	0	0	0	0
Strain B	CG	IC	0	0	0	0
Strain C	CG	CG	+	0	0	0
Strain D	IC	IC	0	0	0	0
Strain E	CG	CG	+	0	0	0
Mode	CG	CG	0	0	0	0

(Table 45.5). The range must therefore be adjusted to give this result during the preliminary exercises. Once the range is set it is seldom necessary to vary it by more than one dilution.

Interpretation

The resistance ratio is found by dividing the MIC of the test strain by the modal MIC of the control strains. When the readings are all CG or IC, this is easy but, when there are tubes showing +, care and experience may be required in interpretation. In general, a resistance ratio of ≤ 2 can be reported as susceptible, 4 as resistant and 8 as highly resistant. A ratio of 3 is borderline except for ethambutol,

when it probably indicates resistance. Mixed susceptible and resistant strains occur. Examples of these findings are shown in Table 45.6.

Automated methods

The radiometric and non-radiometric systems (see p. 384) may be used to obtain rapid results, e.g. within 7 days. Appropriate amounts of each drug are added to the vials followed by the inoculum. For more information see the manufacturers' brochures.

Table 45.6 Interpretation of results of drug susceptibility by the resistance ratio method based on 'modal resistance'

Strain	Drug concentrations in tube no.						Resistance ratio	Interpretation
	1	**2**	**3**	**4**	**5**	**6**		
Mode	CG	IC	0	0	0	0		
Strain A	CG	0	0	0	0	0	0.5	Susceptible
Strain B	CG	GC	0	0	0	0	1	Susceptible
Strain C	CG	IC	+	0	0	0	1	Susceptible
Strain D	CG	CG	CG	0	0	0	2	Susceptible
Strain E	CG	CG	CG	+	0	0	3	Borderline[a]
Strain F	CG	CG	CG	IC	0	0	4	Resistant
Strain G	CG	CG	CG	CG	IC	0	8	Highly resistant
Strain H	CG	CG	CG	CG	CG	CG	16	Highly resistant
Strain I	CG	CG	CG	+	+	+	–	Mixed susceptible and resistant

[a] Probably resistant with ethambutol

CG, confluent growth; IC, innumerable discrete colonies

Pyrazinamide sensitivity tests

Pyrazinamide acts upon tubercle bacilli in the lysosomes, which have a pH of about 5.2. For reliable susceptibility tests the medium should therefore be at this pH. Tubercle bacilli do not grow well on egg medium at pH 5.2 and agar medium (e.g. Middlebrook 7H11) is therefore frequently used. We have had good and consistent results with Yates' (1984) modification of Marks (1964) stepped pH method. This is a Kirchner semisolid medium layered on to butts of Lowenstein–Jensen medium.

Pyrazinamide stock solution

Dissolve 0.22 g dry powder in 100 ml distilled water; sterilize by filtration or steaming.

Solid medium

Use Lowenstein–Jensen medium containing only half the usual concentration of malachite green (at an acid pH, less dye is bound to the egg and the higher concentration of 'free' dye is inhibitory to tubercle bacilli). Adjust 600 ml medium to pH 5.2 with 1 mmol/l HCl. To 300 ml add 9 ml water (control); to the other 300 ml add 9 ml of the stock pyrazinamide solution (test). Tube each batch in 1-ml amounts and inspissate upright, to make butts, at 87°C for 1 h.

Semisolid medium

Add 1 g agar and 3 g sodium pyruvate to 1 litre Kirchner medium. Adjust pH to 5.2 with 5 mol/l HCl. To 500 ml add 15 ml water (control); to the other 500 ml add 15 ml of the stock pyrazinamide solution. Steam both bottles to dissolve the agar. Cool to 40°C and add to each bottle 40 ml OADC supplement.

Final medium

Layer 2 ml of the control semisolid medium on butts of the control Lowenstein–Jensen medium. Do the same with the test media. Store in a refrigerator and use within 3 weeks. (It may keep longer; we have never tried it.)

Bacterial suspensions

Two inocula are required. Use one as described above for other susceptibility tests and also a 1 : 10 dilution of it in sterile water.

Inoculation

With a suitable pipette (e.g. the Jencons MicroRepette and plastic tip) add 20 μl (approximately) of (1) the undiluted and (2) the diluted suspension to pairs of test and control media.

Interpretation

Two concentrations of inoculum are used because:

1. Some strains require a heavy inoculum to give growth at an acid pH, even in the control medium.
2. A heavy inoculum of some other strains may overcome the inhibitory action of pyrazinamide.

Colonies of tubercle bacilli should be distributed throughout the semisolid medium in one or both controls. If there is no growth in the test bottle from either the heavy or light inoculum, report the strain as susceptible. If there is growth in the test bottle that received the heavy inoculum but not in that which received the diluted suspension, report the strain as susceptible. If there is growth in both test bottles, comparable with that in their respective controls, report the strain as resistant.

Note that resistance to pyrazinamide in strains of *M. tuberculosis* from previously untreated patients is uncommon but strains of *M. bovis* are naturally resistant. Other methods for pyrazinamide susceptibility testing, including a radiometric (BACTEC) technique, have been compared by Cutler *et al.* (1997).

APPLICATION OF NUCLEIC ACID-BASED TECHNOLOGY TO DIAGNOSTIC MYCOBACTERIOLOGY

Nucleic acid-based technology has been used in the tuberculosis laboratory for the rapid detection of members of the *M. tuberculosis* complex and other mycobacteria in clinical specimens, the identification of mycobacteria cultivated in various media and systems, the typing or 'fingerprinting' of isolates of *M. tuberculosis* for epidemiological purposes, and the rapid detection of drug (principally rifampicin) resistance.

Detection of mycobacteria in clinical specimens

This is based on DNA or RNA amplification systems, principally the PCR. Various amplification techniques that have been evaluated are reviewed by Drobniewski *et al.* (2001). Some methods are commercially available and a widely used. One is the Cobas Amplicor PCR system (Roche Molecular Systems, USA), which may be automated. The latter system was found to be highly specific (99.7%) in a multicentre evaluation (Bogard *et al.*, 2001) but, although its sensitivity was high (96.1%) when applied to microscopically positive respiratory specimens, it was less sensitive (71.7%) when applied to microscopically negative specimens.

Identification of isolated mycobacteria

DNA probes are commercially available for the identification of commonly encountered mycobacteria in clinical material. Such probes may be used to identify mycobacteria isolated in automated culture systems as well as conventional culture. Currently available systems such as the INNO-LiPA Mycobacteria (Innogenetics, Ghent, Belgium), a reverse-hybridization-based line probe assay, and the AccuProbe assay (Gen-Probe Inc., San Diego, CA, USA) detect species-specific variations in the variable regions of the 16-S rRNA gene or the 16-S to 23-S rRNA spacer region (Scarparo *et al.*, 2001).

Typing or 'fingerprinting' of *M. tuberculosis*

Fingerprinting of *M. tuberculosis* is usually performed by restriction fragment length polymorphism (RFLP) analysis, which detects certain repetitive DNA sequences inserted in the bacterial genome. Up to 25 copies of the insertion sequence IS6110 is found in most isolates of *M. tuberculosis* and is usually found in 5–25 copies throughout the chromosome. After digestion of the mycobacterial DNA by restriction endonucleases, the DNA fragments are separated by electrophoresis and the number and position of the insertion sequences are revealed by DNA hybridization. Standardized methods and criteria for establishing identity of isolates have been developed (Fletcher, 2001; van Soolingen, 2001).

An alternative typing method, useful for strains containing few insertion sequences, is known as spacer oligonucleotide typing (spoligotyping) and detects short DNA sequences situated near the sites of the insertion elements mentioned above. Unlike RFLP, spoligotyping can be performed on DNA amplified by PCR directly from the clinical specimen.

Applications of DNA fingerprinting to epidemiological studies of tuberculosis are discussed in detail by Godfrey-Faussett (1998).

Rapid detection of rifampicin resistance

Rifampicin resistance is associated with various mutations in the *RpoB* gene, which are detectable by use of a commercially available line probe assay system. In a study of 411 strains of *M. tuberculosis* from several countries, this system correctly identified all 145 rifampicin-sensitive and 262 (98.5%) of 266 resistant strains (Traore *et al.*, 2000). As rifampicin resistance is usually accompanied by isoniazid resistance, this method is a good surrogate indicator of multidrug resistance.

The sequencing of the genome of *M. tuberculosis* is facilitating the development of rapid molecular tests for other forms of drug resistance (Caws and Drobniewski, 2001) and suitable kits may soon be available.

REFERENCES

Bogard, M., Vincelette, J., Antinozzi, R. *et al.* (2001) Multicenter study of a commercial, automated polymerase chain reaction system for the rapid detection of *Mycobacterium tuberculosis* in respiratory specimens in routine clinical practice. *European Journal of Clinical Microbiology and Infectious Disease* 20: 724–731.

Canetti, G., Fox, W., Khomenko, P. *et al.* (1969) Advances in techniques of testing mycobacterial drug sensitivity and the use of sensitivity tests in tuberculosis control programmes. *Bulletin of the World Health Organization* 41: 21–43.

Caws, M. and Drobniewski, F. A. (2001) Molecular techniques in the diagnosis of *Mycobacterium tuberculosis* and the detection of drug resistance. *Annals of the New York Academy of Science* 953: 138–145.

Collins, C. H., Yates, M. D. and Grange, J. M. (1982) Subdivision of *Mycobacterium tuberculosis* into five variants for epidemiological purposes: methods and nomenclature. *Journal of Hygiene (Cambridge)* 89: 235–242.

Collins, C. H., Grange, J. M. and Yates, M. D. (1985) *Mycobacterium marinum* infections in man. *Journal of Hygiene (Cambridge)* 94: 135–149.

Collins, C. H., Grange, J. M. and Yates, M. D. (1997) *Tuberculosis Bacteriology. Organization and Practice*, 2nd edn. London: Butterworths.

Cutler, R. R., Wilson, P., Villarroel J. et al. (1997) Evaluating current methods for determination of the susceptibility of myobacteria to pyrazinamide, conventional, radiometric BACTEC and two methods of pyrazinamidase testing. *Letters in Applied Microbiology* 24: 127–132.

Drobniewski, F. A., Caws, M., Gibson, A. *et al.* (2001) Modern laboratory diagnosis of tuberculosis. *Lancet Infectious Diseases* 3: 141–147.

Fletcher, H. (2001) Molecular epidemiology of tuberculosis: recent development and applications. *Current Opinions on Pulmonary Medicine* 7: 154–159.

French, G. L., Chan, C. Y., Cheung, S. W. and Oo, K. T. (1987) Diagnosis of pulmonary tuberculosis by detection of tuberculostearic acid in sputum by using gas chromatography-mass spectrometry with selected ion monitoring. *Journal of Infectious Diseases* 156: 356–362.

Godfrey-Faussett, P. (1998) The use of DNA fingerprinting in the epidemiology of tuberculosis. In: Davies, P. D. O. (ed.), *Clinical Tuberculosius*, 2nd edn. London: Chapman & Hall, pp 53–65.

Grange, J. M., Yates, M. D. and de Kantor, I. N. (1996) Guidelines for speciation within the Mycobacterium tuberculosis complex. WHO/EMC/ZOO/96.4. Geneva: World Health Organization.

Heifets, L. (1986) Rapid automated methods (BACTEC system) in clinical mycobacteriology. *Seminars in Respiratory Infections* 1: 242–249.

Jenkins, P. A. (1981) Lipid analysis for the identification of mycobacteria: an appraisal. *Reviews of Infectious Diseases* 3: 862–866.

Kiehn, T. E., Edwards, F. F., Brannon, P. *et al.* (1985) Infections caused by *Mycobacterium avium* complex in immunocompromised patients: diagnosis by blood culture and fecal examination, antimicrobial susceptibility tests, and morphological and scroagglutination characteristics. *Journal of Clinical Microbiology* 21: 168–173.

Leat, J. L. and Marks, J. (1970) Improvements in drug sensitivity tests on tubercle bacilli. *Tubercle* 51: 68–73.

Marks, J. (1964) A 'stepped pH' technique for the estimation of pyrazinamide sensitivity. *Tubercle* 45: 47–50.

Mitchison, D. A., Allen, B. J. and Manickayasagar, D. (1987) Selective Kirchner medium for the culture of specimens other than sputum for mycobacteria. *Journal of Clinical Pathology* 36: 1357–1361.

Scarparo, C., Piccoli, P., Rigon, A., Ruggiero, G., Nista, D. and Piersimoni, C. (2001) Direct identification of mycobacteria from MB/BacT alert

3D bottles: comparative evaluation of two commercial probe assays. *Journal of Clinical Microbiology* **39**: 3222–3227.

Sisson, P. R., Freeman, R., Magee, J. G. and Lightfoot, N. F. (1991) Differentiation between mycobacteria of the *Mycobacterium tuberculosis* complex by pyrolysis mass spectrometry. *Tubercle* **72**: 206–209.

Traore, H., Fissette, K., Bastian, I., Devleeschouwer, M. and Portaels, F. (2000) Detection of rifampicin resistance in *Mycobacterium tuberculosis* isolates from diverse countries by a commercial line probe assay as an initial indicator of multidrug resistance. *International Journal of Tuberculosis and Lung Disease* **4**: 481–484.

van Soolingen, D. (2001) Molecular epidemiology of tuberculosis and other mycobacterial infections: main methodologies and achievements. *Journal of Internal Medicine* **249**: 1–26.

Vestal, A. L. (1975) *Procedures for the Isolation and Identification of Mycobacteria*, DHEW (CDC) 75-8230. Washington DC: Government Printing Office.

Yates, M. D. (1984) The differentiation and epidemiology of the tubercle bacilli and a study into the identification of other mycobacteria. Master of Philosophy thesis, University of London.

46

Nocardia, Actinomadura, Streptomyces and Rhodococcus

This group contains Gram-positive aerobes that may form filaments, branches and aerial mycelium. Some are partly acid fast.

ISOLATION

Culture pus directly or wash some of it to recover granules as described on p. 404. Treat sputum by the 'soft' method used for culturing mycobacteria (see p. 382). Culture on blood agar and on duplicate Lowenstein–Jensen (LJ) and Middlebrook medium slopes. Incubate at 37°C for several days. Incubate one of the slopes at 45°C at which temperature nocardias may grow whereas other organisms are discouraged.

IDENTIFICATION

Nocardias grow on most standard bacteriological media. Media suitable for primary isolation include brain–heart infusion agar, trypticase–soy agar enriched with blood Lowenstein–Jensen medium, and Sabouraud's dextrose agar containing chloramphenicol as a selective agent.

Subculture aerobic growth on blood agar and incubate at 37°C for 3–10 days. Colonies vary in size and may be flat or wrinkled, or sometimes 'star shaped'. They may be pigmented (pink, red, cream, orange, brown or yellow) and are often covered with a whitish, downy or chalky aerial mycelium. Colonies of nocardias are usually on the surface although they adhere firmly to the medium, whereas those of streptomyces may be embedded in the medium.

Examine Gram- and Ziehl–Neelsen (ZN)-stained films (minimum decolourization with the latter). Do slide culture (see below) and inoculate lysozyme broth: dissolve 20 mg lysozyme in 2 ml 50% ethanol in water and add 50 μl to 2 ml nutrient broth at pH 6.8. Test for the hydrolysis of casein, xanthine, hypoxanthine and tyrosine (Table 46.1). The API ZYM system may be used to distinguish

Table 46.1 *Nocardia, Rhodococcus, Actinomadura* and *Streptomyces* spp.

Species	Acid fast	Aerial mycelium	Lysozyme	Hydrolysis of		
				Casein	Xanthine	Tyrosine
N. asteroides	+[a]	+	R	–	–	–
N. brasiliensis	+[a]	+	R	+	–	+
N. otitidis caviarum	+[a]	+	R	–	+	–
Actinomadura	–	+	S	+	v	+
Streptomyces	–	+	S	v	v	v
Rhodococcus	–	–	S	–	–	–

R, resistant; S, sensitive; v, variable
[a] Rarely complete. Acid-fast elements may be few in number or absent

between nocardias, streptomyces and related organisms.

Slide cultures

Melt a tube of malt agar, dilute with an equal volume of distilled water and cool to 45°C. Draw about 1 ml of this mixture into a Pasteur pipette, followed by a drop of a thin suspension of the organisms. Run this over the surface of a slide in a moist chamber. No cover glass is needed. Examine daily with a low-power microscope until a mycelium is seen. Spores may be observed as powdery spots on the surface of the medium.

Partially acid-fast, Gram-positive mycelia that do not branch and that fragment early in slide cultures may be tentatively identified as nocardias. Non-acid-fast, Gram-positive mycelia that branches with aerial mycelia and conidiophores abstricted in chains suggest streptomyces.

Nocardia species

The species of nocardia that are of medical importance are *Nocardia asteroides*, *N. brasiliensis* and *N. otitidis caviarum*. Less common pathogens include *N. nova* and *N. transvalensis*. Some *Nocardia*-like organisms that are very rare causes of human disease are included in the genera *Gordona*, *Oerskovia*, *Rothia* and *Tsukamurella*.

Acid-fast elements may be rare or occur in cultures at different times and are best observed on Middlebrook media. Nocardias are resistant to lysozyme and may be differentiated by hydrolysis of casein, xanthine and tyrosine (Table 46. 1). Identification of other species and members of related genera is not easy and is usually undertaken in reference laboratories by sequence analysis of the 16S ribosomal RNA (rRNA).

Nocardia asteroides
This is the most common; it causes severe pulmonary infections, brain abscesses and occasionally cutaneous infections.

Nocardia brasiliensis
This is mostly confined to North America and the southern hemisphere, and usually causes cutaneous infections although it may be associated with systemic disease.

Nocardia otitidis caviarum (*N. caviae*)
This is a rare cause of human infections.

Actinomadura and *Streptomyces* species

Actinomadura madurae causes madura foot, and *A. pelleteri* and *S. somaliensis* cause mycetomas. They are difficult to distinguish bacteriologically, although *A. madurae* is said to hydrolyse aesculin whereas *S. somalensis* does not. Unlike the saprophytic streptomyces, they do not produce acid from lactose and xylose. Cell wall analysis or 16S rRNA sequence is necessary to sort out these and associated species.

Rhodococcus species

Several species are found in plant material but none is pathogenic for humans, although one, *R. equi*, formerly known as *Corynebacterium equi*, is a pathogen of horses. Rhodococci are important only in that they may be confused with nocardias, actinomaduras and *S. somaliensis* (see also p. 354).

For more information on the organisms in this chapter, see Collins *et al.* (1988), Gyles (1995), Hay (1995) and Grange (2002), and on these and other aerobic Gram-positive bacilli, see Schaal (1998), and Bortolussi and Kennedy (1999).

REFERENCES

Bortolussi, R. and Kennedy, W. (1999) Aerobic Gram-positive bacilli. In: Armstrong, D. and Cohen, J. (eds), *Infectious Diseases*, Vol. 2. London: Mosby, Chapter 8, Sections 15, pp. 1–20.

Collins, C. H., Uttley, A. H. C. and Yates, M. D. (1988) Presumptive identification of nocardias in a clinical laboratory. *Journal of Applied Bacteriology* 65: 55–59.

Grange, J. M. (2002). Actinomyces and nocardia. In: Greenwood, D., Slack, R. and Peutherer, J. (eds), *Medical Microbiology*, 16th edn. Edinburgh: Churchill Livingstone, pp. 221–224.

Gyles, C. L. (1995) Nocardia; actinomyces; dermatophilus. In: Gyles, C. L. and Thoen, C. O. (eds), *Pathogenesis of Bacterial Infections in Animals*. Ames: Iowa State University Press, pp. 124–132.

Hay, R. J. (1995) Nocardiosis. In: Weatherall, D. J., Ledingham, J. G. G. and Warrell, D. A. (eds), *Oxford Textbook of Medicine*, 3rd edn, Oxford: Oxford University Press, pp. 686–687.

Schaal, K. P. (1998) Actinomycosis, actinobacillosis and related diseases. In: Collier, L., Balows, A. and Sussman, M. (eds) *Topley and Wilson's Microbiology and Microbial Infections*, 9th edn. Vol. 3, *Bacterial Infections*. London: Arnold, pp. 777–798.

47

Actinomyces, Propionibacterium, Bifidobacterium and Tropheryma

This group contains Gram-positive, aerobic, anaerobic or microaerophilic organisms that tend to branch. *Actinomyces* spp. require CO_2 for growth and are catalase negative; *Propionibacterium* sp. does not require CO_2 and is catalase positive; *Bifidobacterium* sp. does not require CO_2 and is catalase negative. Species of medical and veterinary importance occur in pus and discharges as colonies or granules.

A recently described actinomycete, *Tropheryma whippelii*, is the cause of Whipple's disease, a multisystem disease presenting with diarrhoea and malabsorption, often with arthralgia and fever and sometimes with central nervous system (CNS) involvement.

ACTINOMYCES

Isolation

Wash pus gently with saline in a sterile bottle or Petri dish. Look for 'sulphur granules', about the size of a pinhead or smaller. Aspirate granules with a Pasteur pipette and crush one between two slides and stain by the Gram method. Look for Gram-positive mycelia in an amorphous matrix surrounded by radiating large Gram-negative, club-like structures. The club forms are acid fast on Ziehl–Neelsen staining, modified by use of 1% sulphuric acid for decolourization. Sulphur granules and mycelia in tissue sections are also identifiable by use of fluorescein-conjugated specific antisera.

Crush granules with a sterile glass rod in a small sterile tube containing a drop of broth. Inoculate

blood agar plates, brain heart infusion agar and enriched thioglycollate medium. These may be made selective for Gram-positive non spore-bearers by adding 30 μg/ml nalidixic acid and 10 μg/ml metronidazole. Do cultures in triplicate and incubate:

- anaerobically plus 5% carbon dioxide
- in 5% carbon dioxide in air
- aerobically

doing all at 37°C for 2–7 days.

Subculture any growth in the broth media to solid media and incubate under the same conditions. Look for colonies resembling 'spiders' or 'molar teeth'.

Identification

Do catalase tests on colonies from anaerobic or microaerophilic cultures that show colonies of Gram-positive coryneform or filamentous organisms.

Actinomyces are catalase negative but corynebacteria and bifidobacteria are usually catalase positive. It is often difficult to distinguish actinomyces from bifidobacteria and propionibacteria by simple tests. Analysis of the metabolic products of carbohydrate fermentation by gas–liquid chromatography (GLC) may be necessary. The spot test for indole production applied to colonies is useful; *Actinomyces* spp. are negative, unlike *Propionibacterium* spp.

Subculture for nitrate reduction test and test for acid production from mannitol, xylose and raffinose, and aesculin hydrolysis (Table 47.1). The API ZYM system may be useful in distinguishing between actinomyces and related genera.

Table 47.1 *Actinomyces* and *Bifidobacterium* species

Species	Nitrate reduction	Acid from			Hydrolysis of	
		Mannitol	Xylose	Raffinose	Starch	Aesculin
A. bovis	−	−	−	−	+	−
A. israelii	v	v	+	+	−	+
A. naeslundii	+	−	v	+	v	+
A. odontolyticus	+	−	v	−	v	v
A. meyeri	−	−	+	+	−	−
A. pyogenes	v	−	−	−	v	−
B. eriksonii	−	+	+	+	−	v

Species of *Actinomyces*

Actinomyces bovis

Colonies at 48 h on blood agar are pinpoint sized and smooth, later becoming white and shining with an entire edge. Spider and molar tooth colonies are rare. In thioglycollate broth, growth is usually diffuse but occasionally there are colonies that resemble breadcrumbs. Microscopically the organisms are coryneform, and rarely branching.

Nitrates are not reduced, and acid is not produced from mannitol, xylose or raffinose. Starch is hydrolysed; aesculin is not hydrolysed.

This is a pathogen of animals, usually of bovines, causing lumpy jaw.

Actinomyces israelii

Colonies at 48 h on blood agar are microscopic, with a spider appearance, later becoming white and lobulated with the appearance of molar teeth. In thioglycollate broth, there are distinct colonies with a diffuse surface growth and a clear medium. The colonies do not break when the medium is shaken. Microscopically, the organisms are coryneform, with branching and filamentous forms.

Nitrates may be reduced, and acid is produced from xylose and raffinose; aesculin is hydrolysed but not starch. Its identity may be confirmed by staining with specific fluorescent antisera

This species causes human actinomycosis and may be responsible for pelvic infections in women fitted with intrauterine contraceptive devices.

Actinomyces naeslundii

Colonies at 48 h on blood agar are similar to those of A. *israelii*; spider forms are common, but not molar tooth forms. In thioglycollate broth, growth is diffuse and the medium is turbid. Microscopically the organisms are irregular and branched, with mycelial and diphtheroid forms.

Nitrates are reduced, and acid is produced from raffinose; aesculin is hydrolysed; starch hydrolysis is variable. This is a facultative aerobe.

It is not known to be a human pathogen but has been found in human material.

Actinomyces odontolyticus

Colonies at 48 h on blood agar are 1–2 mm in diameter and grey, but they may develop a deep-reddish colour after further incubation. CO_2 is required for growth. The organisms are coryneform although branching is rare. Nitrate is reduced; acid is rarely produced (except occasionally from xylose); starch and aesculin hydrolysis is variable.

The normal habitat is the human mouth but the organisms have been isolated from the tear ducts.

Actinomyces meyeri

Colonies at 48 h on blood agar are pinpoint sized, greyish white and rough. CO_2 is required for both aerobic and anaerobic growth. The organisms are short and coryneform but branching is rare. Nitrate is not usually reduced, and acid is produced from xylose and raffinose; starch and aesculin are not hydrolysed.

The normal habitat is the human mouth but the organism has been isolated from abscesses.

Actinomyces pyogenes

This is involved in suppurative lesions and mastitis in domestic and wild animals. On blood agar it produces pinpoint α-haemolytic colonies at 36–48 h. For more information about actinomyces, see Gyles (1995); Schaal and Lee (1995); and Schaal (1999).

PROPIONIBACTERIUM

These are pleomorphic coryneforms varying from coccoid to branched forms. They are aerobic but may be aerotolerant, non-motile and grow poorly in the absence of carbohydrates. They are catalase positive and some strains may liquefy gelatin. The optimum temperature is 30°C. There are several species of interest to food microbiologists. They occur naturally in the bovine stomach and hence in rennet, and are responsible for flavours and 'eyes' in Swiss cheese. A method for presumptive identification is given on p. 241. It is difficult to identify species.

There is one species of medical interest: *P. acnes*.

Propionibacterium acnes

The bacilli are small, almost coccoid, or about 2.0×0.5 μm, and may show unstained bands. The best growth is obtained anaerobically. The colonies on blood agar are either small, flat, grey–white and buttery, or larger, heaped up and more granular. Both colony forms are β-haemolytic. Acid is not usually produced. The spot test for indole production is positive. It is nitrate reduction negative, catalase positive and liquefies gelatin. Slide agglutination is useful.

It is a commensal on human skin, in hair follicles and in sweat glands. It appears to be involved in the pathogenesis of acne. Oral and systemic infections have been suspected, and it is an occasional cause of delayed-onset postoperative endophthalmitis.

This genus is described by Cummings and Johnson (1986).

BIFIDOBACTERIUM

Bifidobacteria are small Gram-positive rods resembling coryneforms, which predominate in the faeces of breast-fed infants, but they are also commensals in the adult bowels, mouth and vagina. They are anaerobic and may need to be differentiated from anaerobic corynebacteria. Bifidobacteria are catalase negative and nitrate reduction negative, and do not produce gas from glucose. CO_2 is not required but it does improve growth.

There is one pathogen, *B. eriksonii*, formerly *Actinomyces eriksonii*, associated with mixed infections in the upper respiratory tract.

For information about bifidobacteria, see Seardovi (1986).

TROPHERYMA WHIPPELII

Gene amplification techniques have shown that a Gram-positive bacillus detectable microscopically in intestinal tissue from patients with Whipple's disease is an actinomycete with a unique 16S rRNA sequence. This organism, *Tropheryma whippelii*, does not grow on conventional culture media, but replicates within cells of the human fibroblast (HEL) line. Confirmation of the diagnosis of Whipple's disease may be made by polymerase chain reaction (PCR) examination of intestinal biopsies and specimens from other tissues involved in the disease process.

For further details, see Veitch and Farthing (1999) and Puechal (2002).

REFERENCES

Cummings, C. S. and Johnson, J. L. (1986) The genus *Propionibacterium*. In: Starr, M.P., Stolp, H. and Troper, H.C (eds), *The Prokaryotes: A handbook* on *habitats, isolation and identification of bacteria,* Vol. 2. New York: Springer, p. 1864 .

Gyles, C. L. (1995) Nocardia; actinomyces; dermatophilus. In: Gyles, C.L. and Thoen, C.O. (eds), *Pathogenesis of Bacterial Infections in Animals.* Ames, Iowa: Iowa State University Press, pp 124–132

Puechal, X. (2002) Whipple's disease. *Joint, Bone, Spine* **69**: 133–140.

Seardovi, V. (1986) The genus *Bifidobacterium.* In: Sneath, P.H.A., Mair, N.S. and Holt, J.G. (eds), *Bergey's Manual of Systematic Bacteriology,* Vol. 2. Baltimore: Williams & Wilkins, p. 1418.

Schaal, K. P. (1999) Actinomycosis, actinobacillosis and related diseases. In: Collier, L., Balows, A. and Sussman, M. (eds), *Topley and Wilson's Microbiology and Microbial Infections,* 9th edn. Vol. 3, *Bacterial Infections,* CD-ROM. London: Arnold, Chapter 39.

Schaal, K. P. and Lee, H-J. (1995) Actinomycete infections in humans – a review. *Gene* **115**: 201–211.

Veitch, A. M. and Farthing, M. J. G. (1999) Whipple's Disease. In: Armstrong, D. and Cohen, J. (eds), *Infectious Diseases.* London: Mosby, pp. 2.36.1–2.36.2.4.

48

Bartonella and Mobiluncus

These two organisms are considered together here although they are not related. *Bartonella* has aroused interest recently as a result of taxonomic changes. The significance of *Mobiluncus* in the vagina is not well documented.

BARTONELLA

This genus contains at least three species of small, Gram-negative bacilli that are oxidase negative and fastidious in their growth requirements.

Isolation

Perform blood cultures using one of the highly enriched media (automated systems are usually satisfactory, but the organisms may not produce enough CO_2 for detection by some of them). Incubate duplicate cultures at 25–30°C and 35–37°C for at least 4 weeks.

Subculture to lysed blood agar (5–10% rabbit blood). Incubate cultures for at least 14 days in a high humidity and under 5–10% CO_2. Colonies may be dry, of 'cauliflower' appearance, or small, circular and entire with pitting of the medium.

Identification is difficult and it is best to send specimens or cultures to a reference laboratory.

Bartonella bacilliformis

This causes Oroya fever (Carrion fever; bartonellosis) and is transmitted by the sandfly (*Phlebotomus* spp.).

It is best observed in Giemsa-stained blood films and the bacilli are seen in and on red blood cells. They are pleomorphic, coccobacilli, often in chains or arranged like Chinese letters.

Bartonella henseliae

Formerly in the genus *Rochalimaea* this is now recognized as the agent of cat-scratch fever (see Windsor, 2001). It has also been isolated from immunocompromised and other patients with bacteraemia and fever.

Bartonella quintana

This is the agent of trench fever, suffered by military personnel. It is spread by the human body louse.

For more information about *Bartonella*, see Welch and Slater (1995).

MOBILUNCUS

This old species, revived in 1980, is now of interest. It is found in vaginal secretions along with other agents of vaginitis but its significance is not fully documented.

Isolation

Culture vaginal fluids on enriched media containing horse or rabbit blood and incubate at 35–37°C for at least 5 days. Colonies are entire, convex, smooth and translucent. It grows at pH 12.0 (Pålson *et al.*, 1986).

REFERENCES

Pålson, C., Hallén, A. and Forsum, U. (1986) Improved yield of *Mobiluncus* species from clinical specimens after alkaline treatment. *Acta Pathologia, Microbiologia et Immunologia* Section B **94**: 113–116.

Welch, D. F. and Slater, L. N. (1995) *Bartonella*. In: Murray, P. R., Barron E. J., Pfaller, F. A. *et al.* (eds), *Manual of Clinical Microbiology*, 6th edn. Washington DC: American Society for Microbiology.

Windsor, J. J. (2001) Cat scratch fever: epidemiology, aetiology and treatment. *British Journal of Biomedical Science* **58**: 101–110.

49

Spirochaetes

Three genera that are of medical importance are considered briefly here: *Borrelia*, *Treponema* and *Leptospira*.

BORRELIA

There are at least 19 species, classified by their arthropod vectors. Only three are considered briefly here.

Borrelia duttonii and *B. recurrentis* cause relapsing fever, transmitted by lice and ticks. *B. burgdorferi* is responsible for Lyme disease (inflammatory arthropathy), first described in Old Lyme, Connecticut, USA and now known in Europe (Burgdorfer, 1984, 1985). It is a tick-borne zoonosis and small woodland rodents and wild deer are reservoirs. Infection is also transmitted to domesticated and farm animals

Identification

Take blood during a febrile period. For the relapsing fever species examine wet preparations by dark field using a high-power dry objective. Stain thick and thin films with Giemsa stain.

Fix smears in 10% methyl alcohol for 30 s. Stain with 1 part Giemsa stock stain and 49 parts Sorensen's buffer at pH 7 in a Coplin jar for 45 min. Wash off with Sorensen's buffer. Dry in air and examine. Borrelias appear as shallow, coarse, irregular, highly motile coils 0.25–0.5 × 8–16 µm.

In cases of suspected Lyme disease diagnosis may be made by fluorescence antibody methods (Technicon kit), enzyme-linked immunosorbent assay (ELISA) or silver staining of biopsies.

It is not usual practice to attempt culture of borrelias from ticks and host animals, but methods are given by Johnson (1999).

TREPONEMA

There are at least 14 species. In medicine the most important are *T. pallidum*, *T. pertenue*, *T. carateum* and *T. vincentii*.

Identification

In suspected syphilis

Examine exudate (free from blood and antiseptics) from lesion by dark-field or fluorescence microscopy. For detailed methods, see Sequira (1987).

Treponema pallidum

This is the causative organism of syphilis. It forms tightly wound slender coils, 0.1–0.2 × 6–20 µm, with pointed ends, each of which has three axial fibrils. They are sluggishly motile with drifting flexuous movements.

In suspected Vincent's angina

Stain smears with dilute carbol–fuchsin. Large numbers of spirochaetes are seen together with fusobacteria.

Treponema vincentii

This is found in cases of Vincent's angina (ulcerative lesions of the mouth or genitals) and pulmonary infections. It is a loosely wound single contoured spirochaete 0.2–0.6 × 7–18 µm.

Other treponemes

Treponema pertenue

This organism causes yaws and is principally transmitted by direct contact. It is morphologically indistinguishable from *T. pallidum*. Laboratory tests are unhelpful.

Treponema carateum

This causes pinta. The mode of spread is, in common with yaws, by direct contact. It is morphologically indistinguishable from *T. pallidum*. Diagnosis is by silver impregnation of tissue. For further information, see Penn (1999).

LEPTOSPIRA

Identification is important only for epidemiological purposes. *Leptospira interrogans* (23 serogroups and about 200 serotypes) is primarily an animal parasite, but in humans causes an acute febrile illness with or without jaundice, conjunctivitis and meningitis. The more severe manifestations are known as Weil's disease. Animal hosts include rodents, dogs, horses, cattle and pigs. *L. biflexa* is non-pathogenic.

Isolation and identification

Examine blood during the first week of illness and urine thereafter. Add 9 parts blood to 1 part 1% sodium oxalate in phosphate buffer at pH 8.1 and centrifuge 15 min at 1500 rev./min. Examine clear plasma under dark-field microscopy using low-power magnification. If this is negative, centrifuge the remainder of the plasma at 10 000 rev./min for 20 min and examine the sediment. Direct films rarely show leptospiras either in dark-field or Giemsa-stained preparations. Centrifuge urine and examine deposit by dark field within 15 min.

Morphology

Leptospiras are short, fine, closely wound spirals, 0.25 × 6–20 µm, resembling a string of beads. The ends are bent at right angles to the main body to form hooks.

Culture

The methods given here are adapted from those of Waitkins (1985). Add two drops of fresh blood, cerebrospinal fluid (CSF) or urine (at pH 8) to 5 ml EMJ/5FU medium (see p. 75) and make five serial dilutions in the same medium to dilute out antibody. Incubate at 30°C. Examine daily for 1 week, then weekly for several weeks. Use dark-field, phase contrast or fluorescence microscopy.

Identification

If leptospiras are seen. subculture to EMJH medium (se p. 75) and incubate at 30 and 13°C. Subculture also to EMJ medium containing 225 mg azoguanine/l.

Leptospira interrogans grows at 30°C, not at 13°C and not in azoguanine medium. *L. biflexa* grows at 13 and 30°C and also in azoguanine medium.

Serological diagnosis

Macroslide agglutinations may be done with patients' sera and (commercial) genus-specific antigen, but further agglutination tests are best done in reference laboratories.

There is a rapid test kit (Leptese: Bradsure Biologicals). For more information about leptospires, culture, serology and epidemiology, see Waitkins (1985) and Faine (1999).

REFERENCES

Burgdorfer, W. (1984) Discovery of the Lyme spirochaete and its relationship to tick vectors. *Yale Journal of Biology and Medicine* 57: 165–168.

Burgdorfer, W. (1985) *Borrelia*. In: Lennette, E. H., Balows, A., Hauser, W. J. *et al.* (eds), *Manual of Clinical Microbiology*, 4th edn. Washington DC: Association of American Microbiologists, pp. 154–175.

Faine, S. (1999) *Leptospira*. In: Collier, L., Balows, A. and Sussman, M. (eds), *Topley and Wilson's Microbiology and Microbial Infections*, 9th edn. Vol. 2, *Systematic Bacteriology*, CD-ROM. London: Arnold, Chapter 57.

Johnson, R. C. (1999). *Borrelia*. In: Collier, L., Balows, A. and Sussman, M. (eds), *Topley and Wilson's Microbiology and Microbial Infections*, 9th edn. Vol. 2, *Systematic Bacteriology*, CD-ROM. London: Arnold, Chapter 56.

Penn, C. W. (1999). *Treponema*. In: Collier, L., Balows, A. and Sussman, M. (eds), *Topley and Wilson's Microbiology and Microbial Infections*, 9th edn. Vol. 2, *Systematic Bacteriology*, CD-ROM. London: Arnold, Chapter 55.

Sequira, P. J. L. (1987) Syphilis. In: Jephcott, A.E. (ed.), *Sexually Transmitted Diseases*. London: Public Health Laboratory Service, pp. 6–22.

Waitkins, S. A. (1985) Leptospiras and leptospirosis. In: Collins, C.H. and Grange, J.M. (eds), *Isolation and Identification of Micro-organisms of Medical and Veterinary Importance*. Society for Applied Bacteriology Technical Series No. 21, London: Academic Press, pp. 251–296.

50

Yeasts

Yeasts are fungi, the main growth form of which is unicellular; they usually replicate by budding. Many can also grow in the hyphal form and the distinction between a yeast and a mould is one of convention only. Some yeast-like fungi such as *Acremonium* and *Geotrichum* spp. and the 'black yeasts' *Exophiala*, *Aureobasidium* and *Phialophora* spp. are traditionally excluded from texts on yeasts.

The yeasts are not a natural group. Many are Ascomycotina, a few are Basidiomycotina and others are asexual forms of these groups (Deuteromycotina or Fungi Imperfecti).

IDENTIFICATION

The methods given in Chapter 9 yield information about morphology, sugar fermentation and assimilation of carbon and nitrogen sources. The morphological key (Table 50.1) allows most yeasts isolated in clinical and food laboratories to be

Table 50.1 Key to identification of genera of yeasts

Cells reproducing by fission, not budding	*Schizosaccharomyces*
Budding cells only (corn meal and coverslip preparation); no ascospores	
1. Cells small (2–3 μm), bottle shaped, with broad base to daughter bud	*Malassezia*
2. Cells oval with prominent lateral 'spur'; colony pinkish, depositing mirror image of colony on lid of inverted Petri dish	*Sporobolomyces*
3. Cells not as in 1 or 2:	
(a) urease negative	Asexual forms of *Candida Saccharomyces* etc.
(b) urease positive; colonies pink/red; cells large; encapsulated	*Rhodotorula*
(c) urease positive; colonies white or cream; cells large; encapsulated	*Cryptococcus*
Budding cells, some with ascospores	
1. Ascospores liberated from parent cells, kidney-shaped or elongate	*Kluyveromyces*
2. Ascospores liberated, round with or without disc-like flange; nitrate assimilated	*Pichia*
3. Ascospores as in 2; nitrate not assimilated	*Pichia*
4. Ascospores retained in parent cells, round, smooth	*Saccharomyces*
5. Ascospores retained, round, warty and ridged	*Debaromyces*
Budding cells borne on pseudomycelium; no ascospores	*Candida* (most species)
Budding cells born on pseudomycelium which grows out to true mycelium after 3–4 days	
1. Chlamydospores present	*C. albicans/ C. dubliniensis*
2. No chlamydospores	*C. tropicalis*
Extensive true mycelium, fragmenting to form arthrospores	
1. Budding on short side branches	*Trichosporon* or *Blastoschizomyces*
2. No budding	*Geotrichum*

identified to genus level. Specific identification may then be made by reference to Table 50.2 and Figure 50.1 (see also Campbell *et al.*, 1996).

Candida

Candida species are commensals in the human gut. The principal species of medical importance is *Candida albicans*, which can cause superficial, mucosal and systemic infections, particularly in immunocompromised patients.

Candida parapsilosis is a normal skin commensal and may infect nails. Systemic infections are often related to the use of parenteral nutrition and have been particularly noted in paediatric intensive care units. It is also a cause of candida endocarditis.

The other species most commonly implicated in human disease include *C. glabrata*, *C. tropicalis*, *C. krusei*, *C. famata* and *C. lusitaniae*. The increasing use of azole antifungal therapy in both prophylaxis and treatment may have contributed to their increasing incidence. *C. krusei* is innately resistant to fluconazole and *C. glabrata* is less susceptible to fluconazole and other azole drugs than *C. albicans*. Identification to species level is therefore an impor-

tant consideration in the choice of treatment of these infections.

All *Candida* spp. produce creamy, white, smooth colonies on malt or peptone agar after 2 days at 30–37°C, except *C. krusei*, which has a ground-glass appearance (see Table 50.2).

Cryptococcus

Cryptococcus neoformans causes meningitis (which may be fulminant or slowly progressive) and subcutaneous or deep granulomata. It is mainly found in HIV-infected patients. Two varieties are found: *C. neoformans* var. *neoformans*, which has a global distribution, and *C. neoformans* var. *gattii*, found mainly in Australia and central Africa. As with other encapsulated yeasts (*Rhodotorula*, *Trichosporon* spp.), these are the asexual forms of Basidiomycetes, although the sexual forms are not normally encountered.

Colonies on Sabouraud's agar incubated 24–48 h at 37°C are large, white and butyrous, and may 'flow' down a vertical slope.

The most sensitive method of diagnosis of *C. neoformans* infection is the detection of the capsular

Table 50.2 Biochemical differentiation of some common yeasts

Species	Fermentation of				Assimilation of								Nitrate reductase
	glu	ma	su	la	glu	ma	su	la	mn	ra	ce	er	
Candida albicans	+	+	−	−	+	+	+	−	+	−	−	−	−
C. tropicalis	+	+	+	−	+	+	+	−	+	−	+	−	−
C. kefyr	+	−	+	+	+	−	+	+	+	+	+	−	−
C. parapsilosis	+	−	−	−	+	+	+	−	+	−	−	−	−
C. guillermondii	+	−	+	−	+	+	+	−	+	+	+	−	−
C. krusei	+	−	−	−	+	−	−	−	−	−	−	−	−
C. glabrata	+	−	−	−	+	−	−	−	−	−	−	−	−
Cryptococcus neoformans	−	−	−	−	+	+	+	−	+	+^{w}	+^{w}	±	−
C. albidus	−	−	−	−	+	+	+	±	+	+^{w}	+	−	+
C. laurentii	−	−	−	−	+	+	+	+	−	−	−	−	−
Trichosporon beigelii	−	−	−	−	+	+	+	+	±	±	±	±	−
Blastoschizomyces capitus	−	−	−	−	+	−	−	−	−	−	−	−	−
Saccharomyces cerevisiae	+	+	+	−	+	+	+	−	±	±	−	−	−

ce, cellobiose; er, erythritol; glu, glucose; la, lactose; ma, maltose; mn, mannitol; NO$_3$, nitrate; ra, raffinose; su, sucrose; w, weak; +, gas produced or growth occurs; −, no gas produced or no growth occurs

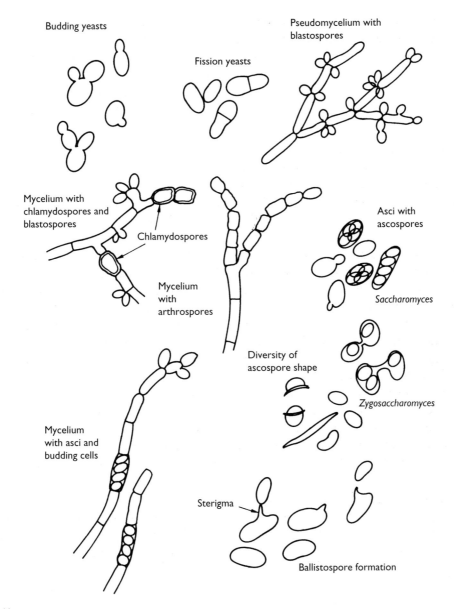

Budding yeasts

Fission yeasts

Pseudomycelium with blastospores

Mycelium with chlamydospores and blastospores

Chlamydospores

Mycelium with arthrospores

Asci with ascospores

Saccharomyces

Diversity of ascospore shape

Zygosaccharomyces

Mycelium with asci and budding cells

Sterigma

Ballistospore formation

Figure 50.1 Yeasts

polysaccharide antigen by latex agglutination using serum or cerebrospinal fluid. Commercial kits are available for this purpose and may also be used to monitor the course of infection following treatment by titration of the samples.

In advanced infections agglutination tests for antibody production may be negative because the capsular material is poorly metabolized by the body, and the accumulation of it causes a high-dose immunological tolerance.

Other species of *Cryptococcus* such as *C. albidus* and *C. laurentii* are sometimes found contaminating solutions but are rarely pathogenic as they do not grow well above 30°C.

Rhodotorula

Members of this genus are frequent skin contaminants. They form red or orange colonies on malt or glucose peptone agar after 2 days' incubation at 30°C. Morphologically they resemble cryptococci and may be capsulated.

Sporobolomyces

These yeasts live on leaf surfaces and reproduce by ballistospores, which are ejected into the air. The salmon-pink colonies of *S. roseus* produce mirror images on the lids of Petri dishes if left undisturbed. They are of no clinical significance.

Trichosporon

Trichosporon beigelli is the one most frequently isolated in clinical work. It is a skin contaminant but occasionally causes deep infections in immunocompromised patients. It is also the cause of 'white piedra', in which masses of yeasts grow on the outer surfaces of axillary and other hair that is kept permanently moist. *T. beigelli* actually encompasses a group of species including *T. asahii*. *T. cutaneum*, *T. inkin*, *T. mucoides* and *T. ovoides*.

Blastoschizomyces

Blastoschizomyce capitatus colonies are whitish, wrinkled, tough, and often slightly 'hairy' on the surface, as a result of the presence of aerial hyphae. It sometimes causes systemic infection in immunocompromised patients.

Pneumocystis

Pneumocystis carinii was originally thought to be a protozoan but was later identified as a fungus on the basis of molecular studies, which place it between the Basidiomycota and Ascomycota. More recently it has been attributed to the Archiascomycetes together with *Schizosaccharomyces* and is described here in the yeast chapter, although it has not been classified as a yeast.

Pneumocystis carinii has emerged as a common cause of pulmonary infection in AIDS patients. The organism cannot be cultivated on routine laboratory media. Diagnosis is therefore based on polymerase chain reaction (PCR) detection or immunofluorescence staining of the thick-walled cysts in samples of broncheoalveolar lavage.

Malassezia species

Malassezia spp. are lipophilic yeasts characterized by monopolar budding from a broad base. They are common skin commensals but may cause disseminated infection in patients, particularly neonates, receiving lipid infusions. Overgrowth of the fungus may cause seborrhoeic dermatitis.

Seven species have been described: *M. furfur*, *M. globosa*, *M. pachydermatis*, *M. obtusa*, *M. restricta*, *M. slooffiae* and *M. sympodialis*. Only *M. pachydermatis* is capable of growth on laboratory culture media without the addition of lipids.

Pityriasis versicolor is a mild skin infection of the chest or back characterized by depigmentation or hyperpigmentation. *M. furfur* is the name usually given to the organism in this condition, although some of the other species listed may also be involved. Diagnosis of pityriasis versicolor is by microscopic examination of infected skin scrapings, which show clumps of yeast cells interspersed by short hyphal fragments.

YEASTS OF IMPORTANCE IN FOOD PROCESSING AND SPOILAGE

Saccharomyces and similar yeasts

This group contains the brewing ('pitching') and baking yeasts. *S. cerevisiae* is the 'top yeast', used in making beer, and *S. carlsbergensis* the 'bottom yeast', used in making lager. There are many special strains and variants used for particular purposes.

Saccharomyces pastorianus

This is a 'bottom yeast' with long sausage-shaped cells, which produces unpleasant flavours in beer.

Saccharomyces cerevisiae

This is sometimes isolated as a non-significant organism from faecal and oral samples but can be a cause of genital or systemic infection.

Zygosaccharomyces

Zygosaccharomyces rouxii and *Z. mellis* are 'osmophilic yeasts' which will grow in concentrated sugar solution (compare with aspergilli of the glaucus group) and spoil honey and jam.

Zygosaccharomyces baillii can grow at low pH in the presence of some preservatives and has become a major problem in some areas of the food and beverage industry.

Pichia spp.

These occur as contaminants in alcoholic liquors. They can use alcohol as a source of carbon, and are a nuisance in the fermentation industry. They grow as a dry pellicle on the surface of the substrate. Some species occur as the 'flor' in sherry and in certain French wines, to which they give a distinctive flavour. They also cause pickle spoilage, particularly in low-salt (10–15%) brines.

Debaromyces spp.

These are also 'pellicle-forming' yeasts. They are found in animal products such as cheese, glue and rennet, and are also responsible for spoilage in high-salt (20–25%) pickle brines.

Apiculate yeasts

These lemon-shaped yeasts are common contaminants in food. Perfect forms are placed in the genus *Hanseniaspora* and imperfect forms in the genus *Kloeckera*.

Yeasts are considered in detail by Kurtzman and Fell (1998).

REFERENCES

Campbell, C. K., Johnson, E. M., Philpot, C. M. and Warnock, D. W. (1996) *Identification of Pathogenic Fungi*. London: Public Health Laboratory Service

Kurtzman, C. P. and Fell, J. W. (1998) *The Yeasts, a Taxonomic Study*. Amsterdam: Elsevier Science.

51

Common moulds

Many common moulds may occur as air-borne contaminants on ordinary culture media if care is not taken. Most of them are saprophytes, causing spoilage of food and other commodities; others are plant pathogens. Some, e.g. *Aspergillus*, are described in this chapter although they are also recognized causes of human disease, especially in immunosuppressed patients, or are of public health importance because they produce mycotoxins (e.g. *Penicillium*). Each strain of a mould isolated from human material must therefore be assessed individually and some reliance must be placed on the observation of hyphae by direct microscopy. The preliminary identification guide, based on colony colour and texture (Table 51.1), includes genera in which some of the pathogens described in Chapter 52 occur.

Table 51.1 Colour guide to mould cultures

Black or dark brown mycelium, spores or both	
Alternaria	*Cladosporium*[a]
Arthrinium	*Exophiala*[a]
Aureobasidium	*Fonsecaea*[a]
Botrytis	*Madurella*[a]
Chaetomium	*Phialophora*[a]
Aspergillus (niger)	*Sporothrix*[a]
	Stachybotrys
Some shade of green predominant	
Trichoderma	*Aspergillus*
	(flavus/parasiticus)
Chaetomium	*Penicillium*
Some shade of red predominant	
Monascus	*Trichophyton*[a]
Paecilomyces (lilacinus)	*Acremonium*
Aureobasidium	*Fusarium* (some strains)[a]
Colonies white or cream	
Chrysosporium	*Coccidioides*[a]
Geomyces	*Blastomyces*[a]
Dermatophytes[a]	Mucoraceous species
Histoplasma[a]	*Fusarium*[a]
Colonies sandy-brown to khaki	
Paecilomyces (variotii)	*Epidermophyton floccosum*[a]
Scopulariopsis	
Aspergillus terreus	

[a] Described in Chapter 52

ALTERNARIA AND *ULOCLADIUM*

These are two large genera of plant pathogens and saprophytes. Without knowledge of the host plant speciation is very difficult (see Ellis, 1971). Rare cases of subcutaneous granulomata in humans and cats have been reported. *Alternaria* species have caused cutaneous infection on the faces of cats.

Colonies on Saboraud's medium after incubation at 20–25°C for 3–4 days are high, densely fluffy mats, white at first, becoming dark-grey, deepest in colour in the centre. On some media the whole colony is black, with less aerial growth. Spores are large, brown pigmented, club shaped and multicellular, with some septa longitudinal or oblique. In *Alternaria* spp. (Figure 51.1) the spores are in chains mostly with an apical beak; in *Ulocladium* spp. they are single or only occasionally joined together. A tape impression mount will demonstrate this difference. *Alternaria* species are sometimes slow to produce spores.

Incubation at room temperature in daylight can encourage sporulation. *Ulocladium chartarum* may be responsible for discoloration of paintwork and wall paper on damp walls. Some species of *Alternaria* may produce mycotoxins in food plant products.

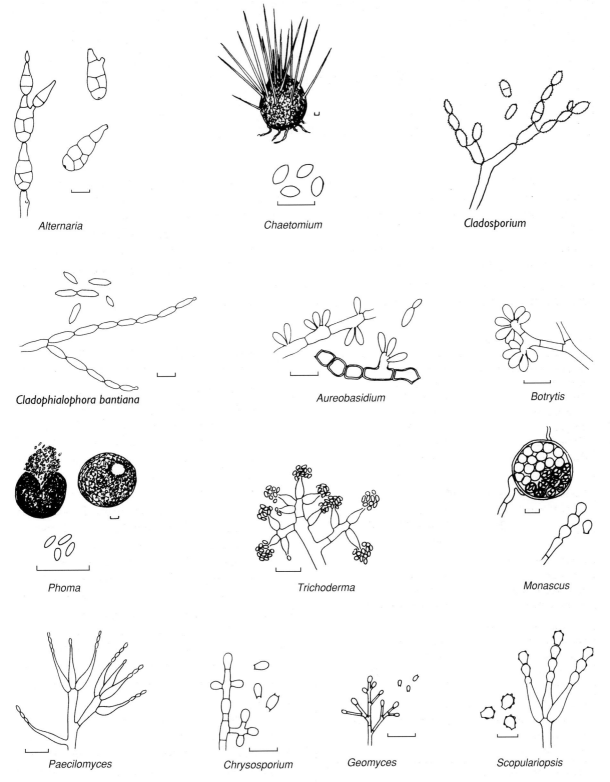

Alternaria

Chaetomium

Cladosporium

Cladophialophora bantiana

Aureobasidium

Botrytis

Phoma

Trichoderma

Monascus

Paecilomyces

Chrysosporium

Geomyces

Scopulariopsis

Figure 51.1 Common moulds (bar = 10 μm)

ARTHRINIUM

These are common contaminants of no medical importance. Most species are from plant material, on which they look quite unlike the *in vitro* growth.

The growth on laboratory media is dense, vigorous and at first pure white in colour. The centres of older cultures become grey because of the abundance of spores, which are unicellular, rounded and black, often with a conspicuous germ slit (Ellis, 1971).

CHAETOMIUM

These are saprophytes on plant material. Some are troublesome pests of paper and cellulosic materials causing discoloration and degradation They have no medical implications, although some species are known to produce toxic metabolites.

Colonies on ordinary laboratory media are flat with low, often sparse aerial growth, with black or greenish discrete bodies standing on the agar surface. Low power examination shows that these bodies are covered with long black spines (Figure 51.1). Under the high power these are seen to be straight, spiral or irregularly branched according to species. Large numbers of globose asci are produced inside the perithecium, each containing dark brown, ovoid ascospores. In *Chaetonium* the ascus walls rapidly break down, releasing the spores and giving the impression that the perithecium itself is full of loose spores.

CLADOSPORIUM

There are many species, most of which are of no clinical significance. Long chains of conidia are produced by acropetal budding and may branch where a conidium develops two buds (Figure 51.1). They differ from the pathogenic *Cladophialophora* spp., which have longer, rarely branching chains of smooth, slender and thin-walled conidia (see Figure 51.1).

AUREOBASIDIUM

This mould occurs on damp cellulosic materials and grows on painted surfaces. Colonies on ordinary laboratory media are at first yeast like, often mucoid, with pink, white and black areas. In time the black areas increase in size and the colony becomes drier and rough surfaced. There may be aerial mycelia. They are often referred to as 'black yeasts'.

Microscopy shows a bizarre mixture of oval to long budding yeasts and pigmented, irregularly shaped, hyphal cells. Under low power the edge of the colony shows yeasts clustered in round masses along the submerged hyphae. This is best seen in a slide culture or needle mount (see Figure 51.1). *A. pullulans* is the most common of the saprophytic species in the group.

BOTRYTIS

This is a group of plant pathogens, of which the common *B. cinerea* has a very wide host range and is also a successful saprophyte. It is well known as the grey mould of fruit such as strawberries.

Growth on ordinary laboratory media is vigorous, producing a high, grey turf which develops discrete black bodies (sclerotia) at the agar surface, especially at the edge of the culture vessel. The large, colourless unicellular spores are produced in grape-like clusters on long, robust black conidiophores (see Figure 51.1).

PHOMA

This is a representative of a large number of pycnidial genera, all plant parasites, any of which may be encountered as contaminants. It is included here because it is probably one of the most common and many others were once classified with it.

Some species produce low, flat colonies on laboratory media with areas of black mixed with pink (spore masses). Others may give dense, fluffy growths with black or grey shading. Microscopically they all show pycnidia – spherical structures containing large numbers of small conidia, often released in a wet, worm-like mass through a round ostiole. These are best seen by examining the growing colony *in situ* under low power (see Figure 51.1).

TRICHODERMA

Growth is white, rapidly spreading and cobweb like, with irregular patches of dark-green spores. These are abundant, small, ovoid, arising in ball-like groups on short right-angled branches (see Figure 51.1). This fungus may be mistaken for a *Penicillium* but the rapidly spreading growth form, often covering the agar surface and even the sides of the Petri dish, would be very unusual for *Penicillium*.

MONASCUS

The colonies of this saprophyte are flat, rather granular with a deep-red pigment. The characteristic ascospores are produced inside spherical ascocarps. These are unusual in that each develops from a single (or a very few) hypha and they have a thin wall (see Figure 51.1). Chains of conidia superficially resembling *Scopulariopsis* may also be present.

PAECILOMYCES

Usually seen as a spoilage fungus, there is one species, *P. variotti*, which has been found in rare, deep infections in immunosuppressed patients.

Growth resembles that of *Penicillium*, producing either pale purple or yellowish olive colonies depending on species. This feature, the distinctly elongated spores, and the long, tapering tips, which are often curved, to the spore-producing cells of the 'penicillus' distinguish it from *Penicillium* (see Figure 51.1).

CHRYSOSPORIUM

These are soil dwellers, many of which specialize in using keratinaceous substrates; some may be capable of causing infection of the skin and nails.

This genus is close to the dermatophytes and, like them, is unaffected by cyclohexamide in the medium. Colonies of most species are white or cream coloured and flat with felt-like surfaces. Spores are similar to those of dermatophyte microconidia but mostly longer than 5 μm (see Figure 51.1).

GEOMYCES

One species, *G. pannorus*, a soil organism, is often mistaken for a dermatophyte and it has been reported on the surfaces of meat carcasses in cold storage at temperatures as low as –6°C (usually under the name *Sporotrichum carnis*).

Colonies are restricted, often heaped up, with a very thin white crust of aerial spores. These resemble dermatophyte microconidia but are smaller than most (< 1.5 μm long) and produced on minute, acutely branched, Christmas tree-like structures (see Figure 51.1).

SCOPULARIOPSIS

One species, *S. brevicaulis*, is a saprophyte, living in soil although it does seem to be adapted to use keratin and is one of the more common invaders of human nail tissue. It is also notable for releasing gaseous arsines from the arsenical pigment Paris green.

Colonies are often folded and heaped up, but they may be flat. When sporing they are cinnamon–brown in colour and have a powdery surface. The spores are large, rounded, with a flat detachment scar and are produced in long chains from a penicillium-like branching structure. In most strains the spore surfaces are distinctly roughened (see Figure 51.1).

STACHYBOTRYS

One species, *S. chartarum* (*S. atra*) may produce very toxic macrocyclic trichothecenes such as the satratoxins. This species is cellulolytic and may occur as black patches on the walls of damp houses where it has been implicated in some forms of damp building illness. It forms effuse, black or blackish green colonies, developing black glistening heads of unicellular phialospores with verrucose walls.

ASPERGILLUS AND PENICILLIUM

These genera are closely related but, although there are several species that appear intermediate in morphology, there is generally no difficulty in placing isolates in one genus or the other. Both produce long, dry chains of conidia by repeated budding through the end of a bottle-shaped cell, the phialide.

In *Penicillium* the phialides are clustered at the tips of tree-like conidiophores and develop unevenly.

In *Aspergillus* they are clustered on a club-shaped or spherical vesicle and develop synchronously (Figure 51.2).

In the subgenus *Aspergilloides* of *Penicillium*, the phialides are attached directly to the tip of the conidiophore (monoverticillate), which may be distinctly swollen, giving the impression of a very simple aspergillus fruiting body. However, the conidiophore of *Penicillium* is septate, and very similar in appearance to the mycelium from which it has branched, whereas that of *Aspergillus* is a more specialized structure, usually non-septate and developing from a distinct foot cell differentiated clearly from the supporting mycelium.

In both groups the mycelium is usually colourless, but may be coloured by the production of pigments

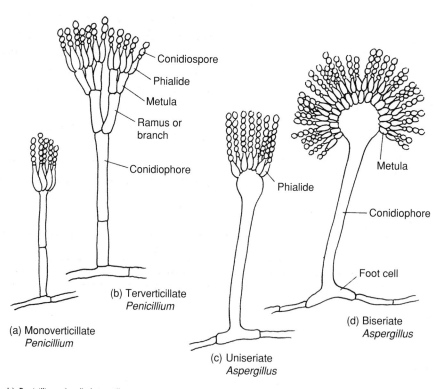

Figure 51.2 (a, b) *Penicillium*; (c, d) *Aspergillus*

retained in the cytoplasm or secreted into the medium; any colour on the obverse is the result of the coloured spores and on the reverse is the result of secreted pigments. Penicillium spores are usually some shade of green or blue. Aspergillus spores may be green, ochre-yellow, brown or black. In both genera there are species with white spores.

Some *Penicillium* spp. are plant parasites but the majority of both genera occur on decaying vegetation. As a result of their wide range of secondary metabolites several are of importance in industrial fermentation and synthetic processes. They are also important in that they may produce mycotoxins in foods.

A single species, *P. marneffii*, is implicated in infections of people who lived in or visited some regions of south-east Asia in which the species is endemic (see Chapter 52).

Aspergillus glaucus group

Members of this group flourish in conditions of physiological dryness (low water activity), e.g. on the surface of jam, on textiles and on tobacco. They are very distinctive in appearance, with bluish-green or grey–green conidia and large yellow cleistothecia of the ascomycete genus *Eurotium*.

Aspergillus restrictus group

Like *A. glaucus*, these will grow on dry materials, such as slightly dry textiles. As their name suggests, they are slow-growing organisms.

Aspergillus fumigatus

This is so called because of its smoky-green colour. It is a common mould in compost, and can grow at temperatures above 40°C; hence it can survive in the tissues of birds, for which it is a pathogen. It spores more readily at 37°C than at 26°C. It is now well-recognized as a human pathogen, underlying pulmonary eosinophilia. In this condition, the fungus may be found in the mucus plugs that are expectorated. It also invades old tuberculous cavities and gives rise to aspergilloma of the lung. When isolated from such cases, the organism often does not produce spores; incubation at 42°C may encourage it to do so. Patients who harbour this organism develop precipitating antibody, which is rather weak in pulmonary eosinophilia but strong in cases of aspergilloma. For methods, see p. 135. If *A. fumigatus* is examined with a lens, the conidiophores are seen to be columnar. It looks like a test-tube brush (see Figure 51.2).

Aspergillus niger

In contrast to *A. fumigatus*, this organism has round heads that are large enough to appear discrete to the naked eye. This, together with their black colour, makes them easy to recognize. It is a common cause of ear infections.

The *flavus–oryzae* group

Members of this group are notorious for the production of aflatoxin in groundnuts, maize, and

Table 51.2 Morphology of *Aspergillus* species

| | *Aspergillus* | | | | | | |
	fumigatus	*flavus*	*niger*	*terreus*	*versicolor*	*nidulans*	*glaucus*
Finely roughened stalks	−	+	−	−	−	−	−
Stalks pale brown	−	−	−	−	−	+	−
Green colony	+	+	−	−	+	+	+
Black colony	−	−	+	−	−	−	−
Sand-brown colony	−	−	−	+	−	−	−
Metulae present	−	+ / −	+	+	+	+	−

several other foods derived from tropical and sub-tropical plants, when badly stored (see Moss *et al.*, 1989). They are also of value commercially, as a source of amylase and protease.

The specific name, which means yellow, is an error; the colony colour in this series is green. The heads are round, but smaller than those of *A. niger*.

Table 51.2 shows the morphological properties of some *Aspergillus* species/groups.

ZYGOMYCOTINA

These are fungi with non-septate, wide hyphae in which sexual fusion results in a thick-walled resting zygospore. They are seldom seen, however, and distinction of genera and species is largely based on the asexual structures. The group contains two main orders. Many of them grow rapidly and produce aerial mycelium resembling sheep wool.

Mucorales

These reproduce by many-spored sporangia. The majority belong to three genera: *Rhizopus, Absidia* and *Mucor*.

Rhizopus

The columella enlarges greatly after rupture of the sporangium and collapses to form a characteristic mushroom shape. Spores are angular and delicately

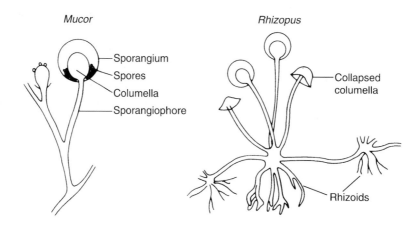

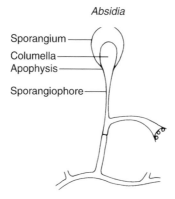

Figure 51.3 Zygomycetes

striated. Root-like rhizoids may occur at the base of the sporangium stalk (Figure 51.3).

Absidia

A trumpet-shaped columella and sporangium stalk are more or less continuous (see Figure 51.3).

Mucor

The absence of features characteristic of the latter two genera suggests *Mucor* spp. The common pathogenic species, all causing mucormycosis, are *Rhizomucor pusillus, Absidia corymbifera* and *Rhizopus arrizhus*. They grow at 37°C, which is useful in identification (see Figure 51.3).

Non-sporing moulds

Some moulds do not readily produce spores by which they may be identified. If they are considered significant on the basis of direct microscopic evidence, sporulation may be encouraged by subculturing on a variety of different media, such as potato sucrose agar or dilute strength cornmeal agar, and on exposure to daylight at room temperature.

For detailed information about these moulds, see Sutton (1980), Pitt and Hocking (1997) and de Hoog *et al.* (2000).

REFERENCES

de Hoog, G. S., Guarro, J., Gené, J. and Figueras, M. J. (2000) *Atlas of Clinical Fungi*, 2nd edn. Utrecht: Centraalbureau voor Schimmelcultures/Universitar Rovira I Virgili.

Ellis, M. B. (1971) *Dematiaceous Hyphomycetes*. Farnham: Farnham Royal, Commonwealth Agricultural Bureau.

Moss, M. O., Jarvis, B. and Skinner, F. A. (1989) Filamentous fungi in foods and feeds. *Journal of Applied Bacteriology* (Symposium Suppl. No. 18) 67: 1S–144S.

Pitt, J. I. and Hocking, A. D. (1997) *Fungi and Food Spoilage*, 2nd edn. London: Blackie Academic.

Sutton, B. C. (1980) *The Coelomycetes*. Farnham: Farnham Royal, Commonwealth Agricultural Bureau.

52

Pathogenic moulds

The following fungi are considered in this chapter: dermatophytes, causing infections of epidermal tissues; some opportunistic deep-tissue pathogens (but see Chapter 51), agents of subcutaneous infections, and the dimorphic systemic pathogens in Hazard Group 3.

DERMATOPHYTES

There are three genera of common dermatophytes, classified according to the types of conidia they produce in culture:

- *Microsporum*: macroconidia spindle shaped, roughened, and microconidia clavate
- *Trichophyton*: macroconidia cylindrical and smooth, and microconidia clavate or round
- *Epidermophyton*: macroconidia clavate, smooth or slightly roughened, no microconidia.

The nature of the disease assists in identification as many of these fungi have characteristic infection patterns (Table 52.1). This is particularly helpful in identifying strains with abnormal morphology.

Identification to the species level depends upon colonial morphology, pigment production and the microscopic study of these conidia. Strains with many microconidia and no macroconidia usually belong to *Trichophyton* species.

Appearance in clinical material

In nail and skin cleared as described in Chapter 9, fungal hyphae may be seen, some of them breaking

Table 52.1 Characteristic infection patterns of dermatophytes

Patient	Body site	Geographical origin	Organism
Adult	Feet, hands, groin	Any	T. rubrum
			T. interdigitale
			E. floccosum
Adult or child	Scalp, face, arms, chest, legs	Any	M. canis
			M. gypseum
			T. verrucosum
			T. mentagrophytes
Adult or child	Scalp, face, arms, chest, legs	N Africa, E and SE Asia	T. violaceum
Child	Scalp	Any	T. tonsurans
Child	Scalp	W and Central Africa, Caribbean	M. audouinii
Child	Scalp	N and E Africa	T. soudanaense
Child	Scalp	N and E Africa, Middle East	T. schoenleinii

into arthrospores, although specific identification cannot be made from their arrangement.

Some dermatophyte-infected hairs fluoresce under a Wood's lamp (examine in a darkened room). Table 52.2 indicates the appearance of infected hairs and Figure 52.1 shows various types of infected hairs, skin and nails.

Table 52.2 Appearance of infected hairs

Spore arrangement	Fluorescence	Organism
Small spore ectothrix	+	M. canis
	+	M. audouinii
	−	T. mentagrophytes
Large spore ectothrix	−	M. gypseum
	−	T. verrucosum
Endothrix	−	T. soudanense
	−	T. violaceum
'Favic' hairs	weak	T. schoenleinii

Identification

Identification to species level depends on colonial and microscopic appearances, both of which are influenced by the growth medium.

Use Sabouraud's dextrose agar which has been shown to give adequate pigmentation and spore production.

Microsporum audouinii

This is a scalp fungus with no animal or soil reservoir. It spreads from child to child in schools, and infection usually resolves at puberty. The organism is of medium growth rate, with a thinly developed, white-to-pale buff, aerial mycelium. The reverse of the colony is typically a pale apricot colour. Macroconidia and microconidia are rare (Figure 52.2).

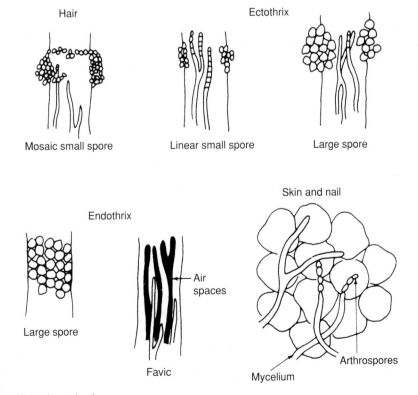

Figure 52.1 Infected hair, skin and nail

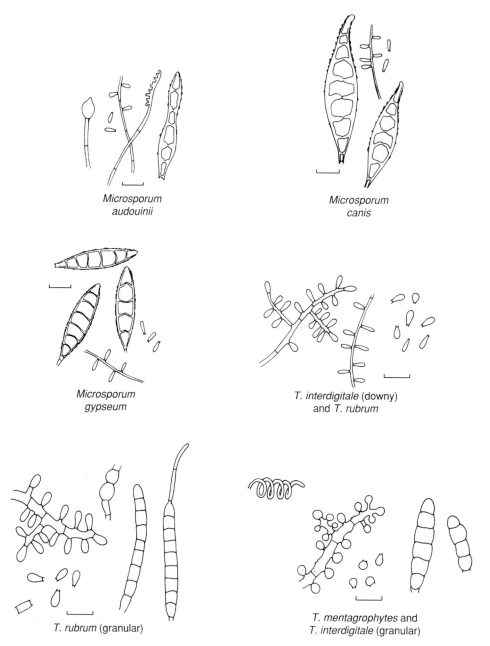

Figure 52.2 Microscopic appearances of some dermatophytes (bar = 10 μm)

Microsporum canis

As its name suggests, this is a common parasite of dogs and cats, where the infection is often difficult to detect. Children are commonly infected from pets. This organism, which is fast growing, usually produces a vivid yellow pigment and fairly abundant macroconidia. Occasionally, strains are non-pigmented and slow growing. To differentiate them from *M. audouinii*, grow on rice (5 g rice and 20 ml water in a 100-ml flask, autoclaved at 115°C for 20 min and cooled). Under these conditions *M. canis* produces aerial mycelium; *M. audouinii* does not (see Figure 52.2).

Microsporum gypseum

This occurs as a saprophyte in soil and has a low virulence for humans. Infections are therefore uncommon and typically present with a solitary lesion, related to contact with the soil. The fungus grows fast, and is soon covered with a buff-coloured granular coating of spores. These tend to be arranged in radial strands on the surface, rather like a spider's web. Macroconidia are abundant (see Figure 52.2).

Trichophyton rubrum

This is the most common dermatophyte seen in dermatology clinics in developed countries, but the less common *T. interdigitale* is probably more widespread in the general population.

There are many varieties of *T. rubrum*; the one seen most often gives a downy, white colony with a red–brown pigment and a narrow white edge on reverse. Microscopy shows only club-shaped microconidia (see Figure 52.2).

Other varieties are 'melanoid' in which a dark-brown pigment diffuses into the medium, 'yellow', which is a non-sporing pale yellow form, and 'granular', which has the usual dark-red reverse but powdery aerial growth with patches of pink. This variety has larger microconidia and usually produces macroconidia (see Figure 52.2).

Non-pigmented strains occur which resemble the downy variety of *T. interdigitale* (see Figure 52.2). To distinguish these, subculture on urea agar. *T. rubrum* does not usually produce the colour change by 7 days but most other dermatophytes do.

Trichophyton mentagrophytes complex

The most common form in this group is the anthropophilic *T. interdigitale*, which causes athlete's foot. Colonies resemble those of the zoophilic *T. mentagrophytes* (see below), with a uniformly powdery, cream-coloured surface made up of an abundance of almost spherical microconidia. Spiral hyphae and macroconidia may be present (see Figure 52.2). The downy variety forms a pure white cottony colony and has elongated microconidia resembling those of *T. rubrum*.

The zoophilic (usually from rodents) *T. mentagrophytes* has coarsely granular colonies with areas of agar surface showing between radiating zones of white or cream sporing growth. Reverse pigmentation often shows dark brown 'veins'.

Trichophyton tonsurans

This is a cause of scalp ringworm. Its colonies are slow growing with a dark red–brown edge, and granular to velvety centre, often with folding. Most strains show a mixture of microconidia and swollen hyphal cells (chlamydospores). Microconidia are large, oval and arranged along wide, often empty hyphae (see Figure 52.3).

Trichophyton verrucosum

This is the causative organism of cattle ringworm. It also causes scalp, beard or nail infections in farm workers, and other people such as slaughtermen and veterinary surgeons who come into contact with cattle. In culture it grows very poorly; suspected cultures should be examined carefully with a lens after incubation, for 14 days, because colonies are often submerged and minute. This, together with the large, thin-walled, balloon-like chlamydospores at the end of a straight hypha and chains of thick-walled cells (see Figure 52.3), is sufficient to identify the organism.

Trichophyton schoenleinii

This is the cause of favus. The colony is slow growing, hard and leathery, with a surface like white suede. The only microscopic features are thick, knobbly hyphae, like arthritic fingers – the so-called 'favic chandeliers' – and chlamydospores.

Trichophyton violaceum

This is another cause of scalp ringworm, microscopically similar to *T. verrucosum*, although the hyphae are usually thinner. The characteristic feature is a deep red–purple pigment, which appears as a spot in the centre of a young colony, and gradually spreads to the edge.

Epidermophyton

There is only one species, *E. floccosum* which is one cause of tinea cruris. It also occasionally infects feet. Its growth on malt agar is khaki in colour with a powdery surface caused by masses of macronidia which have a very characteristic shape (see Figure 52.3)

ENTOMOPHTHORALES

These insect pathogens are seen rarely in clinical material in tropical countries. The asexual spore form is a single, relatively large 'conidium' which is forcefully discharged.

Colonies are wrinkled, cream coloured and waxy. A haze of discharged spores forms a mirror image of the colony on the lid of the Petri dish.

Basidiobolus

This is a cause of tropical subcutaneous zygomycosis.

Conidiobolus coronatus (Entomophthora coronata)

This causes tropical rhinoentomophthoromycosis of the nasal mucosa.

MISCELLANEOUS PATHOGENIC MOULDS

Madurella

Madurella mycetomatis and *M. grisea* cause tropical black grain mycetoma. The former produces flat or wrinkled, yellowish colonies, often releasing melanoid pigments into the medium. The latter grows as domed, densely fluffy, mouse-grey to black colonies. Neither forms conidia.

Exophiala

Colonies are dark grey to black, intensely fluffy or wet and yeast-like depending on degree of conidial budding, strain and growth conditions. All produce conidia from small, tapering nipples which grow in length as more conidia are cut off (Figure 52.4).

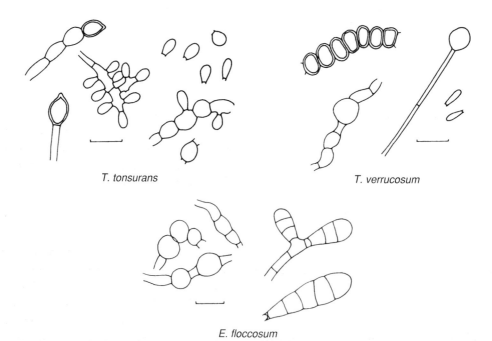

T. tonsurans

T. verrucosum

E. floccosum

Figure 52.3 Microscopic appearances of some dermatophytes (bar = 10 μm)

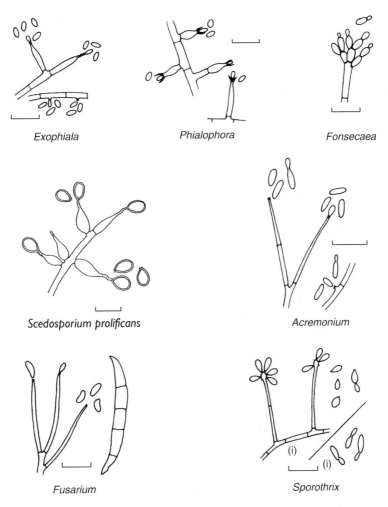

Figure 52.4 Microscopic appearances of some pathogenic fungi (bar = 10 μm)

Exophiala jeanselmei is a cause of mycetoma, *E. werneckii* of tinea nigra, *E. dermatitidis* of chromomycosis and *E. spinifera* of subcutaneous cysts.

Phialophora

Members of this genus are black moulds in which the spore apex has a definite collarette or funnel-shaped cup in which the conidia are formed (see Figure 52.4). *P. verrucosa* is one cause of chromomycosis; *P. parasitica* is occasionally found in subcutaneous cysts.

Fonsecaea

Members of this genus are black moulds in which the conidia are formed in short chains by apical budding of the cell below. They differ from *Cladosporium* and *Cladophialophora* spp. only in that the chains consist of no more than four spores (see Figure 52.4). *F. pedrosi* is the common cause of chromoblastomycosis; *F. compacta* is less common.

Cladophialophora

The most frequent *Cladosporium*-like fungi causing human including cerebral phaeohyphomycosis have been moved from *Cladosporium* to *Cladophialophora*. The common species are *C. carrionii* (subcutaneous chromoblastomycosis) and *C. bantiana* (a pathogen of the brain, now rated in the UK as in Hazard Category 3). They differ from

Cladosporium (see Figure 51.1) in having very long, rarely branching chains of conidia (see Figure 52.4). The conidia are slender, smooth and thin walled in comparison with *Cladosporium*. Some strains also produce phialides of the phialophora type (see Chapter 51).

Scedosporium

Variously named *Allescheria*, *Petriellidium* or *Monosporium* spp. It causes pale grain mycetoma and miscellaneous other infections (otitis externa in the UK). Two species cause a variety of clinical infections, *S. apiospermum* (the asexual form of *Pseudallescheria boydii*) and *S. prolificans*. *S. apiospermum* colonies are fluffy, white to light grey, producing oval, slightly yellowish spores in succession at the tips of long, straight hyphae. Each spore has a basal scar (see Figure 52.4). Occasional strains may produce black ascomata beneath the agar.

Scedosporium prolificans colonies are flatter and slower growing, often with areas of black wet surface. Under the microscope they have short, lateral conidia bearing hyphae, which are swollen at the basal half and taper to a narrow filament-like tip area. A number of fatal cases of deep opportunistic infections by this species have been described. *S. prolificans* is particularly resistant to modern systemic antifungal drugs.

Acremonium

This is a very large group of white or pink moulds with wet heads of conidia produced one by one from the tips of straight hyphae or lateral nipples (see Figure 52.4). In some, the conidia bud secondary spores and the colony becomes yeast like. *A. kilense* is a cause of pale grain mycetoma; other species invade nail tissue.

Fusarium

Colonies are often loose and fluffy with a pink or purple pigment. They are characterized by curved, multiseptate macroconidia, although these may be absent. Many strains produce microconidia similar to those of *Acremonium* spp. but usually somewhat curved (see Figure 52.4).

Fusarium solani and *F. oxysporum* cause a variety of infections including mycetomas and keratitis. They are plant pathogens.

Sporothrix

At low temperatures (25–30°C) colonies of *Sporothrix* are grey to black, dry, membranous and wrinkled. Hyaline or pigmented spores are produced initially as rosettes on minute, teeth-like points on expanded apical knobs (see Figure 52.4). Culture at 37°C when they convert to yeast form facilitates differentiation from non-pathogenic species. Causes sporotrichosis in which it exists as budding, spindle-shaped yeast.

DIMORPHIC SYSTEMIC PATHOGENS

These cause systemic or generalized infections resulting from the inhalation of airborne conidia. Heavily sporing mould cultures are more hazardous than those in the yeast-like phase. These organisms are classified as Hazard Group 3 pathogens and they must be handled in microbiological safety cabinets in Containment Level 3 laboratories.

A cultural feature of these organisms is that they are not inhibited by cycloheximide at 0.5 mg/ml.

Blastomyces dermatitidis

This causes North American blastomycosis. Infection is almost always by inhalation of spores. Skin lesions are usually secondary to a pulmonary infection, even a mild one. The mycelial form, at 26°C, is a moist colony at first, developing a fluffy aerial mycelium with microconidia. At 37°C or in tissue, large oval yeasts are seen (Figure 52.5).

Histoplasma capsulatum

This causes the pulmonary disease histoplasmosis, which is very common in the Ohio River valley in the southern half of the USA, although it is also found in other countries. Infection is associated with exposure to dust containing faecal material in barns and sheds. Most people recover rapidly: only

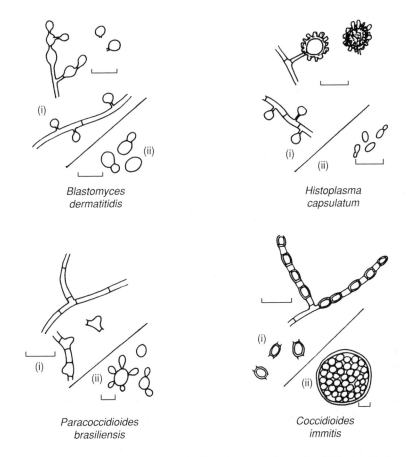

Figure 52.5 Microscopic appearances of *Blastomyces dermatitidis*; *Histoplasma capsulatum*; *Coccidioides immitis*; *Paracoccidioides brasiliensis*: (i) mycelial form; (ii) tissue form (bar = 10 μm)

an occasional patient develops progressive disease. The route of infection is inhalation, and the pathology similar to that of tuberculosis (see Figure 52.5). (*See caution above.*)

At 26°C, the organism grows with a white aerial mycelium, which becomes light brown. The spores are striking and characteristic. At 37°C, small, oval budding yeasts are found. In tissue, these yeasts are intracellular, except in necrotic material.

Paracoccidioides brasiliensis

This causes South American blastomycosis. The primary lesion is almost always in the oral mucosa. The spread is through the lymphatics. The organism is very slow growing, white and featureless at 26°C. At 37°C on rich media, characteristic yeast cells are formed (see Figure 52.5).

Coccidioides immitis

This causes another lung infection, coccidioidomycosis. This disease is limited to a few special areas in the south west of the USA and in Mexico and Venezuela. It is inhaled in desert dust. Again, most people suffer only a mild, transient, 'flu-like' illness. Occasionally, cases progress to generalized destructive lesions. (*See caution above.*)

At 26°C, the colony is at first moist, then develops an aerial mycelium, becoming brown with age and breaking up into arthrospores. There is no second form in artificial culture. In the clinical lesions, however, large, thick-walled cells are seen. When mature, these are full of spores (see Figure 52.5).

Penicillium marneffei

Penicillium marneffei has become one of the most frequent causes of opportunist infection in AIDS patients who have resided in or visited south-east Asia. At lower temperatures (25–35°C) this dimorphic fungus is a hyphal mould with classic penicillium conidial heads (see Chapter 51). The conidia are pale green in mass and contribute this colour to the colony surface. A deep-red diffusing pigment is produced on most media. Microscopically the conidial heads are more spread out and irregular than those of the saprophytic *P. purpurogenum*, which is sometimes encountered as a contaminant and also has a deep-red diffusing pigment.

At blood temperature on nitrogen-rich media (blood agar, brain–heart agar) the fungus converts to a yeast-like colony, which is composed of cylindrical cells reproducing by fission.

In view of the difficulty of identification of *Penicillium* spp., this Hazard Group Category 3 pathogen should be sent to a reference laboratory for confirmation.

For more details about pathogenic fungi, please see Campbell and Stewart (1980), Campbell *et al.* (1996), Midgeley *et al.* (1997) and de Hoog *et al.* (2000).

REFERENCES

Campbell, M. C. and Stewart, J. C. (1980) *The Medical Mycology Handbook.* New York and Chichester, Wiley.

Campbell, C. K., Johnson, E. M., Philpot, C. M. *et al.* (1996) *Identification of Pathogenic Fungi.* London: Public Health Laboratory Service.

de Hoog, G. S., Guarro, J., Gené, J. *et al.* (2000) *Atlas of Clinical Fungi.*, 2nd edn. Utrecht: Reus, Centraalbureau voor Schimmelcultures/Universitat Rovira I Virgili.

Midgley, G., Clayton, Y. and Hay, R. J. (1997) *Diagnosis in Color. Medical mycology.* Chicago: Mosby-Wolfe.

Appendix

Regulatory safety issues that affect work with micro-organisms in the UK

All employers whose work involves the deliberate exposure of their employees to biological agents need to comply with the law. The information given here provides the reader with the framework of the primary legislation in the UK that applies in the microbiology laboratory, thus providing a basis upon which a good health and safety policy can be produced.

HEALTH AND SAFETY AT WORK

The Health and Safety at Work, etc. Act (the Act) 1974 (HSE, 1990) provides the foundations for the health and safety system as we know it today. Before the Act, health and safety legislation was prescriptive, out of date, complex and inflexible, and would not have been able to meet the demands of managing the type of risks encountered in a modern microbiology laboratory. The implementation of the Act established the Health and Safety Commission (HSC) and the Health and Safety Executive (HSE). The Act places on the Commission the responsibility of undertaking revision and replacement of legislation, to recommend new legislation based on developments in technology and to comply with European Union Directives. The Commission's main instrument in operating these responsibilities is the HSE.

The Act is described as goal-setting, leaving employers the freedom to decide how to control the risks that they identify. There are therefore general duties, which employers have towards employees and members of the public, and which employees have to themselves. These duties are qualified by the principle 'so far as is reasonably practicable'.

This means that the degree of risk in a particular job, e.g. the culture of Hazard Group 3 biological agents, needs to be balanced against the time, trouble, cost and physical difficulty of taking measures to avoid or reduce the risk.

REGULATIONS, CODES OF PRACTICE AND GUIDANCE

The Act states that legislation passed before 1974 should be 'progressively replaced by a system of regulations and approved codes of practice'. In fact, the general principle has been that regulations, like the Act itself, should, so far as is possible, express general duties, principles and goals, and that subordinate detail should be set out in Guidance and Approved Codes of Practice (ACoP).

Regulations, e.g. the Control of Substances Hazardous to Health (COSHH) Regulations, are law, approved by Parliament. An ACoP gives advice, but employers are free to take other equally effective measures provided that they do what is reasonably practicable. Some risks are so great, or the proper control measures so costly, that it would not be appropriate to leave employers the discretion to decide what to do about them. Regulations identify these risks and set out specific action that must be taken. These requirements may be absolute and must be complied with, without the qualification of reasonable practicability, e.g. a Containment Level 3 Laboratory must be sealable to permit disinfection.

An ACoP offers practical examples of good practice. Advice is provided on how to comply with the law, by providing a guide to what is 'reasonably practicable', e.g. regulation 6 of the COSHH Regulations requires an 'assessment of health risks created by work involving substances hazardous to health'. This assessment should be 'suitable and sufficient'. The general ACoP accompanying the COSHH Regulations informs that a suitable and sufficient assessment should include:

- an assessment of the risks to health
- consideration of the practicability of preventing exposure to hazardous substances
- the steps that need to be taken to achieve adequate control of exposure where prevention is not reasonably practicable
- identification of other action necessary to comply with the remaining COSHH regulations.

For biological agents, as distinct from other substances hazardous to health, the assessment should reflect the ability they may have to replicate, be transmitted and infect. Additional clarification for biological agents is provided in the main body of the ACoP of the COSHH Regulations. Approved Codes of Practice have *special legal status*: if employers are prosecuted for a breach of health and safety law, and it is proved that they have not followed the relevant provisions of the ACoP, a court can find them at fault unless they can show that they have complied with the law in some other way.

The HSE publishes guidance on a range of subjects, which may be specific to the health and safety problems of an industry or of a particular process used in a number of industries, e.g. the management, design and operation of microbiological containment laboratories (Advisory Committee on Dangerous Pathogens, ACDP, 2001) and successful *health and safety* management (HSE, 1997). The main purposes of guidance are:

- to interpret – helping people to understand what the law says
- to help people comply with the law
- to give technical advice.

Following guidance is not compulsory and employers are free to take other action; however, if they do follow guidance they will normally be doing enough to comply with the law.

MANAGEMENT RESPONSIBILITIES

Successful control of exposure to biological agents depends on having effective arrangements for managing health and safety. An overview of the relevant health and safety legislation and guidance that should be consulted when working with biological agents in microbiological containment laboratories is shown in Figure A.1. Employers need to consider a number of key areas related to work in microbiology laboratories, including:

- risk assessment
- information, instruction and training
- buildings and accommodation
- safe working practices
- health surveillance and immunization
- checking what has been done.

Detailed information on local arrangements and procedures is best provided in standard operating procedures. These should be based on a risk assessment and are a good place to record the findings of assessments. Standard operating procedures need to be written down and kept up to date. All staff need to be aware of the local procedures that apply to them.

HEALTH AND SAFETY LAW: KEY REGULATIONS

The following section provides some information on the Key Regulations that apply to the prevention of infection in the laboratory. The information provided is not intended to be comprehensive and managers must ensure that all other risks are identified and all other legal obligations fulfilled. Examples of additional regulations that will apply in the laboratory include the Manual Handling Operations Regulations 1992 (HSE, 1992), the Display Screen Equipment Regulations 1992 (HSE, 1994a) and the Noise at Work Regulations 1989 (HSE, 1998a).

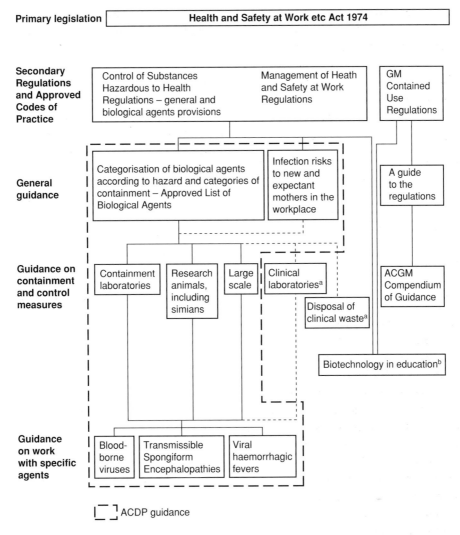

Primary legislation | Health and Safety at Work etc Act 1974

Secondary Regulations and Approved Codes of Practice
- Control of Substances Hazardous to Health Regulations – general and biological agents provisions
- Management of Heath and Safety at Work Regulations
- GM Contained Use Regulations

General guidance
- Categorisation of biological agents according to hazard and categories of containment – Approved List of Biological Agents
- Infection risks to new and expectant mothers in the workplace
- A guide to the regulations

Guidance on containment and control measures
- Containment laboratories
- Research animals, including simians
- Large scale
- Clinical laboratories[a]
- Disposal of clinical waste[a]
- ACGM Compendium of Guidance
- Biotechnology in education[b]

Guidance on work with specific agents
- Blood-borne viruses
- Transmissible Spongiform Encephalopathies
- Viral haemorrhagic fevers

⌐ ⌐ ACDP guidance

Figure A.1 Health and safety legislation and guidance relevant to work with biological agents in microbiological containment laboratories. [a] Health Service Advisory Committee guidance. [b] Education Services Advisory Committee guidance

The Health and Safety at Work, etc. Act

As previously stated, the Act places general duties on employers, employees and others.

Key duties

Employers must:

- protect the health and safety of their employees
- protect the health and safety of others who

might be affected by the way they go about their work (e.g. cleaners, visitors or contractors working in the laboratory)
- prepare a statement of safety policy and the organization and arrangements for carrying it out (if five or more people are employed, this statement must be written down).

Employees must:

- take care of their own health and safety, and that of others
- cooperate with their employer.

The Management of Health and Safety at Work Regulations 1999

The general duties in the Act are developed in the Management of Health and Safety at Work Regulations 1999 (the Management Regulations) and other more specific pieces of law. The *ACoP on the Management of Health and Safety at Work Regulations* provides further guidance (HSE, 2000).

Key duties

Employers must:

- assess risks to staff and others, including young people, and new and expectant mothers (HSE, 1994b; ACDP, 1997)
- make appropriate health and safety arrangements, which must be recorded if five or more people are employed
- appoint competent people to help them comply with health and safety law
- establish procedures to deal with imminent danger
- cooperate and coordinate with other employers and self-employed people who share the workplace.

Employees must:

- work in accordance with training and instruction given by their employer
- report situations, which they believe to be unsafe.

The Control of Substances Hazardous to Health Regulations

The COSHH Regulations 2002 apply to substances that have been classified as being very toxic, toxic, harmful, corrosive, sensitizing or irritant (HSE, 1999a). Biological agents are also treated by the Regulations as substances hazardous to health. Additional requirements of the Regulations, permitting work with biological agents, are described in Schedule 3 of the Regulations (see Table 1.3,

p. 4) which lists some of the containment measures for laboratories that are necessary to ensure compliance with the Regulations. Further practical guidance can be found in Schedule 3 and Appendix 2, 'Additional provisions relating to work with biological agents', of COSHH (HSE, 2002).

Key duties

Employers must:

- carry out risk assessments
- select and use appropriate control measures where there is a likelihood of exposure to biological agents
- maintain, examine and test control measures
- inform, instruct and train employees
- keep records of employees exposed to Group 3 biological agents
- carry out health surveillance of employees.

Reporting of Injuries, Diseases and Dangerous Occurrences Regulations 1995

The Reporting of Injuries, Diseases and Dangerous Occurrences Regulations (RIDDOR) (HSE, 1999b) require employers and others to report accidents and specified diseases, which arise out of or in connection with work, to the HSE.

Key duties

Employers must:

- report any acute illness that requires medical treatment, where there is reason to believe that this resulted from exposure to a biological agent or its toxins or infected material
- report any accident or incident that resulted, or could have resulted, in the release or escape of a Hazard Group 3 or 4 biological agent
- report specified infectious diseases, such as hepatitis and tuberculosis
- report any infection reliably attributable to work in the laboratory.

RISK ASSESSMENT

Risk assessment is not simply a paper exercise. Its purpose is to ensure that there are appropriate precautions in place to prevent and control risk. The process can be broken down into the following five steps (HSE, 1998b):

1. Identify the hazards.
2. Ascertain how the hazards cause harm.
3. Assess the risks – determine what preventive measures or controls are required.
4. Record and implement the findings of the assessment.
5. Monitor, review and revise the assessment.

Step 1: Identify the hazards

The first step in any risk assessment is to look for the hazards. A *hazard* is the potential for something to cause harm. Employers will need to consider:

- which biological agents may be present
- the virulence of these agents
- the route(s) of infection
- their transmissibility.

Step 2: Ascertain how the hazards cause harm

Employers must acknowledge the fact that their staff may all potentially be exposed to infectious agents. Those employees who may be affected, along with the routes of transmission, need to be considered when assessing risks and deciding which precautions are needed.

Step 3: Assess the risks – determine what preventive measures or controls are required

A *risk* is the likelihood of harm occurring. Employers need to evaluate likely risks arising from any exposure to hazardous agents, taking into account:

- the severity of the diseases caused
- prevalence of particular infections in the community
- the nature of any likely contact with a patient specimen
- the likelihood of an accident involving exposure to biological agents
- the likelihood of infection from biological agents present
- the individual susceptibility of staff.

Employers need to devise and implement a mechanism for reporting and responding rapidly and effectively to accidents and incidents of ill-health. An official local record should be made of all incidents and occurrences with infectious or potentially infectious material involving the exposure of individuals. This record should include near misses. It must be remembered that, in addition to local reporting of accidents and exposures, in some cases the HSE must be notified under the RIDDOR Regulations.

When deciding whether existing preventive measures are adequate or whether further controls are required, employers need to consider the laboratory facilities and the activities therein of the staff. Where assessments identify the need for improvements, employers need to plan how to do this and make sure that their plans are implemented.

Step 4: Record and implement the findings of the assessment

The main findings of the risk assessment must be written down in all but the simplest of cases. Employers often incorporate the findings in the standard operating procedures. These procedures are then used as working documents for managers, employees and safety representatives.

Step 5: Monitor, review and revise the assessment

Employers should review the assessment(s) regularly to check that they are still valid. This is best done as part of the day-to-day management of the laboratory. It may include verifying that:

- standard operating procedures are relevant to current practices
- staff have received appropriate information and training about their work
- the system for reporting and responding to accidents, incidents and ill-health is in place and being followed
- equipment, such as microbiological safety cabinets, is working appropriately.

If the review identifies deficiencies in the assessment, employers should revise it and implement any necessary further precautions. Employers also need to make practical checks to ensure that staff are following the standard operating procedures appropriately.

INFORMATION, INSTRUCTION AND TRAINING

All staff need sufficient information, instruction and practical training about the risks associated with their work and on the precautions required to control/prevent the risks. Until staff have received adequate instructions about hygiene and safety at work they should not work with biological agents. Employers need to ensure that staff, contractors and visitors know the relevant safety standards applicable to them while working in or attending the laboratory. Employers are required to identify particular staff needs, including any gaps in knowledge and/or experience, and to provide the necessary information, instruction and training to fill these gaps.

To be useful, instructions and information should:

- take account of the level of training, knowledge and experience that staff and others have

- be easily understood, e.g. by those with language or communication difficulties.

INCIDENT REPORTING

Management needs to have clear procedures in place for dealing with incidents and accidents that could occur to staff or visitors. In addition, employees and other workers should be encouraged to report incidents of ill-health. Standard operating procedures should cover the arrangements for:

- immediate action in the event of an accident, fire or other emergency, especially where there is a risk of infection
- the action to take in the event of a significant spillage, e.g. dropping of culture plates
- reporting, recording and investigating accidents, incidents and ill-health including those with the potential to cause injury or ill-health
- notifying employees and their representatives of the causes of the incident and any remedial measures.

As previously described, the RIDDO Regulations require employers to report and keep records of specified types of accident, incident and diseases. For effective monitoring of health and safety arrangements, the internal reporting system needs to take account of all incidents and accidents that may occur in the laboratory, not just the more serious ones. Following up all such occurrences will help managers, safety representatives, safety managers and others monitor the adequacy of their precautions, check performance, learn from mistakes, identify jobs or activities that cause the greatest number of problems and identify deficiencies in staff training needs. This information is also very useful when carrying out the risk assessment process.

HEALTH SURVEILLANCE

Employers need to advise on the health surveillance requirements and immunization arrangements for their staff. Advice should be obtained from a qualified occupational health practitioner. More detailed information is provided by the HSE (1999c), Health Services Advisory Commission (HSAC, 1984) and Association of National Health Service Occupational Physicians (ANHSOP, 1986).

Pre-employment screening

Routine medical examination before employment is not considered necessary; instead, candidates can complete a questionnaire. One purpose of the questionnaire would be to determine the immune status of employees and offer the appropriate immunization or to identify those who may be more susceptible to infection because of a pre-existing medical condition. Only those whose replies suggest that a problem may exist need to seek advice from an occupational health professional.

Immunization

Employers need to ensure that occupational health arrangements include agreed immunization procedures for all staff. They should base these requirements on guidance from the Joint Committee on Vaccination and Immunisation (Department of Health, 1996). Generally, microbiologists will need to be immunized. Selected immunization procedures should be implemented as soon as possible after the appointment and ideally before starting work. The need for immunization will be determined as part of the risk assessment; invariably protection against hepatitis B, tetanus and tuberculosis will be included. Immunization should, however, be seen only as a useful supplement, reinforcing procedural controls and the use of protective equipment. The COSHH Regulations require employers to make effective vaccines available to employees exposed to biological agents. As this is a specific requirement under the Regulations, employers cannot charge their employees for such vaccines. In providing vaccines, employers should

ensure that employees are made aware of the advantages and disadvantages of immunization and its limitations. Adequate records should be kept of any vaccines given and checks made to ensure that protection has been provided.

Monitoring

This involves checking employees' health to detect workplace illness by, for example, following up sickness absence or explaining symptoms of infection to employees so that they can monitor their own health. In some cases, it may be useful to provide medical contact cards to alert medical practitioners about the nature of the work in the event of sudden unexplained illness.

Record keeping

Accurate health records are needed for all staff. Records that are more detailed should be kept for those who may be exposed regularly to a high risk of infection. It is important that, where hazards exist, employers monitor the health of their staff and note and act upon occurrences of work-related illness, e.g. sickness absence. Active health surveillance will be required for needlestick injuries and following exposure to significant pathogens, e.g. *Mycobacterium tuberculosis*. The health record is different from a clinical record and may be kept with other confidential personnel records. Records that include medical information arising from clinical examination should be held in confidence by the occupational health doctor and can be released only with the written consent of the individual.

Employers are required to keep the following details about employees exposed to a Group 3 biological agent:

- the type of work that the employee does
- the biological agent(s) to which he or she has been exposed (where this is known)
- records of exposures, accidents and incidents.

Health records will need to be maintained in most cases for a period of 40 years after the last known exposure.

CHECKING WHAT HAS BEEN DONE

Measuring performance and review

Employers need to measure how effectively they are implementing their standard operating procedures and assess how effectively they are controlling risks in the laboratory. Actively monitoring systems before something goes wrong is a key part of a successful monitoring regimen and part of a manager's responsibility. Information that is more detailed can be found in HSE guidance – *Successful Health and Safety Management* (HSE, 1997).

Bio-security

Historically, security measures in laboratories maintaining or researching pathogenic cultures have focused on the containment of the pathogen and the safety of workers. The supply of pathogens to such laboratories requires those laboratories to have suitable facilities for the containment of those organisms and, in order to be supplied with pathogens, to have notified the HSE about their intent. It is recognized that the choice of containment measure is largely determined by the categorization of the biological agent (ACDP, 1995).

Recent events, e.g. 11 September 2001, have focused attention on the threat posed by the terrorist use of biological agents as 'weapons of mass destruction' (WMD). Consequently, there is an increasing requirement for greater scrutiny of distribution of biological agents and making it more difficult to access areas where such agents are stored and used. A full discussion on the implications of bioterrorism and subsequent impact on the microbiology laboratory is beyond the scope of this book, and the reader is referred to other publications (McGinness *et al.*, 2001; Beeching *et al.*, 2002) (see also p. 10).

REFERENCES

Advisory Committee on Dangerous Pathogens (ACDP) (1995) Second supplement to: *Categorisation of Biological Agents According to Hazard and Categories of Containment*, 4th edn. Sudbury: HSE Books.

ACDP (1997) *Infection Risks to New and Expectant Mothers in the Workplace: A guide for employers.* Sudbury: HSE Books.

ACDP (2001) *The Management, Design and Operation of Microbiological Containment Laboratories.* Sudbury: HSE Books.

Association of National Health Service Occupational Physicians (ANHSOP) (1986) *Health Assessment for Employment in the NHS: Physician's guidance Note 9.* Bath: ANHSOP, Royal United Hospital.

Beeching, N. J., Dance, D. A. B., Miller, A. R. O. and Spencer, R. C. (2002) Biological warfare and bioterrorism. *British Medical Journal* **324**: 336–339.

Department of Health (1996) *Immunisation Against Infectious Disease.* Department of Health Joint Committee on Vaccination and Immunisation. London: HMSO.

Health Services Advisory Committee (HSAC) (1984) *Guidelines on Occupational Health Services in the Health Service.* Sudbury: HSE Books.

Health and Safety Executive (HSE) (1990) *A Guide to the Health and Safety at Work etc Act 1974 (L1).* Sudbury: HSE Books.

HSE (1992) *Manual Handling. Manual Handling Operations Regulations 1992 Guidance on Regulations (L23).* Sudbury: HSE Books.

HSE (1994a) *VDUs: An easy guide to the regulations (HSG90).* Sudbury: HSE Books.

HSE (1994b) *New and Expectant Mothers at Work: A guide for employers (HSG 122).* Sudbury: HSE Books.

HSE (1997) *Successful Health and Safety Management (HSG65).* Sudbury: HSE Books.

HSE (1998a) *Reducing Noise at Work: Guidance on The Noise at Work Regulations 1989* (L108). Sudbury: HSE Books.

HSE (1998b) *Five Steps to Risk Assessment* (INDG163). Sudbury: HSE Books.

HSE (1999a) *Approved Guide to the Classification and Labelling of Substances Dangerous for Supply: Chemicals (hazard information and packaging for supply) regulations 1994: guidance on regulations* (L100). Sudbury: HSE Books.

HSE (1999b) *A Guide to the Reporting of Injuries, Diseases and Dangerous Occurrences Regulations 1995: Guidance on regulations* (L73). Sudbury: HSE Books.

HSE (1999c) *Health Surveillance at Work* (HSG 61). Sudbury: HSE Books.

HSE (2000) *Management of Health and Safety at Work, Management of Health and Safety at Work Regulations 1999 Approved Code of Practice and guidance* (L21). Sudbury: HSE Books.

HSE (2002) *Control of Substances Hazardous to Health: Approved Codes of Practice* (L5), 4th edn. Sudbury: HSE Books.

McGinness, S., Sleator, A and Youngs, T. (2001) The anti-terrorism, crime and security bill Parts VI & VII: pathogens, toxins and weapons of mass destruction. Parliamentary Library Research Paper 01/94. www.parliament.uk

Index

Page numbers in **bold type** refer to tables.